AF577291

Benchmark Papers in Geology

Series Editor: Rhodes W. Fairbridge
Columbia University

27 PERIGLACIAL PROCESSES/*Cuchlaine A. M. King*
28 LANDFORMS AND GEOMORPHOLOGY: Concepts and History/ *Cuchlaine A. M. King*
29 METALLOGENY AND GLOBAL TECTONICS/*Wilfred Walker*
30 HOLOCENE TIDAL SEDIMENTATION/*George deVries Klein*
31 PALEOBIOGEOGRAPHY/*Charles A. Ross*
32 MECHANICS OF THRUST FAULTS AND DÉCOLLEMENT/*Barry Voight*
33 WEST INDIES ISLAND ARCS/*Peter H. Mattson*
34 CRYSTAL FORM AND STRUCTURE/*Cecil J. Schneer*
35 OCEANOGRAPHY: Concepts and History/*Margaret B. Deacon*
36 METEORITE CRATERS/*G. J. H. McCall*
37 STATISTICAL ANALYSIS IN GEOLOGY/*John M. Cubitt and Stephen Henley*
38 AIR PHOTOGRAPHY AND COASTAL PROBLEMS/*Mohamed T. El-Ashry*
39 BEACH PROCESSES AND COASTAL HYDRODYNAMICS/*John S. Fisher and Robert Dolan*
40 DIAGENESIS OF DEEP-SEA BIOGENIC SEDIMENTS/*Gerrit J. van der Lingen*
41 DRAINAGE BASIN MORPHOLOGY/*Stanley A. Schumm*
42 COASTAL SEDIMENTATION/*Donald J. P. Swift and Harold D. Palmer*
43 ANCIENT CONTINENTAL DEPOSITS/*Franklyn B. Van Houten*
44 MINERAL DEPOSITS, CONTINENTAL DRIFT AND PLATE TECTONICS/ *J. B. Wright*
45 SEA WATER: Cycles of the Major Elements/*James I. Drever*
46 PALYNOLOGY, PART I: Spores and Pollen/*Marjorie D. Muir and William A. S. Sarjeant*
47 PALYNOLOGY, PART II: Dinoflagellates, Acritarchs, and Other Microfossils/*Marjorie D. Muir and William A. S. Sarjeant*
48 GEOLOGY OF THE PLANET MARS/*Vivien Gornitz*
49 GEOCHEMISTRY OF BISMUTH/*Ernest E. Angino and David T. Long*

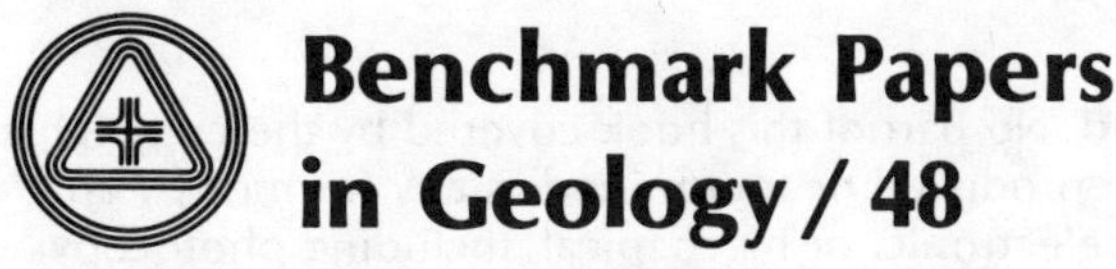

Benchmark Papers in Geology / 48

A BENCHMARK® Book Series

GEOLOGY OF THE PLANET MARS

Edited by

VIVIEN GORNITZ

Columbia University
and
Institute for Space Studies

STROUDSBURG, PENNSYLVANIA

Benchmark Papers in Geology, Volume 48
Library of Congress Catalog Card Number: 78-13589
ISBN: 0-87933-339-1

81 80 79 1 2 3 4 5
Manufactured in the United States of America

LIBRARY OF CONGRESS CATALOGING IN PUBLICATION DATA

Main entry under title:
Geology of the planet Mars.
(Benchmark papers in geology ; v. 48)
Includes bibliographical references and index.
1. Mars (Planet)—Geology. I. Gornitz, Vivien.
QB641.G46 559.9'23 78-13589
ISBN 0-87933-339-1

Distributed world wide by Academic Press, a subsidiary of Harcourt Brace Jovanovich, Publishers.

SERIES EDITOR'S FOREWORD

The philosophy behind the "Benchmark Papers in Geology" is one of collection, sifting, and rediffusion. Scientific literature today is so vast, so dispersed, and, in the case of old papers, so inaccessible for readers not in the immediate neighborhood of major libraries that much valuable information has been ignored by default. It has become just so difficult, or so time consuming, to search out the key papers in any basic area of research that one can hardly blame a busy man for skimping on some of his "homework."

This series of volumes has been devised, therefore, to make a practical contribution to this critical problem. The geologist, perhaps even more than any other scientist, often suffers from twin difficulties—isolation from central library resources and immensely diffused sources of material. New colleges and industrial libraries simply cannot afford to purchase complete runs of all the world's earth science literature. Specialists simply cannot locate reprints or copies of all their principal reference materials. So it is that we are now making a concerted effort to gather into single volumes the critical materials needed to reconstruct the background of any and every major topic of our discipline.

We are interpreting "geology" in its broadest sense: the fundamental science of the planet Earth, its materials, its history, and its dynamics. Because of training and experience in "earthy" materials, we also take in astrogeology, the corresponding aspect of the planetary sciences. Besides the classical core disciplines such as mineralogy, petrology, structure, geomorphology, paleontology, and stratigraphy, we embrace the newer fields of geophysics and geochemistry, applied also to oceanography, geochronology, and paleoecology. We recognize the work of the mining geologists, the pertroleum geologists, the hydrologists, the engineering and environmental geologists. Each specialist needs his working library. We are endeavoring to make his task a little easier.

Each volume in the series contains an Introduction prepared by a specialist (the volume editor)—a "state of the art" opening or a summary of the object and content of the volume. The articles, usually some twenty to fifty reproduced either in their entirety or in significant extracts, are selected in an attempt to cover the field, from the key papers of the last century to fairly recent work. Where the original works are in foreign languages, we have endeavored to locate or commission translations. Geologists, because of their global subject, are often acutely aware of the oneness of our world. The selections cannot, therefore, be restricted to any one country, and whenever possible an attempt is made to scan the world literature.

To each article, or group of kindred articles, some sort of "highlight commentary" is usually supplied by the volume editor. This commentary should serve to bring that article into historical perspective and to emphasize its particular role in the growth of the field. References, or citations, wherever possible, will be reproduced in their entirety—for by this means the observant reader can assess the background material available to that particular author, or, if he wishes, he, too, can double check the earlier sources.

A "benchmark," in surveyor's terminology, is an established point on the ground, recorded on our maps. It is usually anything that is a vantage point, from a modest hill to a mountain peak. From the historical viewpoint, these benchmarks are the bricks of our scientific edifice.

RHODES W. FAIRBRIDGE

PREFACE

The planet Mars has always stimulated mankind's imagination, from the ancient myths to countless science fiction stories, such as H. G. Wells's *War of the Worlds*. The true picture of Mars may appear less dramatic than in fiction, but is extraordinary nevertheless. The Viking Landers, which transmitted to Earth the first detailed pictures of the martian surface, represent an extension of man's eyes, ears, and hands over a distance of 80 million km of empty space.

Mars is in many ways the most Earth-like of the planets. It has seasons, ice caps, a thin atmosphere, volcanoes, rift valleys, sand dunes, rocky deserts, and water-worn channels. Mars lies between the Moon and Earth in its geological evolution, and its relatively primitive nature may help elucidate the more complex processes that have shaped the Earth. In the words of James C. Fletcher, former director of NASA: " . . . every new insight in comparative planetology may be the greatest of treasures sent home by the Viking explorers."

The geologic study of Mars began with space exploration. Each succeeding space mission drastically altered our preconceptions about Mars. The now historic initial results of the U.S. Mariner and Viking and Soviet Mars programs appear in several papers that have been reproduced in this volume. The new discoveries in planetary exploration have produced an information explosion and the field is still in a state of flux. There exists a need to place the flood of data on Mars into some perspective and to assemble the source papers that have strongly influenced our current understanding. This volume in the Benchmark series represents an attempt to fulfill that need.

Papers were selected to cover a broad range of topics in martian geology: geomorphology, degradation processes, geochemistry, geophysics, volcanism and tectonics, and cratering among others. Because of the relevance to planetary evolution, more space is allotted to the polar caps and the controversial channels than, for instance, to eolian activity (which is described in the Introduction). The composition and evolution of the martian atmosphere are closely related to the extent of outgassing, which points to the level of geological development. The amount of available information in geophysics and internal structure remains sparse. Although strictly speaking, the Viking biology experiments lie outside the realm of geology, their inclusion was considered for several reasons: the primary motivation behind the Viking program was to establish whether life exists on Mars; biological processes (wherever present) interact with and modify geological processes; and the curious responses actually observed in all the biology experiments may have an inorganic explanation and therefore be part of geochemistry.

Most papers on martian geology are published in a limited number of journals. Nearly 70 percent of Mars articles in my files as of September 1977 came from only three journals: *Icarus, Science,* and *Journal of Geophysical Research.* This bias was prevalent in spite of a thorough literature search, including astronomy and foreign language journals.

In conclusion, I would like to thank Dr. M.H. Carr, Professor R.B. Hargraves, Dr. W.K. Hartmann, Dr. J.F. McCauley, Professor T.A. Mutch, and Professor C. Sagan for their helpful comments and kind advice during the early stages of the preparation of this volume.

VIVIEN GORNITZ

CONTENTS

PART III: POLAR REGIONS

PART IV: REMOTE SENSING OF THE SURFACE

PART V: GEOCHEMISTRY—IN SITU ANALYSES

PART VI: EROSION AND WEATHERING

PART VII: VOLCANISM AND TECTONICS

PART VIII: GEOPHYSICS AND INTERNAL STRUCTURE

PART IX: CRATERING

PART X: THE MARTIAN ATMOSPHERE

PART XI: CLIMATIC CHANGES ON MARS

PART XII: THE CANALS OF MARS

PART XIII: BIOGEOCHEMISTRY

PART XIV: THE MOONS OF MARS

CONTENTS BY AUTHOR

INTRODUCTION

HISTORICAL PERSPECTIVE

Mars, fourth planet from the sun, revolving in an orbit between that of the Earth and Jupiter, has attracted man's attention ever since prehistoric times because of its reddish color. The ancients associated the unusual color with warfare and bloodshed, which thus lead the Babylonians, Greeks, and Romans to name the planet after their gods of war respectively: Nirgal, Ares, and Mars. [a] The first recorded observation of Mars was made by Ptolemy in 272 B.C.

The planet Mars has occupied a significant place in astronomy, since Kepler's study of its motion led him directly to the discovery (published in 1609) that planetary orbits are elliptical, with one focus of the ellipse centered at the sun, rather than circular, thereby confirming the Copernican theory of a heliocentric Solar System. Among the early maps of Mars, those by Dawes (1864) and Schiaparelli (1877-1881) were the most detailed. Schiaparelli is best known, however, for his popularization of the so-called martian canals, features that generated intense controversy for many years. In fact, some of the presumed canals were observed previously by Beer and Maedler in 1840, and Secchi in 1869 had used the Italian term "canale" (channel) to describe them. Schiaparelli believed that the channels were a natural phenomenon: "It is not necessary to suppose them the work of intelligent beings, and notwithstanding the almost geometrical appearance of all of their system, we are now inclined to believe them to be produced by the evolution of the planet, just as on the Earth we have the English Channel and the Channel of Mozambique."

[a] The classical names for Mars still survive as the names for the large channels discovered by Mariner 9: Nirgal Vallis, Ares Vallis, Ma'adim Vallis, and so forth (DeVaucouleurs et al. 1975). Mars still appears as the name of the month *March* (*mars,* in Fr., *März* in Ger.) and the day of the week: *mardi* (Fr.), while *Tuesday* is after the Germanic god of war.

P. Lowell, however, in his popular books *Mars and Its Canals* (1906) and *Mars as the Abode of Life* (1908), developed the idea that the canals were artificial waterways constructed by intelligent beings who lived on a dehydrating planet and were desperately tapping water from the polar caps to supply their needs. Lowell described the canals in 1906 as follows: "Suggestive of a spider's web seen against the grass of a spring morning, a mesh of fine reticulated lines overspreads it, which with attention proves to compass the globe from one pole to the other." The modern viewpoint on this once controversial subject is discussed in Paper 33.

Mars—The View from Earth-Based Telescopes

Partly as a reaction to Lowell's excesses, planetary science declined during the first half of the twentieth century, a position from which it emerged only during the last twenty years, largely due to space exploration. Nevertheless, data about Mars continued to accumulate, and a picture gradually developed of a cold, dry planet, with a very thin atmosphere lacking essential oxygen—in other words, an environment extremely hostile to life as we know it.

Mars travels around the sun in an orbit whose plane is inclined to the ecliptic by 1.85°. The orbit of Mars has a rather high eccentricity (0.093; a value of 0.00 would indicate perfect circularity); thus its perihelion distance of 206.66 million kilometers differs significantly from its aphelion distance of 249.22 million kilometers. Since its axis of rotation is inclined by 23°59′ from the perpendicular to the plane of its orbit, Mars experiences seasons just like Earth (the corresponding value for the tilt of the Earth's axis is presently 23°27′). However, because of the orbital eccentricity, the duration of the seasons on Mars is unequal; for example, spring in the northern hemisphere (or autumn in the southern hemisphere) is 199 days, summer 183, autumn 147, and winter 158 days. All four seasons on earth are of nearly equal length—approximately 90 days each. The martian year—that is, the period to complete one revolution of Mars around the sun—is 687 days, as compared with 365.26 days for the Earth. Another consequence of the eccentric orbit is that the southern hemisphere experiences a hotter summer than the northern one, because its summer (axis tilted *toward* sun) occurs during perihelion. Southern hemisphere summers are not only hotter but also shorter than their northern counterparts. Other planetary properties are listed in Appendix A.

Through the telescope, Mars appears as a yellowish-ochre planet with light and dark markings. The bright areas—orange to yellowish-brown—constitute around 70 percent of the surface and are located mainly in the northern hemisphere. The belief that the dark areas were vegetated persisted as recently as the 1960s. Opik (1966) felt that the seasonal "regenerative properties" of the dark areas implied vegetation. He referred to the *wave of darkening:* With the beginning of spring in each hemisphere, a dark band appears to form around the polar caps and to advance toward the equator as the ice retreats. This phenomenon was generally attributed to the spring blooming of vegetation as the ice melts. Alternatively, darkening was ascribed to soil moisture (Focas 1962). Kuiper (1952) considered the presence of lichens a distinct possibility. Yet by 1957, Kuiper had abandoned the vegetation hypothesis in favor of wind-scoured lava fields. Spectral measurements kept the controversy alive for several more years. Sinton (1957, 1959) detected absorption bands around 3.4 μ, which he interpreted as absorption by organic molecules. Colthup (1961) more specifically attributed the band at 3.43 μ to carbohydrates or plant protein and the band at 3.67 μ, to the aldehyde group. But according to Rea et al. (1965) the absorption bands were caused by the presence of the heavy hydrogen isotope in terrestrial atmospheric water vapor. This paper marked a turning point, from which time geological interpretations prevailed for the wave of darkening and other phenomena.

McLaughlin (1954, 1956) was one of the earliest to suggest a geological approach to the study of another planet. He explained the dark markings as windblown volcanic ash. Seasonal shifts in wind circulation pattern could thus account for the observed changes in the dark areas. This theme was further elaborated by Pollack and Sagan (1967) and Sagan and Pollack (1967, 1969). Frost-heaving as a cause for seasonal darkening was proposed by Otterman and Bronner (1966) but was considered unlikely on a large scale since the amount of liquid water was strictly limited (Anderson et al. 1967). As discussed below, with new data from satellite imagery, the eolian hypothesis has gained increasing favor (Sagan et al. 1972, 1973a; Greeley et al. 1974; Arvidson 1974).

Further understanding of the nature of the martian surface has come from measurement of optical properties and reflectance spectrophotometry. Polarization measurements (Dollfus 1957a, b; 1961; 1969) suggested the presence of limonite. Support for this viewpoint was contributed by Sharonov (1961), Moroz (1964), Draper et al. (1964), and Tull (1966). Alternatively, the spectral properties of Mars could be caused by a thin veneer of iron oxides or limonite coating silicate grains. Sensors in the visible to near-IR penetrate only ~ 1 μ and thus respond entirely to such surface coatings. The extent of iron oxide staining in terrestrial rocks often depends more on the amount of weathering than on iron oxide concentration. Data acquired by Van Tassel and Salisbury (1964), Younkin (1966), Binder and Cruikshank (1966), and Salisbury and Hunt (1968, 1969) apply this hypothesis to the martian surface. Plummer and Carson (1969) proposed that the color of Mars was caused by carbon suboxide, C_3O_2, derived from a photochemical reaction between CO_2 and CO. Perls (1973) noted that carbon suboxide would not be stable under martian conditions below 20 km in the absence of solar flares. Furthermore, a 2550 Å absorption feature, originally assigned to carbon suboxide, is probably caused by ozone. Therefore the presence of carbon suboxide was considered inprobable. (It may be of interest to note that carbon suboxide has recently been proposed as a reagent in the martian soil, in conjunction with the Viking biology experiments [Oyama et al. 1977]).

As mentioned above, like the Earth, Mars experiences seasons. The most prominent seasonal changes occur with the regular recession (and growth) of the polar caps whose composition will be reviewed below. As each cap recedes, a dark band or collar, originally seen by Beer and Von Madler in 1830, forms at its edge, as though melting ice moistened the soil. Each winter, a "polar hood" forms over the caps. This phenomenon, not to be confused with the "collar," is caused by white clouds of frozen water or carbon dioxide.

White clouds were already reported by Secchi in 1858 and Dawes in 1864. They occur in spring and summer mornings (probably as ground fog); also at sunset. Localized late afternoon clouds form orographically over prominent volcanoes in Amazonis and Tharsis (Masursky et al. 1972; Smith and Smith 1972; Leovy et al. 1973). Yellow dust clouds have been observed to form more commonly at perihelion, when surface heating is maximum (Antoniadi 1909; Capen 1971; Golitsyn 1973). Dust storms appear to originate in geographically restricted locations; for example, the Hellas basin. De Vaucouleurs (1967) has estimated wind velocities up to 100 km/hr. Kiess et al. (1960) proposed that yellow clouds might consist of NO_2, traces of which were reported by Gaiduk (1971). However, the estimated abundance is too low to account for the clouds.

Another phenomenon is the blue haze, which strictly speaking is not haze, but rather refers to the absence of contrast when the martian surface is observed with blue filters (below 4500 Å). It appears as though the "haze" has absorbed blue light and reflected red. At times, a "clearing" has been noted. Ice crystals, dry ice crystals, dust particles, and low surface contrast in the blue (as deduced from reflectance spectra) have all been advanced as possible explanations. Earth-based spectral and polarization measurements support the low surface contrast theory (Evans 1965; Dollfus et al. 1966; Pollack and Sagan 1967; Adams and McCord 1969) as do early Mariner results (Leighton et al. 1969; Leovy et al. 1971). However, the presence of reddish dust particles suspended in the atmosphere remains a possiblility, now supported by recent Viking observations (Prokof'eva et al. 1975; Mutch et al. 1976a).

While it had been known for a long time that Mars possessed a thin atmosphere, its exact composition and pressure were uncertain until quite recently. In 1947, Kuiper detected CO_2, with an estimated partial pressure of 0.3 mb (Kuiper 1952). Grandjean and Goody in 1955 revised this value to 2 mb. At that time, the total atmospheric pressure was estimated at $\sim$80 mb, and N_2 was assumed to be a major constituent (De Vaucouleurs 1954; Kellog and Sagan 1961). In the 1960s, new data gradually forced a downward revision of the total pressure to about 15–20 mb and that of CO_2 to approximately 4 mb (Chamberlain and Hunten 1965; Belton and Hunten 1966; Vasil'chenko and Moroz 1967). Proof of water vapor in the martian atmosphere was furnished by Spinrad et al. (1963). However, the amount was extremely low: about 14 ppt. μ[b]. By contrast, the atmosphere over a terrestrial desert contains 1000 ppt.μ. Traces of CO and O_2 were also detected (Kaplan et al. 1969; Belton and Hunten 1968). A more accurate determination of the atmospheric composition of Mars was provided by the Mariner space missions to Mars beginning in 1965.

THE SPACE EXPLORATION OF MARS

The geological examination of Mars really begins with the era of space exploration. The most powerful earth-based optical telescopes have a resolution of 100 km on the martian surface, although narrower high contrast features can be detected. On the other hand, Mariner 4 TV imagery had a nominal resolution of about 3 km at closest approach, and camera B TV imagery on Mariner 9 could resolve details as small as 100 m across. It is not at all

[b] A precipitable micron is a unit of thickness of water, if all of the moisture on Mars contained in a column one cm^2 at the base were condensed.

surprising that our knowledge of Mars has undergone an exponential increase beginning with Mariner 4 and culminating in the dramatic soft landings of the two Vikings in the summer of 1976. The major findings of the various space missions are summarized below and in Appendix B.

The Mariner 4 spacecraft, arriving in the vicinity of Mars in July 1965, obtained 22 TV pictures covering 1 percent of the surface, from a distance of 17,000 to 12,000 km. Over 70 craters were clearly observed, ranging from 4 to 120 km in diameter. Leighton et al. (1965) reported a crater density similar to that of the lunar uplands, which by analogy to the moon, was inferred to be several billion years old. The old surface age was challenged, since due to the proximity to the asteroid belt, the cratering rate on Mars should be 10 to 25 times greater than on the Moon. The intensified cratering rate would thus reduce the age of the martian surface to between 300 and 800 million years (Baldwin 1965; Witting et al. 1965; Anders and Arnold 1965). Although Hartmann (1966) and Binder (1966) accepted a high asteroid flux, they still argued that the martian surface was more comparable in crater density to the lunar maria, whose estimated age lies between 2 and 4 billion years old.

A very weak magnetic field was detected by the Mariner 4 magnetometer, which gave an upper limit to the martian magnetic dipole between 3×10^{-4} that of the earth (Smith et al. 1965). The occultation experiment provided the first direct measurement of Mars' atmosphere (Kliore et al. 1965). It found a surface temperature of $180° \pm 20°$ K and total pressure ranging between 4.1 and 7.0 mb. The results from Mariner 4 generated the impression of an ancient, heavily cratered planet, which had been geologically inactive since its earliest days, whose surface always had lacked flowing water, and which never had a thicker atmosphere. In other words, Mars appeared almost to be a duplicate of our Moon.

The impression of a Moon-like planet was initially reinforced by Mariners 6 and 7, which flew past Mars in 1969 at a closest approach of some 3500 km from the surface. The two spacecraft acquired a total of 201 complete TV pictures, with a best resolution (camera B) at closest approach of 0.3 km and over 10 percent of the surface sampled with the low resolution A cameras. Mariner 6 and 7 TV images showed that craters were the dominant landform on Mars (Leighton et al. 1969). However, two distinct crater morphologies were recognized: small, bowl shaped (diameter less than 15 km) and large, flat-bottomed ($>$15 km). The depth-diameter ratios for the large, flat-floored craters were considerably lower than for equivalent-sized lunar craters, thereby implying erosional modification. Small craters appeared fresh and unmodified (Chapman et al. 1969; Murray et al. 1971). Previous asteroidal impact rates were now believed "gross overestimates," and the true ratio of objects striking Mars and Moon was considered to be closer to unity (Murray et al. 1971). The conclusion was that the large, flat craters dated from the final stages of planetary accretion nearly 4.6 billion years ago. However, the fact that the small craters always had a "fresh" morphology suggested an erosional episode strong enough to erase craters smaller than 15 km in diameter [c] after the terminal meteorite bombardment had ceased, yet no significant erosion during the interval of accumulation of the small bowl-shaped craters. A break in slope in the crater frequency-diameter curve suggestive of episodic erosion had already been noted on Mariner 4 imagery (Loomis 1965; Binder 1966; Hartmann 1966; Opik 1966; Chapman et al. 1969). This question subsequently became a major controversy in interpreting Mariner 9 data.

Other clues to the diversity of geological processes on Mars came from the discovery of *chaotic terrain*. This landform, centered at about 10°S, 325°E, consists of a rough, irregular jumble of blocks, ridges, and troughs formed by subsurface collapse. The exact cause of collapse was uncertain, however. The leading hypotheses included sudden thawing of buried permafrost, or volcanotectonic subsidence (Sharp et al. 1971). Whatever the origin, the existence of chaotic terrain implied a certain degree of internal activity and possibly a sign of "geothermal maturing of the planet." To quote Sharp et al. (1971): ". . . as earlier hinted by Urey, the most interesting time for Mars may still be ahead."

Mariners 6 and 7 detected other new landforms: the "featureless" (at a resolution of 500 m) surface of Hellas, an ancient impact basin 1700 km across, and "quasi-linear features" and "etch pits" at the edge of the south polar cap. Sharp et al. (1971) originally thought that a thick layer of dust had buried craters in the Hellas basin, but since recent Viking TV pictures reveal a rugged morphology, the apparent smoothness was probably caused by haze or suspended dust. (On the other hand, Dollfus [1970] reports polarization curves for cloud-free Hellas that match the other bright areas of Mars. The absence of craters, in his view, is not an atmospheric phenomenon.) Dunes, moraines, ridges were all proposed as possible explanations for the "quasi-linear" features, and wind

[c] Equivalent to an initial depth of approximately 2000 m.

deflation basins for the "etch pits" (Sharp et al. 1971). Frost thicknesses at the edge of the south polar cap were estimated to be some tens of meters.

The occultation experiment (Kliore et al. 1969) measured a surface atmospheric pressure between 7 and 6 mb. The IR spectrometer established that the chief atmospheric constituent was CO_2 (Herr et al. 1972) and also detected solid CO_2 absorption bands near 3 μ at the south polar cap (Herr and Pimental 1969). The IR radiometer found temperatures low enough to be consistent with dry ice (condensation T of CO_2 at 6.5 mb is 148°K) as the chief component of the south polar cap (Neugebauer et al. 1969).

The Mariner 9 spacecraft, launched in May 1971, became the first artificial satellite to orbit another planet. It reached its destination in November 1971, during a major dust storm that completely obscured the surface of Mars. About a week later, the first of two Russian space probes began orbiting the dusty planet (Appendix B). Mars 2 and 3 each consisted of an orbiter and lander pair. Although the Mars 3 capsule landed successfully, its TV camera transmitted only 20 seconds before stopping, either because of a mechanical malfunction or because the capsule may have been overturned by the high winds (Marov and Petrov 1973). However, the Mars orbiters detected a weak magnetic field about 1/4000 that of the Earth (Dolginov et al. 1973, 1975). Meanwhile, as the dust settled out, Mariner 9's two TV cameras began transmitting data to the Goldstone tracking station in California. During one year, the Mariner 9 TV cameras obtained over 7300 pictures that covered nearly the entire surface of Mars at low resolution (about 1 km) and about 10 percent of the surface at high resolution (0.1 km). The Mariner 9 UVand IR spectrometers and IR radiometer measured atmospheric pressure, temperature profiles, composition, and topography, while the S-band occultation experiment provided additional data on atmospheric pressure, topography, and the shape of Mars, and radio tracking data supplied information on the gravity field (Steinbacher et al. 1972; Masursky et al. 1972; Hanel et al. 1972a; Lorrell et al. 1972).

The Dynamic Mars—Mariner 9 Findings

While the previous flybys of Mars had detected a cratered surface similar to the Moon, Mariner 9 revealed a planet not merely pock-marked with craters, but also displaying entirely unsuspected landforms such as towering volcanoes; enormous canyons; channels that seemed to have been carved by torrential floods; and jumbled, collapsed terrain; and layered deposits at the polar caps. It soon became apparent that the surface of Mars was shaped by processes familiar to the geologist, although often on a scale unknown on earth.

Among the features that could be seen during the 1971 dust storm were the south polar cap and four dark spots. One of the spots was identified with the classical feature, Nix Olympica. The dark spots were giant volcanoes that projected above the dust layer, capped by complex, coalescing craters characteristic of collapse calderas (Masursky et al. 1972). Olympus Mons (formerly Nix Olympica) stands 23 km high, is 500 km wide at its base, and has a summit caldera 60 km in diameter. Thus when compared with Mauna Loa (Hawaii)—200 km diameter at its base and 9 km high from the ocean floor—Olympus Mons appears to be the largest known volcano in the Solar System. The other three "spots," aligned roughly NE in Tharsis are Arsia Mons, Pavonis Mons, and Ascraeus Mons, going south to north. Each volcano is about 20 km high, and 400 km in diameter.

Over a dozen large volcanic piles are now known (Carr 1973, 1974a; Greeley 1973). They occur on or near plains units in Tharsis, in Elysium, and around the ancient impact basin Hellas (Carr 1975; Potter 1976; Scott and Allingham 1976). The Tharsis Mountains have been classified as shield volcanoes by their morphology, but domes, calderas, crater chains (maars) and features with radiating channels and ridges (highly eroded volcanoes?) have also been observed. Only one possible stratovolcano, Elysium Mons, has been recognized (Scott and Allingham 1976; Malin 1977), although it has also been classified as a shield eruption (Carr 1976). Cinder cones are rare, perhaps because of the reduced gravity and thinner atmosphere (McGetchin 1973), although West (1974) documents presumed pyroclastic features.

The volcanoes span a wide range in age, with the Tharsis volcanics being the youngest and the increasingly older and more degraded constructs being in Elysium and Hellas. A minimum age for Olympus Mons is 130 million years, based on the same rate of accumulation as the Hawaiian Islands (Carr 1973). Hartmann (1973) and Soderblom et al. (1974) obtain a similar age, by crater density measurements, but Blasius (1976), refining the cratering analysis, finds an age closer to 1 billion years, while Neukum and Wise (1976), using different assumptions of impact fluxes, determine that Olympus Mons is at least 2.5 million years old. Resolution of the true age of the youngest volcanics remains a high priority, since the time of cessation of volcanism places constraints on the thermal history of Mars. The presence of wrinkle ridges and lobate flow structures suggests that much of the smooth, sparsely cratered plains of the northern hemisphere of Mars may consist of lava sheets blanketed by a veneer of eolian deposits (Carr et al. 1973; Carr 1973). These wrinkle or mare ridges are sinuous arches that occur in the lunar maria and also on Mars. They have been interpreted as either viscous lava flows extruded from fissures, or as compressional structures or overthrusts related to the cooling of mare lavas.

The existence of major volcanic features raises the question of planetary differentiation. The shield morphology, extensive lava channels and flow fronts suggest, by analogy, a similarity in composition to terrestrial low viscosity basalt. Initial results from the IR spectrometer analysis of atmospheric dust during the 1971 storm indicated a SiO_2 composition of 60 ± 10 percent, which is characteristic of intermediate igneous rocks, thus supporting the contention of geochemical differentiation (Hanel et al. 1972b).

A striking aspect of martian volcanoes is their giant size. Carr (1973, 1974a) has proposed that the size stems from a long period of accumulation of lava over a stationary magma source (i. e., a "hot spot" or "mantle plume"), with no relative motion between lithosphere and asthenosphere. By scaling the hydrostatic head needed to erupt magma from its source to the summit, Carr (1973) estimates that the magma chamber is at least 130 km deep on Mars, as compared with 60 km for Hawaii. Alternatively, Vogt (1974) suggests that volcano height is determined by plate thickness and therefore the lithosphere on Mars is believed to be 2 or 3 times as thick as around Hawaii.

While the evidence from Mariner 9 points to pervasive signs of extensional faulting, fracturing, and upwarping, no corresponding signs of crustal compression or horizontal displacement have been found on a comparable scale (excepting wrinkle ridges, if they originate by compression; however, they are of minor significance in size and number). Lyttleton (1965) has proposed that global expansion would result from a phase change of forsterite from high-pressure spinel structure to ordinary olivine, as internal temperatures increased by radioactive decay. Solomon and Chaiken (1976) also favor global expanison.

Mariner 9 images revealed for the first time a vast chasm in the Tithonius Lacus region, over 2500 km long, 200 km wide, and 3–4 km deep. The main canyon, named Valles Marineris after the spacecraft, superficially resembles a rift valley, such as the Red Sea (Masson 1977) but probably did not form by a spreading process, although Allègre and Courtillot (1974), Courtillot et al. (1975), and Sengör and Jones (1975) favor this hypothesis. More probably, the Valles Marineris graben belongs to the radial fracture system centered on Tharsis (Carr 1974a; Mutch et al. 1976b). Structural control is evident in the parallelism of the canyon walls, grabens, crater chains, and smaller, closed elongate basins, all trending roughly E 15°S (Sharp 1973b; Blasius et al. 1977). The canyon walls have receded mainly by mass wasting—although in several places, the branched side valleys resembling an incipient dendritic drainage system that are observed could have formed by ground ice sapping along fractures (Sharp 1973b). Mariner 9 and Viking imagery clearly show massive landslides and slumps along the canyon walls, but no trace of stream channels on canyon floors (Carr et al. 1976).

Crater frequency-diameter curves have been widely interpreted to indicate evidence for episodic erosion. A kink in the crater frequency-diameter curve, between 8 and 30 km, in the heavily cratered province represents a steady-state curve, where the rate of cratering is balanced by the rate of crater destruction (Hartmann 1973). The upswing for craters less than 10 km represents accumulation of smaller craters since the erosional episode occurred some 600 million to 3.5 billion years ago, if the martian cratering rate is six times that of the Moon. (Recent studies, however, cast doubt on episodic erosion as the cause for the observed break in slope. Instead, the slope change may be produced by the process of crater generation itself; see Oberbeck et al. 1977; Woronow 1977.)

Soderblom et al. (1974) assume that Mars and the Moon have had a similar cratering chronology and that the cratering flux has declined exponentially on both bodies. The comparable crater density of fresh, bowl-shaped craters on both heavily cratered martian highlands and cratered plains indicates that a short time intervened between the end of the intense bombardment, erosion, and formation of the oldest cratered plains unit around 4 billion years ago. Neukum and Wise (1976) also date the erosional event before or at the time of deposition of the oldest plains unit, around 4 billion years ago.

Jones (1974) finds that the period of modification coincides with the time of formation of the cratered plains, but places this period later in martian history at 450 million years ago, with a duration of 25 million years. The end of the preriod of intense surface modification marks an important time-stratigraphic horizon on Mars by providing a maximum age for the plains units. Martian channels and fretted terrain have comparable crater densities as the Lunae Planum unit (cratered plains; see Malin 1976a; Arvidson and Coradini 1975; Arvidson 1974a). It is tempting to relate the crater obliteration event to the onset of mare-type volcanism, outgassing, a thicker atmosphere, and brief period of fluviatile erosion (Jones 1974). If the Neukum and Wise (1976) chronology is valid, the bulk of the geological evolution of Mars could be compressed into the first half of its existence.

Eolian erosion and deposition are currently the most active geological processes shaping the surface of Mars. Even prior to Mariner 9, the redistribution of unconsolidated material by winds was inferred to contribute to seasonal and long-range albedo changes (Rea 1964; Pollack and Sagan 1967; Sagan and Pollack 1969; Cutts et al. 1971).

Mariner 9 has shown that the darkening of the classical low albedo feature, Syrtis Major, arises from the growth and coalescence of dark crater streaks, while the darkening of Lunae Palus results from gradual depletion of bright dust deposited during the 1971 storm (Sagan et al. 1973a). Streaks have been variously interpreted as deposition of bright dust in the lee of craters or removal of bright material everywhere but in the lee of craters to produce dark "tails" (Greeley et al. 1974; Sagan et al. 1973; Arvidson 1974b; Veverka 1975). The dark blotches have been identified as dune fields in several high resolution images (Cutts and Smith 1973). "Wind vane" maps of crater "tail" directions show close agreement with actual winds in the equatorial belt (Sagan et al. 1973). Dark splotches occur predominantly in large, flat-floored craters, while bright streaks are associated with small, bowl-shaped craters (Arvidson 1974). The latter association suggests that the high crater rims impede the flow of dust-laden air and force it to deposit its load downwind, in agreement with wind tunnel simulations (Greeley et al. 1974). Flat-floored craters act as sand traps, and where crater rims are low, the dark material spills over and forms a dark plume. The mobility of the dark material appears greater than the bright (Sagan et al. 1972, 1973a). If dark blotches generally represent sand dunes, the sands may lie closer to the optimum size for particle motion than the brighter, more stable streaks, which may consist of extremely fine-grained dust. This discussion implies that particle size rather than mineralogy or weathering is responsible for the albedo variation, although the latter causes have not been disproven.

Examples of wind sculpturing have been documented in equatorial and mid-latitude regions (McCauley 1973) and near the poles (Cutts 1973b; Sharp 1973c). Products of wind erosion include "exhumed" craters that had been buried under a dust blanket, longitudinal grooves and ridges (yardangs), deflation hollows, and pitted terrain.

A rather high value for the rate of wind erosion—1/cm/year—has been computed (Sagan 1973). A more realistic rate of 5×10^{-4}

to 10^{-3} m/year has been determined from recession of ejecta blankets (Mutch et al. 1976; Arvidson et al. 1975). Although wind scouring is still effective, erosion today is considerably reduced from former times.

Wind velocities were originally determined from the rate of motion of dust clouds across the surface. Leovy et al. (1972) estimated a minimum wind velocity of 55 m/sec from behavior of wave clouds. Gierasch and Sagan (1971) predicted wind speeds of 80m/sec on a 1 percent slope due to temperature differences generated by topographic relief ("slope" winds). Because of the thinner atmosphere and lower gravity on Mars, higher threshold drag velocities are required to initiate particle movement on Mars than on Earth. Minimum[d] threshold velocities on Mars are about 2m/sec for 300 μ-sized particles (Arvidson 1972; Hess 1973; Iverson et al. 1976), as compared with only 1/10 this velocity on Earth. Equivalent wind speeds are 50–100 m/sec, which lie close to the maximum wind speeds observed (Pollack et al. 1976; Arvidson 1972).

Once fine-grained particles have come into suspension, the rate at which they settle out is given by the Stokes-Cunningham equation. Unlike the large difference in velocities between Mars and Earth required to initiate particle motion, the settling velocities for both planets are remarkably similar (Arvidson 1972). In particular, once fine particles are aloft, they can be transported over long distances before being redeposited. Values for particle sizes of a few hundred microns have been derived from optical properties during the dust storm (Leovy et al. 1972). In order for such large particles to remain in suspension for so long, strong winds must have prevailed for several weeks.

Eolian deposits occur in the laminated strata of the polar regions (Cutts 1973b), as dune fields (Cutts and Smith 1973) and as a debris mantle. The deposits are thickest near the poles but gradually disappear between 30°N and 30°S (Soderblom et al. 1973a). The polar strata probably represent accumulations of windblown dust, possibly admixed with frozen volatiles. The layering probably records depositional discontinuities resulting from variation in the ability of the atmosphere to transport dust to the polar regions. These depositional breaks and unconformities have been linked to climatic changes (Cutts, 1973a; Cutts et al. 1976).

Perhaps the most unanticipated result of the Mariner 9 mission has been the discovery of river-like channels on Mars, which

[d] A plot of threshold drag velocity versus particle diameter displays a U-shape curve (Bagnold 1941). Large particles require higher velocities to overcome increased inertia while for very fine-grained particles, friction and intra particle cohesion must be overcome.

exhibit geomorphological features strongly suggestive of aqueous erosion, such as dendritic drainage, braided channels with sand bars, levees, terraces and a downslope gradient (McCauley et al. 1972; Milton 1973; Baker and Milton 1974). The braided channels of Mangala Vallis in Amazonis provide the most convincing evidence in favor of water erosion (Milton 1973), although nonfluvial explanations have also been advanced for these features (Maxwell et al. 1973; Carr 1974b; Schumm 1974).

The presence of river-like channels indicates that the environment of Mars must have been significantly different in the past. Liquid water is not stable under the present low atmospheric pressure (average surface pressure 5-8 mb; see Hord et al. 1972; Hanel et al. 1972b) and low temperatures except in very localized depressions where the partial pressure of water vapor might exceed 6.1 mb at 0°C (triple point pressure). About 95 percent of the total atmosphere is CO_2, and water vapor forms merely a trace constituent with maximum concentrations of 70–80 precipitable microns near the poles during spring and summer (Owen and Mason 1969; Barker et al. 1970; Farmer and Laporte 1972; Moroz and Nadzhip 1975; Farmer 1976). Water ice clouds have also been detected (Curran et al. 1973).

Climates in the past may have varied considerably. The layered deposits near the poles are generally assumed to contain a record of previous climate alterations. Variations in maximum insolation due to changes in martian orbital parameters have been cited as possible cause of climatic modification (Ward 1973, 1974; Murray et al. 1973; Sagan et al. 1973b). The eccentricity of the martian orbit displays a 95,000-year cycle, superimposed on a larger cycle of 2×10^6 years, during which the eccentricity ranges between 0.01 and 0.14 (Murray et al. 1973). The *average* polar insolation varies by only 1 percent, but the insolation of the subpolar point at perihelion can increase by 30 percent. Cyclical oscillations in the obliquity (axial tilt), which range between 15° and 35°, over 100,000- and million-year cycles must produce even larger changes. The annual solar insolation can double between the extremes of the obliquity range and in turn affect the intensity of planetwide dust storms and thus the ability of the atmosphere to transport dust to the polar regions. Long-term variations in solar luminosity would also affect climate (Sagan and Young 1973; Sagan 1977; Hartmann 1974a), although the terrestrial record speaks strongly against any major changes during the last 3×10^9 years.

As a consequence of orbital variations at times in the past, martian temperatures could have risen by as much as 30°K, and pressures could have attained 1 atm. Mutch et al. (1976b) point out, however, that the proposed climatic changes appear insufficient to maintain the liquid water required for rainfall and stream runoff. However, if the atmospheric pressure were sufficiently high, water could traverse long distances and erode large channels before evaporating.

Another problem is to establish the time of channel formation. If the channels are linked to periods of milder climate caused by obliquity or eccentricity changes, they should occur throughout martian history, right up to the present time. The polar layered deposits, which are believed to contain a climate record, rank among the youngest terrains of Mars and show few, if any, impact craters (Murray et al. 1972; Soderblom et al. 1973b). Estimates of the time required to accumulate these sediments range from a few million years (Murray et al. 1972) to 500 million years (Cutts 1973a). The channels, on the other hand, appear to date back several billion years to the formation of the cratered plains (Malin 1976a; Arvidson and Coradini 1975; Pieri 1976).

An alternative hypothesis relates the channels to the early presence of a thick atmosphere (Jones 1974). A case can be made for early outgassing on Mars, by analogy, to the differentiation inferred for Earth, Moon, and Mercury prior to 4 billion years ago (Hanks and Anderson 1969; Anderson 1972; Murray et al. 1975). Rapid outgassing of a thick atmosphere was a probable byproduct of the early melting on Earth (Fanale 1971a; Schwartzmann 1973; Ozima 1975) and may have also occurred on Mars (Fanale 1971b).

Many of the large martian channels may not even require an atmospheric source of water. A number of channels, especially those in the Chryse depression, emanate from chaotic terrain, which may have been formed by the sudden melting of subsurface ice (Belcher et al. 1971; McCauley et al. 1972). The rapid release of large volumes of water may have caused catastrophic floods that could have carved out the Chryse channels. Baker and Milton (1974) point out many resemblances between the martian outflow channels and the Channeled Scablands of eastern Washington state that were produced when an ice-dammed lake burst and drained in a few weeks (Bretz 1969). Floods may have occurred repeatedly (Hartmann 1974b). Melting of subsurface ice, however, is an inadequate explanation for the sinuous, winding furrowed channels that occur in a broad subequatorial belt (Pieri 1976). The widespread distribution of these small channels argues for a period of atmospheric precipitation. Preliminary crater counts suggest that the small channels are also very old, and their presence supports the concept of an early thicker atmosphere (Pieri 1976).

Major proposed sinks of volatiles include adsorption of H_2O

and CO_2 on the regolith (Fanale and Cannon 1971, 1974; Pimental et al. 1974), in hydrated minerals (Huguenin 1974; 1976), in laminated deposits (Cutts 1973a), at the polar caps (Murray and Malin 1972), and as buried ice underlying chaotic terrain. These sources could yield up to 4×10^{21} g (Appendix C). Anders and Owen (1977) estimate some 10^{21}g.

The polar caps constitute the most visible reservoir for volatile deposits, but their composition and thickness have aroused considerable controversy over the years. Past observers of Mars assumed that the polar caps were frozen water by analogy to the earth (De Vaucouleurs 1954; Kuiper 1952) although A.R. Wallace in 1907, in attacking Lowell's notions of an inhabited Mars, suggested that because of extremely low temperatures, the caps should consist of dry ice. This idea became more popular after Leovy (1966) demonstrated that polar temperatures in winter were indeed low enough (less than 145°K) to allow condensation of dry ice, and Leighton and Murray (1966) developed a thermal model that explained the seasonal behavior of the caps in terms of CO_2 ice. While thermal and spectral data from Mariners 6 and 7 initially appeared to support the CO_2 theory (Neugebauer et al. 1969; Herr and Pimentel 1969), evidence from Mariner 9 and especially Viking has demonstrated that the *residual* polar caps are water ice (Murray et al. 1972; Soderblom et al. 1973b; Murray and Malin 1973b; Kieffer et al. 1976).

The Viking Mission

The most ambitious and complex unmanned planetary exploration to date has been the highly successful Viking mission to Mars. Two identical spacecraft, each consisting of an Orbiter and Lander capsule, were launched on August 20 and September 9, 1975, respectively.

Viking Lander instruments probed the inorganic soil chemistry (X-ray fluorescence), measured atmospheric composition (mass spectrometer), seismology, meteorology, surface magnetism, and provided ground imaging (two Lander facsimile scanning cameras). From orbit, a pair of TV cameras surveyed the ground, while the Infrared Thermal Mapper and Atmospheric Water Detector respectively measured surface temperatures and atmospheric water vapor content (Corliss 1975; Soffen and Synder 1976).

The Viking Landers were aimed at targets where the prospects for life appeared greatest. Viking 1 was directed toward the "mouth" of the large channel systems in Chryse Planitia (Plains of Gold) at about 20°N latitude in apparent alluvial deposits. Traces of soil moisture might still have been present and might have supported a population of microorganisms, or preserved fossils from an earlier, wetter epoch. The second Lander was pointed to 44°N latitude, where the atmospheric moisture was highest. Safety considerations ruled out any terrain that was not topographically smooth and hazard free. Furthermore, the first landing was restricted to a zone ±20° latitude in order to be accessible to Earth-based radar.

Viking Lander 1 successfully reached the surface of Mars at 8:12 AM EDT, on July 20, 1976, at 22.4°N, 47.5°W (Masursky and Crabill 1976a). After an even more extensive search, Viking 2 touched down on Mars, in Utopia Planitia on September 3, 1976, 7 PM EDT, at 47.89°N, 225.86°W (Masursky and Crabill 1976b).

Orbiter

A preliminary survey of obital TV images has led to an increased awareness of the role of water and buried ice in shaping the martian landscape. The pictures removed any remaining doubts that fluvial erosion had incised a widespread network of channels, streamlined teardrop "islands," and plucked and scoured channel floors in vast floods in Chryse and elsewhere. Although the floods were not recent, as implied by the number of impact craters superposed on the channel beds (Carr et al. 1976), they may have spanned a wide range in time to 0.5 billion years (Masursky et al. 1977).

Ejecta blankets of many impact craters display flow lobes and other indications of fluidization rather than the ballistic deposition characteristic of the Moon and Mercury. One interpretation has been that the projectiles crashed into a buried layer of permafrost that melted or vaporized on impact (Carr et al. 1976, Carr et al. 1977a).

Polygonal fractures have been observed near the Viking 2 landing site and have been compared to Arctic "patterned ground." However, the fissures also resemble cooling cracks in congealed lava flows, or dessicated mud-cracks (Masursky and Crabill 1976b; Lachenbruch 1962; Neal et al. 1968). Other possible permafrost features have been documented by Carr and Schaber (1977). One striking image of a 120 km long channel originating in a triangular collapsed depression strongly supports the relation between formation of chaotic terrain and sudden melting of subsurface ice (Carr et al. 1976).

The Viking mission has conclusively established the composition of the permanent polar caps to be water ice (Kieffer et al. 1976; Farmer et al. 1976). The Viking Orbiter pictures have also shown that much of what appeared as sparsely cratered "smooth plains" on Mariner 9 images actually consists of heterogeneous terrain affected by a complex sequence of deposition, erosion, cratering, volcanism, and fracturing. This is particularly true for the regions surrounding the two Viking landing sites originally selected for their geological "blandness" (Greeley et al. 1977; Guest et al. 1977).

Lander

Among the earliest Lander results have been the chemical analyses of the atmosphere. The mean surface pressure at the Viking 1 site was 7.65 m bars and 7.75 m bars at Utopia Planitia (Hess et al. 1976).

The presence of about 3 percent N_2 and 1.6 percent Ar has been determined for the first time (Owen and Biemann 1976; Clark et al. 1976a). The low value for Ar contrasts sharply with earlier estimates of up to 25–30 percent from the Soviet Mars 6 space probe (Istomin et al. 1975; Moroz 1976; Istomin and Grechnev 1976; Kaplan et al. 1975). Traces of xenon and krypton have been detected, with the latter being more abundant as on Earth (Owen et al. 1976, 1977). The xenon is adsorbed on terrestrial shales, and its deficiency in the martian atmosphere may arise from similar causes. The upper atmosphere of Mars contains traces of CO, O_2, O, N_2, and NO in addition to CO_2 (Nier et al. 1976a).

In view of the geological diversity of terrain and the great distance separating both lander sites, it is rather surprising that the chemical composition of the soil, as determined by X-ray fluorescence, should be so similar for both sites (Toulmin et al. 1976; Clark et al. 1976b; Baird et al. 1976, 1977). A low SiO_2 content (below 45 percent) is present; the total Fe (as Fe_2O_3) is near 20 percent; and lesser amounts of alumina, magnesium, and calcium oxides occur. Potassium is rather low (3.1 percent). The large concentration of sulfur (6.5–9.5 percent) and significant amounts of chlorine were unexpected. The mafic character of the soil contrasts with the fairly silicic composition of atmospheric dust initially deduced from the IR spectrometer data during the Mariner 9 mission (Hanel et al. 1972; Aronson and Emslie 1975; Toon et al. 1976). On the other hand, montmorillonite had also been proposed as a possible dust component, which agrees better with the tentative soil mineralogy (Hunt et al. 1973).

The yellowish-brown surface color is probably produced by

a thin coating ($\ll 0.25\mu$) of hydrated iron oxides (Clark et al. 1976) or nontronite (Huck et al. 1977). Some magnetite (Fe_3O_4) or maghamite (γ-Fe_2O_3) could be present, since magnetic particles have been detected (Hargraves et al. 1976). The homogeneity of the surface materials suggests that the soil may be a product of chemical weathering and efficient mixing. The vesicular texture of many martian surface rocks, and the morphology of the giant volcanoes and lava flows point to a widely distributed basaltic parent rock, which implies that the overall degree of magmatic differentiation on Mars is less than on Earth (Carr et al. 1977b).

The Viking 2 seismometer detected only one probable seismic event during the first 146 sols (martian days) of operation (Anderson et al. 1977). However, the short operating period of the instrument and paucity of Mars quakes prevent final conclusions about the level of martian seismicity.

Biogeochemistry

A major Viking objective has been to establish whether or not life exists on the Red Planet, and four experiments were expressly designed for this purpose. These include the pyrolytic release (a test of carbon assimilation, or photosynthesis), gas exchange (a test of metabolism), labelled release experiments (tests of metabolism and growth), and the gas chromatograph–mass spectrometer (to identify organic molecules) (Horowitz et al. 1972; Oyama 1972; Levin 1972; Klein et al. 1972).

The Viking biology experiments assume a carbon-based life, although other exotic chemistries are conceivable (Pimentel et al. 1966). The biology experiments have yielded ambiguous results. However, the test for the presence of organic molecules has been essentially negative (Biemann et al. 1976b). The gas chromatograph–mass spectrometer, whose sensitivity is on the order of 1 ppb, failed to detect anything beyond chemically bound water, adsorbed CO_2, and traces of solvent impurities. If fewer than 10^6 microorganisms were theoretically present in a sample, the organic matter could go undetected. Alternative explanations are that martian microorganisms recycle efficiently any organic debris ("scavengers") or that the results of the remaining experiments have a nonbiological origin.

The pyrolytic release (carbon assimilation) experiment was designed to measure biological conversion of CO_2 and CO into organic matter (Horowitz et al. 1972). While the Utopia results are somewhat less clear, the PR analyses ". . . suggest that an organic synthesis from CO or CO_2 occurs in the martian surface material.

The synthesis is weak compared to that found in biologically active terrestrial soils and, unlike the latter, is inhibited by small amounts of water. It resembles a biological reaction in being thermolabile, although there is room for doubt that the synthesis was completely abolished in [the sterilized] sample, which had been heated to 180°C for 3 hours" (Horowitz et al. 1976). A more probable nonbiological explanation involves the photochemical conversion of CO_2 into carbon suboxide polymer, in the presence of dust particles (Oyama 1977). Radioactive decay of C^{14} from the "tagged" gases provides enough energy to activate the polymer and incorporate it into the "second peak." Alternatively, organic compounds could be synthesized by reaction of UV light on CO under martian conditions (Hubbard et al. 1971).

The gas exchange experiment attempted to monitor changes in the gas composition, as byproducts of biological metabolism (Oyama 1972). The soil sample may either be in contact with (humid) or directly wetted by the nutrient medium. In the humid mode, quantities of O_2 and CO_2 were evolved, reaching a plateau after 50 hours (Oyama 1977). When immersed in nutrient, the rate of evolution decreased. The gasses O_2 and CO_2 are believed to be chemically bound to iron-oxide coated silicate soil particles. Ultra-violet radiation has formed calcium (or other alkali, alkaliearth) peroxides or superoxides, which release O_2 and H_2O_2 (hydrogen peroxide) in the presence of water vapor.[e] Adsorbed CO_2 can be removed by reaction with Ca^{++}.

The labelled release experiment tested for metabolism during incubation of a sample moistened with C^{14}-tagged nutrient solution (Levin 1972; Levin and Straat 1976, 1977). Results demonstrate a rapid increase in radioactivity upon injection of nutrient to a level indicating that only one of the nutrients was consumed. The responses show good reproducibility at both Lander sites. The "active" response is eliminated by sterilization for 3 hours at 160°C and significantly reduced by "cold" sterilization for 3 hours at 50°C, in two separate runs. The nonbiological explanation requires the presence of maghamite (γ-Fe_2O_3) in the soil, to catalyze the formation of hydrogen peroxide, which then reacts with formate and all of the other organic nutrients (except glycine) to yield $C^{14}O_2$. However, two major difficulties remain for the peroxide theory. Although the pyrolytic release "second peaks" were sharply curtailed for a sample collected from beneath a large rock (protected from UV-induced photochemical reactions), the label-

[e] $Ca(O_2)_2 + 2H_2O \rightleftarrows Ca(OH)_2 + H_2O_2 + O_2$

$Ca(OH)_2 + CO_2 \rightleftarrows CaCO_3 + H_2O$

led release experiment showed the *same* level of radioactivity for the Utopia "under-the-rock" sample as the Chryse surface specimens. The sharp reduction of the response at 50°C places constraints on possible chemical oxidants. Furthermore the positive "under-the-rock" result casts doubts on theories requiring direct and recent UV reactions. Levin and Straat (1976) conclude: "As yet, however, no chemical experiment has quantitatively reproduced the LR Mars data. Thus, despite all hypotheses to the contrary, the distinct possibility remains that biological activity has been observed on Mars." While the present consensus of opinion favors a nonbiological explanation for all the observed results, the labelled release experiment is the only one that could still be considered biologically "positive" (Klein 1977), although Klein (1978) now feels that because of the kinetics of the reaction and the amount of nutrient consumed, that a biological interpretation is improbable. Further Viking analyses are not expected to resolve this ambiguity.

The Moons of Mars

The two moons of Mars were discovered by Asaph Hall in 1877, and were named Phobos (Fear) and Deimos (Terror) after the mythological attendants of Ares. Curiously, Kepler in 1610, had predicted that Mars should have two moons. However his reasoniong was based more on numerology than science. He noted the geometrical progression: Venus (zero satellites), Earth (one), and Jupiter (four Galilean moons). Thus Mars, situated between Earth and Jupiter, should have two. Evidently, the British writer Jonathan Swift was familiar with Kepler's prediction, which he incorporated into *Gulliver's Travels* (1726).[f]

Phobos and Deimos travel in nearly circular orbits that lie in the plane of Mars' equator. Phobos completes a revolution about Mars in 7 hours and 39 minutes while Deimos, which is farther from Mars, requires 30 hours, 18 minutes. Mariner 9 TV images have revealed both moons as irregular lumpy objects, saturated with craters (Masursky et al. 1972; Pollack et al. 1973). Their shapes

[f] Swift gives the distance of the orbit of the innermost moon as 3 diameters from the center of Mars, that for the outermost moon as 5 diameters. The corresponding periods from Kepler's Third Law are 10 and 21.5 hours, respectively. However, the mass of Mars when calculated from this data turns out to be about 6 times the actual mass, as determined from the orbital parameters of Phobos and Deimos reported by Hall. Richardson and Bonestell (1964) point out that if Swift had meant "radius" instead of "diameter," the values for the mass of Mars would be much closer to the true value—another uncanny coincidence!

are best approximated by triaxial ellipsoids, whose radii are listed in Appendix A. Both satellites display synchronous rotation, with their longest axis pointing toward Mars, and shortest axis normal to the orbital plane. Calculations indicate that Phobos would need 10,000 to one million years to achieve synchronous rotation while Deimos would require 100 million years (Pollack et al. 1973). The orbital velocity of Phobos is slowly increasing, which brings the satellite ever closer to Mars. At the present rate, Phobos is expected to crash into Mars in 100 million years. The phenomenon, termed secular acceleration, was originally noted by A.B. Sharpless in 1945 and confirmed by Smith and Born (1976). This puzzling phenomenon led Shklovskii to the ingeneous suggestion that Phobos must have a very low density and thus must be hollow and artificial (Shklovskii and Sagan 1966). Since Mariner 9 and Viking clearly show the nonartificial appearance of Phobos, a more probable explanation for the acceleration is tidal drag.

TV imagery shows that both moons are heavily cratered. The irregular shape of Phobos suggests that not only is it a solid object (as opposed to a gravitationally bound aggregate of planitesimals), but that pieces must have been knocked off by large impacts. Crater chains and striations (visible in Viking images) suggest structural damage in a rigid body. However, the surface is covered by a regolith, several hundred meters deep. The presence of a regolith has been inferred by polarization, photometric, and thermal measurements (Zellner 1972; Noland and Veverka 1977).

While the escape velocities for Phobos and Deimos are quite low, and ejecta might be expected to escape, some fraction of the ejecta may have less than escape velocity. Furthermore, the debris might orbit Mars and be recaptured by the two moons (Pollack et al. 1973). Phobos and Deimos strongly backscatter light as though the surface were covered by a fine, texturally complex powder consistent with the idea of a regolith. The two satellites reflect only 6 percent of visible light as compared with 11 percent for our Moon, while the color is dark gray, as compared with orange-yellow for Mars.

The composition of Phobos has been uncertain until quite recently. Exact knowledge of the density would place constraints on its bulk composition. The two leading possibilities are carbonaceous chondrites (ρ=2.3g/cm^3) or basalt (ρ=2.9g/cm^3). Discrimination between these two alternatives would also narrow down choices as to the origin of the objects. A basaltic composition would imply a melting history, hence origin within a larger, now shattered asteroid; if chondritic, the moons could represent residue of the planetary accretion process (Veverka 1977).

Measurement of the mass of Phobos from Viking 1 close encounter suggests a density near 2.0 g/cm^3 (Christensen et al. 1977; Tolson et al. 1978). The low density indicates a carbonaceous chondrite material. Furthermore, the spectral reflectance of Phobos more closely resembles spectral curves of carbonaceous chondrites and several asteroids than powdered basalt or our Moon, thus supporting the former alternative (Pang et al. 1978; Pollack et al. 1978).

Part I

INITIAL RESULTS OF SPACE EXPLORATION

Editor's Comments on Papers 1 Through 4

1 **LEIGHTON et al.**
Mariner IV Photography of Mars: Initial Results

2 **LEIGHTON et al.**
Excerpt from *Mariner 6 and 7 Television Pictures: Preliminary Anlaysis*

3 **MOROZ**
Preliminary Results of Studies Conducted on the Soviet Automatic Stations Mars 4, Mars 5, Mars 6 and Mars 7

4 **CARR et al.**
Preliminary Results from the Viking Orbiter Imaging Experiment

Space exploration has made the geological study of Mars possible. The following set of four papers records some of the historic, early discoveries of the space missions.

In 1965 Mariner 4 was the first unmanned spacecraft to obtain TV images of the surface of another planet. The TV cameras on this mission, as recounted by R. B. Leighton et al. (Paper 1), transmitted only 22 pictures to earth, yet fundamentally altered our conception of Mars. Although by the 1960s, earlier speculations about the existence of advanced forms of life on Mars had been largely dismissed, spectral observations and the seasonal color changes were still cited in support of a primitive vegetation cover (Sinton 1957, 1959; Colthup 1961; Opik 1966). The Mariner IV photographs of a heavily cratered Moon-like planet, with no signs of aqueous erosion and a very low (<10 mb) atmospheric pressure, indicated a geologically inactive body and an environment even more hostile to the evolution of life than had been previously anticipated.

The shortcomings of extrapolating from a 1 percent sample became evident from subsequent missions. The second paper presents the initial results of the Mariner 6 and 7 missions. While craters were still seen as the dominant landform on Mars, other as yet unknown processes had created "chaotic" and "featureless" terrain, and unusual morphologies near the south polar cap. Mariner 9 and later, Vikings 1 and 2, revealed that the martian surface exhibits a far greater diversity than the preceding missions had led to expect, and that Mars represents a body intermediate between the Earth and Moon in its geological development. (The Mariner 9 findings are reproduced in Paper 5 in Part II.)

The Viking orbital TV cameras documented convincing evidence for fluvial activity, volcanism, permafrost, and extensive surface modification (Paper 4). A dramatic moment was the successful soft landing of Viking 1 in Chryse Planitia on July 20, 1976, followed by the transmission of the first photographs from the surface of Mars showing a deep rust-colored, rock-strewn desert (Mutch et al. 1976).

During the 1970s, the Soviet Union launched a series of spacecraft toward Mars. "Mars 2" and "Mars 3" were the first artificial devices to land intact on the martian surface (Marov and Petrov 1973). Results of the 1973 "Mars" 4 through 7 missions are summarized in Paper 3. Some significant findings of these missions include confirmation of the presence of a weak magnetic field (Dolginov et al. 1975), the highest pre–Viking report of atmospheric water vapor content—up to 100 microns of precipitable water (Moroz and Nadzhip 1975)—and gamma-ray spectrometer measurements from orbit, which suggest a mafic surface composition (see Paper 15, Part IV).

The most controversial result to emerge from the "Mars" program was the estimate of up to 35 percent atmospheric argon, based on the behavior of a malfunctioning mass spectrometer ion pump (Istomin and Grechnev 1976). However, the Viking experiments show that the correct value lies closer to 2 percent (Owen and Biemann 1976; Clark et al. 1976; see also Paper 30, Part X).

1

Reprinted from *Science* **149**:627–630 (1965)

MARINER IV PHOTOGRAPHY OF MARS: INITIAL RESULTS

Robert B. Leighton, Bruce C. Murray, Robert P. Sharp
California Institute of Technology

J. Denton Allen and Richard K. Sloan
Jet Propulsion Laboratory, California

The Mariner IV spacecraft successfully acquired and transmitted to earth 22 pictures of the planet Mars taken from a distance of 17,000 to 12,000 km just before its closest approach to Mars at approximately 00:30 UT on 15 July 1965. This first report describes the performance of the television camera system, the resultant picture quality, and the more prominent surface features present in the picture. We feel that some of these features are so striking that certain physical and geological inferences can be drawn even at this early date.

The complete set of pictures in their current state of processing was released to the scientific community and the public on 29 July 1965. The completely processed pictures, the relevant calibration data, and a more detailed analysis of surface features will be presented later.

In regard to the design and performance of the TV system, one of the most difficult problems associated with the Mariner photographic mission to Mars was the wide illumination range and the low surface contrast to be expected. Since the camera would be viewing the surface from the bright limb to, and beyond, the evening terminator, the camera was called upon to respond to brightness ranging from full solar illumination near the subsolar point to near total darkness as the terminator was crossed. The slow-scan vidicon chosen was capable of handling a 30-to-1 range of illumination with fixed operating voltages. The low communications rate from the planet required a digital transmission system. In order to effectively utilize a high signal-to-noise ratio the video signal was divided into 64 equal increments. Since the video signal level would decrease as the photo path approached the terminator, automatic video gain control was incorporated. The control was designed to maintain a video level which would contain at least 15 of 64 increments. The camera telescope was of the Cassegrain type with 12-inch (30-cm) focal length, $f/8$ focal ratio, and 0.2-second shutter time.

The camera operated at the mini-

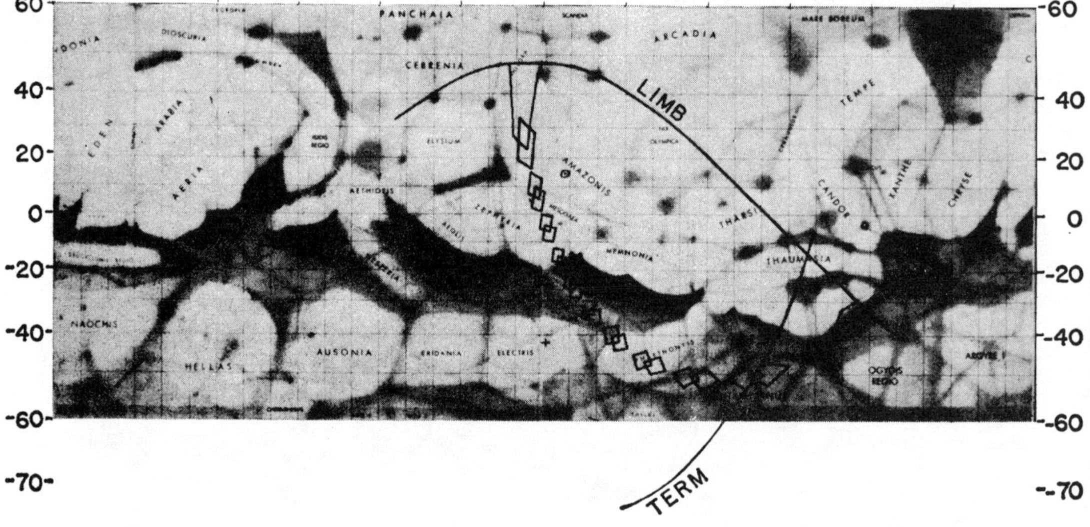

Fig. 1. Diagram of locations on the planet of the Mariner IV photographs. The pictures were taken in overlapping pairs, an orange filter being used for one member of a pair and a green filter for the other. *TERM*, terminator.

mum gain until picture No. 18, in which the video level fell below the minimum allowed (Fig. 1). The system then increased gain for picture No. 19 which was taken 96 seconds later. As the terminator was approached, the gain increased a second time for picture No. 20 and reached its maximum for pictures No. 21 and No. 22. The automatic adjustment of video signal level was not able to fully cope with the unexpectedly low light intensity, and pictures No. 15 through 20 show decreasing signal-to-noise ratios. In addition, a spurious background brightness was detected on picture No. 1 apparently at a distance of more than 100 km from the limb of the planet. Preliminary analysis has indicated this spurious background to be approximately one-fourth the brightness of the planet itself. Similar analysis of pictures Nos. 21 and 22, which were taken on the dark side of the terminator, reveals considerably lower levels of brightness, about 1/8th and 1/25th that of picture No. 1, respectively. This background is tentatively attributed to an instrumental defect of an optical nature which developed during the 7½-month

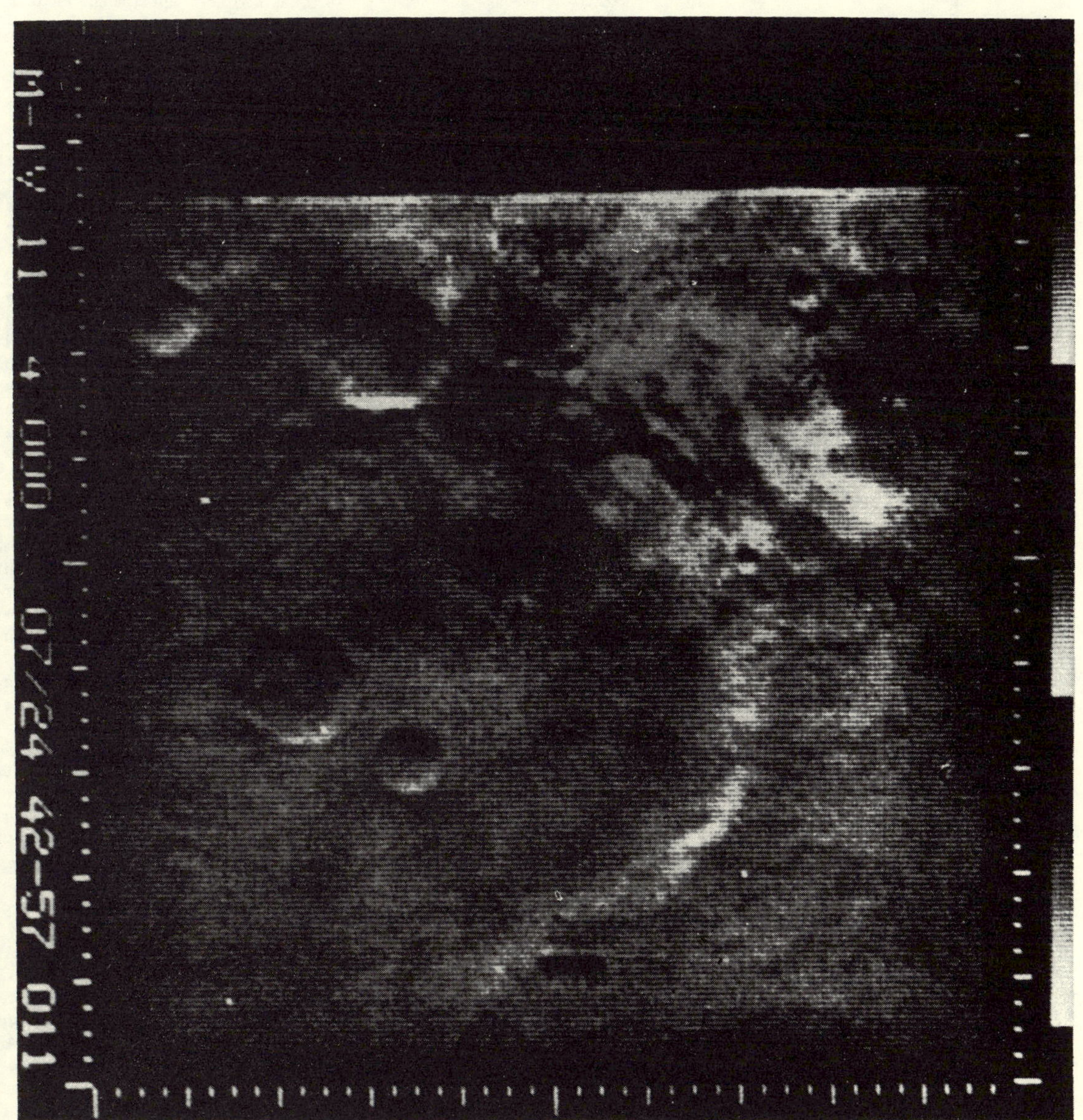

Fig. 2. Picture No. 11 of the Mariner IV sequence. This picture shows twelve craters ranging from five to about 120 km in diameter. The area of Mars covered by the picture is about 238 by 275 km.

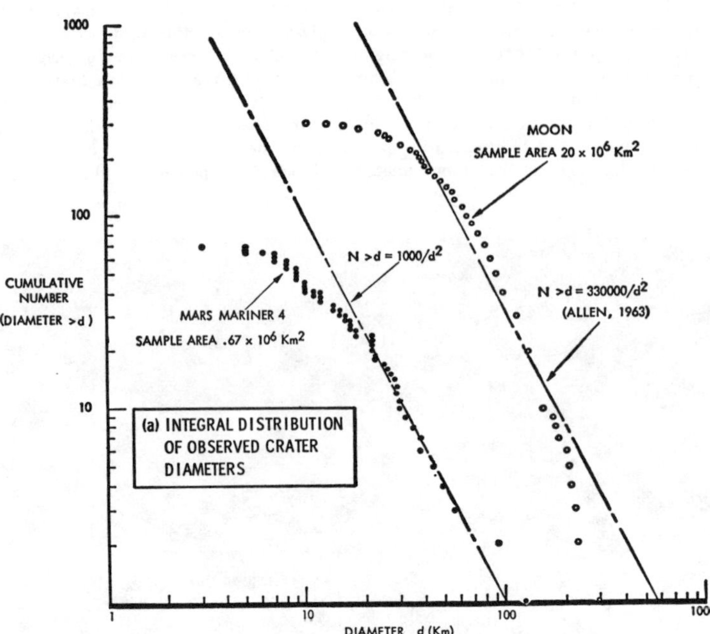

Fig. 3. Integral size distribution of crater diameters on Mars and the Moon: (*a*) as observed; (*b*) the same data, normalized to an area of 10^6 km^2; the lunar crater diameters were taken from Baldwin (*2*); the maria and uplands were not differentiated in the analysis.

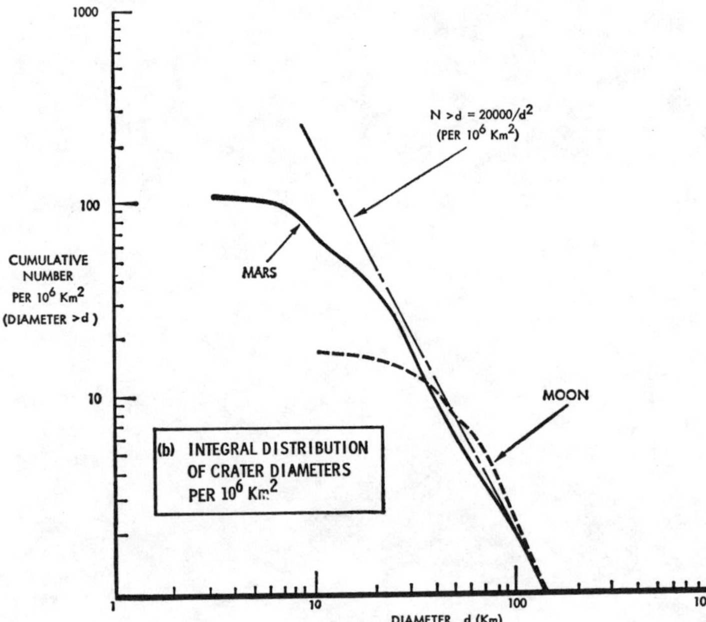

space flight. All other camera characteristics such as resolution and geometrical fidelity appear normal. Results of the first tape playback have indicated that the tape machine and communications equipment operated as designed.

In pictures No. 1 through No. 4, the very high solar illumination of the terrain viewed by the camera significantly reduced the visibility of surface features, as had been anticipated. Pictures No. 5 through No. 14, however, present a view of a densely cratered surface, closely comparable to bright upland areas of the Moon (Fig. 2).

We have observed more than 70 clearly distinguishable craters ranging in diameter from 4 to 120 km. It seems likely that smaller craters exist; there also may be still larger craters than those photographed, since Mariner IV photographed, in all, only about 1 percent of the Martian surface.

The observed craters have rims rising to about 100 m above the surrounding surface and depths of many hundred meters below the rims. Crater walls so far measured seem to slope at angles up to about 10°. The number of large craters present per unit area on the Martian surface and the size distribution of those craters resemble remarkably closely the lunar uplands, as illustrated in Fig. 3.

If the Mariner sample is representative of the Martian surface, the total number of craters of the sizes so far observed is more than 10,000 compared to a mere handful on Earth. In appearance, the Martian craters closely resemble impact craters on Earth, both artificial and natural, and the craters of the Moon. Craters of widely different degree of preservation and, presumably, age are distinguished. A few elongate markings of diffuse nature are present on the Mariner photos but at this early stage of analysis no conclusions can be offered concerning them. On frame No. 13, one such feature looks like a part of the edge of a very large crater and, perhaps significantly, lies near the border of a Martian dark area. In southern subpolar latitudes, where the season is now late midwinter, some craters appear to be rimmed with frost, particularly those in frame No. 14.

Some mention must be made of features looked for, but not seen, on the Mariner photos. Although the line of flight crossed several "canals" sketched from time to time on maps of Mars, no trace of these features was discernible. It should be remembered in this respect that the visibility of many Martian surface features, including the "canals," is variable with time. No Earth-like features, such as mountain chains, great valleys, ocean basins, or continental plates were recognized. Clouds were not identified, and the flight path did not cross either polar cap.

Although it may be difficult to ever arrive at an unambiguous identification and interpretation of all the features recorded on the Mariner photographs, we feel that the existence of a lunar-type cratered surface, even in only a 1-percent sample, has profound implications about the origin and evolution of Mars and further enhances the uniqueness of Earth within the solar system. By analogy with the Moon, much of the heavily cratered surface of Mars must be very ancient—perhaps 2 to 5 $\times$ 10^9 years old (*1*). The remarkable state of preservation of such an ancient surface leads us to the inference that no atmosphere significantly denser than the present very thin one has characterized the planet since that surface was formed. Similarly, it is difficult to believe that free water in quantities sufficient to form streams or to fill oceans could have existed anywhere on Mars since that time. The presence of such amounts of water (and consequent atmosphere) would have caused severe erosion over the entire surface.

The principal topographic features of Mars in the areas photographed by Mariner have not been produced by stress and deformation originating within the planet, in distinction to the case of Earth. Earth, of course, is internally dynamic, giving rise to mountains, continents, and other such features, whereas Mars has evidently long been inactive. The lack of internal activity is also consistent with the absence of a significant magnetic field on Mars, as determined by the Mariner magnetometer experiment.

As we had anticipated, Mariner photos neither demonstrate nor preclude the possible existence of life on Mars. Terrestrial geological experience would suggest that the search for a fossil record appears less promising if Martian oceans never existed. On the other hand, if the Martian surface is truly "near pristine," that surface may prove to be the best—perhaps the only—place in the solar system still preserving clues to primitive organic development, traces of which have long since disappeared from Earth.

References

1. E. M. Shoemaker, R. J. Hackmon, R. E. Eggleton, *Advan. Astronaut. Sci.* **8**, 70 (1963).
2. R. B. Baldwin, *The Measure of the Moon* (Univ. of Chicago Press, Chicago, 1963).

29 July 1965

2

Reprinted from *Science* **166**:49, 54–67 (1969)

Mariner 6 and 7 Television Pictures: Preliminary Analysis

R. B. Leighton, N. H. Horowitz, B. C. Murray, R. P. Sharp, A. H. Herriman, A. T. Young, B. A. Smith, M. E. Davies, C. B. Leovy

Before the space era, Mars was thought to be like the earth; after Mariner 4, Mars seemed to be like the moon; Mariners 6 and 7 have shown Mars to have its own distinctive features, unknown elsewhere within the solar system.

The successful flyby of Mariner 4 past Mars in July 1965 opened a new era in the close-range study of planetary surfaces with imaging techniques. In spite of the limited return of data, Mariner 4 established the basic workability of one such technique, which involved use of a vidicon image tube, on-board digitization of the video signal, storage of the data on magnetic tape, transmission to the earth at reduced bit rate by way of a directional antenna, and reconstruction into a picture under computer control. Even though the Mariner 4 pictures covered only about 1 percent of Mars's area, they contributed significantly to our knowledge of that planet's surface and history (*1, 2, 19, 21*).

The objectives of the Mariner 6 and 7 television experiment were to apply the successful techniques of Mariner 4 to further explore the surface and atmosphere of Mars, both at long range and at close range, in order to determine the basic character of features familiar from ground-based telescopic studies; to discover possible further clues as to the internal state and past history of the planet; and to provide information germane to the search for extraterrestrial life.

The Mariner 6 and 7 spacecraft successfully flew past Mars on 31 July and 5 August 1969, respectively; first results of the television experiment, based upon qualitative study of the uncalibrated pictures, have been reported (*3, 4*). The purpose of this article is to draw together the preliminary television results from the two spacecraft; to present tentative data concerning crater size distributions, wall slopes, and geographic distribution; to discuss evidences of haze or clouds; to describe new, distinctive types of topography seen in the pictures; and to discuss the implications of the results with respect to the present state, past history, and possible biological status of Mars.

The data presented here and in the two earlier reports were obtained from inspection and measurement of a partial sample of pictures in various stages of processing. As such, the results must be regarded as tentative, subject to considerable expansion and possible modification as more complete sets, and better-quality versions, of the pictures become available over a period of several months. They are offered at this time because of their unique nature, their wide interest, and their obvious relevance to the forthcoming Mariner 1971 (orbiter) and Viking 1973 (lander) missions.

Drs. Leighton, Horowitz, Murray, and Sharp are affiliated with the California Institute of Technology, Pasadena; Mr. Herriman and Dr. Young, with the Jet Propulsion Laboratory, Pasadena; Mr. Smith, with New Mexico State University, Las Cruces; Dr. Davies, with the RAND Corporation, Santa Monica, California; and Dr. Leovy, with the University of Washington, Seattle.

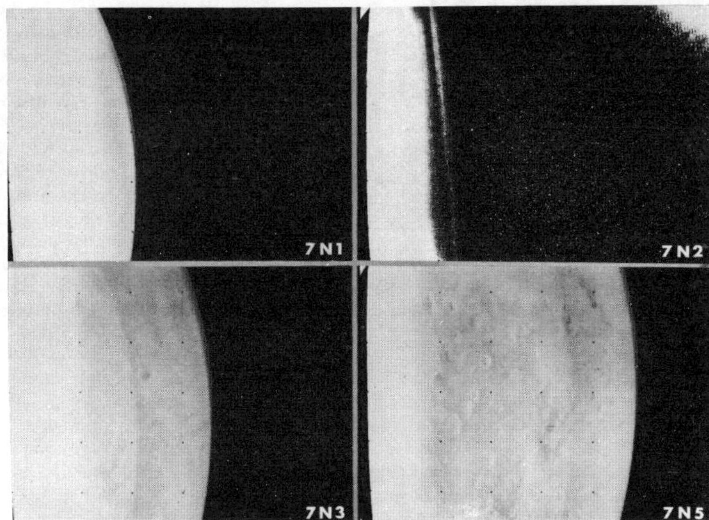

Fig. 4. Mariner 7 limb frames 7N1, 2, 3, and 5. Note the sharp haze layer adjacent to the limb in frames 7N1, 3, and 5, and the magnified view (tenfold magnification) in 7N2. The prominent, cratered dark feature in frame 7N5 is Meridiani Sinus. North is approximately toward the right.

[*Editor's Note:* Material has been omitted at this point.]

Observed Atmospheric Features

Aerosol scattering. Clear-cut evidence for scattering layers in the atmosphere is provided by the pictures of the northeastern limb of Mariner 7. The limb appears in frames 7N1, 2, 3, 5, and 7, and in a few real-time digital A-camera frames received immediately prior to frame 7N1. The limb appears again in frame 7N21 after the platform slew which began the track across Hellas. Thus the limb coverage includes each of the filters of the A camera and one B-camera frame.

Several characteristics of the scattering layer shown in Fig. 4 are evident even at this early stage. (i) The scattering is distinctly stratified in horizontal layers, just as scattering from aerosol layers in the earth's atmosphere is. (ii) The intensity of the scattering varies substantially over distances of a few hundred kilometers and is more intense toward the west or toward earlier local times of day. (iii) The thickness of the scattering layer is about 10 kilometers. (iv) The height of the layer is difficult to determine because of the difficulty of locating the true planetary limb, but it is estimated to be between 15 and 25 kilometers in the region covered by frames 7N1 to 7N7 and up to 40 kilometers in frame 7N21. (v) The layer is about 50 percent brighter in the blue-filter pictures than in the red or green. This is less difference in intensity than would be expected for Rayleigh scattering, but corresponds more closely to λ^{-2} wavelength dependence.

The relationship between this scattering layer and the martian tropopause is obviously of great interest and will be studied carefully as more refined data become available.

The normal-incidence optical depth, isotropic scattering being assumed, is estimated as 0.01 in the red and about 0.03 in the blue. A λ^{-2} dependence suggests that scattering should be predominantly forward, so these very small values should be underestimates.

The real-time digital data reveal an apparent limb haze near the south polar cap, and over the regions of Mare Hadriaticum and Ausonia just east of Hellas. The haze over these regions is not as bright as the haze discussed above, so it is unlikely that it is sufficiently dense to obscure surface features seen at NE viewing angles. A faint limb haze may also be present in the Mariner 6 limb frames.

The "blue haze." Despite these evidences of very thin aerosol hazes, visible tangentially on the limb, there is no obscuring "blue haze" sufficient to account for the normally poor visibility of dark surface features seen or photographed in blue light and for their occasional better visibility—the so-called "blue-clearing" phenomenon (*11*, *12*).

The suitability of the Mariner blue pictures for "blue haze" observations was tested by photographing Mars through one of the Mariner blue filters on Eastman III-G plates, whose response in this spectral region is similar to that of the vidicons used in the Mariner camera. Conventional blue photographs on unsensitized emulsions and green photographs were taken for comparison. A typical result is shown in Fig. 5; the simulated TV blue picture is very similar to the conventional blue photographs. The effective wavelength of the actual blue TV pictures should be even shorter, owing to a lower ambient temperature and to the absence of reddening due to the earth's atmosphere.

The blue pictures taken by Mariners 6 and 7 clearly show craters and other surface features, even near the limb and terminator, where atmospheric effects are strong. Polar cap frame 7N17 shows sharp surface detail very near the terminator. The blue limb frame 6N1 shows surface detail corresponding to that seen in the subsequent overlapping green frame 6N3. Figure 6 includes blue, green, and red pictures in the region of Sinus Meridiani. Although craters show clearly in all three colors, albedo variations, associated both with craters and with larger-scale features, are much more pronounced in green and red than in blue. Blue photographs obtained from the earth

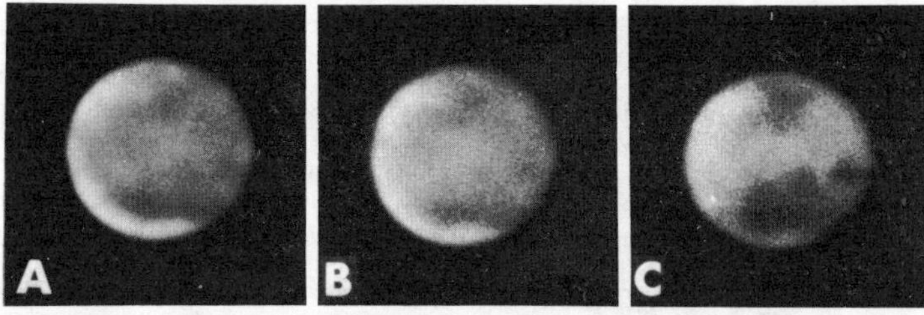

Fig. 5. Photographs of Mars from the earth, taken to compare Mariner-type blue-filter pictures with "standard" green and blue pictures of Mars. The pictures were taken 24 May 1969 at New Mexico State University Observatory. (A) "Standard" blue (0915 U.T.); (B) Mariner blue (0905 U.T.); (C) standard green (0844 U.T.). North is at the top.

during the Mariner encounters show the normal "obscured" appearance of Mars.

South polar cap shading. Another possible indication of atmospheric haze is the remarkable darkening of the south polar cap near both limb and terminator in the FE pictures (Fig. 7). This darkening is plainly *not* due to cloud or thick haze since, during near-encounter, surface features are clearly visible everywhere over the polar cap. It may be related to darkening seen in NE Mariner 7 frames near the polar cap terminator, and to the decrease in contrast with increasing viewing angle between the cap and the adjacent mare seen in frame 7N11 (Fig. 8b). The darkening may be due to optically thin aerosol scattering over the polar cap, or possibly to unusual photometric behavior of the cap itself. In either case, it may be complicated by systematic diurnal or latitudinal effects.

North polar phenomena. Marked changes seem to have occurred, between the flybys of Mariners 6 and 7, in the appearance of high northern latitudes. Some of these changes are revealed by a comparison of frames 6F34 and 7F73, which correspond to approximately the same central meridian and distance from Mars (Fig. 7). A large bright tongue (point 1 in frame 73) and a larger bright region near the limb (point 2) appear smaller and fainter in the Mariner 7 picture, despite the generally higher contrast of Mariner 7 FE frames. Much of the brightening near point 2 has disappeared entirely between the two flybys; in fact, it was not visible at all on pictures taken by Mariner 7 during the previous Mars rotation, although it was clearly visible in several Mariner 6 frames taken over the same range of distances. The bright tongue (point 1) increases in size and brightness during the martian day, as may be clearly seen from a comparison of frames 7F73 and 7F76 (Fig. 7).

The widespread, diffuse brightening covering much of the north polar cap region (point 3) apparently corre-

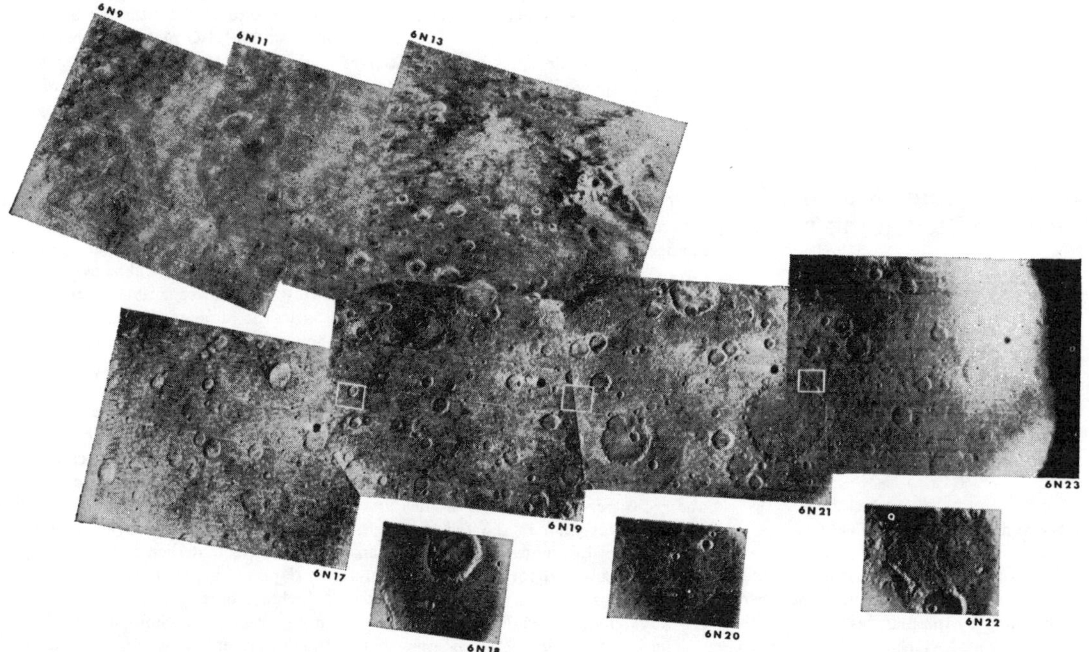

Fig. 6. Composite of ten Mariner 6 pictures showing cratered terrain in the areas of Margaritifer Sinus (top left), Meridiani Sinus (top center), and Deucalionis Regio (lower strip). Large-scale contrasts are suppressed by AGC and small-scale contrast is enhanced (see text). Craters are clearly visible in blue frames 6N9 and 6N17, but albedo variations are subdued. Locations of three camera-B frames are marked by rectangles. North is approximately toward the top, and the sunset terminator lies near the right edge of 6N23.

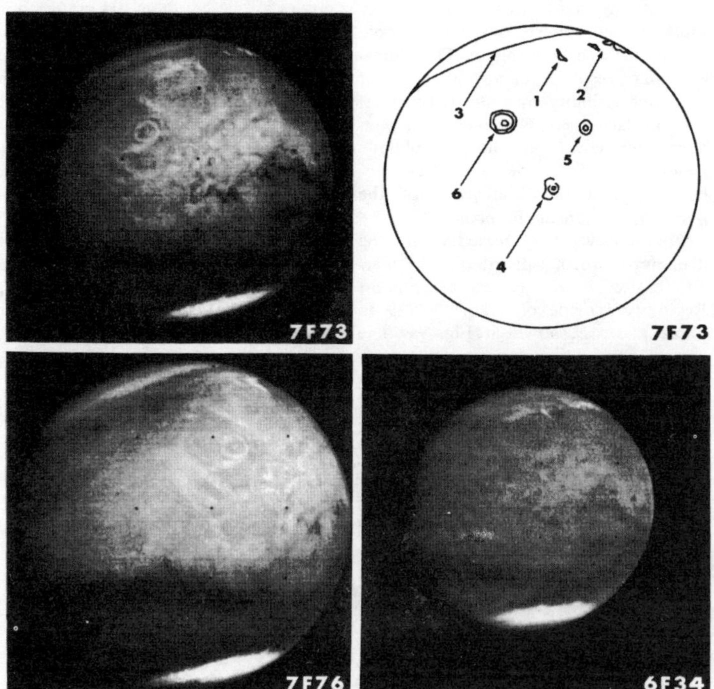

Fig. 7. Far-encounter pictures showing atmospheric and atmosphere-surface effects. Picture shutter times were as follows: 6F34, 30 July 0732 U.T.; 7F73, 4 August 1115 U.T.; 7F76, 4 August 1336 U.T.

sponds to the "polar hood" which has been observed from the earth at this martian season (northern early autumn). The extent of this hood is smaller in Mariner 7 than in Mariner 6 pictures; the region between, and just north of, points 1 and 2 appears to be covered by the hood in the Mariner 6 frames, but shows no brightening in the Mariner 7 frames.

The different behaviors of the discrete bright regions and the hood suggest different origins for these features, although both apparently are either atmospheric phenomena or else result from the interaction of the atmosphere and the surface. The discrete bright regions have fixed locations suggesting either surface frost or orographically fixed clouds. The fluctuation in the areal extent of the diffuse hood suggests cloud or haze. An extensive cloud or haze composed of either CO_2 or CO_2 and H_2O ice would be consistent with the atmospheric temperature structure revealed by the Mariner 6 occultation experiment (*13*).

Diurnal brightening. Other variable bright features which may be indicative of atmospheric processes appear throughout the Tharsis, Candor, Tractus Albus, and Nix Olympica regions (Fig. 7, frame 7F73; see also *3*). The brightness of these areas is observed to develop during the forenoon and increase during the martian afternoon, both in the Mariner pictures and in photographs taken from the earth. The structure and locations in the FE pictures do not change over the 6-day time span during which the region was observed. Particularly striking are several long light streaks near Nix Olympica (point 6 in frame 7F73) and numerous circular features resembling craters, which have bright centers and dark edges. Several of the circular features exhibit one or more concentric circles similar to, but less striking than, those near Nix Olympica. Two features of this type form two of the westernmost points of the classical "W-cloud" (points 4 and 5). No morphology associated with clouds—such as waves, billows, or cirriform streaks—appear in this region, although at the 30-kilometer resolution of these pictures such features would be visible in terrestrial clouds.

Search for local clouds and fog. All NE frames from both spacecraft were carefully examined for evidences of clouds or fog. Away from the south polar cap there are no evidences of such atmospheric phenomena. Over the polar cap and near its edge a number of bright features which may be atmospheric can be seen, although no detectable shadows are present and no local differences in height can be detected by stereoscopic viewing of overlapping regions whose stereo angles lie between 5° and 12°. Little or no illumination is evident near and beyond the polar cap terminator. On the other hand, frames 7N11, 12, and 13 (Fig. 8) show several diffuse bright patches suggestive of clouds near the polar cap edge. Also, on the cap itself a few local diffuse bright patches are present in frames 7N15 (green) and 7N17 (blue). Unlike most polar cap craters, which appear sharp and clear, a few crater rims and other topographic forms appear diffuse (frames 7N17, 18, and 19). In frames 7N17 (blue) and 7N19 (green), remarkable curved, quasi-parallel bright streaks are visible near the south pole itself. While these show indications of topographic form or control, including some crater-like shapes, their possible cloud-like nature is suggested by lack of shading. Also frames 7N17 and 7N19 show faint but definite streaks and mottles very near the terminator, superimposed on a sparsely cratered surface.

Observed Surface Features

A primary objective of the Mariner 6 and 7 television experiment was to examine, at close range, the principal types of martian surface features seen from the earth.

Mariners 6 and 7, while confirming the earlier evidence of a Moon-like cratered appearance for much of the martian surface, have also revealed significantly different terrains suggestive of more active, and more recent, surface processes than were previously evident. Preliminary analyses indicate that at least three distinctive terrains are represented in the pictures, as well as a mixture of permanent and transitory surface features displayed at the edge of, and within, the south polar cap; these terrains do not exhibit any simple correlation with the light and dark markings observed from the earth.

Cratered terrains. Cratered terrains

are those parts of the martian surface upon which craters are the dominant topographic form (Fig. 6). Pictures from Mariners 4, 6, and 7 all suggest that cratered terrains are widespread in the southern hemisphere.

Knowledge of cratered terrains in the northern hemisphere is less complete. Cratered areas appear in some Mariner frames as far north as latitude 20°. Nix Olympica, which in far-encounter photographs appears to be an unusually large crater, lies at 18°N. Numerous craters are visible in the closer-range FE frames. These are almost exclusively seen in the dark areas lying in the southern hemisphere, few being visible in the northern hemisphere. This difference may result from an enhancement of crater visibility by reflectivity variations in dark areas. However, poor photographic coverage, highly oblique views, and unfavorable sun angles combine to limit our knowledge of the northern portion of the planet.

Preliminary measurements of the diameter-frequency distribution of martian craters in the region Deucalionis Regio were made on frames 6N19 to 6N22 and are shown in Fig. 9a. The curves are based upon 104 craters more than 0.7 kilometer in diameter seen on frames 6N20 and 22, and upon 256 craters more than 7 kilometers in diameter seen on frames 6N19 and 21. The most significant result is the existence of two different crater distributions, a dichotomy also apparent in morphology. The two morphological crater types are (i) large and flat-bottomed and (ii) small and bowl-shaped. Flat-bottomed craters are most evident on frames 6N19 and 6N21. The diameters range from a few kilometers to a few hundred kilometers, with estimated diameter-to-

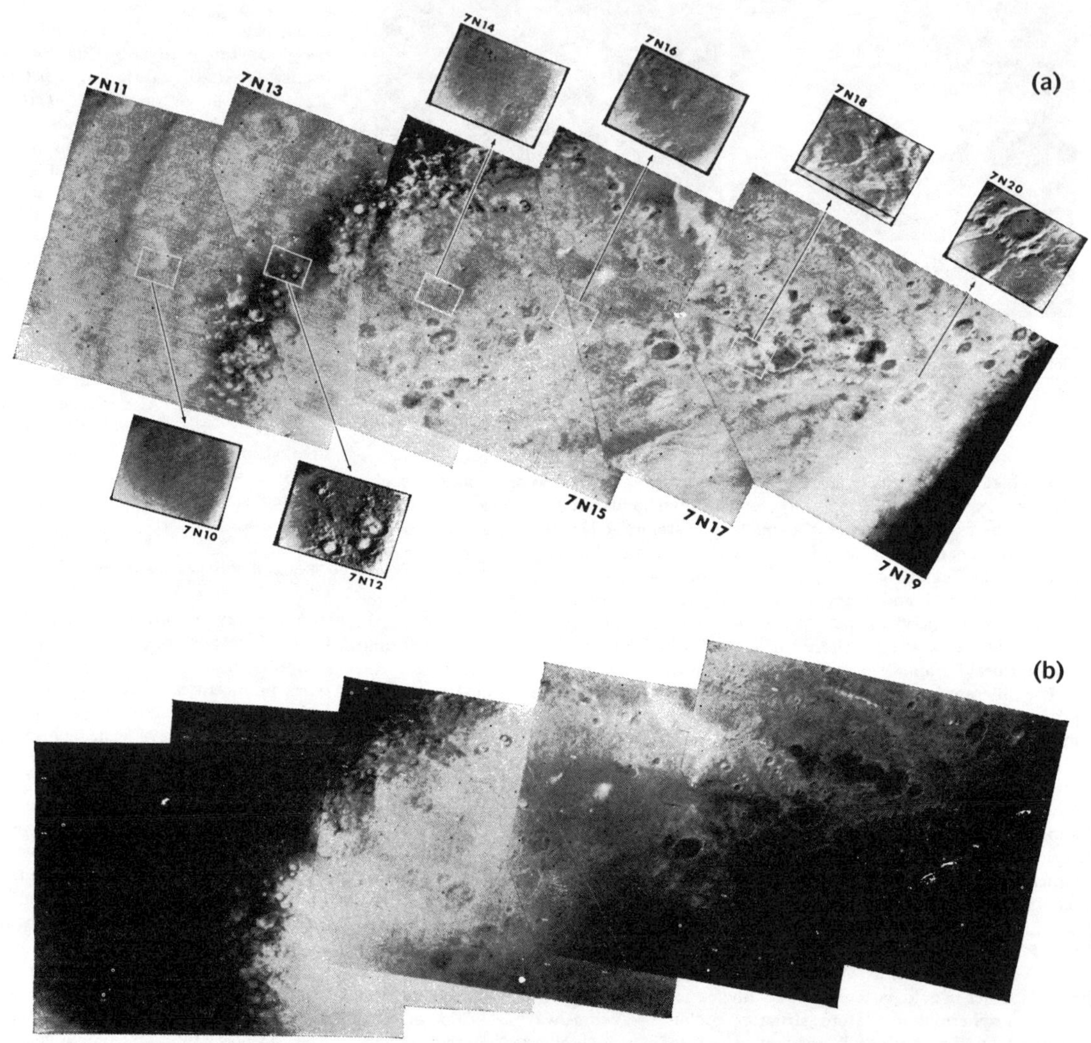

Fig. 8. (a) Composite of polar cap frames 7N10 to 7N20. Effects of AGC are clearly evident near the terminator (right) and at cap edge. (b) Composite of poplar cap camera-A frames 7N11 to 7N19. The effects of AGC have been partially corrected, but contrast is enhanced. The south pole lies near the parallel streaks in the lower right corner of frame 7N17.

depth ratios on the order of 100 to 1. The smaller, bowl-shaped craters are best observed in frames 6N20 and 6N22 and resemble lunar primary-impact craters. Some of them appear to have interior slopes steeper than 20 degrees.

The diameter-frequency distribution of large flat-bottomed craters is compared in Fig. 9b with the distribution of craters in the uplands on the far side of the moon near Tsiolkovsky. This particular lunar region was chosen for comparison because it is evidently a primordial surface, devoid of large post-upland features which might have modified the original crater distribution. The distribution of the small bowl-shaped craters is compared in Fig. 9c with the distribution of craters on the lunar maria (*14*). The distribution curve for small martian craters larger than about 1 kilometer in diameter has a slope of about −2, which is similar to the curve for primary craters larger than 3 kilometers in diameter on the lunar maria.

There are large variations in crater morphology among different cratered terrains. The craters of the dark area Meridiani Sinus (Fig. 6) have more marked polygonal outlines and more central peaks than craters in some other martian areas; the especially distinctive lighter marking in the northwest portions of the crater floors is also, but less clearly, seen in some FE frames. Many of the primary and secondary features associated with large lunar impact craters can be found on the martian terrain (see, for example, frame 6N18); however, certain others, such as rays and secondary crater swarms, appear to be absent. Also, ejecta blankets appear to be much less well developed. The missing features are generally those most easily removed or hidden by erosion or blanketing—a pattern consistent with the observation that the martian craters are generally shallower and more smooth than lunar ones.

On frame 6N20 there are low irregular ridges similar to those seen on the lunar maria. However, no straight or sinuous rills have been identified with confidence. Similarly, no Earth-like tectonic forms possibly associated with mountain building, island-arc formation, or compressional deformation have been recognized.

Chaotic terrains. Mariner frames 6N6, 14, and 8 (Fig. 10a) show two types of terrain—a relatively smooth cratered surface that gives way abruptly to irregularly shaped, apparently lower areas of chaotically jumbled ridges. This chaotic terrain seems characteristically to display higher albedo than its surroundings. On that basis, we infer that significant parts of the overlapping frames 6N5, 7, and 15 may contain similar terrain, although their resolution is not great enough to reveal the general morphological characteristics. As shown in Fig. 10a, frames 6N6, 14, and 8 all lie within frame 6N7, for which an interpretive map of possible chaotic terrain extent has been prepared (Fig. 11).

About 10^6 square kilometers of chaotic terrain may lie within the strip, 1000 kilometers wide and 2000 kilometers long, covered by these Mariner 6 wide-angle frames. Frames 6N9 and 10 contain faint suggestions of similar features. This belt lies at about 20°S, principally within the poorly defined, mixed light-and-dark area between the dark areas Aurorae Sinus and Margaritifer Sinus.

Chaotic terrain consists of a highly irregular plexus of short ridges and depressions, 1 to 3 kilometers wide and 2 to 10 kilometers long, best seen in frame 6N6 (Fig. 10a). Although irregularly jumbled, this terrain is different in setting and pattern from crater ejecta sheets. Chaotic terrain is practically uncratered; only three faint possible crate[illegible] recognized in the 10^6-square-kilo[illegible] area. The patches of chaotic terrain are not all integrated, but they constitute an irregular pattern with an apparent N to N 30°E grain.

Featureless terrains. The floor of the bright circular "desert," Hellas, centered at about 40°S, is the largest area of featureless terrain so far identified. Even under very low solar illumination the area appears devoid of craters down to the resolution limit of about 300 meters. No area of comparable size and smoothness is known on the moon. It may be that all bright circular "deserts" of Mars have smooth featureless floors; however, in the present state of our knowledge it is not possible to define any significant geographic relationship for featureless terrains.

The Mariner 7 traverse shows that the dark area Hellespontus, lying west of Hellas, is heavily cratered. The 130- to 350-kilometer-wide transitional zone is also well cratered and appears to slope gently downward to Hellas, interrupted by short, en echelon scarps and ridges (Fig. 12). It gives way abruptly along an irregular foot to the flat floor of Hellas. Craters are observed within the transitional zone but abruptly become obscured within the first 200 kilometers toward the center of Hellas.

The possibility that a low haze or fog may be obscuring the surface of Hellas, and that the featureless images are therefore not relevant to the true surface, has been considered. However, in frame 7N26 the ridges of the Hellas-Hellespontus boundary are clearly visible, proving that the surface is seen; yet there are virtually no craters within that frame. Thus the absence of well-defined craters appears to be a real effect.

South polar cap features. The edge of the martian south polar cap was visible at close range over a 90° span of longitude, from 290°E to 20°E, and the cap itself was seen over a latitude range from its edge, at −60°, southward to, and perhaps beyond, the pole itself. Solar zenith angles ranged from 51° to 90° and more; the terminator is clearly visible in one picture. The phase angle for the picture centers was 35°. The superficial appearance is that of a clearly visible, moderately cratered surface covered with a varying thickness of "snow." The viewing angle and the unfamiliar surface conditions make quantitative comparison with other areas of Mars difficult with respect to the number and size distributions of craters. Discussion here is therefore confined to those qualitative aspects of the polar cap which seem distinctive to that region.

The edge of the cap was observed in the FE pictures to be very nearly at 60°S, as predicted from Lowell Observatory measurements (*15*); this lends confidence to Earth-based observations concerning the past behavior of the polar caps.

The principal effect seen at the cap edge is a spectacular enhancement of crater visibility and the subtle appearance of other topographic forms. In frames 7N11 to 7N13, where the local solar zenith angle was about 53°, craters are visible both on and off the cap. However, in the transition zone, about 2 degrees of latitude in width, the population density of visible craters is several times greater, and may equal any so far seen on Mars. This enhancement of crater visibility results mostly from the tendency, noted in Mariner 4 pictures 14 and 15, for snow to lie preferentially on poleward-facing slopes.

In frame 7N12 the cap edge is seen in finer detail. The tendency mentioned above is here so marked as to cause confusion concerning the direction of the illumination. There are several tiny craters as small as 0.7 kilometer in diameter, and areas of fine mottling and sinuous lineations are seen near the larger craters. The largest crater shows interesting grooved structure, near its center and on its west inner wall, which appears similar to that in frame 6N18.

On the cap itself, the wide-angle views show many distinct reflectivity variations, mostly related to moderately large craters but not necessarily resulting from slope-illumination effects. Often a crater appears to have a darkened floor and a bright rim, and in some craters having central peaks the peaks seem unusually prominent. In frames 7N17 and 7N19 several large craters seem to have quite dark floors.

In contrast, the high-resolution polar cap frames 7N14 to 7N20 suggest a more uniformly coated surface whose brightness variations are mostly due to the effects of illumination upon local relief. Several craters are visible in each frame, some quite small; but, unlike most other areas of Mars, regions of positive relief are also visible, notably in frames 7N14 and 7N16. Also distinctive in these frames are areas of fine, irregular, quasi-parallel bright boundaries and irregular, shallow depressed regions. In frame 7N14, three such regions lie on the floor of a crater. Other irregular, shallow depressions apparently unrelated to craters appear in frames 7N15 and 7N17. Some of these are tens of kilometers in diameter and have no known counterparts elsewhere in the picture series of either spacecraft.

Finally, in frame 7N19 a curved, scalloped escarpment is seen, forming a boundary between a well cratered area on its convex side and a relatively crater-free area on its concave side; this feature suggests the large circular structures associated with the mare basins on the moon.

Relationship of terrain to light and dark markings. The contrast of light and dark markings on Mars varies with wavelength, a fact long known from telescopic photography. In violet light, "bright" and "dark" areas are essentially indistinguishable, as they have approximately the same reflectivity. With increasing wavelength, contrast is enhanced as redder areas become relatively brighter. Frame 6N13 (Fig. 6) shows craters in a dark area that have partially bright rims and floors, while craters in bright areas have rim and floor reflectivities similar to the reflectivity of the surroundings. These differences tend to increase the visibility of craters in dark areas, but only in photographs taken in red or green light. West of Meridiani Sinus (frames 6N9 and 6N11), and in some other parts of the planet, there are a number of dark-floored craters within bright areas, which, although otherwise relatively conspicuous, are difficult to see in blue light. The floors of these craters thus exhibit the same general dependence of reflectivity on wavelength that the larger dark areas do.

The distinction between bright and dark areas on the martian surface is generally more obvious in FE than in NE views. At higher resolution, the boundaries tend to become dispersed and indistinct. Exceptions are the sharp northeast and southern boundaries of Meridiani Sinus (frame 6N13) and the east edge of Hellespontus. The clearest structural relationship seen between a dark and a bright area is that of Hellespontus and Hellas (Fig. 12).

Chaotic terrain appears to be lower and to have a somewhat higher reflectivity than adjacent cratered areas. Whether chaotic terrain is extensive enough to comprise any previously identified bright areas remains to be determined.

Some of the classical "oases" observed from the earth have now been identified with single, large, dark-floored craters (such as Juventae Fons, see *4* and Fig. 4) or groups of such craters (such as Oxia Palus, frame 7N5). At least two classical "canals" (Cantabras and Gehon) have been found to coincide with quasi-linear alignment of several dark-floored craters, shown also in frame 7N5 (Fig. 13). As reported elsewhere (*4*), other canals are composed of irregular dark patches. It is probable

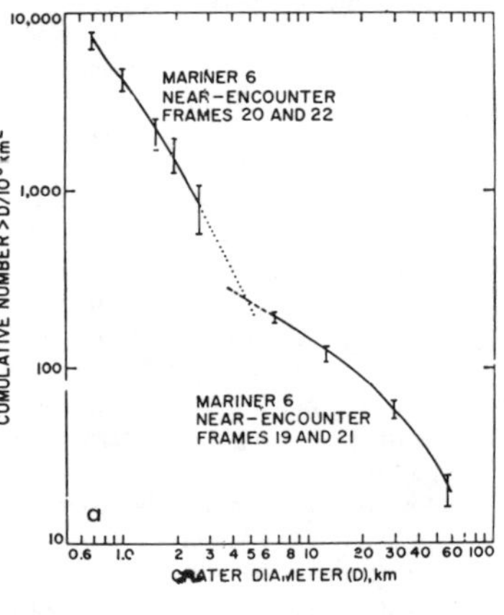

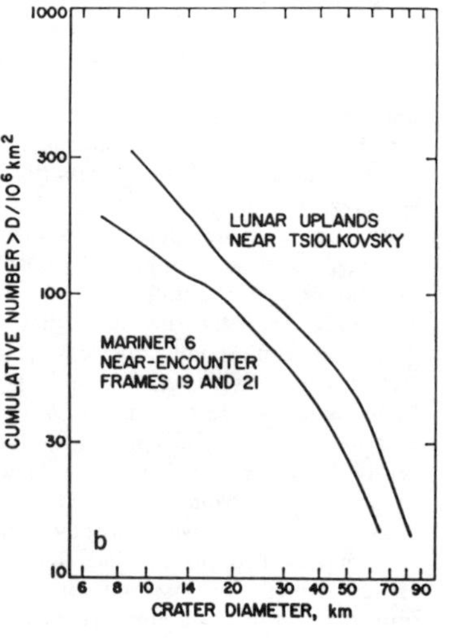

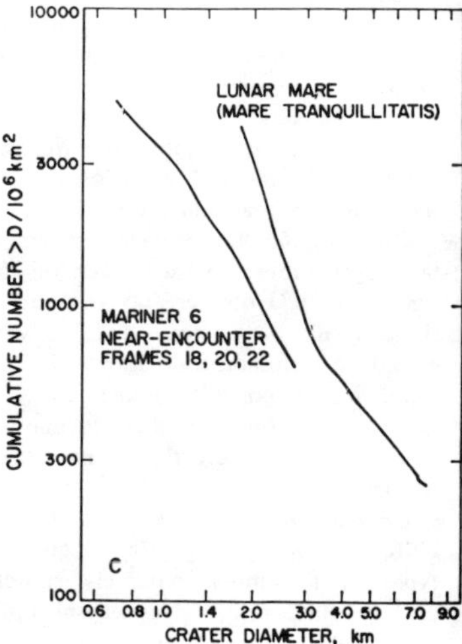

Fig. 9. (a) Preliminary cumulative distribution of crater diameters. Solid curve at right is based upon 256 counted craters in frames 6N19 and 6N21 having diameters $\geqslant$ 7 kilometers. The solid curve at left is based upon 104 counted craters in frames 6N20 and 6N22. The error bars are from counting statistics only ($N^{1/2}$). (b) Comparison of size distribution of large craters on Mars and on the lunar uplands. (c) Comparison of size distribution of small craters on Mars and on a lunar mare.

that most canals will, upon closer inspection, prove to be associated with a variety of physiographic features, and that eventually they will be considered less distinctive as a class.

Some early drawings and "maps" of Mars show a circular bright area within the dark area south of Syrtis Major and east of Sabaeus Sinus, very nearly in the place occupied by a large crater (*4*, fig. 3; *16*). Further comparison of the Mariner pictures with early maps and photographs may prove fruitful in revealing long-term aspects of topographic associations of dark-area boundaries.

It is possible to make a few rough comparisons between the Mariner 6 and 7 pictures and estimates of Mars topographic elevations based on Mariner occultations (*13*), Earth-based radar measurements (*17*), or CO_2 equivalent-width spectral measurements (*18*).

The long stretch of cratered terrain in Deucalionis Regio (Fig. 6) is located on what Earth-based radar and CO_2 measurements both suggest to be a very gradual slope rising westward. Here, at least, cratered terrain is not restricted to regionally high or to regionally low areas; like dark and light areas, it may not exhibit any particular correlation with planetary-scale relief.

The chaotic terrain viewed in Mariner 6 NE frames occurs in what appears to be the topographically lowest area surveyed by the spacecraft, if Earth-based CO_2 measurements are correct. Thus, it is at least possible that chaotic terrain is related in some consistent way to planetary-scale relief.

Mariner 7's occultation point occurred near Hellespontica Depressio, near latitude 55°S, longitude 20°E. A very low surface pressure was measured there, suggesting highly elevated terrain. If Hellespontica Depressio is elevated, Hellespontus may be also. This relationship would be consistent with the impression gained from the Mariner pictures that the floor of Hellas, with its featureless terrain, is a local basin rimmed by higher areas.

Inferences concerning Processes and Surface History

The features observed in the Mariner 6 and 7 pictures are the result of both present and past processes; therefore, they provide the basis of at least limited conjecture about those processes and their variations through time. In this section we consider the implications of (i) the absence of Earth-like tectonic features; (ii) the erosion, blanketing, and secondary modification evidenced in the three principal terrains; and (iii) the probable role of equilibrium between CO_2 solid and vapor in the formation of features of the south polar cap. We also consider the possible role of equilibrium between H_2O solid and vapor as an explanation of the diurnal brightenings observed in the FE photographs and biological implications.

Significance of the absence of Earth-like forms. The absence of Earth-like tectonic features on Mars indicates that, for the time period represented by

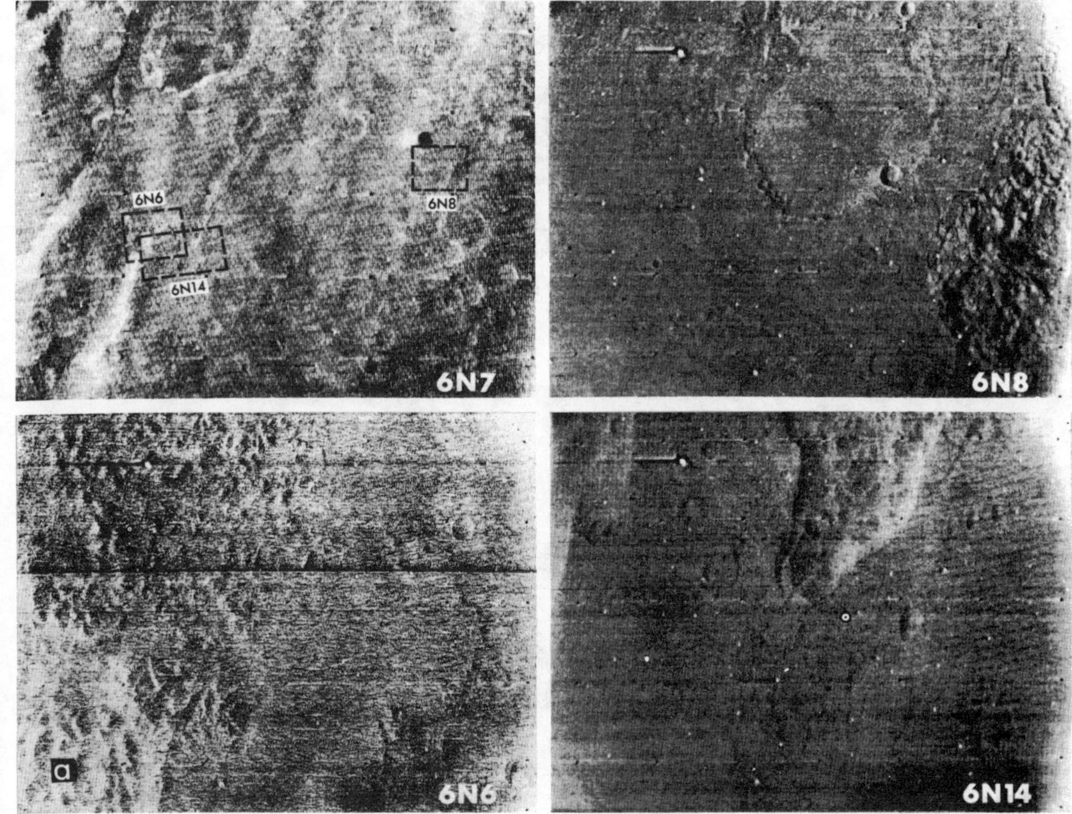

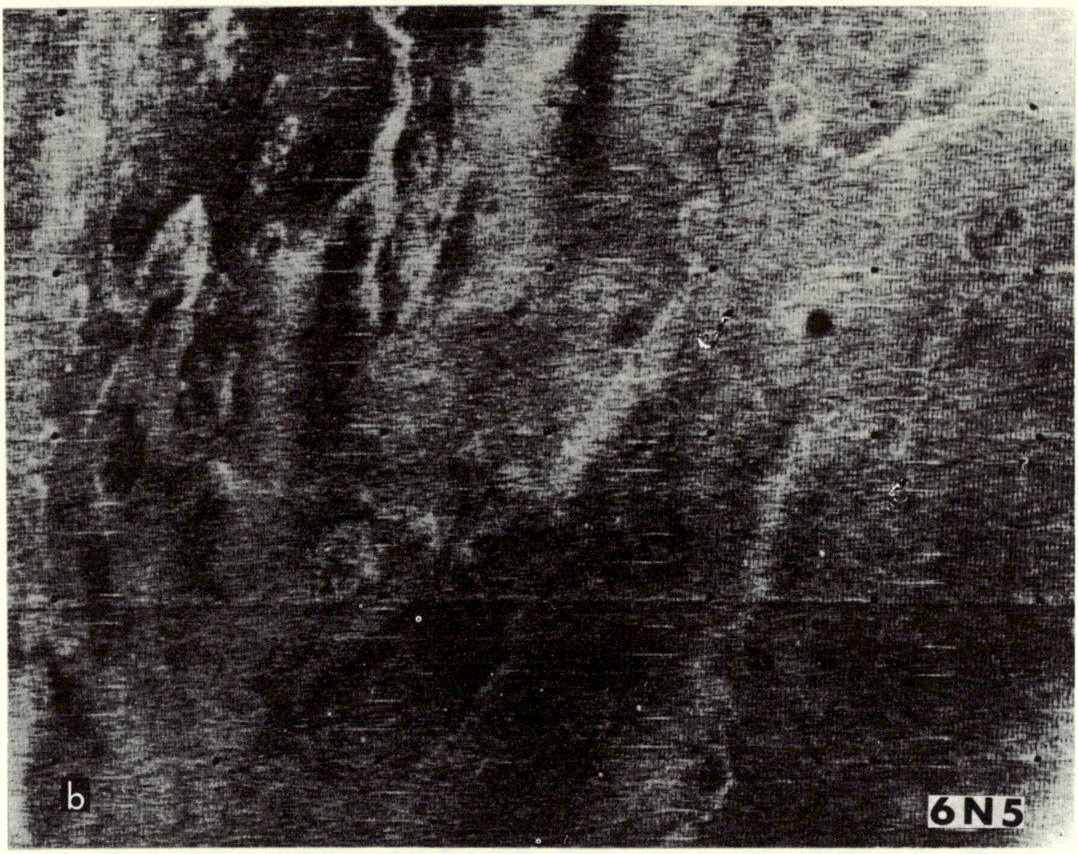

Fig. 10. (a) Examples of chaotic terrain. The approximate locations of the camera-B views inside camera-A frame 6N7 are shown by the dashed rectangles. North is approximately at the top. (b) Example of possible chaotic terrain. The lighter color and the absence of craters suggest that large parts of the right-hand half of this camera-A view may consist of chaotic terrain. (c) Example of chaotic terrain. The location of frame 6N14 inside frame 6N15 is shown by the solid rectangle.

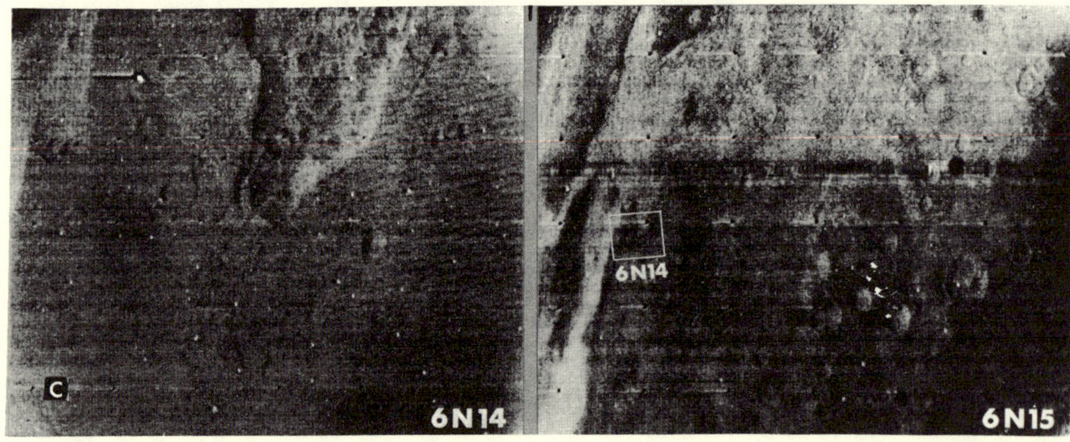

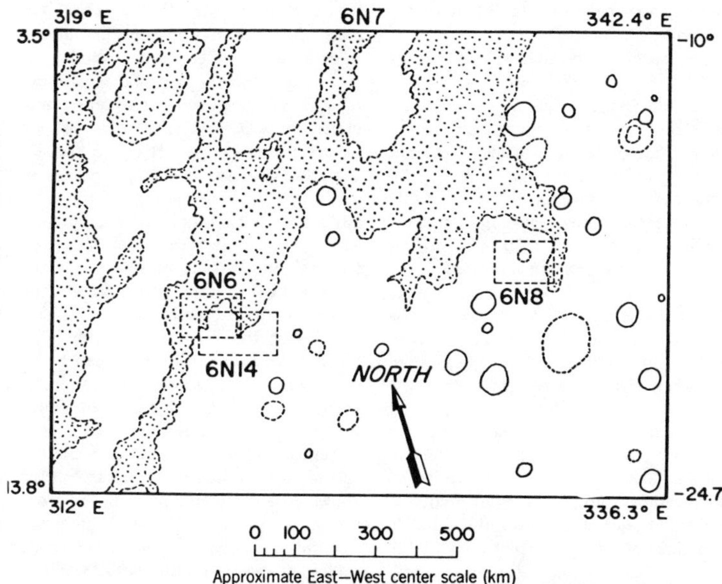

Fig. 11. Interpretive drawing showing the possible extent of chaotic terrain in frame 6N7.

the present large martian topographic forms, the crust of Mars has not been subjected to the kinds of internal forces that have modified, and continue to modify, the surface of the earth.

Inasmuch as the larger craters probably have survived from a very early time in the planet's history, it is inferred that Mars's interior is, and probably has always been, much less active than the earth's (*19*). Furthermore, a currently held view (*20*) is that the earth's dense, aqueous atmosphere may have formed early, in a singular event associated with planetary differentiation and the origin of the core. To the extent, therefore, that surface tectonic features may be related in origin to the formation of a dense atmosphere, their absence on Mars independently suggests that Mars never had an Earth-like atmosphere.

Age implications of cratered terrains. At present, the ages of martian topographic forms can be discussed only by comparison with the moon. Both the moon and Mars exhibit heavily cratered and lightly cratered areas, which evidently reflect in each case regional differences in the history of, or the response to, meteoroidal bombardment over the total life-span of the surfaces. The existence of a thin atmosphere on Mars may have produced recognizable secondary effects in the form and size distribution of craters, by contrast with the moon, where a significant atmosphere has presumably never been present. To the extent that relative fluxes of large objects impinging upon the two bodies can be determined, or a common episodic history established, a valid age comparison may be hoped for, except in the extreme case of a saturated cratered surface, where only a lower limit to an age can be found.

It is a generally accepted view that the present crater density on the lunar uplands could not have been produced within the 4.5-billion-year age of the solar system had the bombardment rate been no greater than the estimated present rate; that is, the inferred minimum age is already much greater than is considered possible. Indeed, it is found that even the sparsely cratered lunar maria would have required about a billion years to attain their present crater density. Unless this discrepancy is somehow removed by direct measurements of the crystallization ages of returned samples of lunar upland and mare materials, the previously accepted implication of an early era of high bombardment followed by a long period of bombardment at a drastically reduced rate will presumably stand.

In the case of Mars, a bombardment rate per unit area as much as 25 times that on the moon has been estimated (*21*). However, even this would still seem to require at least several billion years to produce the density of large craters that is seen on Mars in the more heavily cratered areas (*19*). Thus these areas *could also be primordial.* Further, were these areas to have actually been bombarded at a constant rate for such a time, at least a few very recent, large craters should be visible, including secondary craters and other local effects. Instead, the most heavily cratered areas seem relatively uniform with respect to the degree of preservation of large craters, with no martian Tycho or Copernicus standing out from the rest. This again suggests an early episodic history rather than a continuous history for cratered martian terrain, and increases the likelihood that cratered terrain is primordial.

If areas of primordial terrain do exist on Mars, an important conclusion follows: these areas have never been subject to erosion by water. This in turn reduces the likelihood that a dense, Earth-like atmosphere and large, open bodies of water were ever present on the planet, because these would almost surely have produced high rates of planet-wide erosion. On the earth, no topographic form survives as long as 10^8 years unless it is renewed by uplift or other tectonic activity.

Implications of modification of terrain. Although erosional and blanketing processes on Mars have not been strong enough to obliterate large craters within the cratered terrains, their effects are easily seen. On frames 6N19 and 6N21 (Fig. 6), even craters as large as 20 to 50 kilometers in diameter appear scarce by comparison with the lunar uplands [a feature originally noted by Hartmann (*19*) on the basis of the Mariner 4 data], and the scarcity of smaller craters is marked. The latter have a relatively fresh appearance, however, which suggests an episodic history of formation, modification, or both. Such a history seems particularly indicated by the apparently bimodal crater frequency distribution of Fig. 9.

Marked erosion, blanketing, and other surface processes must have been operating almost up to the present in the areas of featureless and chaotic terrains; only this could account for the absence of even small craters there. These processes may not be the same as those at work on the cratered terrains, because large craters have also been erased. The cratered terrains obviously have *never been* affected by such processes; this indicates an enduring geographic dependence of these extraordinary surface processes.

The chaotic terrain gives a general impression of collapse structures, suggesting the possibility of large-scale withdrawal of substances from the underlying layers. The possibility of permafrost some kilometers thick, and of its localized withdrawal, may deserve further consideration. Magmatic withdrawal or other near-surface disturbance associated with regional volcanism might be another possibility, but the apparent absence of extensive volcanic terrains on the surface would seem to be a serious obstacle to such an interpretation. It may also be that chaotic terrain is the product either of some unknown intense and localized erosional process or of unsuspected local sensitivity to a widespread process.

Carbon dioxide condensation effects. The Mariner 7 NE pictures of the polar cap give no direct information concerning the material or the thickness of the polar snow deposit, since the observed brightness could be produced by a very few milligrams per square centimeter of any white, powdery material. However, they do provide important indirect evidence as to the thickness of the deposit and, together with other known factors, may help to establish its composition.

The relatively normal appearance of craters on the polar cap in the high-resolution frames, and the existence on these same frames of topographic relief unlike that so far recognized elsewhere on the planet, suggest that some of the apparent relief may be due to variable thicknesses of snow, perhaps drifted by wind. If it is, local thicknesses of at least several meters are indicated.

The structure of the polar cap edge shows that evaporation of the snow is strongly influenced by local slopes—that is, by insolation effects rather than by wind. On the assumption that the evaporation is entirely determined by the midday radiation balance, when the absorbed solar power exceeds the radiation loss at the appropriate frost-point temperature, one may estimate the daily evaporation loss from the cap. We find the net daily loss to be about 0.8 gram per square centimeter in the case of CO_2, although the loss is reduced by overnight recondensation. In the case of H_2O, the loss would be about 0.08 gram per square centimeter, and it would be essentially irreversible because H_2O is a minor constituent whose deposition is limited by diffusion.

Since the complete evaporation of the cap at a given latitude requires many days, we may multiply the above rates by a factor between 10 and 100, obtaining estimates for total cap thickness of tens of grams per square centimeter for CO_2 and several grams per square centimeter for H_2O, on the assumption that the cap is composed of one or the other of these materials. The estimate for CO_2 is quite acceptable, but that for H_2O is unacceptable because of the problem of transporting such quantities annually from one pole to the other at the observed vapor density (*22*). For the remainder of this discussion we assume the polar cap to be composed of CO_2, with a few milligrams of H_2O per square centimeter deposited throughout the layer.

Several formations have been observed which suggest a tendency for snow to be preferentially removed from low areas and deposited on high areas, contrary to what might be expected under quiescent conditions (*23*). These formations include craters with dark floors and bright rims, prominent central peaks in some craters, and irregular depressed areas (frames 7N14, 15, and 17). While such effects might result simply from wind transport of solid material, it is also possible that interchange of solid and vapor plays a role.

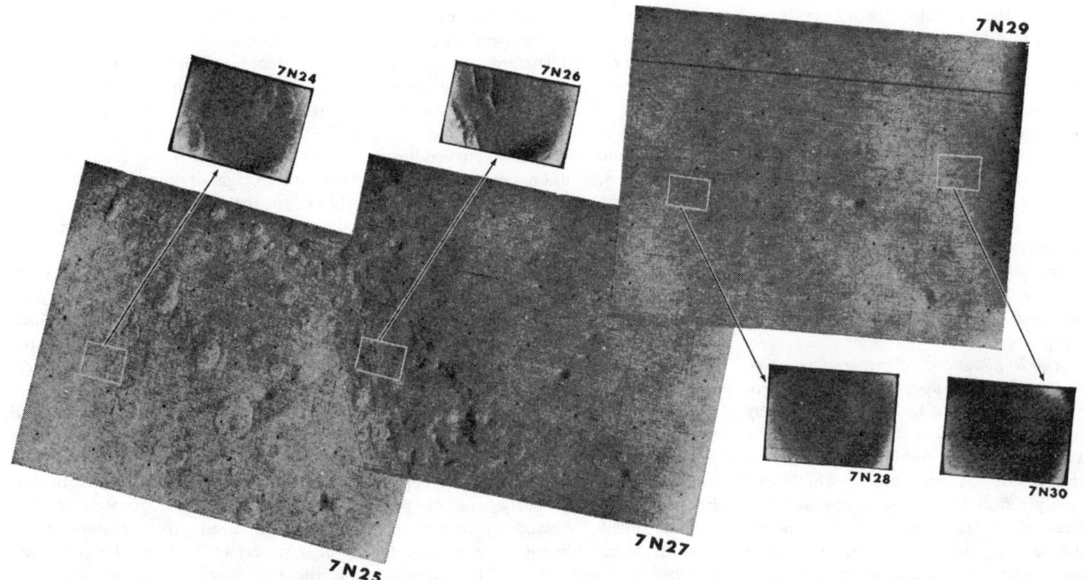

Fig. 12. Composite of seven Mariner 7 frames, showing the cratered dark area Hellespontus, the ridged, broken boundary between Hellespontus and Hellas, and the featureless terrain of the bright, circular "desert" Hellas. Large-scale variations in contrast are suppressed by AGC. Lighting conditions are similar to those of Fig. 6, frames 7N17 to 7N21. North is approximately toward the top.

The adiabatic lapse rate of the polar atmosphere is about 6°K per kilometer, as compared to a frost-point lapse rate of about 1°K per kilometer. Therefore, adiabatic heating or cooling of martian air blowing across sloping terrain may result in evaporation of low-lying solid and precipitation over high areas. The rate is determined by atmospheric density, latent heat, difference in elevation, lateral scale of relief, wind speed, turbulent-layer thickness, interfacial heat transfer, and net adiabatic gradient. For example, a wind of 20 meters per second blowing across a crater 20 kilometers in diameter and 0.5 kilometer deep, with a turbulent layer only 100 meters thick, could differentially deposit about 0.1 gram per square centimeter per day. Since this process would be effective during the entire time that snow is on the ground, several tens of grams per square centimeter could become systematically redistributed in the course of a few hundred days. A speculative possibility is that such a process, or perhaps merely a sustained accumulation of snow in a particular topographic "trap," occasionally results in sufficient accumulation during the winter to survive summer evaporation. The increased albedo might then lead to a permanent accumulation of CO_2 snow within that topographic feature—that is, to formation of a martian "ice field." The very prominent snow-covered central peaks of some of the small craters located at high latitude might conceivably be the sites of such permanent deposits of solid CO_2. The permanent part of the north polar cap is presumably such a structure (*22*).

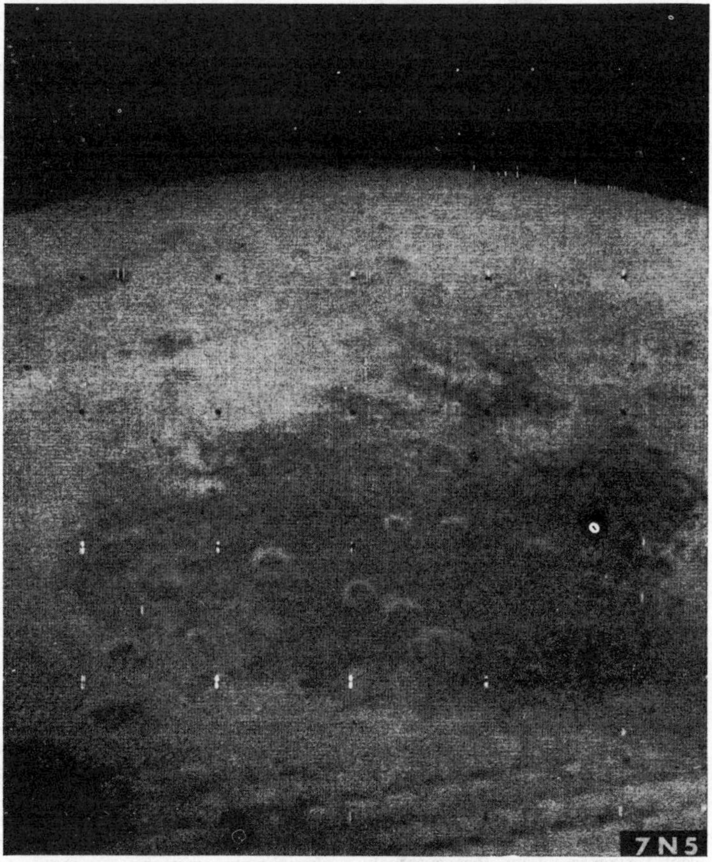

Fig. 13. Partially reconstructed frame 7N5, showing Meridiani Sinus in the foreground. The "oasis" Oxia Palus projects into the picture from the left edge, near the limb. Note the asymmetric shading in several craters in Meridiani Sinus; the isolated, dark-floored craters in surrounding bright areas; and the crater complex in Oxia Palus. The view is approximately toward N 20°W. The banded appearance of the sky area is an artifact of the picture processing.

Water: processes suggested by brightening phenomena. Several of the brightening and haze phenomena described above could be related either to formation of H_2O frost on the surface or to formation of H_2O ice clouds in the atmosphere. In most of these instances, however, the phenomena could equally well be explained by condensation of CO_2. This is true of the bright tongues and polar hood in the north polar region, of the cloud-like features observed over and near the south polar cap, and of the limb hazes observed in tropical latitudes and over the Mare Hadriaticum and Ausonia regions.

On the other hand, the brightenings in the Nix Olympica, Tharsis, Candor, and Tractus Albus regions cannot be explained by CO_2 condensation because their complete topographic control requires that they be on or near the surfaces where temperatures are well above the CO_2 frost point. An explanation of these phenomena in terms of H_2O condensation processes also faces serious difficulties, however. Most of the region is observed to brighten during the forenoon, when the surface is hotter than either the material below or the atmosphere above, so that water vapor could not diffuse toward the surface and condense on it, either from above or below. Thus a surface ice-frost is very unlikely. A few features in the area, parts of the "W-cloud," for example, are observed to brighten markedly during the late afternoon, where H_2O frost could form on the surface if the air were sufficiently saturated. These features are not observed, from the earth, to be bright in the early morning, but a thin layer of H_2O frost persisting through the night would evaporate almost immediately when illuminated by the early morning sun, provided the air were then sufficiently dry. Under these conditions, the behavior of the "W-cloud" could be due to frost.

The diurnal behavior of the bright regions throughout this part of Mars is consistent with a theory of convective H_2O ice clouds, but the absence of any cloud-like morphology and the clear topographic detail observed at the highest resolution available (frame 7F76) render this explanation questionable. Even very light winds of 5 meters per second would produce easily observable displacements of the order of 100 kilometers in the course of the more than one-fourth of the Mars day during which these regions were continuously observed by each spacecraft. Since condensation and evaporation processes are slow at Mars temperatures and pressures, some observable distortion and streakiness due to these displacements should be seen in clouds, even if they are orographically produced. No such distortions or streakiness are observed.

An additional difficulty with an explanation of these phenomena in terms of H_2O condensation lies in the relatively rapid removal of water from the local surface. Water vapor evolved from the surface during the daytime would quickly be transported upward through a deep atmospheric layer by thermal convection, and most of it would be removed from the source region. Local permafrost sources should be effectively exhausted by this mechanism within a few hundred years at most, unless somehow replenished. Since most of this region lies near the equator, where seasonal temperature variations are small, it is difficult to see how any significant seasonal replenishment from the atmosphere could take place. The possibility of replenishment from a subsurface source of liquid water is not considered here.

In summary, in our examination of the data thus far, we see no strong indications of H_2O processes involving vapor and ice. The brightenings seen in the tropics and subtropics at far-encounter are not easily explained by a mechanism involving H_2O. On the other hand, we have no satisfactory alternative explanation for these phenomena. Perhaps detailed exploration of these regions by the Mariner '71 orbiters will provide the answer.

Biological inferences. No direct evidence suggesting the presence of life on Mars has been found in the pictures. This is not surprising, since martian life, if any, would probably be microbial and undetectable at a resolution of 300 meters. Although inconclusive on the question of martian life, the photographs are informative on at least three subjects of biological interest: the general nature of the martian maria, the present availability of water, and the availability of water in the past.

One of the most surprising results so far of the TV experiment is that nothing in the pictures suggests that the dark areas, the sites of the seasonal darkening wave, are more favorable for life than other parts of the planet. On the contrary, it would now appear that the large-scale surface processes implied by the chaotic and featureless terrains may be of greater biological interest than the wave of darkening. We reiterate that these are preliminary conclusions; it may be that subtle physiographic differences between dark and bright regions will become evident when photometrically corrected pictures are examined.

With regard to the availability of water, the pictures so far have not revealed any evidence of geothermal areas. We would expect such areas to be permanently covered with clouds and frost, and these ought to be visible on the morning terminator; no such areas have been seen. A classically described feature of the polar cap which has been interpreted as wet ground—the dark collar—has likewise not been found. Other locales which have been considered to be sites of higher-than-average moisture content are those which show diurnal brightening. A number of such places have been observed in the pictures, but on close inspection the brightening appears not to be readily interpretable in terms of water frosts or clouds. Pending their definite identification, however, the brightenings should be considered possible indications of water.

The results thus reinforce the conclusion, drawn from Mariner 4 and ground-based observations, that scarcity of water is the most serious limiting factor for life on Mars. No terrestrial species known to us could live in the dry martian environment. If there is a permafrost layer near the surface, or if the small amount of atmospheric water vapor condenses as frost in favorable sites, it is conceivable that, by evolutionary adaptation, life as we know it could use this water and survive on the planet. In any case, the continued search for regions of water condensation on Mars will be an important task for the 1971 orbiter.

The past history of water on Mars is a matter of much biological interest. According to current views, the chemical reactions which led to the origin of life on the earth were initiated in the reducing atmosphere of the primitive earth. These reactions produced simple organic compounds which were precipitated into the ocean, where they underwent further reactions that eventually yielded living matter. The pictorial evidence raises the question of whether Mars ever had enough water to sustain an origin of life. If the proportion of water outgassed relative to CO_2 is the same for Mars as for the earth, then, from the mass of CO_2 now in the martian atmosphere, it can be estimated that Mars has produced sufficient water to cover the planet to a depth of a few meters. The question is whether anything approaching this quantity of water was ever present on Mars in the liquid state.

The existence of cratered terrains and the absence of Earth-like tectonic forms on Mars clearly implies that the planet has not had oceans of terrestrial magnitude for a very long time, possibly never. However, we have only very rough ideas of how much ocean is required for an origin of life, and of how long such an ocean must last. An upper limit on the required time, based on terrestrial experience, can be derived from the age of the oldest fossils. $>3.2 \times 10^9$ years (*24*). Since these fossils are the remains of what were apparently highly evolved microorganisms, the origin of life must have taken place at a much earlier time, probably during the first few hundred million years of the earth's history. While one cannot rule out, on the basis of the TV data, the possibility that a comparably brief, aqueous epoch occurred during the early history of the planet, it must be said that the effect of the TV results so far is to diminish the a priori likelihood of finding life on Mars. However, it should be noted that if Mars is to be a testing ground for our notions about the origin of life, we must avoid using these same notions to disprove in advance the possibility of life on that planet.

Potentialities of the Data

Careful computer restoration of the pictures, starting with data recovered from six sequential playbacks of the near-encounter analog tapes, will be carried out over the next several months. This further processing will greatly enhance the completeness, appearance, and quantitative usefulness of the pictures. While it is not yet certain whether the desired 8-bit relative photometric accuracy can be attained, there are reasonable grounds for thinking that much new information bearing on the physiography, meteorology, geography, and other aspects of Mars will ultimately be obtained from the pictures. Some of the planned uses of the processed data are as follows.

Stereoscopy. Most of the NE wide-angle pictures contain regions of two-picture overlap, and a few contain regions of three-picture overlap. These areas can be viewed in stereoscopic vision in the conventional manner of aerial photography. Preliminary tests on pictures of the south polar cap (frames 7N17 and 7N19) indicate that measurement of crater depth, central-peak height, and crater-rim height is possible. However, accuracy can be estimated for the elevation determinations at this time.

Planetary radii. Geometric correction of the FE photographs should make it possible to determine the radius of Mars as a function of latitude, and possibly of longitude. The geometric figure of Mars has been historically troublesome because of inconsistencies between the optical and the dynamical oblateness, a discrepancy amounting to some 18 kilometers in the value for the difference of the equatorial and polar radii. It is possible that the darkening of the polar limb observed by Mariners 6 and 7, if it is a persistent phenomenon, might have systematically affected the earlier telescopic measurements of the polar diameter more than irradiation has, giving too large a value for the optical flattening. However, this cannot explain the large flattening obtained from surface-feature geodesy (*25*). Although a fairly reliable figure for the polar flattening may be obtained from the Mariner data, it is unlikely that the actual radii will be determined with an accuracy greater than several kilometers because of the relatively low picture-element resolution in these frames and the difficulty in locating the limb.

Cartography. The large number of craters found on the surface of Mars makes it feasible to establish a control net which uses topographic features as control points, instead of surface markings based on albedo differences. This net should provide the basic locations for compiling a new series of Mars charts. The NE pictures, which cover 10 to 20 percent of the area of the planet, will constitute the basic material for detailed maps of these areas.

Satellites. We hope to detect the larger of Mars's satellites, Phobos, in two of the Mariner 6 FE pictures taken when Phobos was just beyond the limb of the planet. The satellite should have moved between the two frames by about ten picture elements, and should appear as a "defect" that has moved by this amount between the two pictures. If Phobos itself is not visible, its shadow (again detectable by its motion) should be. The shadow will be some five picture elements across and will have a photometric depth of about 10 percent. If the photometric depth of the shadow can be measured accurately, we can determine the projected area (and hence the diameter) of the satellite. A similar method has been used to measure the diameter of Mercury during solar transits.

Photometric studies. We expect to derive the photometric function for each color, combining data from the two spacecraft. Observations by the current Mariners were made near 25°, 35°, 45°, and 80° phase. Since data obtained from the earth can be used to establish the absolute calibration at the smaller phase angles, we will also be able to relate the 80°-phase data to Earth-based observations, thus doubling the range over which the phase function is determined. This information should then make possible the determination of crater slopes. Agreement for areas of overlap between different filters and between A-camera and B-camera frames can be used to check the validity of the results and possibly to measure and correct for atmospheric scattering.

The reciprocity principle may be useful in testing quantitatively for diurnal changes in the FE pictures. Such changes might include dissipation of frost or haze near the morning terminator and formation of afternoon clouds near the limb.

Overlap areas in NE pictures can be used to obtain approximate colors, even though these areas are seen at different phase angles in each color. In addition, color-difference or color-ratio pictures may be useful in identifying local areas of anomalous photometric or colorimetric behavior. Camera-A digital pictures obtained by Mariner 7 in late far-encounter will be very useful for making color measurements.

Comparison of pictures with radar-scattering and height data. The reflection coefficient of the martian surface for radar waves of decimeter wavelength shows marked variations at a given latitude as a function of longitude. Even though few of the areas of Mars so far observed by radar are visible at close range, some correlation of topography with radar reflectivity may become apparent upon careful study. Clearly, the Mariner pictures will become steadily more valuable in this connection as more radar results and other height data become available.

Effects on Mariner '71

The distinctive new terrains revealed in the Mariner 6 and 7 pictures, the relatively small fraction (10 to 20 percent) of the surface so far viewed even at moderate (A-camera) resolution, and the tantalizing new evidence of afternoon-brightening phenomena all emphasize the importance of an exploratory, adaptive strategy in 1971 as opposed to a routine mapping of geographic features. The fact that each of three successive Mariner spacecraft has revealed a new and unexpected topography strongly suggests that more surprises (perhaps the most important ones) are still to appear.

A primary objective should be to view nearly all of the visible surface at A-camera resolution (1-kilometer pixel spacing), and to inspect selected typical areas at higher resolution, very early in the 90-day orbiting period. The true extent and character of cratered, chaotic, and featureless terrains, and of any new kinds of terrain, can thus be determined and correlated with classical light and dark areas, with regional height data, and so on.

A second objective should be to search for and examine, in both spatial and temporal detail, those areas which suggest the local presence of water, through the afternoon-brightening phenomena, morning frosts or fogs, or other behavior not now recognized. Certainly the known "W-cloud" areas, Nix Olympica, and other, similar areas known from Earth observation take on a new interest by virtue of the Mariner 6 and 7 results.

The complex structure found in the south polar cap calls for further examination, particularly with respect to separation of its more permanent features from diurnally or seasonally varying ones. The sublimation of the cap should be carefully followed, so as to detect evidence of variations in thickness of the deposit and especially evidence of the possible existence of permanent deposits. Study of the north polar cap at close range should also be exceedingly interesting.

Effects on Viking '73

If the effects of the Mariner 6 and 7 results on Mariner '71 are substantial, they at least do not require a change of instrumentation, only one of mission strategy. This may not be true of the effects on Viking '73. The discovery of so many new, unexpected properties of the martian surface and atmosphere adds a new dimension to the problem of selecting the most suitable landing site and may make Viking even more dependent on the success of Mariner '71 than has been supposed. Furthermore, since so much new information is revealed through the tenfold step in resolution afforded by the B-camera frames, a further substantial increase in resolution, not available to Mariner '71, may have to be incorporated in Viking in order to examine even more closely the fine-scale characteristics of various terrain types before a landing site is chosen.

Summary and Conclusions

Even in relatively unprocessed form, the Mariner 6 and 7 pictures provide fundamental new insights concerning the surface and atmosphere of Mars. Several unexpected results emphasize the importance of versatility in instrument design, flexibility in mission design, and use of an adaptive strategy in exploring planetary surfaces at high resolution.

The surface is clearly visible in all wavelengths used, including the blue. No blue-absorbing haze is found.

Thin, patchy, aerosol-scattering layers are present in the atmosphere at heights of from 15 to 40 kilometers, at several latitudes.

Diurnal brightening in the "W-cloud" area is seen repeatedly and is associated with specific topographic features. No fully satisfactory explanation for the effect is found.

Darkening of the polar cap in a band near the limb is clearly seen in FE pictures and is less distinctly visible in one or two NE frames. Localized, diffuse bright patches are seen in several places on and near the polar cap; these may be small, low clouds.

Widespread cratered terrain is seen, especially in dark areas of the southern hemisphere. Details of light-dark transitions are often related to local crater forms. Asymmetric markings are characteristic of craters in many dark areas; locally, these asymmetries often appear related, as if defined by a prevailing wind direction.

Two distinct populations of primary craters are present, distinguished on the basis of size, morphology, and age. An episodic surface history is indicated.

In addition to the cratered terrain anticipated from Mariner 4 results, at least two new, distinctive topographic forms are seen: chaotic terrains and featureless terrains. The cratered terrain is indicative of extreme age; the two new terrains both seem to require the present-day operation of especially active modifying processes in these areas. When seen at closer range, the very bright, streaked complex found in the Tharsis-Candor region may reveal yet another distinctive topographic character. Because of the afternoon-brightening phenomena long known here, this area provides a fascinating prospect for further exploration in 1971.

No tectonic and topographic forms similar to terrestrial forms are observed.

Evidences of both atmosphere-surface effects and topographic effects are seen on the south polar cap. At the cap edge, where the "snow" is thinnest, strong control by solar heating, as affected by local slopes, is indicated. Crater visibility is greatly enhanced in this area.

On the cap itself, intensity variations suggestive of variable "snow" thickness are seen. These may be caused by wind-drifting of the snow or by differential exchange of solid and vapor, or by both.

Snow thicknesses here of several grams or several tens of grams per square centimeter are inferred if the snow material is H_2O or CO_2, respectively. The possibility that the material is H_2O seems strongly ruled out on several grounds.

Variable atmospheric, and atmosphere-surface, effects are seen at high northern latitudes; these effects include the polar "hood" and bright, diurnally variable circumpolar patches.

Several classical features have been successfully identified with specific topographic forms, mostly craters or crater remnants.

The findings are inconclusive on the question of life on Mars, but they are relevant in several ways. They support earlier evidence that scarcity of water, past and present, is a serious limiting factor for life on the planet. Nothing so far seen in the pictures suggests that the dark regions are more favorable for life than other parts of Mars.

References and Notes

1. R. B. Leighton, B. C. Murray, R. P. Sharp, J. D. Allen, R. K. Sloan, *Science* **149**, 627 (1965).
2. ———, "Mariner IV Pictures of Mars," *Tech. Rep. Jet Propul. Lab. Calif. Inst. Technol. No. 32–884* (1967), pt. 1.
3. R. B. Leighton, N. H. Horowitz, B. C. Murray, R. P. Sharp, A. G. Herriman, A. T. Young, B. A. Smith, M. E. Davies, C. B. Leovy, *Science* **165**, 684 (1969).
4. ———, *ibid.*, p. 787.
5. The 1/7 digital TV data for the central 20 percent of each line were replaced by encoded data from other on-board experiments. In this region, coarser, 1/28 digital data (6-bit-encoded for every 28th pixel), stored on the analog tape recorder, were available (see Fig. 2).
6. D. G. Montgomery, in preparation.
7. This procedure is semiautomatic, subject to hand correction by the computer operator as necessary.
8. G. E. Danielson, in preparation.
9. For each spacecraft, this must be done for each filter of each camera and for all calibration temperatures, and the results must be corrected to the observed flight temperature.
10. Most of the real-time FE A-camera digital pictures were of no value because little or none of the image projected outside the central 20 percent blank area.
11. E. C. Slipher, *Publ. Astron. Soc. Pacific* **49**, 137 (1937).
12. J. B. Pollack and C. Sagan, *Space Sci. Rev.* **9**, 243 (1969).
13. A. J. Kliore, G. Fjeldbo, B. Seidel, in preparation.
14. M. J. Trask, *Tech. Rep. Jet Propul. Lab. Calif. Inst. Technol. No. 32–800* (1966), p. 252.
15. G. E. Fischbacher, L. J. Martin, W. A. Baum, "Martian Polar Cap Boundaries," final report under Jet Propulsion Laboratory contract 951547, Lowell Observatory, May 1969.
16. E. Burgess, private communication.
17. R. M. Goldstein, private communication; C. C. Councilman, private communication.
18. M. J. S. Belton and D. M. Hunten, *Science*, in press.
19. W. K. Hartmann, *Icarus* **5**, 565 (1966).
20. D. L. Anderson and R. A. Phinney, in *Mantles of the Earth and Terrestrial Planets*, S. K. Runcorn, Ed. (Interscience, New York, 1967), pp. 113–126.
21. E. Anders and J. R. Arnold, *Science* **149**, 1494 (1965).
22. R. B. Leighton and B. C. Murray, *ibid.* **153**, 136 (1966).
23. B. T. O'Leary and D. G. Rea, *ibid.* **155**, 317 (1967).
24. A. E. J. Engel, B. Nagy, L. A. Nagy, C. G. Engel, G. O. W. Kremp, C. M. Drew, *ibid.* **161**, 1005 (1968); J. W. Schopf and E. S. Barghoorn, *ibid.* **156**, 508 (1967).
25. R. J. Trumpler, *Lick Obs. Bull.* **13**, 19 (1927).

26. We gratefully acknowledge the support and encouragement of the National Aeronautics and Space Administration. An undertaking as complex as that of Mariners 6 and 7 rests upon a broad base of facilities, technical staff, experience, and management, and requires not only money but much individual and team effort to be brought to a successful conclusion. It is impossible to know, much less to acknowledge, the important roles played by hundreds of individuals. We are deeply appreciative of the support and efforts of H. M. Schurmeier and the entire Mariner 1969 project staff. With respect to the television system, responsibility for the design, assembly, testing, calibration, flight operation, and picture data processing lay with the Jet Propulsion Laboratory. We gratefully acknowledge the contributions of G. M. Smith, D. G. Montgomery, M. C. Clary, L. A. Adams, F. P. Landauer, C. C. LaBaw, T. C. Rindfleisch, and J. A. Dunne in these areas. L. Malling, J. D. Allen, and R. K. Sloan made important early contributions. We are indebted to V. C. Clarke, C. E. Kohlhase, R. Miles, and E. Greenberg for their help in exploiting the flexibility of the spacecraft to achieve maximum return of pictorial data. We are especially appreciative of the broad and creative efforts of G. E. Danielson as Experiment Representative. The able collaborative contributions of J. C. Robinson in comparing Mariner pictures with Earth-based photographs and of L. A. Soderblom and J. A. Cutts in measuring craters are gratefully acknowledged.

3

Reprinted from *Cosmic Res.* **13**:1–6 (1975)

PRELIMINARY RESULTS OF STUDIES CONDUCTED ON THE SOVIET AUTOMATIC STATIONS MARS 4 MARS 5, MARS 6, AND MARS 7

V. I. Moroz

Until recently, studies of the reflected solar radiation and the thermal radiation in various spectral ranges by astrophysical methods were the sole source of information about the physical characteristics of the atmospheres and surfaces of other planets. Orbital spacecraft significantly extended the possibilities of astrophysical methods, since they permitted the distance to the object under study to be sharply reduced and the spatial resolution to be improved by tens and hundreds of times in comparison to that which the largest terrestrial telescopes can give. Moreover, the interference from the earth's atmosphere is completely removed, and those parts of the planet which are not visible from the earth become accessible for study.

Astrophysical methods of planet study by their radiation characteristics with equipment placed in orbital craft and direct studies on descent craft mutually supplement each other. A descent craft allows one to conduct detailed direct studies at one point of a planet; an orbital craft gives results of a more indirect nature, but which, on the other hand, extend over sizeable areas and which can cover the whole planet. The optimum consists of a reasonable combination of both methods. The idea of such an optimal combination was established in the expedition which arrived near Mars in February-March, 1974 and consisted of four spacecrafts: Mars 4, Mars 5, Mars 6, and Mars 7.

Mars 4 reached the planet first (Feb. 10, 1974); it passed by at a distance of 2200 km from its surface and conducted photography from the flyby trajectory. Mars 5 arrived second (Feb. 12), and entered the orbit of an artificial satellite of the planet. The orbit has the following basic parameters:

Osculating period	$T = 24^h 52^m 50^s$
Inclination	$i = 35°20'$
Semimajor axis	$a = 20{,}570$ km
Eccentricity	$e = 0.74974$
Planetary center distance at pericenter	$r_p = 5150$ km
Planetary center distance at apocenter	$r_a = 35{,}950$ km

The orbit was selected in such a way that Mars 5 passed into the daylight time of the Martian day over the region chosen for the landing of the descent crafts which Mars 6 and Mars 7 carried. Mars 7 approached the planet first (March 9), but as a result of the failure of one of the systems on board its descent craft passed by at a height of 1300 km above the surface and did not make a landing. On Mars 6 the launch operation of the descent trajectory went flawlessly, and on March 12 it was lowered onto the planet's surface in a region with the coordinates $\varphi \simeq -24°$, $\lambda = 20°$.

This region bears the name Pyrrhae in the International Astronomical Union nomenclature. Direct studies of the Martian atmosphere were conducted for the first time in the history of mankind. Studies of the atmospheric composition (the group directed by V. G. Istominyi) showed that the atmosphere of Mars contains a sizeable amount of inert gases (35 ± 15%). This is most probably argon. Such an amount of argon indicates that the average rate of gassing on Mars is comparable with the terrestrial rate, and the lower density of the Martian atmosphere is most probably explained in that its main part is condensed in the polar caps. This in turn supports the hypotheses that

---

Translated from Kosmicheskie Issledovaniya, Vol. 13, No. 1, pp. 3-8, January-February, 1975. Original article submitted June 28, 1974.

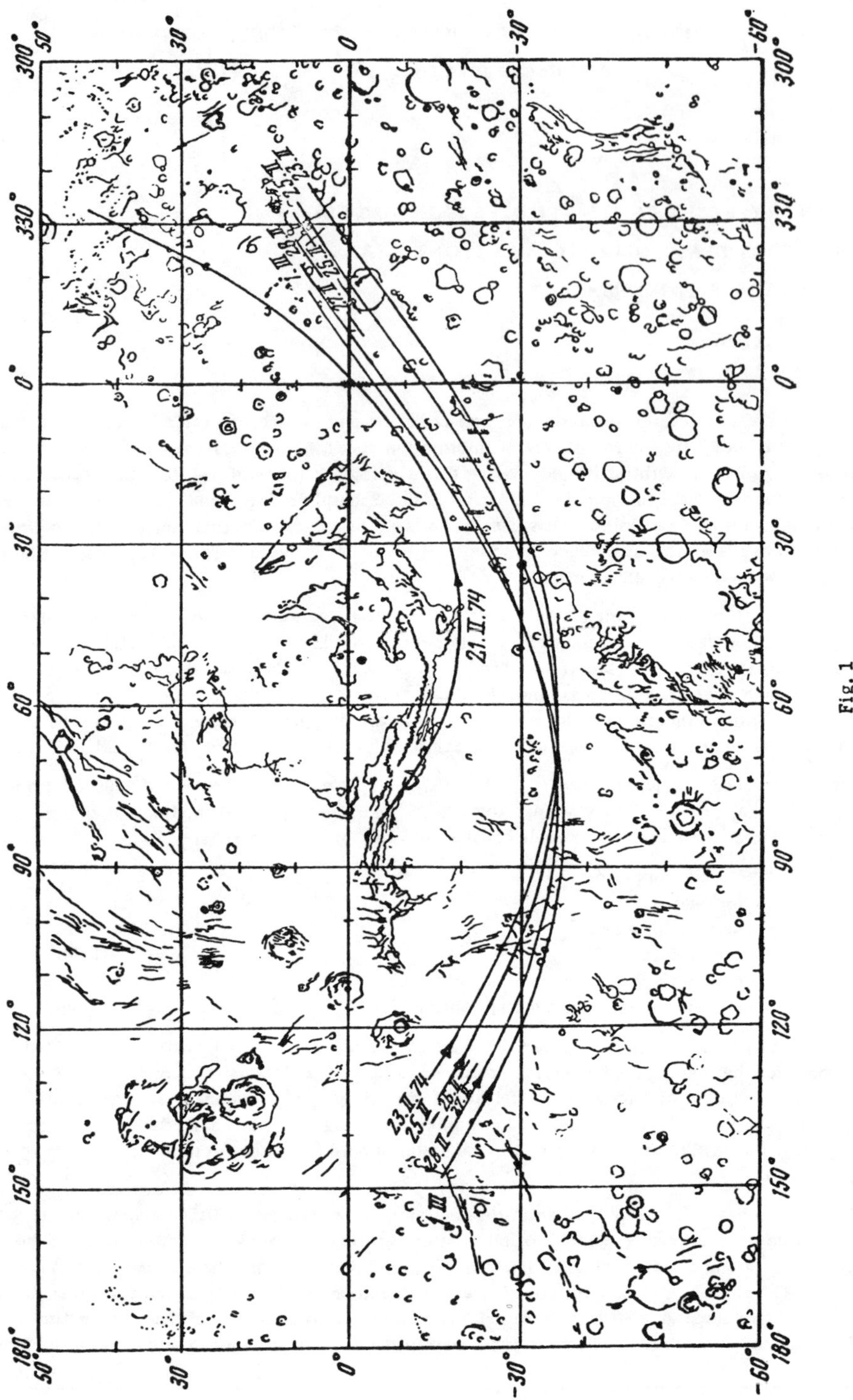

Fig. 1

50°
30
0
-30
-60°
180°
150
120
90
60
30
0
330
300°
Euxinus L.
Amazonis
Mesogaea
Memnonia
Sirenum M.
Phaethontis
Arcadia
Nix Olympica
Biblis F.
Daedalia
Icaria
Ascraeus L.
Tharsis
Araxes
Claritus
Aonius S.
Ceraunius
Noctis L.
Syria
Solis L.
Thaumasia
Dia
Tempe
Nilus
Lunae L.
Tithonius L.
Sinai
Ophir
Bosporus
Ogygis R.
Xanthe
Jamuna
Ganges
Aurorae S.
21.II.1974 г.
Erythraeum M.
Argyre I
Oceanidum M.
Acidalium M.
Niliacus L.
Oxia
Chryse
Margaritifer S.
Aram
Thymiamata
Eden
Cydonia
Ismenius L.
Protonilus
Arabia
Moab
Aeria
Edom
Sabaeus S.
Deucalionis R.
Pandorae F.
Noachis
Hellespontica D.
Hellespontus
Yaonis R.
Hellas
Serpentis M.
Deltoton S.

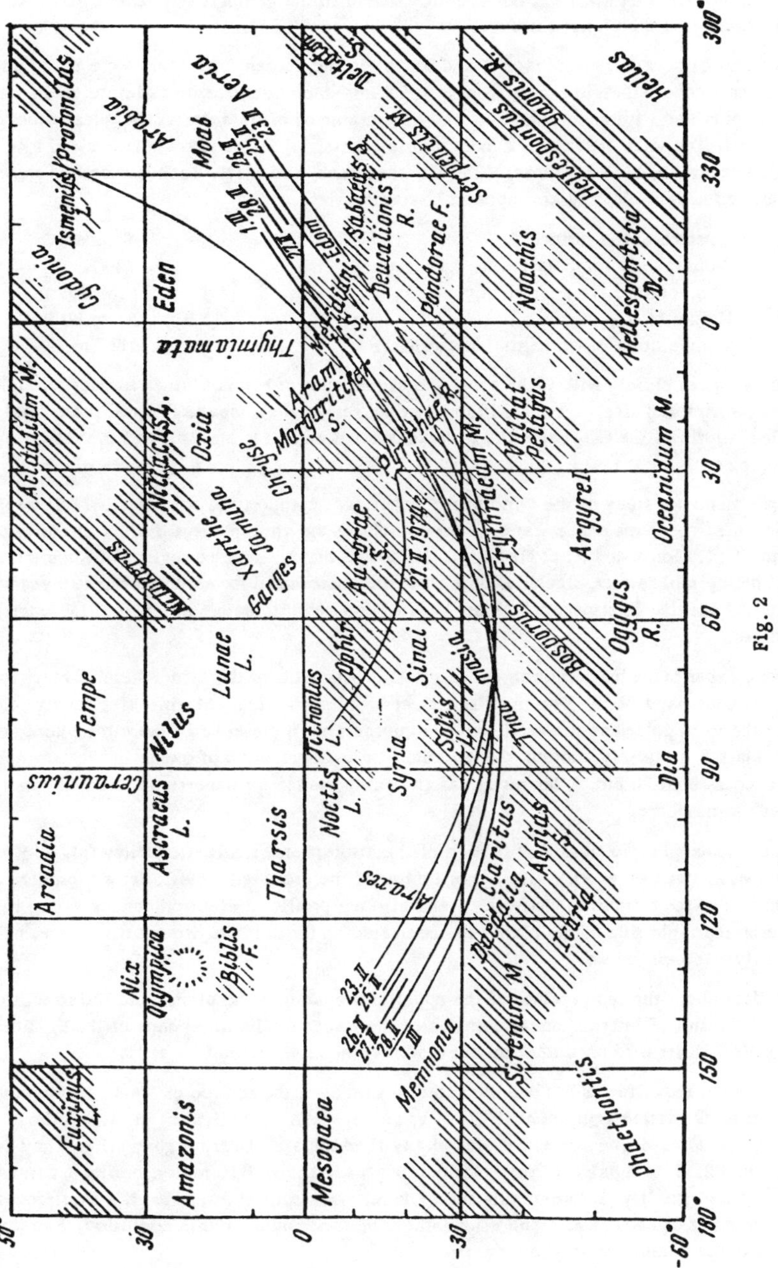

Fig. 2

suggest that the atmosphere on Mars was considerably denser in the geologically recent past than now, and open bodies of water existed on the planet's surface.

Measurements of pressure, temperature, and height on the descent trajectory were made on the descent craft. These measurements cover a height region from 0 to 20 km. Moreover, for an estimate of the basic atmospheric parameters, data obtained with accelerometers and measurements of the relative Doppler velocity along the descent craft to orbital craft line were drawn on. A compatible analysis of all the data (the work of a large group of authors including V. S. Avduevskii, M. Ya. Marov, V. V. Kerzhanovich, and M. K. Rozhdestvenskii) showed that they all could be explained for the following atmospheric characteristics:

Pressure at the surface	6 mbar
Atmospheric temperature at the surface	230°K
Temperature gradient at tropopause	2.5° K/km
Height of tropopause	25 to 30 km
Temperature of isothermal stratosphere	150° to 160°K

This model agrees closely with ideas about the atmosphere of Mars obtained earlier through analysis of the planet's radiation characteristics. Pressures in the Pyrrhae region were measured from the orbital craft Mars 5 by the equivalent widths of the CO_2 bands (L. V. Ksanfomaliti and V. I. Moroz), and the results agree well with the direct measurements.

Two experiments for study of the atmosphere's chemical composition were carried out on the orbital craft. One of these was measurements of the water-vapor content in the atmosphere from the intensity of the 1.38-μ absorption bands (V. I. Moroz and A. É. Nadzhip). It showed that in isolated regions of Mars the H_2O content can reach 100 μ of precipitable water, sizeably more than was observed in the same season two years ago on Mars 3. Moreover, it turned out that in regions which are several hundred kilometers apart, the H_2O content can differ by three to four times.

The second experiment for measurement of small constituents of the atmosphere of Mars was a photometer for the λ 2600 Å ozone band (V. A. Krasnopol'skii et al.). The American Mariner stations had earlier detected ozone bound in the solid polar cap material, but the question of its presence in the atmosphere remained open. The experiment on Mars 5 definitely showed the presence of small amounts of ozone in the atmosphere. The height of the ozone layer is about 20 km. This result has great importance for understanding of the photochemical processes in the planet's atmosphere.

Two experiments on radio transillumination of the atmosphere, single-frequency (M. A. Kolosov et al.) and dual frequency (M. B. Vasil'ev et al.), allowed estimates of the pressure in the lower atmosphere to be obtained from the magnitude of the refraction phase shift, and also the profile of electron concentration in the ionosphere. The ionosphere on the night side of the planet has been studied for the first time; earlier here, only an upper limit of electron density was well known.

Detailed data about the temperature of the upper atmosphere of the planet and the structure of its exosphere were obtained as a result of intensity measurements of the resonance luminescence of the L_α line (the group under the direction of V. G. Kurt with participation of French scientists G. Blamon et al.).

A large series of experiments has been devoted to studies of the surface of Mars. Photography of the planet was done with photo-television equipment (PTE) of various types (A. S. Selivanov et al.). About 60 photographs were obtained on the Mars 4 and Mars 5 crafts, many of which are of very high quality. They cover the region shown in Figs. 1 and 2. It is significant that Mariner 9 photographed this region during a dust storm and could not provide such a high quality picture-taking survey here. Two cameras were used, a short-focus one with a resolution near pericenter of about 1 km and a long-focus one with about 100 m resolution. Moreover, images were obtained with scanning photoelectric photometers.

The photographs obtained are being studied by geologists (see the article of K. P. Florenskii et al.), and also a photogrammetric analysis of them is being made (B. V. Nepoklonov et al.). There are traces of water erosion shown on certain photographs whose ages are estimated conservatively at less than a billion years. This is an independent corroboration of the hypothesis of cyclic density fluctuations of the Martian atmosphere.

The study of properties of the surface and soil by their radiation characteristics was conducted over a broad spectral range, starting from radio waves and ending with gamma radiation.

The devices for these measurements are ridigly bound to the case of the Automatic Mars Station (AMS) and their orientation in a constant direction during measurements is usually provided by the sun—star orientation system of the AMS. Also the photometers for H_2O and ozone were oriented. During approach to pericenter the devices are turned on by a special optical sensing element for several minutes before crossing the limb. The optical axes usually intersect the planet along nearly a great circle and passage from limb to limb takes about 30 min. The terminator is crossed within 22 min after the bright limb. From now on we shall call the track of the optical axis on the planet's surface the measurement path. By a preliminary estimate the accuracy with which the measurement path is determined comes to 1 to 2° in areographic coordinates.

The measurement paths of the orbital craft Mars 5 from Feb. 21 to March 1are shown in Figs. 1 and 2 for all the devices that are oriented parallel to the PTE.

The paths from Feb. 23 to March 1 pass through the Araxes and Claritas region, to the south of the Solis Lacus (they touch it from the south), then across Thaumasia, Mare Erythraeum, and end in Pyrrhae, where the descent craft Mars 6 made the landing. The paths are shifted in longitude by approximately 3° per day; they are almost superposed in the central part, which allows one to conduct mutual control of measurements on different paths. The path of Feb. 21 lies to the north of the remaining ones and has a different shape, as it was obtained with supplementary turn angles. In all, seven full-valued measurement sequences were conducted and results have been obtained for the seven paths.

One of the devices, the infrared radiometer (L. V. Ksanfomaliti and V. I. Moroz), measured the brightness temperature of the soil in the interval from 8 μ to 26 μ. These measurements show that the thermal inertia of the soil is in the range of 0.004 to 0.008 $cal \cdot deg^{-1} \cdot cm^{-2} \cdot sec^{-1/2}$. From this one can estimate grain soil size at from 0.1 to 0.5 mm.

On the other hand, the photometric and polarimetric measurements show that these grains have a microstructure of a finer scale (on the order of a micron). Photometry of the planet in the wavelength range 0.3 μ to 0.8 μ was conducted with several photoelectric devices (L. V. Ksanfomaliti). In just this wavelength range measurements with polarimeters (L. V. Ksanfomaliti, V. I. Moroz, and A. Dollfus, France) were made. Photometric and polarimetric observations from spacecraft are valuable not only in that they provide high spatial resolution, but they also provide the possibility to observe Mars at phase angles that are inaccessible from the earth.

The radio emission of the planet at 3.4 cm wavelength was measured with a radio telescope (a group of authors who consist of A. E. Basharinov, N. N. Krupenio, A. D. Kuz'min, V. S. Troitskii et al.). These measurements allow data about the temperature at a large depth (several tens of centimeters), and also estimates of the dielectric constant to be obtained. The dielectric constant depends on the soil density and thus measurements of radio emission give us the possibility to obtain an idea about the density, in addition, at a fairly sizeable depth below the surface.

The composition of the soil and its structure determine the planet's albedo in the wavelength range from 0.3 to 4 μ. The long wavelength section of this interval was studied with an infrared spectrophotometer (V. I. Moroz and N. A. Parfent'ev). Several hundred spectra have been obtained in the 2 μ to 5 μ interval. The most characteristic detail of them is the presence of a crystallized-water band near 3.2 μ.

A special device, a CO_2 altimeter (L. V. Ksanfomaliti and V. I. Moroz), measured the equivalent widths of the CO_2 band near two microns. From them the profiles of pressures and heights on the measurement paths were determined. A high region with characteristic pressure magnitudes of 3 to 4 mbar is located in the western part of the paths, in the east it is 5-6 mbar. The paths intersect two ridges with heights up to 8 to 10 km above the reference level (6.1 mbar).

The gamma ray spectrometer (Yu. A. Surkov, O. P. Shcheglov et al.) on Mars 5 allowed the gamma radiation spectra of the Martian rocks to be obtained, from which one possibly will obtain an idea about their characteristic composition.

A group of experiments was devoted to studies of the interplanetary plasma and the magnetic field in the vicinity of Mars.

Measurements of the magnetic field (Sh. Sh. Dolginov et al.) and of the interplanetary plasma (K. I. Gringauz et al.) have demonstrated new, real arguments in favor of the existence of an intrinsic magnetic field of the planet. Studies of the plasma in interplanetary space were also conducted by O. L. Waisberg's group. This same group made measurements of electric fields.

Two last experiments which remain to be mentioned are connected with solar studies. These are the measurement of solar cosmic ray fluxes (S. N. Vernov et al.) and the study of solar radio emission in the meter wavelength range by means of simultaneous observation from the spacecraft and from the earth (E. M. Vasil'ev et al. and Steinberg et al., France). The results of the first of these showed interesting peculiarities of the behavior of solar cosmic rays at the time of flares.

Summing up, one may draw the conclusion that the latest expedition to Mars brought science much new, interesting data. Not everything that was planned was carried out; however, a whole series of fundamentally new results has been obtained. The reduction of measurements of the majority of experiments has only been started, and it is a certainty that later on these results will be significantly supplemented. Unfortunately, not all the articles have been published in the present collection (B. V. Nepoklonov et al., A. E. Basharinov et al., Yu. A. Surkov et al., E. M. Vasil'ev, Steinberg, et al.) and will be published later.

4

Reprinted from *Science* **193**:766–776 (1976)

PRELIMINARY RESULTS FROM THE VIKING ORBITER IMAGING EXPERIMENT

Michael H. Carr, Harold Masursky, William A. Baum, Karl R. Blasius, Geoffrey A. Briggs, James A. Cutts, Thomas Duxbury, Ronald Greeley, John E. Guest, Bradford A. Smith, Laurence A. Soderblom, Joseph Veverka, John B. Wellman

This report is a preliminary assessment of pictures acquired from the Viking 1 orbiter during its first 30 orbits (designated as revs). During this period, attention was focused on the selection of landing sites. However, the pictures acquired have broad scientific interest, both for geology and for studies of the martian atmosphere.

The Viking visual imaging system (VIS) (*1*) consists of two high-resolution, slow-scan television framing cameras. Conceptually similar to the Mariner camera systems used in previous Mars, Mercury, and Venus missions, the VIS incorporates improvements designed to increase both spatial resolution and coverage. Each camera employs a 475-mm diffraction limited telescope and a 37-mm-diameter vidicon, the central region of which is scanned with a raster format of 1056 lines by 1182 samples and produces a 1.54° by 1.69° field of view. The optical axes of the cameras are offset by 1.38°. Cameras are shuttered alternately, resulting in contiguous swaths of images 80 km wide, with resolution better than 100 m near periapsis. Six color filters are available to restrict the image spectral bandpass to limited portions of the cameras' near-visual response characteristics.

The orbiter imaging experiment started acquiring calibration data 50 days before Mars orbit insertion (MOI); acquisition of scientific Mars data did not, however, begin until 120 hours before MOI when red and violet picture pairs were acquired every 4 hours. Beginning at MOI − 56 hours, three-color pictures were taken every 2 hours through MOI − 25 hours. A series of pictures taken with the red, the minus blue, and the violet filters completed the approach imaging. These early frames allayed any fears that the planet's atmosphere would interfere with photographing the surface. In several regions, particularly Hellas, Argyre, and Memnonia, they also revealed local brightenings interpreted as surface frost or ice clouds low in the atmosphere. One surprise was the visibility of the surface in the regions of the south pole, where a hood of clouds had been anticipated. After MOI, the orbiter cameras were devoted to finding suitable sites for the Viking landers.

Prior to insertion, detailed plans had been formulated to evaluate potential hazards at the landing site (A1) at 19°N, 34°W (Fig. 1), near the mouth of the large Chryse channels at the southern edge of the Chryse basin. The plan involved a calibration sequence on rev 1, extended coverage of the A1 site area on rev 3, and stereo coverage of the specific landing area on revs 4, 6, and 8. Because of a delayed arrival at the planet, the rev 1 observations were not made; the first high-resolution frames were acquired on rev 3. These revealed surprising detail in the A1 area (discussed below), sufficient to cause apprehension about its suitability as a landing site. Consequently, it was decided to look elsewhere for a safer site. The region to the northwest of the original site appeared most likely to yield a smooth area because it is farther from the mouths of the large channels, close to the deepest part of the Chryse basin, and might therefore be a site of fluvial deposition. This area (A1NW) did appear smooth in the pictures. Radar data (*2*) acquired later, however, indicated adverse conditions and resulted in searches on revs 20 and 22 for additional sites to the west, where a suitable site was eventually found. During this period two sites for the second lander were also examined (Fig. 1). These are the B site at 10°W, 44°N (revs 9 and 26) and the C site at 44°W, 6°S (revs 12 and 14). While the low-altitude coverage was being acquired, high-altitude observations continued to monitor atmospheric activity.

The Chryse Planitia region. Regional and local geological analyses (*3*) on the basis of pre-Viking data show that the area consists of relatively smooth plains of Chryse Planitia near the terminus of three large channel systems (Ares Vallis, Tiu Vallis, and Simud Vallis) that originate in chaotic terrain of Margaritifer Sinus and drain northward into the Chryse basin (Fig. 1). These channels and their associated distributary networks are considered to be primarily fluvial in origin and to have been modified by aeolian processes; however, the degree of modification by wind has not been established.

The predominant feature in the A1 area is lightly cratered plains typified by ridges similar to those on the lunar maria, and which increase in frequency and prominence to the northwest. By analogy to lunar geology, presence of the ridges suggests that the plains are lava flows with low viscosity in the melting range and are probably of basaltic composition. The second most extensive unit in the area forms plateaus which stand topographically above the plains. These are probably remnants of a surface that is older than the lava plains. Almost everywhere in this region streamlined plateau forms indicate sculpturing by fluid flow. In the transition zone between the Chryse basin and the cratered uplands, individual impact craters formed effective barriers to flow; downstream from the barriers the plateaus form islands with teardrop shapes (Fig. 2). Many plateau remnants are regularly terraced, possibly as a result of differential erosion of strata in the plateau material, or they could represent terraces from progressively lower fluvial levels. Toward the northwest the outliers of plateau material are less affected by the channeling process and appear to have undergone erosion by mass wasting along the margin, producing hummocky terrain.

In the southwest part of the A1 area, the lightly cratered plains have been etched, or stripped away in angular patches, to reveal a light toned, topographically lower, and presumably stratigraphically older unit (Fig. 3). Two possible agents of removal are wind and water. The exact stripping process is not known but the removal of material has been influenced by local structure. The general trend of the etched zone is parallel to Tiu Vallis and suggests a genetic relation to the formation of the channel. The northern end of the etched zone grades into incipient chaotic terrain, indicating that the etching may have been initiated or enhanced by ground-sapping processes.

At about 37.5°W and 22.5°N, 250 km northwest of the original A1 site, a channel segment possibly related to Tiu Vallis cuts through plateau material. The floor of the channel in this area is characterized by a network of fractures that are several hundred meters wide and several kilometers long. The general trend of the fracture system is parallel with the channel; a less pronounced set of fractures is transverse to the channel axis.

Knobs occur throughout the plains unit and range in size from several hundred meters to several kilometers across. They appear to be of diverse origin—some have distinct summit craters and are interpreted as volcanic in origin (perhaps cinder cones), others appear to be erosional remnants of plateau material, and still others are undoubtedly remnants of degraded and nearly buried crater rims. Small pancake-shaped features superimposed on both plains and etched terrain are interpreted to be constructional features possibly of volcanic origin (Fig. 3). Central craters and knobs, and possible dikes, support this interpretation.

On rev 10, 80 frames were taken of an area centered at 23°N, 43°W, close to the center of the Chryse basin (Fig. 1). It was anticipated that the region would be free of the fluvial features, etched terrain, and upland remnants that characterized the original A1 area. This prediction was largely fulfilled. Most of the area consists of lightly cratered plains, which in the western part are typified by north-south trending marelike ridges. Two islandlike, streamlined remnants of plateau material occur within the plains. Each remnant is topped with a 5-km-diameter, partly filled crater. Knobs as much as 1 km across occur throughout the area, but the density is greatest in the southeast. Overlying the smooth plains are craters up to 18 km in diameter.

In the search for a landing site on the west side of the Chryse basin, two areas were photographed. The first, centered on 49°S, 22°N, was photographed on rev 20; the second region, centered at 55°W, 22°N, was photographed on rev 22 (Fig. 1). Both regions, being to the west of the center of the Chryse basin, are at a higher elevation than the area photographed on rev 10.

As anticipated, both these areas show evidence of channeling related to Kasei Vallis and other channels farther south. The rev 20 coverage (which includes the final landing site) can be divided into two halves according to the degree of channeling. In the western half, channel-sculptured terrain is common (Fig. 4); in the east are mainly plains that resemble the lunar maria, and which are a continuation of those described above. They are crossed by ridges which, like those on the lunar mare, usually consist of two parts—a low broad linear rise on top of which is a narrower, more steeply sloping ridge. The ridges are mostly north-south trending and are spaced approximately 40 km apart. The plains surface is less cratered than the lunar maria and several ghost craters suggestive of filling are present. In common with other areas observed, the ejecta around large craters are sharply delineated and appear to consist of several flow lobes. The gradual transition from chaotic ejecta outward into secondary crater fields, characteristic of lunar craters, is rarely seen.

The western half of the area photographed on rev 20 has been extensively modified, apparently by fluvial action. Channels appear to have originated to the west and flowed eastward through the area toward the center of the Chryse basin. The whole region is sculptured by linear channels that wind through the

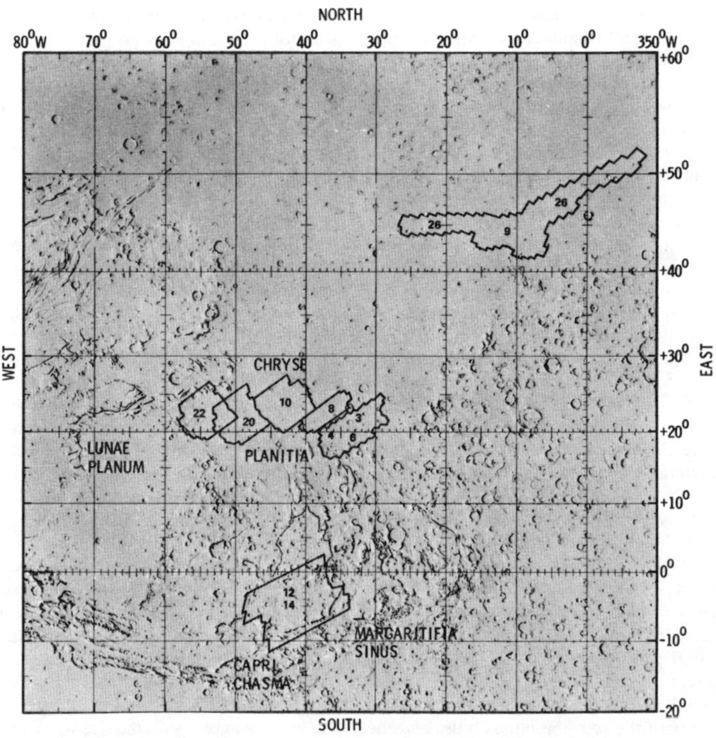

Fig. 1. Index map to show Viking orbital photographic coverage prior to the Viking 1 landing. Orbit numbers for the coverage are indicated. The Viking landing site is in rev 20 coverage.

area and cut across one another. Lenticular bars are common, as are convergent and divergent flow lines. Immediately upstream from the mare ridges sculpturing is often absent as though the fluid pooled there and had little erosive capability. Where gaps occur in the mare ridges the flow lines converge, indicating funneling of the flow through the gap (Fig. 4). Deeply incised channels occur downstream of the gaps. At the southern end of the channeled area the flow appears to have been contained and diverted by a large ridge. The sculptured area over which flow has presumably occurred is approximately 150 km wide. The length of the channels will not be known until more photographs are taken to the southwest. The general impression is of a flood, fairly evenly spread over a large area with very shallow slopes. The northern end of the rev 20 coverage includes part of Kasei Vallis. Again, as in the south Chryse region, long, lenticular, plateau remnants occur and linear striae attest to east-west flow. A low escarpment with a rectilinear outline, which seems to be unsculptured by flow, marks the edge of Kasei Vallis. The scarp consists mostly of intersecting alcoves that give it a serrated outline, although there are some linear sections. In general, it lacks streamline forms. Isolated areas of plains, surrounded by an escarpment, occur within the Kasei Vallis. Large craters superposed on these escarpments suggest a very old age (Fig. 5).

South of Kasei the region photographed on rev 22 can be divided east-west into two broad areas. In the east is a continuation of the lightly cratered plains. They resemble lunar maria and in the south are extensively sculptured by channel processes. The western half of the region is distinctively different from any region yet photographed. From the Mariner 9 data, it appears that a narrow strip of ancient cratered terrain separates Lunae Planum from Chryse Planitia. Numerous large craters within the strip typically have subdued rims and indistinct ejecta patterns, somewhat similar to those of the craters in the C site area described below. The plains between craters are, however, extremely hummocky and are crossed by several channels that are distinctly different from those at the C site. A section of one channel, which runs approximately east-west at 21°N, resembles some of the larger lunar rilles, such as Schröters Valley. Other channels farther south have rounded cross sections and numerous tributaries. All appear to empty onto the plains to the east. These channels are quite old with numerous superposed impact craters. The hummocky deposits in the west on the lunarlike mare plains also appear to be relatively old, since ejecta from several large craters covers both upland and plains.

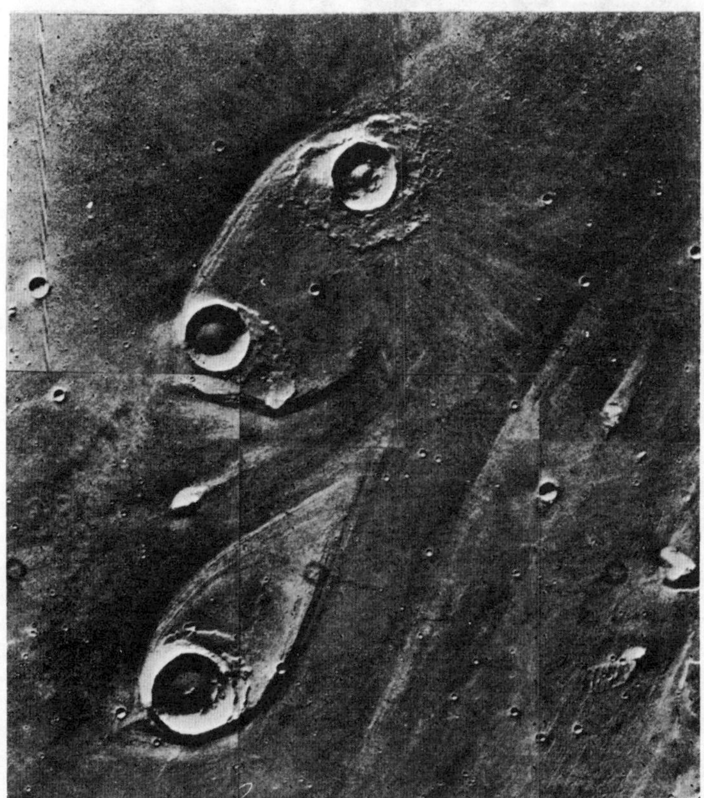

Fig. 2. Photomosaic to show teardrop "islands" on the southern side of the Chryse basin. The islands consist of remnants of the plateau material that forms a more continuous outcrop farther south. Here the plateau material has been largely eroded, apparently by fluvial action; craters on the upstream ends of the islands have protected the plateau material downstream of them from erosion. Each island is about 40 km long (frame numbers 4A50 to 4A54).

Cydonia region (B1 site). The Cydonia site (revs 9 and 26; 44°N, 12°W), as mapped from Mariner 9 (*4*), lies near the boundary between the mottled cratered plains and the smooth plains and includes the transitional boundary between mantled and unmantled terrain (*5*). The area shows a complex history of erosion and deposition. To the west and north, the surface is cut by a complex of curvilinear fractures typically 1 km wide and 10 km long that in plan view forms as a set of roughly polygonal-to-circular forms, most of which are 5 to 20 km across (Fig. 5). Southeast, the fractures seem to be buried by a younger unit that has undergone subsequent erosional stripping. This younger plains unit has a relatively featureless surface apart from superposed degraded craters.

Isolated angular-to-rounded mesas in the south are either (i) the remains of a once-continuous unit which overlay the fractured plains or (ii) the remnants of an ancient eroded surface projecting through the fractured plains from below. The mesas and fractured plains clearly underlie a younger plains unit to the southeast, which indicates that at least one phase of erosion occurred to produce the mesa landforms before deposition and erosion of the plains. Craters on surfaces below the plains are flat floored and have smooth rims surrounded by low, outward-facing scarps.

The youngest unit is a mantle of relatively bright material, probably of aeolian origin, that buried preexisting small craters. Other subsequently formed craters deposited ejecta from lower units on top of the mantle. During a later phase of erosion most of the mantle in the southern and western part of the area was stripped away, presumably by aeolian activity, leaving material trapped inside small bowl-shaped craters, and under and within ejecta deposits. These remnants of bright materials throughout most of the region give the mottled appearance observed in Mariner 9 and Viking pictures.

The origin of the fractured plains is not immediately apparent. Although the fractures could be the result of cooling contraction of extensive, thick lava, the scale of the pattern is larger by an order of magnitude than is normally found in such rocks. Another possibility is that they are of tectonic origin, but the lack of a marked regional trend to the pattern does not support such an origin. The possibility that they are related to a deep permafrost layer is an attractive alternative. However, this layer of permafrost would have to be abnormally thick. The region shows an extremely complex history of volcanic, tectonic, aeolian, and, possibly, periglacial processes, which yield a complex variety of landforms and materials.

Capri Chasma region (C1 site). On revs 12 and 14, frames were taken of the C site at 6°S, 43°W, one of the alternative sites for the second lander. The site is adjacent to Capri Chasma, a branch of the equatorial canyon system. The area outside the canyon is characterized by numerous large, flat-floored, subdued craters, between which are areas of relatively smooth intercrater plains. North of the site are several areas of chaotic terrain in which several large channels appear to originate. The channels drain northward, converge with other channels, and ultimately debouch into the Chryse basin.

The views of the canyon are some of the most spectacular pictures yet acquired. Landslides (Fig. 6) are clearly visible on both walls. The walls are as much as 2 km high and display several stratigraphic units that erode differentially. The uppermost layer breaks into large blocks while the lower layers seem to have poor cohesion and exhibit more fluid flow. There are a few low hills or knobs on the canyon floor, some of which may be remnants of coherent materials that have slumped into the canyon. Much of the canyon floor is featureless, devoid even of craters at the limiting resolution of the cameras. This suggests that the canyon floor is relatively young, certainly younger than any other surface yet observed on Viking pictures. The presence of bright streaks and dune fields (Fig. 6) indicates an active aeolian regime. These observations imply that the canyon is enlarged by collapse of the canyon walls to produce debris flows and removal of the material by wind. Although the causes of slope failure are uncertain, groundwater sapping or undercutting by aeolian action (or both) may be contributing processes (*6*).

Most of the large (> 50-km-diameter) craters in the region are flat-floored and have low rims; their ejecta appears to be covered by intervening plains material, indicating that the craters are older than the plains. However, some large craters are younger than the plains and have well-developed albeit subdued ejecta. Strings of secondary craters are abundant in some areas. Since Mariner 9, the origin of the intercrater plains has been a subject of some controversy. It has been variously ascribed to ballistic and base surge phenomena associated with impacts (*7*), atmospherically redistributed impact debris and weathering products (*8*), and volcanic activity (*9*). The layers in the canyon wall provide a cross-sectional view of the plains, and the morphology of the landslides provides an impression of the mechanical properties of the near-surface materials. It is possible that coherent layers resistant to erosion form the surface and that they are underlain by more easily erodible, less coherent materials. An inter-

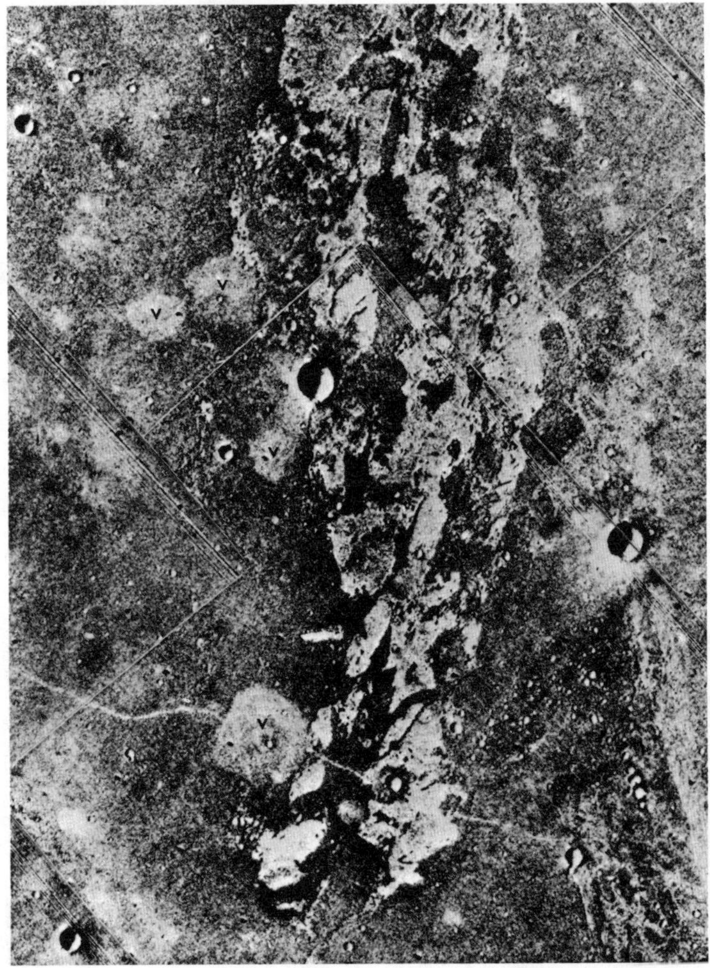

Fig. 3. Photomosaic to show the light-toned, "etched" terrain on the southern side of the Chryse basin. Also shown are light, domical features (*v*) thought to be volcanic shields. The mosaic covers an area about 60 km from top to bottom (frame numbers 4A78 to 4A81).

pretation of the general terrain of the C1 area is that it is composed of highly brecciated rocks of the ancient cratered terrain overlain by plains-forming deposits. The origin of the plains is not clear but the volcanic hypothesis is supported by the presence on the surface of numerous ridges and scarps similar to those on the lunar maria.

The close relation between fluvial features and chaotic terrain, noted on the basis of Mariner 9 observations, is clearly seen in the northern part of the site (Fig. 7). An area of chaotic terrain 50 km across is at the head of a series of fluvial-like features. The surface at the head of the channel appears to have collapsed into a jumbled, chaotic mass of debris, as though the underlying material had been removed. The area to the west has been sculptured and shaped into streamlined forms, suggesting flow to the west. The channel-like feature can be traced about 400 km to the west, where it passes off the edge of the photographic coverage into the region of the Hydrocates Chaos from which the channels of the Simud Vallis originate. These latter channels lead to Chryse Planitia, where the first Viking has landed. It thus appears that some of the materials that were carried out of the chaotic terrain in the C site ultimately may have been deposited far to the north, close to the present Viking landing site. The origin of the chaos and the channels is still unclear, but massive removal of subsurface materials, including a transporting fluid, is amply demonstrated.

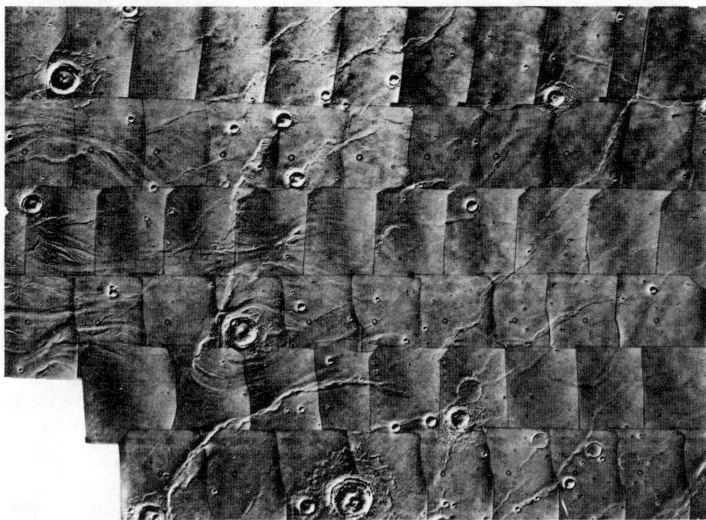

Fig. 4. Photomosaic of the landing site region in Chryse. Channeling is seen to the west (left). To the middle and right the plains have ridges similar in appearance to lunar mare ridges, implying that the plains were formed by extensive lava flows. The distance across the mosaic is about 240 km.

Fig. 5. Photomosaic of the fractured plains (*pf*) and overlying plains (*p*) in Cydonia. Eroded mesas of plateau material (*pl*) appear to have been exhumed by scarp retreat of the plains material. The area shown is about 120 km across (frame numbers 9A62 to 9A64).

Craters. Craters with fresh ejecta blankets and associated secondary craters were rarely seen in Mariner 9 pictures. It was predicted that, because the surface gravity on Mars is about the same as it is on Mercury, martian craters would have a similar morphology to those on Mercury (*10*) with the secondary crater fields much closer to the crater rim than on the moon, where the surface gravity is less. Viking pictures, however, show that fresh craters do exist on Mars and that, at least in the areas photographed, their morphology (Figs. 8 and 9) is dissimilar to that on Mercury or on the moon. Fresh craters are surrounded by lobate flow scarps and ridges, outside of which, in some cases, are bright rays and secondary crater clusters. The ejecta were apparently emplaced largely by flow. On arrival at the surface from ballistic throwout, the ejecta may have transformed to a fluidized sheet, possibly as a result of melted and vaporized ice or entrapped atmospheric gas. The lack of this distinctive ejecta pattern on the moon and Mercury may be explained by their lack of atmospheres and subsurface ice.

Not all martian craters have the same form. Many are degraded to smooth forms with flat floors and are encircled by an erosional scarp that faces outward. However, other apparently fresh craters, such as those near Kasei (Fig. 9), have closely spaced radial lineations on the continuous ejecta and few or no lobate forms. Whether such differences between fresh craters result from different states of target material or from other parameters remains to be determined.

Another striking characteristic of the areas photographed thus far is the presence of fresh crater clusters and irregularly shaped craters similar in form to lunar secondary craters. Some of the smaller clusters can be related to nearby fresh craters but there are other, often extensive clusters, particularly at the C site, which are not obviously related to large fresh craters. It is possible that some craters resembling secondary craters result from the impact of meteoroid showers produced by breakup of a large meteoroid on entry into the martian atmosphere.

Variable features. Four types of variable features (*11*) are prominent in the Viking orbiter pictures: (i) bright streaks associated with craters, (ii) bright streaks associated with small hills, (iii) dark streaks associated with craters, and (iv) sand dune fields. These were compared with the available Mariner 9 coverage to determine the changes that have occurred since 1972. As a result of the increased resolution of the Viking photography, many more small craters and streaks are visible in the Viking images. Comparison with the Mariner 9 data shows that the bright streaks are generally unchanged in direction and outline since 1972, although in a few places they have increased in size or new bright streaks have appeared.

In the Chryse region dark streaks are less prominent than bright streaks and trend southwest to northeast, opposite to the bright streak direction. The dark streak direction coincides with the regional wind flow expected at the present season (northern summer). The bright streak pattern defines wind flow from northeast to southwest, the direction of strong winds that are expected during southern summer dust storms. The directions and outlines of bright crater streaks are unchanged since the Mariner 9 coverage in 1972, thereby confirming the speculations that bright streaks, unlike some dark streaks, are stable over many martian seasons and are unaffected by the weaker winds that occur during the present season. Evidently, the winds responsible for the formation of the dark streaks did not significantly modify the bright ones.

Westward across the Chryse basin the bright streak direction shifts from an azimuth of 220° to 230° near 34°W to 255° near 58°W, a pattern consistent with that observed by Mariner 9. Wind streaks appear to be concentrated in regions that look generally smooth at orbital picture resolution. This correlation of streak density with terrain type may be valid down to roughness scales of tens of centimeters since, for example, in the rev 20 coverage the only large area that appears to be smooth but which is devoid of streaks appear rough to radar. There is supporting evidence, based on Mariner 9 data, that wind streaks are best developed in regions where the areas between craters are relatively smooth.

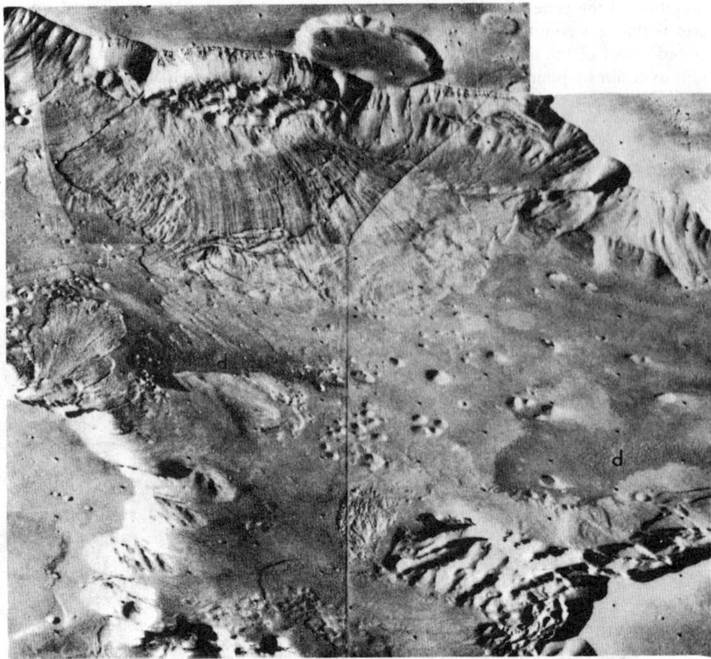

Fig. 6. Oblique view across the Capri Chasma to show dark fields of sand dunes (*d*) in the floor of the canyon, and landslides on the walls. Distance from foreground to background is about 150 km; the canyon is about 2 km deep (frame numbers 14A29 to 14A32).

No wind directions in the Cydonia region can be mapped because the area lacks streaks that are associated with craters and other evident wind markers. At least during this season (northern summer), wind transportable material appears not to move; either the texture of the surface is too coarse, or the fine particles are effectively bound together. The annual dust storm fallout that probably blankets the area after perihelic dust storms could be swept up effectively into the numerous troughs which occur in this region. Significantly perhaps, some of the troughs show definite signs of infilling by high albedo material.

In the Capri region, bright streaks trend northeast to southwest and are unchanged in orientation and outline since 1972 (Mariner 9 data). Low albedo dune fields occur on the floor of neighboring Gangis Chasma (Fig. 6). Bright, streaklike markings behind hills on the canyon floor are common in the vicinity of dune deposits and may represent protected areas where no sand has accumulated. The directions of these bright streaks associated with hills show that the dominant regional northeast to southwest wind flow is channeled by the canyon walls into a general east-west direction.

High-altitude pictures of the Oxia Palus region show that the general albedo boundaries and bright streak patterns are unchanged since 1972. Nevertheless, conspicuous albedo changes have occurred in some localized areas, such as within and around the crater Galilaei. A new bright streak, trending south from a 4-km crater, has appeared since 1972, and a streaklike bright area emanating from a channel has grown significantly during the past 4 years. These areas will be studied at high resolution later in the mission.

Atmospheric phenomena. Numerous atmospheric phenomena have been observed in the orbiter camera images, both on approach and from high altitude in orbit. Approach images indicated that the atmosphere of Mars was then relatively clear in the southern hemisphere but obscuring hazes were present at all longitudes in the north. After Mars orbit insertion, most of the high-altitude images were acquired in order to monitor a broad area around the planned landing site for dust activity. In this coverage, taken from a range on the order of 30,000 km, the site is seen in morning hours and each pixel spans about 800 m.

Figure 10 shows a mosaic of a typical high-altitude, five-frame sequence taken through a red filter by the Viking 1 orbiter cameras and processed without spatial frequency discrimination. Although some differences are seen from day to day, the general haziness of the morning sky in the northern hemisphere has looked much like Fig. 10 since Viking 1 arrived. The midday and afternoon skies are not visible for comparison, but the finding of diffuse morning cloudiness is consistent with long-term photography of Mars from Earth. On some days, high-altitude pentads have been obtained through violet and green filters as well as through red, and the overall differences in haziness are not very great.

In regions of Fig. 10 where surface features such as crater rims can be detected, the atmospheric extinction coefficient γ can be estimated from contrast measurements. Where craters are near the threshold of detection, the local values of γ for blue light turn out to be comparable with Earth's atmosphere on a relatively clear day, but values for red light are somewhat greater than on Earth. When expressed in terms of equal air paths, however, this morning scattering above threshold craters is fully two orders of magnitude greater than in clear Earth air. In regions of similar emission angles where craters exist but cannot be detected, the martian atmosphere is optically still thicker.

The morning haziness in the north that is shown in Fig. 10 would not, on the basis of photography from Earth, be expected to typify other latitudes and times of day. Indeed, Fig. 11a, which was obtained on 11 July 1976, shows a much clearer atmosphere around 50° to 60°S in midafternoon, where the extinction coefficient of red light was found to be $\gamma \sim 0.1$ per air mass. It was approximately winter solstice ($L_s = 93°$) in the southern hemisphere, thus the terminator lies only a few degrees below the bottom of Fig. 11a. In Viking approach images, there were bright patches in this region that did not obscure underlying topography and that contrasted more strongly with their surroundings in violet images than in red, but their nature is not yet clear and further observations are planned for distinguishing between fog and frost.

Figure 11a, taken at a range of about 18,000 km, also illustrates well the layered structure of the atmosphere. From the foreshortening of craters, we infer the true surface limb to lie about 3 km under the bright low-altitude haze, while upper layers extend to nearly 40 km above the surface. At the limb, the low-altitude haze is optically thick, but the uppermost layer has an edge-on extinction of only about 0.3. This implies that the uppermost layer has a vertical extinction coefficient less than 0.01 per air mass, thus it plays very little role in the total atmospheric obscuration.

The height of the limb haze in the tropics has been estimated in an image from rev 4. Landmarks in the scene were used to locate the surface limb, which was more heavily obscured than in Fig. 11a. In the rev 4 image, the highest visible haze layer occurs at 25 km, and the height of unit optical depth on the limb (normal optical depth ~ 0.02) occurs at 15 km. Since the surface optical depth in the same region as judged from crater contrasts is ~ 0.5, the scattering power

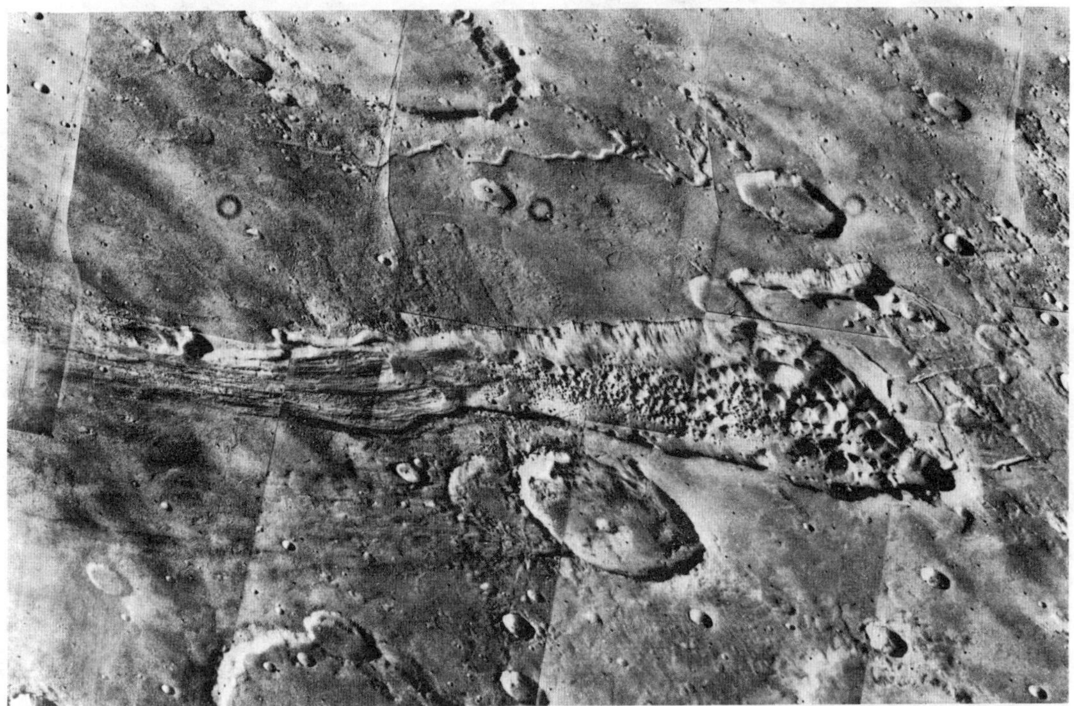

Fig. 7. Photomosaic of a 120-km-long channeled area near Capri Chasma. The apparent source of the fluid that cut the channels is a depression enclosing chaotic terrain (right) which appears to have been formed by collapse (frame numbers 14A67 to 14A69)

of the atmosphere increases more rapidly with depth less than 15 km than does the atmospheric density. Infrared thermal mapper temperature measurements, which have a very broad weighting function but which center near 20 km, show that the highest tropical haze layers cannot be CO_2 ice.

Discrete clouds, as distinct from diffuse regional haze, seem to have several morphological forms. The classical white clouds of Tharsis, known for decades to observers on Earth, are seen in Fig. 11b, which was obtained through a violet filter at a range of 300,000 km as Viking 1 initially approached Mars. Images taken when this region was clear to the morning terminator do not show these clouds, and we hope that orbital images will provide additional information about their growth phase. They were observed by Mariner 9 only when fully formed (*12*).

Several diffuse bright clouds of similar size have been seen within the morning haze in the north. Figure 11c shows one covering about 40 square degrees that appeared near the final A1 landing site on rev 28. It contrasted more with its surroundings in this violet image than in a corresponding red one. Images 6 minutes apart revealed no motion with respect to surface features; an upper limit on its motion is estimated to be 10 m/sec.

Equatorial clouds seen thus far are much less diffuse and are made up of many patches with dimensions of a few kilometers. Figure 12a shows one seen on rev 4 at 23°W, 1°S. It is typical of condensate clouds and showed much higher contrast in violet light than in red. Its motion was 46 ± 3 m/sec westward. Since no shadows were identified, the height of the cloud above the surface is unknown. However, such clouds appear to have a convective structure and are inferred to be within a few kilometers of the surface. Westward cloud motions at speeds ranging from 15 to 45 m/sec are expected theoretically in this region and at this season (*13*). This east-west flow is part of an expected anticyclonic circulation around the Chryse basin.

Similar clouds were also detected in low-altitude vertical imaging of candidate landing sites, as illustrated in Fig. 12b, which shows a portion of two C1 site mosaics centered at about 40°W 4°S. Frames in the second mosaic were taken two martian days after those in the first mosaic under similar viewing conditions. Both sets of frames were acquired through a clear filter and have been processed in such a way as to enhance high spatial frequencies. Sequences designed to provide stereoscopic coverage of the landing sites yield repeated coverage of those regions over time intervals of about 2 minutes. Changes in position and appearance of these cloudlike features over such short time intervals confirms them to be atmospheric phenomena.

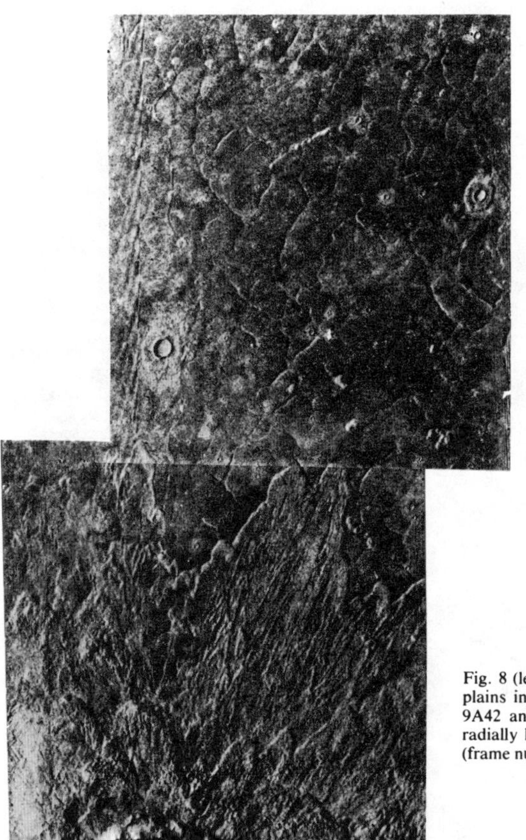

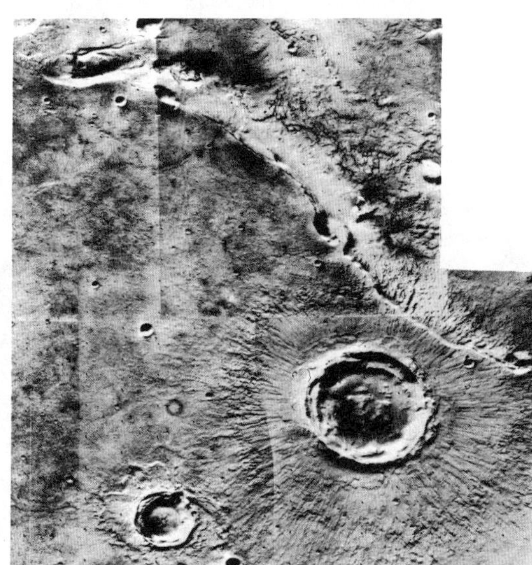

Fig. 8 (left). Part of a relatively fresh 30-km-diameter crater overlying fractured plains in Cydonia. The lobate ejecta flows are well developed (frame numbers 9A42 and 9A43). Fig. 9 (right). A 20-km-diameter crater near Kasei with radially lineated ejecta surface. The ejecta overlies the scarp on the valley side (frame numbers 22A52 to 22A54).

Wave clouds have been observed in midmorning images of the equatorial region. A wave pattern can be seen in the lower left corner of Fig. 12b, but some better examples are found in Fig. 12c. A comparison with frames of the same area taken on another day indicates that many discrete albedo features in Fig. 12c, such as bright splotches and crater tails, are

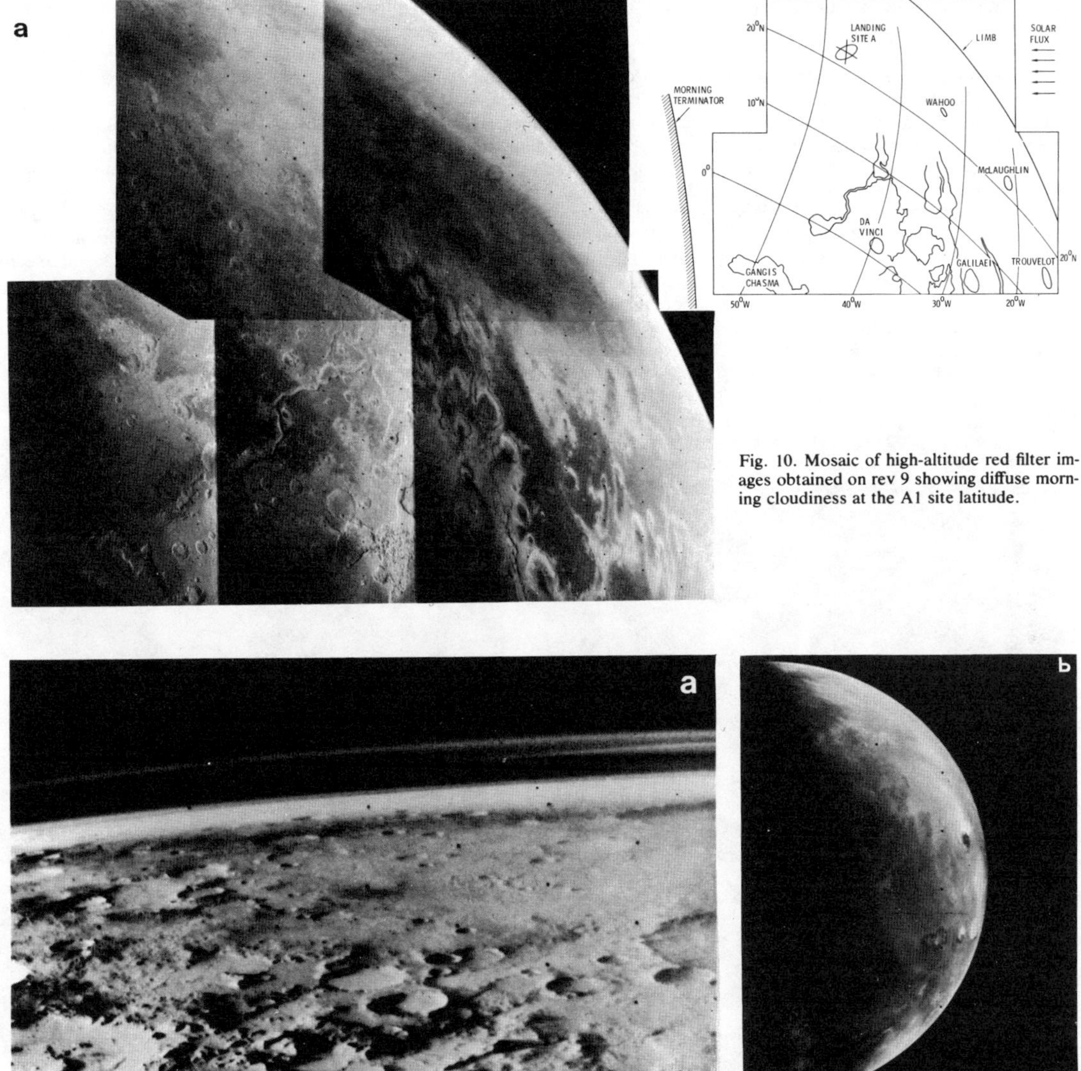

Fig. 10. Mosaic of high-altitude red filter images obtained on rev 9 showing diffuse morning cloudiness at the A1 site latitude.

Fig. 11. (a) Southern hemisphere winter limb in the Noachis region imaged in red light from medium altitude on rev 22. (b) Large white clouds on the western flanks of Olympus Mons (upper) and Ascraeus Mons (lower), imaged through a violet filter during approach. (c) White cloud near the A1 landing site imaged through a violet filter on rev 28.

on the surface, but that the whole area is overcast with a nonuniform haze. The presence of wave clouds, presumably composed of water ice, provides an indication of the wind direction and of the static stability of the atmosphere (*14*). Observed wavelengths are on the order of 10 km.

At the lower right of the images in Fig. 12d is a bright patch seen in Capri Chasma from high altitude on rev 4. It does not obscure or diffuse surface detail, therefore it evidently lies closer to the surface than the resolution limit (about 2 km). It may be either a fog of water ice or possibly a thin frost patch. The former seems rather more likely in view of the smaller amount of water required to produce the observed brightness.

No direct evidence for dust clouds has been seen in early Viking 1 orbiter images, although the presence of micron-size dust particles could well be postulated in some of the haze without contradicting the observed characteristics.

On the basis of preliminary examination of about 1000 frames obtained from Viking 1 orbiter, the following conclusions are reached.

1) Most of the surfaces examined are old. Crater frequencies on the various plains range from one-tenth that of the lunar maria to approximately the same as the lunar maria. Only the floor of Vallis Marineris is significantly younger, on the basis of crater frequencies.

2) Despite the seemingly old age, of almost all the surfaces so far photographed, small craters are preserved, thereby suggesting that aeolian erosion is extremely slow.

3) Abundant new evidence of catastrophic floods has been revealed in the southern and western margins of the Chryse basin; however, no evidence for a thick accumulation of sediments was found in the middle of the Chryse basin.

4) One mechanism for growth of the equatorial canyon system is slumping of the canyon walls into the canyon and subsequent removal of the slumped debris by wind.

5) In most of the areas examined, crater ejecta morphology is distinctively different from either the moon or Mercury; the principal mechanism of ejecta emplacement appears to be surface flow rather than ballistic deposition.

6) At 44°N numerous intersecting cracks on the surface of the plains form polygonal patterns reminiscent of patterned ground in the Arctic regions of Earth.

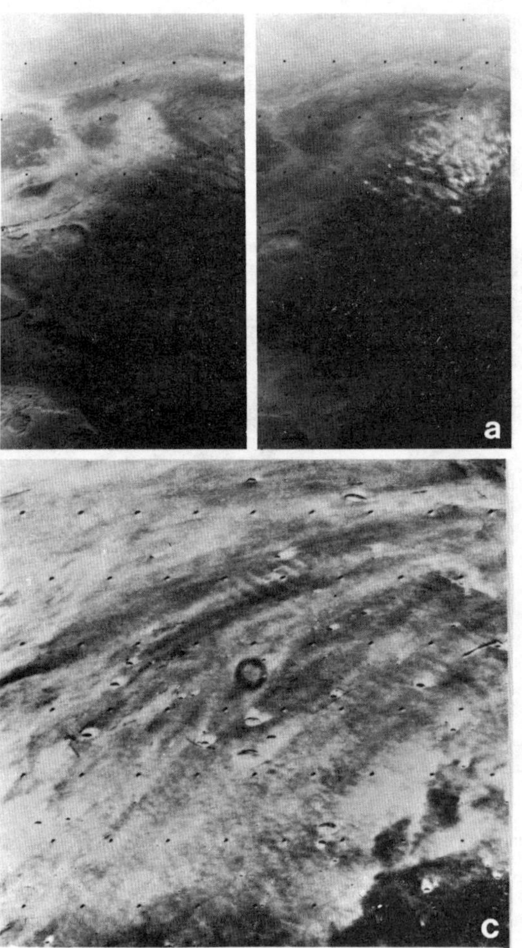

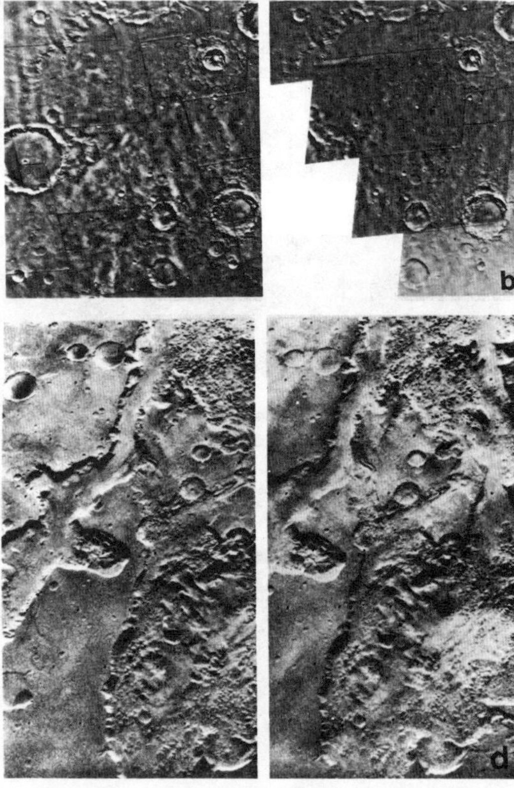

Fig. 12. (a) Equatorial cloud (upper right) imaged from high altitude on rev 4. Red filter is on the left and violet filter version on the right. (b) Equatorial clouds seen on C site photomosaics of low-altitude images obtained through a minus-blue filter. One set was obtained 2 rev after the other. (c) Wave clouds at 7°N, 84°W seen in red light from medium altitude on rev 12. (d) Bright patch (lower right) imaged from high altitude on rev 4. Red filter version is on the left and violet filter version on the right.

7) Variable features, when compared to Mariner 9 pictures taken 4 years ago, show relatively little change.

8) Several types of clouds were observed, including diffuse morning hazes in the northern hemisphere, discrete equatorial white clouds, and extensive wave clouds.

9) A direct measurement of wind velocity from cloud motion was achieved for the first time.

References and Notes

1. J. B. Wellman, F. P. Landauer, D. D. Norris, T. E. Thorpe, *J. Spacecraft Rockets*, in press; T. E. Thorpe, *Icarus* **27**, 229 (1976).
2. L. Tyler, personal communication.
3. D. H. Wilhelms, *U.S. Geol. Surv.* (1976), p. I-895; H. Masursky, N. J. Trask, M. E. Strobell, A. L. Dial, G. W. Colton, *U.S. Geol. Surv. Map*, in press; D. Milton, *U. S. Geol. Surv.* (1974), p. I-894.
4. M. H. Carr, H. Masursky, R. S. Saunders, *J. Geophys. Res.* **78**, 4031 (1973); D. Scott, *U.S. Geol. Surv.*, in press.
5. L. A. Soderblum, T. J. Kreidler, H. Masursky, *J. Geophys. Res.* **78**, 4117 (1973).
6. R. P. Sharp, *ibid.*, p. 4073.
7. M. Malin, thesis, California Institute of Technology (1976).
8. B. C. Murray, L. A. Soderblum, R. P. Sharp, J. A. Cutts, *J. Geophys. Res.* **76**, 313 (1971).
9. D. E. Wilhelms, *ibid.* **79**, 3933 (1974).
10. D. E. Gault, J. E. Guest, J. B. Murray, D. Dzurisin, M. Malin, *ibid.* **80**, 2444 (1975).
11. Variable features were first defined during the Mariner 9 mission to mean surface albedo patterns which changed with time; the term has since been used to include all surface forms that are commonly attributed to aeolian or wind processes [C. Sagan *et al.*, *J. Geophys. Res.* **78**, 4163 (1973)].
12. C. B. Leovy *et al.*, *ibid.*, p. 4252.
13. J. B. Pollack, C. B. Leovy, Y. Mintz, W. Van Camp, *Geophys. Res. Lett.*, in press; P. W. Webster, in preparation.
14. J. A. Pirraglia, *Icarus* **27**, 517 (1976).

15. We thank the following people for their untiring help during the early hectic stages of this mission: J. Boyce, P. S. Butterworth, K. W. Farrell, E. A. Flinn, H. Ferguson, A. Jankevics, K. P. Kaasen, C. Leavy, B. Lucchitta, J. MacQueen, T. E. Poe, Jr., G. Schaber, A. Spruck, E. E. Theiling, D. T. Thompson, and T. E. Thorpe. Financial support for the work of team members was provided by NASA Viking Project Office, NASA Office of Planetary Geology (R.G.), and U.K. Natural Environment Research Council (J.E.G.).

26 July 1976

Part II

SURFACE MAPPING AND GEOMORPHOLOGY

Editor's Comments on Papers 5 and 6

5 **McCAULEY et al.**
Excerpt from *Preliminary Mariner 9 Report on the Geology of Mars*

6 **CARR, MASURSKY, and SAUNDERS**
A Generalized Geologic Map of Mars

Our current understanding of martian geology has been largely shaped by Mariner 9 results. Familiar geological processes such as volcanism, tectonic activity, and eolian and aqueous erosion have modified the once densely cratered surface in many areas, but on a scale and in combinations unlike anything seen on Earth. J. F. McCauley and others (Paper 5) describe some of the diverse landforms of Mars and discuss possible modes of origin of controversial features such as the broad channels and equatorial canyon system. M. H. Carr, H. Masursky, and R. S. Saunders (Paper 6) have refined the geological map of the equatorial belt that originally appeared in the paper by McCauley et al.

The preliminary results of the Viking missions published to date have reinforced the basic conclusions reached in these two papers. Some of the dilemmas raised by McCauley et al. seem a little less perplexing today. For instance, the removal of material out of the canyons is no longer a serious problem, since recent pictures suggest that the canyons are in fact grabens formed by collapse (Blasius et al. 1977). The scarcity of water to carve the giant channels is also less of a problem, since permafrost features appear widespread (Carr and Schaber 1977), and water ice is stored at the permanent polar caps (Farmer et al. 1976; see also Paper 9, Part III).

5

Reprinted from *Icarus* **17**:289, 325–327 (1972)

Preliminary Mariner 9 Report on the Geology of Mars[1]

J. F. McCAULEY[2], M. H. CARR[2], J. A. CUTTS[3], W. K. HARTMANN[4], HAROLD MASURSKY[2], D. J. MILTON[2], R. P. SHARP[5], AND D. E. WILHELMS[2]

Received June 7, 1972

Mariner 9 pictures indicate that the surface of Mars has been shaped by impact, volcanic, tectonic, erosional and depositional activity. The moonlike cratered terrain, identified as the dominant surface unit from the Mariner 6 and 7 flyby data, has proven to be less typical of Mars than previously believed, although extensive in the mid- and high-latitude regions of the southern hemisphere. Martian craters are highly modified but their size-frequency distribution and morphology suggest that most were formed by impact. Circular basins encompassed by rugged terrain and filled with smooth plains material are recognized. These structures, like the craters, are more modified than corresponding features on the Moon and they exercise a less dominant influence on the regional geology. Smooth plains with few visible craters fill the large basins and the floors of larger craters; they also occupy large parts of the northern hemisphere where the plains lap against higher landforms. The middle northern latitudes of Mars from 90 to 150° longitude contain at least four large shield volcanoes each of which is about twice as massive as the largest on Earth. Steep-sided domes with summit craters and large, fresh-appearing volcanic craters with smooth rims are also present in this region. Multiple flow structures, ridges with lobate flanks, chain craters, and sinuous rilles occur in all regions, suggesting widespread volcanism. Evidence for tectonic activity postdating formation of the cratered terrain and some of the plains units is abundant in the equatorial area from 0 to 120° longitude. Some regions exhibit a complex semiradial array of graben that suggest doming and stretching of the surface. Others contain intensely faulted terrain with broader, deeper graben separated by a complex mosaic of flat-topped blocks. An east–west-trending canyon system about 100–200 km wide and about 2500 km long extends through the Coprates–Eos region. The canyons have gullied walls indicative of extensive headward erosion since their initial formation. Regionally depressed areas called chaotic terrain consist of intricately broken and jumbled blocks and appear to result from breaking up and slumping of older geologic units. Compressional features have not been identified in any of the pictures analyzed to date. Plumose light and dark surface markings can be explained by eolian transport. Mariner 9 has thus revealed that Mars is a complex planet with its own distinctive geologic history and that it is less primitive than the Moon.

[*Editor's Note:* Material has been omitted at this point.]

Major Conclusions

Mariner 9 has shown that Mars is geologically far more heterogeneous than previously suspected from the earlier flyby missions. The analyses to date indicate convincingly that it has a geological style of its own different from that of either the Earth or Moon, and suggest that it represents a body intermediate in its evolutionary sequence somewhere between the Earth and Moon. It is now tempting to consider Mars as a planet that has partly made the transition from a relatively primitive impact-dominated (but not primordial) body like the Moon to an orogenically mobile, volcanically active, water-dominated planet like the Earth. Phobos and Deimos (Pollack *et al.*, 1972) are clearly the most primitive solar system bodies closely investigated to date.

Like the Moon, Mars shows extensively cratered regions as well as numerous large circular basins. The basins, some of which are larger than any lunar basin, seem to exercise less control on the regional topography and distribution of the volcanic units than they do on the Moon.

The crater and basin terrains and the plains material that fills depressions in it are reminiscent of the Moon. But over much of the planet, the later parts of the Martian geological record are punctuated by huge and spectacular tectonic and volcanic features, not moonlike and only partly earthlike, that have destroyed or covered its earlier crater and basin aspect.

Extensive tectonic activity has occurred in huge regions of Mars. Much of this can be ascribed to circumferential tension in the upper parts of the lithosphere and to local doming. No shear or compressional features have been identified to date.

Volcanism has also played an important role in shaping the surface of Mars and probably has contributed in great part to its tenuous atmosphere. Martian volcanism is dramatically more varied and may span a larger part of the planet's history than lunar volcanic activity. Preliminary crater frequency studies point to the possibility that the major shield volcanos, calderas, plains, and other volcanic features could be relatively young.

Erosion and sedimentation, neither of them related to impact cratering, has occurred on a planetwide scale and surface modification processes are more widespread than previously envisioned. Erosion channels and depositional features abound, commonly related geographically and probably genetically to terrain that has collapsed chaotically. Extensive transport of materials has occurred in these channels, and moving fluids probably in episodic surges seem to be the only possible mechanism by which this can be explained, that is consistent with the present observational data.

Mars has clearly undergone a different proportionate mix of major surface-shaping processes than the Earth; the interplay between impact, volcanism, tectonism, and various erosion and sedimentation processes is clearly distinctive. Elucidation of these relations certainly will be the major fruit of the Mariner 1971 mission and should contribute significantly to a better understanding of the Earth.

Most markings observed by Mariner 9 seem to be surficial and of probable eolian origin. They are partly controlled by topographic features such as craters and scarps and appear to be excellent indicators of both past and recent wind regimes.

Acknowledgments

G. W. Colton and Joseph Veverka reviewed the manuscript and made numerous helpful suggestions. Colton also helped review and expedite production of the geologic maps. Joseph Boyce also helped in the original compilation of the maps. Special thanks are extended to James Van Divier, Carl Zeller, Roger Carroll, and William Miller for their efficient and timely help in photo reproduction and the drafting of the various illustrations. The original base materials were prepared under the direction of R. M. Batson. The work contained in this report was done under the auspices of the Mars Mariner 1971 Project, Jet Propulsion Laboratory, California Institute of Technology mostly under contract WO-8122.

References

CUTTS, J. A., SODERBLOM, L. A., SHARP, R. P., SMITH, B. A., AND MURRAY, B. C. (1971). The surface of Mars: 3. Light and dark markings. *J. Geophys. Res.* **76**, 343.

HARTMANN, W. K. (1966). Martian cratering. *Icarus* **5**, 565.

LEIGHTON, R. B., AND MURRAY, B. C. (1966). Behavior of carbon dioxide and other volatiles on Mars. *Science* **153**, 136.

LEOVY, C., BRIGGS, G. A., YOUNG, A. T., SMITH, B. A., POLLACK, J. B., SHIPLEY, E. N., AND WILDEY, R. L. (1972). The Martian atmosphere: Mariner 9 television experiment progress report. *Icarus* **17**, 373.

MASURSKY, H. *et al.* (1971). Mariner 9 television reconnaissance of Mars and its satellites: Preliminary results. *Science* **175**, 294.

MCCAULEY, J. F., AND WILHELMS, D. E. (1971). Geological provinces of the near side of the Moon. *Icarus* **15**, 363.

MURRAY, B. C., SODERBLOM, L. A., SHARP, R. P., AND CUTTS, J. A. (1971). The surface of Mars: 1. Cratered terrains. *J. Geophys. Res.* **76**, 313.

MURRAY, B. C., SODERBLOM, L. A., CUTTS, J. A., SHARP, R., MILTON, D., AND LEIGHTON, R. (1972). Geological framework of the south polar region of Mars. *Icarus* **17**, 328.

POHN, H. A., AND OFFIELD, T. W. (1970). Lunar crater morphology and relative-age determination of lunar geologic units—pt. 1. Classification. *In* Geological Survey Research 1970: *U.S. Geol. Survey Prof. Paper* 700-C, C153.

POLLACK, J., VEVERKA, J., NOLAND, M., SAGAN, C., HARTMANN, W., DUXBURY, T., BORN, G., MILTON, D., AND SMITH, B. (1972). Mariner 9 television observations of Phobos and Deimos. *Icarus* **17**, 394.

SAGAN, C. (1971). The long winter model of Martian biology. *Icarus* **15**, 511.

SAGAN, C., AND POLLACK, J. B. (1969). Windblown dust on Mars. *Nature, London* **223**, 791.

SAGAN, C., VEVERKA, J., FOX, P., DUBISCH, R., LEDERBERG, J., LEVINTHAL, E., QUAM, L., TUCKER, R., POLLACK, J., AND SMITH, B. (1972). Variable features on Mars: Preliminary Mariner 9 television results. *Icarus* **17**, 346.

SHARP, R. P., SODERBLOM, L. A., MURRAY, B. C., AND CUTTS, J. A. (1971). The surface of Mars 2. Uncratered terrains. *J. Geophys. Res.* **76**, 331.

SIMKIN, T., AND HOWARD, K. A. (1970). Caldera collapse in the Galapagos Islands, 1968. *Science* **169**, 429.

WILHELMS, D. E. (1970). Summary of lunar stratigraphy—Telescopic observations. *U.S. Geol. Survey Prof. Paper* 599-F, F1.

WILHELMS, D. E., AND MCCAULEY, J. F. (1971). Geologic map of the near side of the Moon. *U.S. Geol. Misc. Geol. Inv. Map* I-703.

6

Reprinted from *J. Geophys. Res.* **78**:4031–4036 (1973)

A Generalized Geologic Map of Mars

M. H. CARR

U.S. Geological Survey, Menlo Park, California 94025

HAROLD MASURSKY

U.S. Geological Survey, Flagstaff, Arizona 86001

R. S. SAUNDERS

Jet Propulsion Laboratory, California Institute of Technology, Pasadena, California 91103

A geologic map of Mars has been constructed largely on the basis of photographic evidence. Four classes of units are recognized: (1) primitive cratered terrain, (2) sparsely cratered volcanic eolian plains, (3) circular radially symmetric volcanic constructs such as shield volcanoes, domes, and craters, and (4) tectonic erosional units such as chaotic and channel deposits. Grabens are the main structural features; compressional and strike slip features are almost completely absent. Most grabens are part of a set radial to the main volcanic area, Tharsis.

A generalized geologic map of Mars has been constructed largely on the basis of differences in the topography of the surface. The success with which the geology can be deduced from surface topography depends on how distinctively the original topography of a feature reflects its mode of origin and the extent to which subsequent modification can be recognized and assessed. We are fortunate in having a number of topographic features on Mars whose forms are highly diagnostic of their origin. Of particular note are the shield volcanoes and lava plains. In some areas the original features have been considerably modified by subsequent erosional and tectonic processes. These have not, however, resulted in homogenization of the planet's surface but rather have emphasized its variegated character by leaving a characteristic imprint in specific areas. The topography of the planet therefore lends itself well to remote geologic interpretation.

The map (Figure 1) is an outgrowth of an earlier version of the equatorial belt [*McCauley et al.*, 1972]. The techniques and conventions used are similar to those used for the moon and have been fully described elsewhere [*Wilhelms*, 1972]. The surface has been divided into several units, each of which has a specific range of topographic characteristics. In the map explanation (Figure 2) the units are arranged according to their age as inferred from superposition and transection relations, crater counts, and so forth. The map also provides a generalized indication of tectonic deformation. Most of the units represent materials of a specific origin deposited within a restricted period of time. Other units, such as the chaotic and knobby materials, are not strictly geologic deposits but are modifications of preexisting materials. The modifications, however, have been so drastic as to, in effect, create new geologic units, and they are mapped as such.

For this early version of the map, data from the various spectral instruments have been largely ignored because of the difficulty in obtaining the data in a form readily correlatable with the visual image. However, we would not expect that, at the scale at which the map is depicted here, the spectral data would significantly affect delineation of the various units. The scale was dictated by journal format and should not be taken as indicative of a present level of knowledge. The map is a very coarse generalization of the information available. A more detailed map that will do justice to the wealth of information in the Mariner photography will be published in the not too distant future.

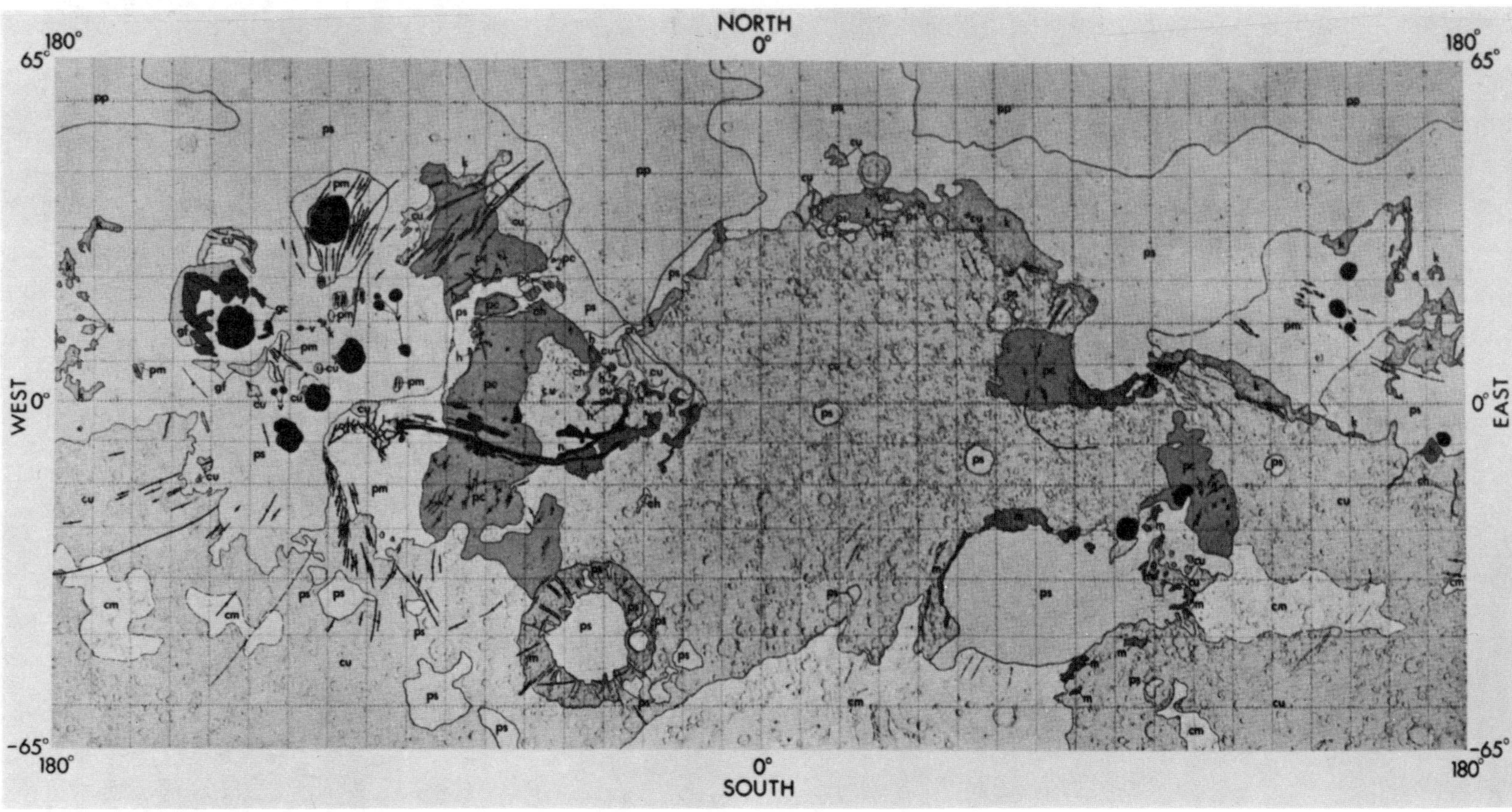

Fig. 1. Generalized geologic map of the region of Mars between 65°N and 65°S.

AGE RELATIONS

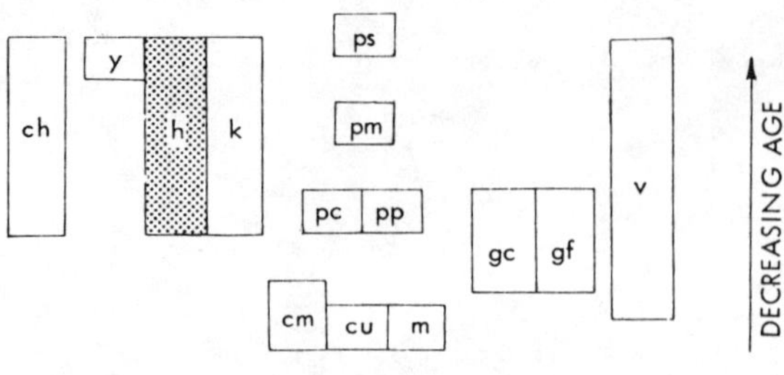

STRUCTURAL SYMBOLS

FAULTS, BAR AND BALL ON DOWNTHROWN SIDE.

GRABEN, WHERE FAULTS MAPPED SEPARATELY, BALL ON DOWNTHROWN BLOCK. WHERE GRABEN IS NARROW, SHOWN BY SINGLE LINE AND BALL.

ROUNDED ESCARPMENT. CARET POINTS DOWNHILL.

LOW RIDGE, RESEMBLING THOSE ON LUNAR MARIA

LINEAMENT

LIST OF UNITS

ch	CHANNEL DEPOSITS	ps	SPARSELY CRATERED PLAINS
y	CANYON DEPOSITS	pm	MODERATELY CRATERED PLAIN
h	CHAOTIC DEPOSITS	pc	HEAVILY CRATERED PLAINS
k	KNOBBY DEPOSITS	pp	MOTTLED CRATERED PLAIN
gc	GROOVED TERRAIN MATERIAL, COARSE	m	MOUNTAINOUS DEPOSITS
gf	GROOVED TERRAIN MATERIAL, FINE	cm	CRATERED DEPOSITS, MANTLED
v	VOLCANIC DEPOSITS	cu	CRATERED DEPOSITS, UNDIVIDED

Fig. 2. Age relations and explanations of map symbols.

DESCRIPTION OF UNITS

Densely Cratered Units

Densely cratered terrain covers approximately one half of the planet's surface, including most of the central and southern parts of the map and the south polar regions. The cratered area was photographed by Mariner 4, 6, and 7 in 1964 and 1969 and has been described in detail [*Murray et al.*, 1971]. The surface is almost saturated with large (>20 km) flat-floored craters; the density of smaller craters, which are mostly bowl shaped, falls short of saturation by a factor of 10. Several large ring structures resembling lunar basins occur within the densely cratered terrain [*Wilhelms*, 1973]. The character of the cratered terrain varies both locally and on a regional scale. Around the large impact basins, positive relief features are more common than elsewhere, providing the basis for discriminating a mountainous unit (m). In other areas the cratered terrain appears to be partly mantled, so that fewer intermediate and small (<20 km) craters are present and the larger craters appear subdued. A unit termed 'cratered deposits, mantled' (cm) has therefore been designated; the rest of the densely cratered deposits are left undivided (cu).

Cratered deposits, undivided (*cu*). This unit forms the primitive accretionary surface of the planet. It occurs primarily in the southern midlatitudes but extends to 40°N around the

330°W meridian. Isolated areas occur in the volcanic province. The unit is saturated with large flat-floored commonly rimless craters. Intercrater areas are flat and featureless except for scattered small bowl-shaped craters. Locally they may be partly covered by younger deposits, so that the topography has a muted appearance. Ridges, resembling those on the lunar maria, occur in some of the muted areas. The unit probably consists mostly of reworked impact breccias but locally includes younger volcanic and eolian deposits.

Cratered deposits, mantled (*cm*). This unit is mapped only where extensive areas appear mantled. Fewer intermediate and small craters (<20 km) occur than in unit cu. The unit occurs primarily in high southern latitudes and is interpreted as cratered deposits mantled by various thicknesses of younger material. Lobate flow fronts indicate volcanic materials in some areas, but the mantling material almost certainly includes significant portions of eolian debris.

Mountainous deposits (*m*). This unit forms the rugged parts of the rims of the three largest recognized impact basins, Argyre, Libya, and Hellas. Argyre (50°S, 43°W) is surrounded by rugged terrain cut by graben. The Libya basin (15°N, 270°W) rim is preserved as a distinct mountainous unit only at the south side of the basin. The Hellas (45°S, 295°W) rim is subdued. Mountainous terrain with relief comparable to Argyre is preserved only in isolated areas, particularly to the east where there is a complex array of isolated mountains. The mountainous unit is interpreted as remnants of parts of the primitive crust uplifted during the formation of the impact basins. Remnants of basin ejecta are also included.

Plains-Forming Materials

Plains, showing various degrees of cratering, occur over most of the planet not covered by the densely cratered units. The plains-forming materials are thought to be largely volcanic and eolian in origin. They have been divided into three units on the basis of the number and character of the superimposed craters.

Heavily cratered plains materials (*pc*). This unit is the most heavily cratered of the plains units but has fewer large craters than the densely cratered units described. It is equivalent to unit mc of *McCauley et al.* [1972] and occurs mainly in the areas north and south of Lunae Palus and in Hesperia. Ridges resembling those on the lunar maria are common. This material is interpreted as old lava plains similar to the lunar maria. (Small areas of material similar to this are present widely throughout the densely cratered area; they are not mapped because of scale limitations.)

Moderately cratered plains materials (*pm*). This unit has fewer craters in the size range 2–20 km than unit pc but more than unit ps [*Carr*, 1973]. It occurs in Elysium around the major shield volcanoes, in the Arcadia-Tharsis region, where it is exposed mainly as islands surrounded by unit ps, and in the region of Phoenicus Lacus. In the Arcadia-Tharsis region the unit is almost everywhere intensely fractured. It is interpreted as volcanic lava plains intermediate in age between units ps and pc.

Sparsely cratered plains materials (*ps*). This unit occurs mainly in the Amazonis-Tharsis region and in the large impact basins in the densely cratered province. Unit ps is the least cratered and presumably the youngest of the plains units. At wide-angle resolution it is relatively featureless, except in places close to its contact with the densely cratered terrain, where indistinct ridges and low rounded hills are common. At narrow-angle resolution irregular lobate scarps suggestive of flow fronts are common, especially in the Tharsis region. Low hills, islands of highly fractured terrain, sinuous channels, and polygonal fractures occur in other areas. More rarely, the narrow-angle pictures are featureless. In the Tharsis region the plains appear to be composed mainly of volcanic flows, since lobate flows are detectable on nearly all narrow-angle pictures. In the Amazonis region and in the large impact basins, Argyre, Hellas, and Libya, volcanic features are rare, and the eolian component probably dominates. The unit embays all other units, confirming the young age inferred from the crater counts.

Mottled cratered plains materials (*pp*). This unit occurs only at high northern latitudes where it forms an annulus around the pole. The density of large (>20 km) craters is comparable to that of unit pc, but the density of smaller craters is substantially lower. Large craters are subdued and apparent mainly because of an albedo contrast between the light crater floors and the dark surrounding materials. The local albedo contrasts give the unit a mottled appearance. The albedo of the unit as a whole is lower than that of unit ps, with which it is generally in contact to the south. It is interpreted as cratered plains material (pc) overlain by a mantle of eolian debris.

Volcanic Units

Included under this heading are all volcanic units associated with roughly circular volcanic structures, as distinct from the extensive plains units. Because of limitations of scale, only two categories are depicted: a general unit v and another unit consisting of two facies (gc and gf), which form some very distinctive terrain around Nix Olympica.

Volcanic materials (*v*). This unit includes materials associated with shield volcanoes, volcanic domes, and volcanic craters (units vs, vd, and vc of *McCauley et al.* [1972]). Most of these features are circular and radially symmetric and have a central crater and gently sloping flanks. They occur primarily outside the densely cratered region, although two features within the densely cratered province are indicated on the map. The volcanic deposits have a wide range of ages. Those associated with Nix Olympica are relatively young [*Hartmann*, 1973]; those within the cratered province are relatively old [*Carr*, 1973].

Grooved terrain materials (*gf and gc*). Two facies of grooved terrain are recognized: a unit gc with a coarse surface topography that occurs close to Nix Olympica and a unit gf with a finer surface texture that occurs farther away. The coarse unit gc is characterized by linear mountains 1–5 km wide and typically 100 km long. The mountains are commonly separated by valleys that have flat floors with a fine striation parallel to the length of the valleys. The unit appears to be broken into blocks along arcuate faults that tilt the blocks gently inward toward Nix Olympica. The fine-textured unit gf is characterized by closely spaced equidimensional mountains whose horizontal and vertical dimensions decrease outward from Nix Olympica until the mountains merge with the surrounding plains. Both units are complex embayed by the surrounding plains deposits.

The grooved terrain deposits may represent old volcanic materials derived from the Nix Olympica center and since complexly fractured by the continual tectonic activity associated with the formation of the central shield. An alternative hypothesis is that the grooved terrain represents the outer remnants of a once much larger Nix Olympica that has been reduced in size by whatever process has formed the bounding scarp.

Other Units

In several areas erosional and tectonic processes have resulted in the formation of distinctive geologic units. Four categories have been identified.

Channel deposits (*ch*). These have all the characteristics of terrestrial stream deposits. They occur in long, linear, sometimes sinuous channels that commonly have well-developed tributaries. Narrow-angle pictures show terraces, bars, finely braided networks of channels, and superimposed meanders. The most prominent channels head in the area of chaotic terrain around 2°N, 30°W and run northwest to the Chryse basin. Most, but not all, other channels occur in the densely cratered terrain near its contact with the plains. The unit is interpreted as materials deposited by a fluid, presumably water [*Milton*, 1973].

Canyon deposits (*y*). This unit includes materials that are exposed on the floor of the major rift system that extends 4000 km across the surface close to the equator between 30° and 100°W. The talus on the canyon walls is excluded. In most places the floor appears smooth, but locally jumbled blocks are present. At the eastern end of the canyon the unit merges with the chaotic terrain. At the western end the unit is pinched out as the canyon grades into a zone of branching troughs.

Chaotic deposits (*h*). Chaotic deposits have been described in detail previously [*Sharp et al.*, 1971]. The surface of the unit is crossed by numerous intersecting cracks that break the surface into blocks that have a wide range of sizes and that may be tilted slightly in different directions. The chaotic deposits normally occur in locally low areas, occasionally in completely closed basins. The outcrop areas are usually surrounded by an inward-facing scarp. The main area of occurrence is a broad region around 5°S latitude, 35°W longitude. The origin of the unit is obscure. Some process of sapping from below and subsequent collapse is required.

Knobby deposits (*k*). The unit is characterized

by irregular rounded hills and intervening plains. The hills are all sizes, from several hundred kilometers down to the limit of resolution (<1 km). The unit occurs mostly in a zone up to 500 km wide between the cratered terrain and the plains units. In general, the individual hills are larger and have more rectilinear outlines and flatter tops closer to the cratered terrain. Farther from the cratered terrain they become rounded and smaller in both horizontal and vertical dimensions until they merge with the surrounding plain. The unit includes the unit kt of *McCauley et al.* [1972] and the fretted terrain of *Sharp* [1973].

STRUCTURAL FEATURES

Mars is characterized by an abundance, if not a great variety, of structural features. The most common are graben, typically 1–5 km wide. They may occur as closely spaced parallel arrays, as in the Arcadia region, or as isolated fractures several thousand kilometers long, as in the Mare Sirenum region. The distribution of graben is markedly nonuniform. Most grabens are part of a system of faults approximately radial to the Tharsis ridge. The western part of the Coprates canyon also appears to be part of this set. The focal point of this vast system of fractures is the Phoenicus Lacus area, which is also the highest part of the ridge. The pattern appears to have formed as a result of the extension associated with the broad domical uplift of the Tharsis region.

Elsewhere fractures occur concentric and radial to the large impact basin of the densely cratered province [*Wilhelms*, 1973]. Fractures also occur locally around the large shield volcanoes [*Carr*, 1973] and at and parallel to the margin of the densely cratered terrain to form the knobby and fretted terrains [*Sharp*, 1973].

Mare ridges are the other common structural feature. They occur primarily on the most heavily cratered plains unit (pc). West of the Tharsis ridge they are aligned along directions concentric with the center of the Tharsis ridge.

Acknowledgment. Publication authorized by the Director, U.S. Geological Survey.

REFERENCES

Carr, M. H., Volcanism on Mars, *J. Geophys. Res.*, *78*, this issue, 1973.

Hartmann, W. K., Martian cratering, 4, Mariner 9 initial analysis, *J. Geophys. Res.*, *78*, this issue, 1973.

McCauley, J. F., M. H. Carr, J. A. Cutts, W. K. Hartmann, H. Masursky, D. J. Milton, R. P. Sharp, and D. E. Wilhelms, Preliminary Mariner 9 report on the geology of Mars, *Icarus*, *17*, 289, 1972.

Milton, D. J., Water and processes of degradation in the Martian landscape, *J. Geophys. Res.*, *78*, this issue, 1973.

Murray, B. C., L. A. Soderblom, R. P. Sharp, and J. A. Cutts, The surface of Mars, 1, Cratered terrains, *J. Geophys. Res.*, *76*, 313, 1971.

Sharp, R. P., Mars: Fretted and chaotic terrain, *J. Geophys. Res.*, *78*, this issue, 1973.

Sharp, R. P., L. A. Soderblom, B. C. Murray, and J. A. Cutts, The surface of Mars, 2, Uncratered terrains, *J. Geophys. Res.*, *76*, 331, 1971.

Wilhelms, D. E., Geologic mapping of the second planet, *Astrogeology 55*, interagency report, 36 pp., U.S. Geol. Surv., Washington, D.C., 1972.

Wilhelms, D. E., Comparison of lunar and Martian multiring basins, *J. Geophys. Res.*, *78*, this issue, 1973.

Part III

POLAR REGIONS

Editor's Comments on Papers 7 Through 10

7 LEIGHTON and MURRAY
Behavior of Carbon Dioxide and Other Volatiles on Mars

8 MURRAY and MALIN
Polar Volatiles on Mars—Theory versus Observation

9 KIEFFER et al.
Martian North Pole Summer Temperatures: Dirty Water Ice

10 CUTTS et al.
North Polar Region of Mars: Imaging Results from Viking 2

The composition of the martian polar caps has been the subject of vigorous debate for many years, with dry ice and water ice being the two leading contenders. Earlier workers advocated water ice, in part by analogy to the Earth, and in part because of spectral data (Kuiper 1952; Dollfus 1961). Leighton and Murray (Paper 7) present a major discussion in support of the dry-ice theory. Evidence in favor includes the observed atmospheric CO_2 abundance, the low polar cap temperatures (Leovy 1966), and the agreement between the seasonal growth of the polar caps and the proposed model. The presence of a permanent CO_2 reservoir was considered important at the time, because it permitted estimates of the total amount of CO_2 outgassing on Mars relative to the Earth and to Venus. Furthermore, the vulnerability of frozen CO_2 to climatic instability could have led to the complete sublimation of the polar caps, with a consequent increase in mass of the atmosphere and buildup of sufficient pressure to stabilize the formation of liquid water (Sagan et al. 1973). The Leighton-Murray model initially received support from the Mariner 6 and 7 missions (Neugebauer et al. 1969; Herr and Pimentel 1969). However, Mariner 9 images show *residual* polar caps, whose very persistence contradicts the Leighton-Murray model (Murray et al. 1972; Soderblom et al 1973b).

Murray and Malin (Paper 8) recognized that the residual polar caps were probably water ice. On the other hand, they believed that a solid reservoir of CO_2 was in equilibrium with the atmosphere and was probably located at the north polar cap. Ingersoll (1974) argued against a permanent CO_2-ice reservoir, whether exposed or buried. Briggs (1974) also predicted that the residual polar caps should be water ice.

The Viking thermal infra-red and water vapor mappers conclusively establish the presence of water ice near the north polar cap, and the IR thermal mapper results by H. Kieffer et al. are reprinted here as Paper 9. The authors conclude that because of the relatively high summer temperatures, it is unlikely that any permanent frozen CO_2 could survive in the area proposed by Murray and Malin: "All of the polar condensate at this season [northern hemisphere summer] is H_2O." The high observed water vapor concentrations are consistent with temperatures well above 200°K (the condensation temperature of CO_2 is 148°K: see Farmer et al. 1976).

The Viking 2 orbital imagery has shed important new light on polar processes. These preliminary results are reviewed by J. A. Cutts and other members of the Viking imaging team (Paper 10). One significant discovery has been the recognition of unconformities in the sequence of polar layered deposits. Since the strata are widely considered to represent eolian deposition, regulated by cyclical (perhaps seasonal) variations in wind velocities, the unconformities may indicate more drastic climatic changes (resulting in periods of enhanced erosion or nondeposition). At present, the polar regions are experiencing active eolian erosion (Cutts 1973a; Sharp 1973c). A possible consequence of the current erosional cycle may have been the accumulation of a "collar" of dark sand dunes surrounding the stratified sequence. Present wind patterns are remarkably persistent over wide distances, judging by the unchanging dune spacing and direction. The new imagery raises many questions concerning the nature of the cyclic climate changes, the extreme youthfulness of the terrain, and the relation of the layered deposits to the evolution of the martian atmosphere and the storage of volatile materials at the poles.

7

Reprinted from *Science* **153:**136–144 (1966)

Behavior of Carbon Dioxide and Other Volatiles on Mars

Robert B. Leighton and Bruce C. Murray

The nature of the Martian polar caps has been a subject of speculation for many decades. They have been variously conjectured to be composed of frozen water, carbon dioxide, or oxides of nitrogen (*1*, p. 362; *2*, *3*). The weight of scientific opinion, based upon a variety of evidence, currently favors water ice as the dominant substance comprising the polar caps. However, much of the relevant observational evidence has undergone significant change and improvement in recent years, so that it seems appropriate to reexamine the question in the light of the more complete and reliable measurements now available.

The purpose of this article is to report results of a new study of the problem based on a consideration of the heat balance of the planet. We believe that this study points strongly toward frozen CO_2 as the dominant substance comprising the Martian polar caps.

Further, we are led to suggest that (i) the total amount of CO_2 on Mars may exceed the amount present in the atmosphere by a considerable factor, the excess being present as solid CO_2 and CO_2 permafrost in and under the permanent north polar cap; (ii) the partial pressure of CO_2 in Mars's atmosphere may be regulated by the north polar cap; and (iii) the total pressure of the Martian atmosphere may change semiannually by a significant amount because of the freezing out of much of the atmospheric CO_2 at either pole. Considerable quantities of water-ice permafrost may also be present in the subsurface of the polar regions and perhaps of more temperate regions as well. Many organic compounds of biological interest, however, are more volatile than water at low temperatures and might be easily evaporated from permeable soil and become trapped as minor constituents of the permanent polar cap.

We propose that selective trapping of CO_2 (and H_2O) in the solid phase may help explain the anomalously high concentration of CO_2 (and probably H_2O) on Mars relative to nitrogen.

Dr. Leighton is professor of physics at California Institute of Technology, Pasadena; Dr. Murray is associate professor of planetary science at California Institute of Technology.

Surface and Soil Temperatures

In order to understand the behavior of volatiles on Mars more fully, we have investigated the expected diurnal and annual temperature variations at various latitudes, using a thermal model of the Martian surface in which each surface element is treated as a horizontal plane, at some instantaneous temperature T_1, which absorbs a fraction F of the incident solar radiation, emits blackbody radiation at temperature T_1 with an emissivity E, exchanges heat with underlying layers by thermal conduction (with constant conductivity), and absorbs or releases latent heat if a condensed volatile is present. Radiative thermal exchange with the atmosphere is assumed negligible except insofar as the atmosphere may affect E by blocking certain wavelengths emitted by the surface. Horizontal heat transport by wind is also neglected, although some form of planetary circulation is presumed to replenish condensed CO_2 and H_2O.

The latter two assumptions may constitute a significant oversimplification of the actual situation on Mars. In particular, Sinton has suggested the possibility of some horizontal heat transport, on the basis of temperature measurements derived from infrared brightness (*3a*). It is also possible that condensed H_2O or CO_2 particles in the atmosphere, particularly at the poles during the autumn and winter, may provide a significant local blanketing effect. We recognize that both effects may be important, but we have purposely confined our attention to a simple model in order to determine how well such a model can account for the observed properties of Mars. We find, in fact, surprisingly good agreement between the observed properties and the properties predicted on the basis of this model.

Typical values adopted for the various numerical parameters were as follows: $F = 0.85$, corresponding to an average Bond albedo $A = 0.15$; $E = 0.85$, a value which allows for the blocking effect of 5000 centimeter-atmosphere of CO_2 at pressure of 5 millibars [based on the tables of calculated CO_2 transparency of Stull, Wyatt, and Plass (*4*)] and an intrinsic emissivity of 0.95 to 1.00 (characteristic of powdered silicate minerals in the 10- to 50-micron wavelength range observed under natural conditions); $K = 2.5 \times 10^{-4}$ watt cm^{-1} $(°K)^{-1}$; $C = 3.3$ j g^{-1} $(°K)^{-1}$; $\rho = 1.6$ g cm^{-3}. The above values for soil conductivity (K), specific heat (C), and density (ρ) are similar to those derived by Leovy (*5*) from analysis of the infrared observations of Sinton and Strong (*6*).

The thermal history of such a surface was followed for 2 to 5 Martian years at 19 latitudes from $-90°$ to $+90°$, the equations of radiative and thermal exchange being evaluated (by a 7094 computer) for each 1-hour period during every 5th day to a depth of 3 meters, well below the thermal "skin depth" for annual temperature changes. The depth coordinate x was treated in three parts: (i) the top surface, $x_1 = 0$; (ii) nine equal layers 1.5 cm thick centered at $x_2 = 0.75, \ldots x_{10} = 12.75$ cm; and (iii) ten equal layers 30 cm thick, centered at $x_{11} = 15, \ldots x_{24} = 2.85$ cm. The boundary conditions on the temperature at a given latitude were as follows.

1) The surface temperature T_1 was taken as a linear extrapolation to the surface of the values for the upper two layers: $T_1 = (3T_2/2) - (T_3/2)$.

2) The net heat gain or loss, Q_2, for the top layer was taken as

$$Q_2 = (-E\sigma T_1^4 + S_0 \cos Z + K(T_3 - T_2)/\Delta x)\,\Delta t$$

where S_0 ($= 0.06$ W cm^{-2}) is the solar constant at Mars; σ is the Stefan-Boltzmann constant; cos Z is the cosine of the zenith angle of the sun; $\Delta t =$ 3600 seconds, and $\Delta x = 1.5$ cm.

3) The quantity Q_2 was used either to change the temperature of the top layer or (as latent heat) to condense or evaporate CO_2 or H_2O, as required by the ambient conditions.

4) The layer x_{10} was assumed to exchange heat with the upper layers in proportion to the temperature gradient $(T_{10} - T_9)/\Delta x$, and with the lower layers in proportion to the gradient $(T_{12} - T_{11})/\Delta x'$, with $\Delta x' = 30$ cm.

5) The temperature T_{11} was set equal to T_{10}, as described below.

6) The layer x_{20} was assumed to exchange heat only with overlying layers.

The initial temperatures at each latitude were constant with depth and equal to the mean annual temperatures found in pilot calculations. In order to speed the convergence to a steady state, the temperatures T_i of all layers were adjusted by amounts ΔT_i at the end of each annual calculation according to the equation

$$\Delta T_i = (T_{av} - T_{20})\,(2x_i/x_{20} - x_i^2/x_{20}^2)$$

where T_{av} is the calculated yearly average surface temperature at the corresponding latitude.

For each diurnal calculation, made on an hourly basis, only the temperatures of the upper nine layers were changed; at the end of this calculation five successive iterations on the lower ten layers were made, T_{11} being held constant and equal to the final value of T_{10}.

Representative curves for the daily temperature variations at the equator are shown in Fig. 1 for three different thermal conductivities; these are in adequate agreement with the measurements of Sinton and Strong (*6*). At first, the possibility of condensation of CO_2 was ignored. The annual variation of daily minimum temperature found for this moonlike case are plotted in Fig. 2, which shows that the minimum night-time temperatures would remain above 145°K at all seasons at the equator and at low temperate latitudes, but would drop considerably below 145°K in winter time at subpolar and polar latitudes. Since the condensation temperature of CO_2 at the currently accepted pressure of about 4 millibars is 145°K, it is immediately clear from Fig. 2 that CO_2 should precipitate out and accumulate at the higher latitudes during local winter.

Carbon Dioxide Relationships

A number of calculations which permitted the condensation of CO_2 were made. These were made progressively more complete and realistic until ultimately even the effects of orbital eccentricity and depletion of CO_2 by freezing were included. Various albedos for solid CO_2 were used, ranging from 0.60 to 0.75, and different initial amounts of atmospheric CO_2 were assumed. All these calculations are in agreement with respect to the following major effects.

1) Large amounts of CO_2 were always found to precipitate near each pole during the Martian winter, as shown in Fig. 3. Typical maximum amounts were 100 to 150 g cm^{-2}.

2) Carbon-dioxide precipitation occurred above latitude + 50° and below − 45°; this is about the range observed for the actual Martian polar caps.

3) At latitudes near the boundary of a polar cap, the calculations indicated, CO_2 would precipitate at night and evaporate during the day; the "frost" would usually disappear before noon, unless the cap was growing rapidly.

4) The amount of CO_2 precipitated in the polar regions was so great as to appreciably affect the total atmospheric pressure. In a typical case, the atmospheric pressure varied semiannually by almost ± 1 millibar from the mean (Fig. 4).

In addition, the calculated rate of disappearance of a CO_2 frost cap agrees rather well with the observed rate as illustrated in Fig. 5.

The total radio (thermal) emission from Mars has been measured at various wavelengths (*7–9*) and constitutes an important constraint upon the soil temperature derived from a model. Of course, only the very surface of the soil undergoes diurnal and annual temperature variations as great as those shown in Figs. 1 and 2. Within a few centimeters' depth the diurnal variation is damped out, and within a depth of a few tens of centimeters the annual variation is no longer significant. The average temperature over the planetary disk as seen from the direction of the sun was calculated for each depth below the surface for a number of models. Representative results for the CO_2 model are shown in Fig. 6. Since dry silicates are good insulators and therefore somewhat transparent to radio waves, radio brightness temperatures refer effectively to emission from some region within the soil at, perhaps, a depth of three to ten wavelengths. The curve of Fig. 6 for a depth of 15 centimeters can therefore be taken as an approximate guide to the radio emission of 3- to 4-centimeter wavelength. In Fig. 7, the temperature at this depth is compared with recent radio observations (*8*) for two models.

The calculations brought out another expected consequence of CO_2 solid-vapor equilibrium: if the total abundance of CO_2 is greater than about 15 g cm^{-2}, the north polar cap will not disappear during the summer, as is in fact the case on Mars. The equilibrium partial pressure of CO_2 for this (or any greater) abundance is 3 to 5 millibars. Thus, the observed partial pressure of CO_2 is itself strong evidence for the existence of a permanent CO_2 ice cap and, probably, CO_2-saturated permafrost in the north polar region. This important point merits elaboration.

Let us assume an initial amount of atmospheric CO_2 sufficient to allow some CO_2 to remain condensed in the north polar region throughout the summer. In this case the permanent cap is, on the average, the coldest place on the planet, and, furthermore, it is of nearly constant temperature T_0 throughout the year. Therefore, the rate R_r at which heat is radiated throughout the year is constant and is

$$R_r = E\sigma T_0^4$$

where E is the infrared emissivity of solid CO_2, corrected for the effects of atmospheric blanketing. On the other hand, the yearly average insolation rate I_{av} at the north pole is fixed by the size and eccentricity of Mars's orbit and the inclination of the rotation axis. Thus,

$$I_{av} = \frac{1}{2\pi}\int_{-\pi/2}^{\pi/2} S_0 \sin\delta_0 \cos\Phi\,(1 - 2\epsilon\cos\Phi) \times (1 - \beta)^{\sec Z}\, d\Phi$$

where S_0 is the mean intensity of solar radiation at Mars's orbit; δ_0 is the inclination of Mars's equator with respect to its orbital plane; Φ is the longitude of the sun, measured from the north summer solstice; ϵ is the orbital eccentricity

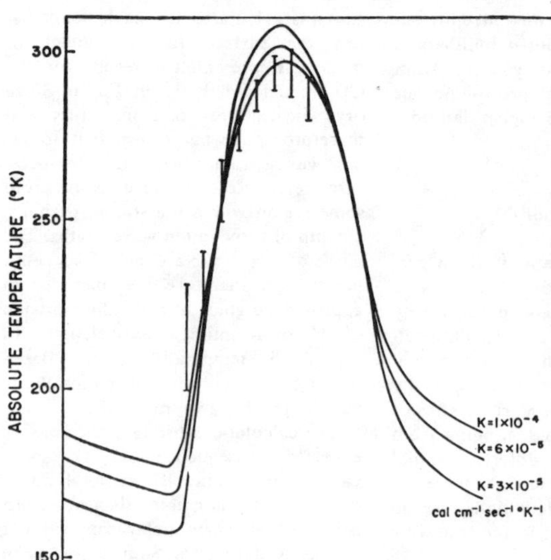

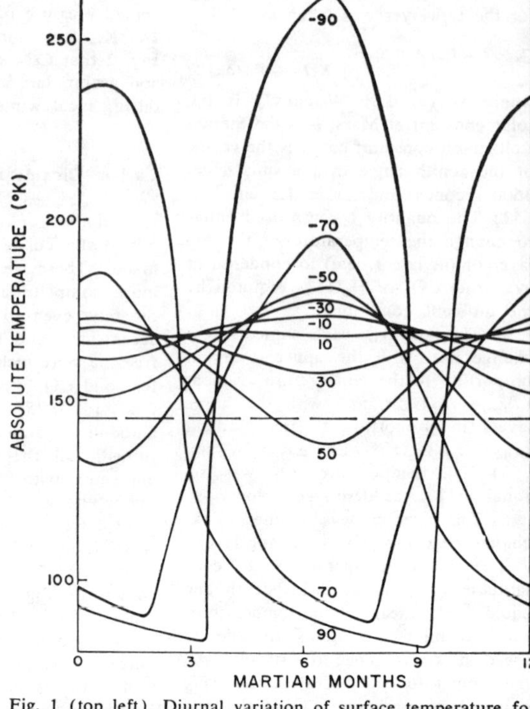

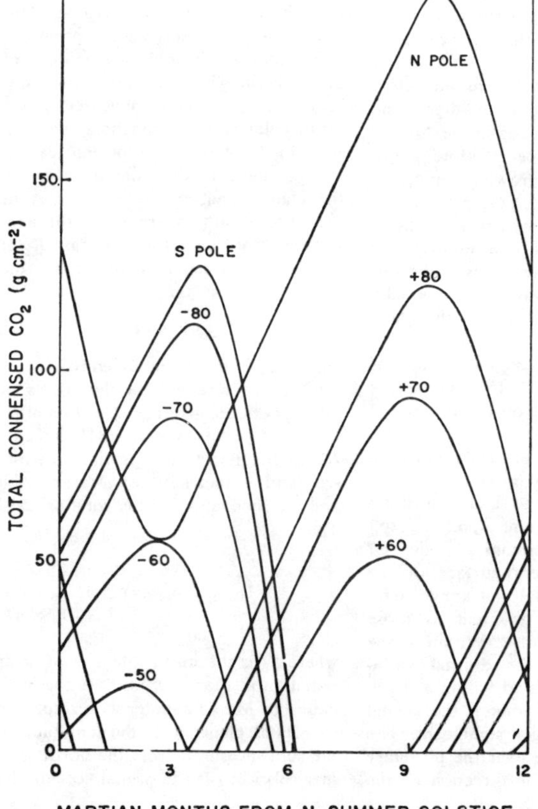

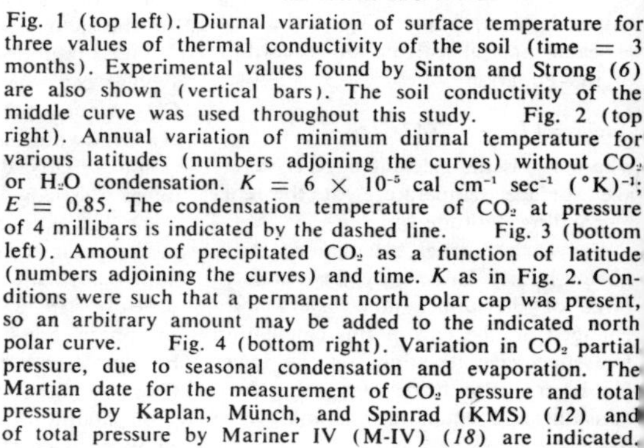

Fig. 1 (top left). Diurnal variation of surface temperature for three values of thermal conductivity of the soil (time = 3 months). Experimental values found by Sinton and Strong (*6*) are also shown (vertical bars). The soil conductivity of the middle curve was used throughout this study. Fig. 2 (top right). Annual variation of minimum diurnal temperature for various latitudes (numbers adjoining the curves) without CO_2 or H_2O condensation. $K = 6 \times 10^{-5}$ cal cm^{-1} sec^{-1} (°K)$^{-1}$; $E = 0.85$. The condensation temperature of CO_2 at pressure of 4 millibars is indicated by the dashed line. Fig. 3 (bottom left). Amount of precipitated CO_2 as a function of latitude (numbers adjoining the curves) and time. K as in Fig. 2. Conditions were such that a permanent north polar cap was present, so an arbitrary amount may be added to the indicated north polar curve. Fig. 4 (bottom right). Variation in CO_2 partial pressure, due to seasonal condensation and evaporation. The Martian date for the measurement of CO_2 pressure and total pressure by Kaplan, Münch, and Spinrad (KMS) (*12*) and of total pressure by Mariner IV (M-IV) (*18*) are indicated.

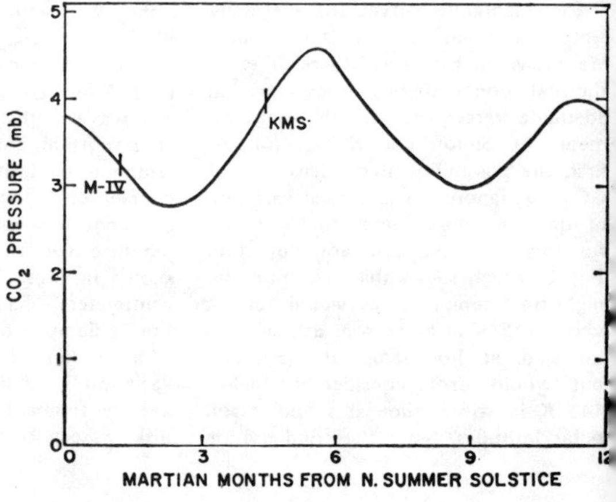

(aphelion is assumed to coincide with the north summer solstice); $(1 - \beta)$ is the fraction of incident solar radiation which is assumed to reach the surface when the sun is at the zenith; and sec Z $(= 1/\sin \delta_0 \cos \Phi)$ is the secant of the zenith angle of the sun. Numerically, $S_0 = 0.060$ watt cm^{-2}, $\sin \delta_0 = 0.407$; $\epsilon = 0.096$; and $\beta \approx 0.02$ (as a rough estimate). These values yield

$$I_{av} \approx 0.0060 \text{ watt cm}^{-2}$$

Now, if $R_r > I_{av}(1 - A)$, the polar cap will gain CO_2 at an average rate

$$M = (R_r - I_{av})/L$$

where L (≈ 450 j g^{-1}) is the latent heat of vaporization of CO_2 at temperature T_0. This removal of CO_2 from the atmosphere will then reduce the partial pressure of CO_2, thereby decreasing the condensation temperature T_0 and R_r. On the other hand, if $R_r < I_{av}(1 - A)$, the polar cap will lose CO_2 at an equivalent rate, a loss resulting in an increase in T_0 and R_r. An equilibrium will therefore exist such that

$$R_r = I_{av}(1 - \bar{A})$$

This condition fixes T_0, which in turn fixes the mean CO_2 partial pressure P_0 through the relation of vapor pressure to temperature. If $A = 0.65$ and $E = 0.85$, we find

$$T_0 = (I_{av}(1 - A)/\sigma E)^{1/4} = 145°\text{K}$$

At this temperature, $P_0 = 4.0$ millibars, in good agreement with current experimental values. If A and E differ from the above values (or if we have incorrectly estimated I_{av}), P_0 will, of course, differ also. The dependence of P_0 upon A and E is shown in Fig. 8.

The *rate* at which the above-described equilibrium will be approached may be estimated. Let the minimum radius of the polar cap be r radians of Martian latitude, and let $P = P_0 e^{\alpha(T - T_0)}$ for T near T_0. Then, if $T = T_0 + \Delta T$, a linear approximation to ΔR_r and ΔP yields a time constant

$$\tau = T_0 P_0 \alpha L / r^2 R_r g \approx 0.6 \times 10^{10} \text{ sec } (= 200 \text{ yr})$$

In spite of its small radius of only 3 degrees, the cap should thus be quite effective in maintaining a constant partial pressure of CO_2.

Several unsuccessful attempts were made to simulate the very small residual north polar cap in the model before it was recognized that the observed behavior of the cap (it regularly shrinks to a diameter of a few hundred kilometers but always resists complete evaporation) does not imply that the cap is on the verge of disappearing when it is at minimum size. In fact, a permanent polar cap will tend to become smaller in diameter and correspondingly thicker because of the variation in insolation near the pole. Calculation shows that the insolation varies approximately according to the formula

$$I_{av}(\lambda) = I_{av}(90°)[1 + (5 \times 10^{-4})(90 - \lambda)^2]$$

At latitude 80°, for example, the mean insolation is 5 percent greater than it is at the pole. However, all parts of the cap are at the same temperature, since the cap is in equilibrium with the CO_2 partial pressure, and thus radiates equally everywhere (*10*). The differential insolation therefore leads to a differential rate of growth or loss of CO_2 at the two latitudes in question until the permanent CO_2 all resides at the pole. This transfer rate is relatively rapid. In the example cited, the differential rate of loss at latitude 80° would be about 7 g cm^{-2} yr^{-1}.

The approximate annual "snowfall" of CO_2 at the north pole may also be estimated to be about half the amount that would be gained during the Martian year at the steady rate R_r/L. This amounts to

$$M = I_{av}(1 - A)t_y/2 \approx 140 \text{ g cm}^{-2}$$

where t_y $(= 687 \times 86{,}400)$ is the length of the Martian year, in seconds. This is in good agreement with the results of the more precise calculations shown in Fig. 3.

In summary, on a rather general basis, CO_2 can be expected to be the major constituent of the polar caps on Mars, and this conclusion is consistent with (i) the size and rate of disappearance of the observed caps, (ii) the observed permanence of the north polar cap, (iii) the observed partial pressure of CO_2 on the planet, and (iv) the disk brightness temperature at radio wavelengths.

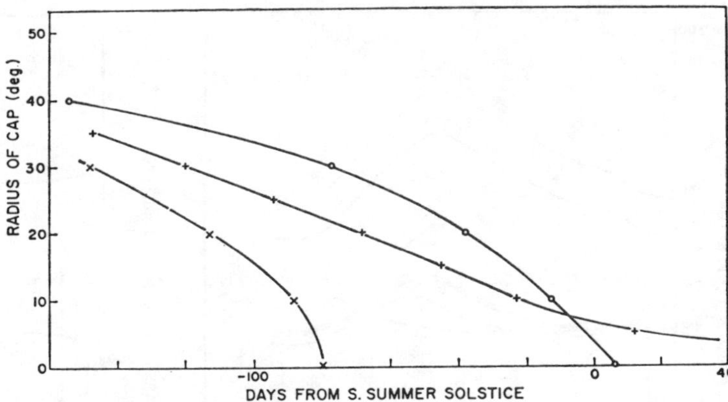

Fig. 5. Observed radius of south polar cap as a function of time (*21*) compared with predictions of models in which the caps are composed of CO_2 alone or H_2O alone. (+) Observed radius; (○) model with cap of CO_2 alone; (×) model with cap of H_2O alone. Local topographic irregularities are probably responsible for the slow recession of the cap after summer solstice, since the remanent southern cap is situated 6.5° from the pole. The recession of the cap in the H_2O model is closely in step with the solar declination because of the negligible mass of H_2O precipitated.

Water Relationships

We have found, in the preceding section, that the properties of a rather simple thermal model of the Martian surface are in good agreement with many of the observed properties of Mars and suggest that the Martian polar caps consist largely of frozen CO_2. However, this result is in conflict with observations which seem to indicate that water ice, and not CO_2, is the substance comprising the polar caps. We have sought to understand the extent of this conflict and to find ways of resolving it. We see three possibilities.

1) The observations which indicate the presence of water ice, or the conclusions derived therefrom, may be in error.

2) Effects not taken into account in

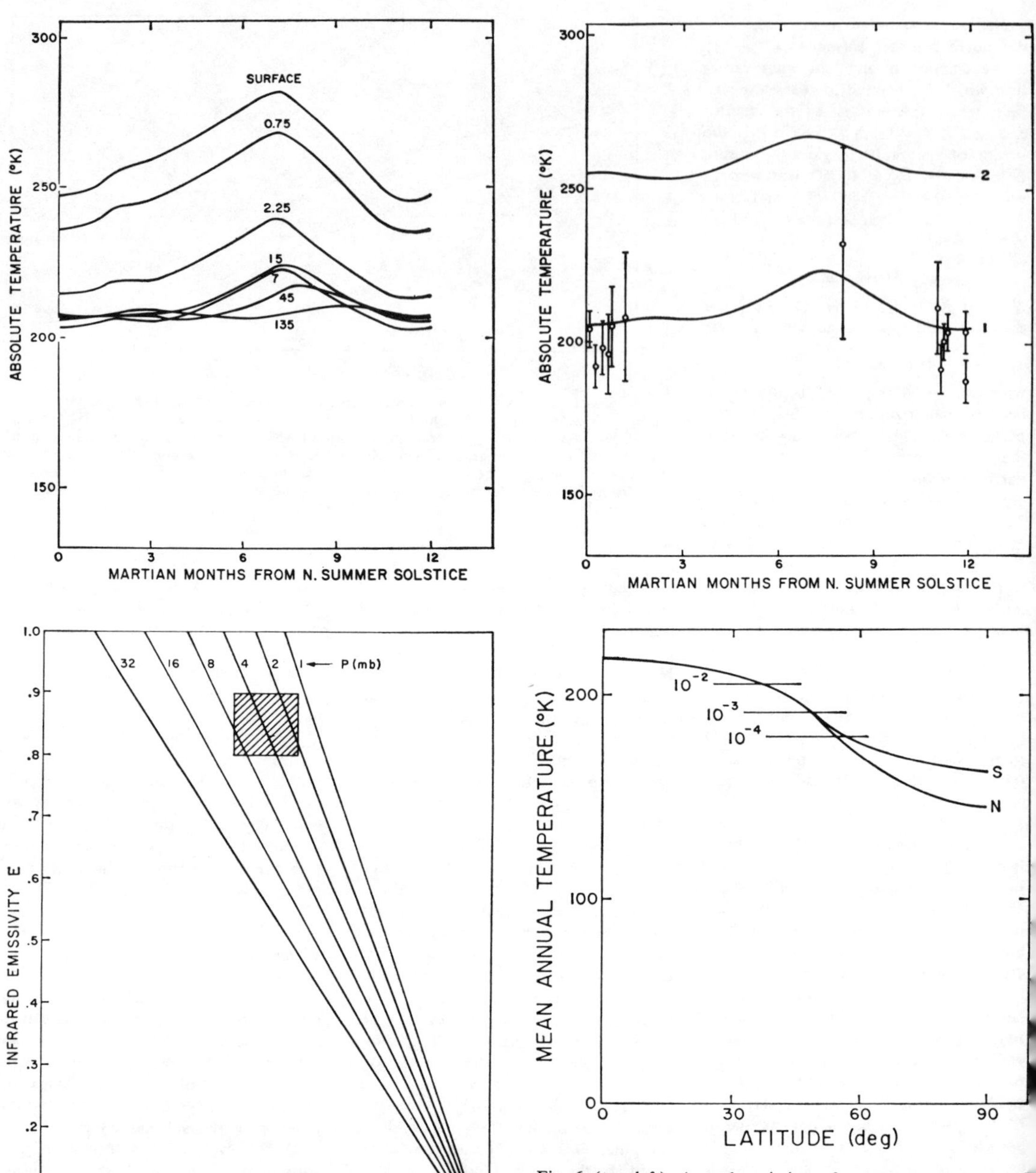

Fig. 6 (top left). Annual variation of average temperature of the disk as viewed from the sun for various depths, in centimeters (numbers adjoining curves), for the CO_2 model. The damping out of the diurnal and annual variations with depth are well shown in these curves. Fig. 7 (top right). Comparison of average disk temperature at depth of 15 centimeters for two thermal models with radio observations (vertical bars), at wavelength of 3 to 4 centimeters. The radio brightness temperatures have been multiplied by a factor 1.1 to correct approximately for an observed reflectivity of about 0.10 (*24*). The two models are (curve 1) the CO_2 model and (curve 2) a model having lower (and variable) infrared emissivity adjusted so as to prevent the precipitation of CO_2 but permit the condensation of H_2O. Fig. 8 (bottom left). Dependence of partial pressure *P* of CO_2 in the Martian atmosphere upon visual reflectivity *A* and effective infrared emissivity *E* of CO_2 frost. Fig. 9 (bottom right). Mean annual temperature as a function of latitude (CO_2 model). Saturation temperatures corresponding to three values for precipitable water vapor (in grams per square centimeter) are indicated.

the study discussed here may render our conclusions invalid.

3) A small amount of water ice may be present together with a large amount of frozen CO_2 and in some way affect the observational results, giving a magnified estimate of the abundance of water.

Water in solid form on the frost caps of Mars has been reported by Kuiper (*1*) and Dollfus (*11*), and water vapor has been reported by Kaplan, Münch, and Spinrad (*12*) and by Dollfus (*13*). Kuiper compared the infrared spectral reflectivity of the polar cap with similar laboratory spectra of ice and dry ice, while Dollfus made a study of the polarization of the polar cap. Moroz (*14*) has recently confirmed the existence of the near-infrared spectral feature and, like Kuiper, attributes it to the presence of water ice. However, we are not aware of a sufficiently thorough published study of the reflection spectrum of *both* solid H_2O and CO_2 deposited under simulated Martian conditions to justify the identification.

Dollfus' discussion of his polarization measurements was restricted to a consideration of the physical state of the condensed material and not its composition. He found difficulty in producing a form of water ice that would exhibit the very small amount of polarization that was observed, and he did not study the properties of CO_2 frost. There is at present no evidence, from polarization studies of Mars, which would enable one to distinguish between a frost cap of H_2O and one of CO_2.

There are also difficulties in reconciling the optical properties and seasonal behavior of supposed water ice polar caps with the extremely minute quantities of water that spectroscopic studies have revealed. The Kaplan, Münch, and Spinrad observation was sensitive enough to permit an estimate of the total precipitable water above the Martian surface: $14 \pm 7 \times 10^{-4}$ g cm^{-2}. Comparable measurements of water vapor have also been obtained independently by Spinrad over a number of years (*15*). Although Heyden *et al.* (*16*) have recently pointed out that heretofore unrecognized weak H_2O absorption lines originating in the earth's atmosphere might complicate the interpretation of H_2O absorption features of the Mars spectrum, it seems unlikely that all observations would be similarly affected because of the variable Doppler displacements of the Martian lines. A further consideration regarding Martian water vapor has been raised by Johnson (*17*). He argues that very low atmospheric temperatures are indicated by the Mariner IV occultation experiment (*18*) and that these would be incompatible with a water-vapor concentration of as much as 10^{-3} g cm^{-2}, because ice-crystal cloud formation should set in at a concentration lower by perhaps an order of magnitude. On the other hand, Gross, McGovern, and Rasool (*19*) believe that higher atmospheric temperatures are more likely, in which case there is not necessarily a contradiction. Any such discrepancy between theory and observation evidently will not be resolved until more is known about the details of Martian meteorology, so it seems reasonable to conclude for the present that there may well be 10^{-3} g of water vapor per square centimeter commonly present in the Martian atmosphere but that an upper limit can perhaps be set around 3×10^{-3} g cm^{-2}. Thus we see that water constitutes at most 0.03 percent, by weight, of the Martian atmosphere, and probably less than 0.01 percent. It is, therefore, a minor constituent, and its condensation and evaporation will differ markedly from that of the major constituent, CO_2, in several ways.

First, the basic problem of the transfer of H_2O in an evaporation-condensation cycle from a solid state at one place on the planet to a solid state at another is vastly more difficult than the corresponding problem for CO_2, since, at the minimum, 10^4 times as much CO_2 as H_2O must be transferred by the wind system. If all the H_2O is not removed during a single cycle (because of insufficient vertical mixing, for example) the ratio becomes correspondingly larger, and an unreasonably strong wind system would be required for the effective transport of water vapor. For example, one might assume, as an extreme upper limit, that during the disappearance of one cap and the growth of the other the entire 10^{-3} g of water vapor above each square centimeter of the disappearing cap is transported to the growing cap and deposited there *daily* as frost. (Such a process would require net north-south winds of about 500 kilometers per hour blowing steadily for perhaps 100 days, combined with complete vertical mixing daily at each end.) Even in such an extreme case only about 0.1 g cm^{-2} woud be transferred from one cap to the other—a thickness of a few millimeters at most. Since the Martian surface undoubtedly has some microrelief, this tiny coating of frost would probably accumulate only in the shadowed areas, such as on the colder, poleward slopes of topographic features and on the poleward sides of individual rock fragments. A frost cap only a few millimeters thick very probably would not be observable at all from the direction of the sun, yet this is the direction from which the planetary frost caps are observed from Earth at opposition.

A second, and equally serious, difficulty with the hypothesis that water ice is the dominant constituent of the Martian caps is the fact that, unlike the case for CO_2, the atmosphere near the surface can become depleted in H_2O and the depletion causes the condensation rate to become extremely small and dependent upon the vagaries of the vertical and horizontal circulation system of the planet. Inasmuch as that circulation system is governed by the planetary heating and reradiation, and perhaps by rotation, and, unlike the case for Earth, is not significantly influenced by the atmospheric content of water vapor, such redistribution of water as does take place is the result of purely accidental effects.

These considerations make us doubt that water could be the dominant constituent of the Martian polar caps. To place the analysis in more quantitative form, however, we have investigated two specific hypothetical cases which have also led us into some interesting considerations concerning the possible occurrence of permafrost on Mars.

First, we have studied what is required if the frost caps are in fact H_2O and if they always remain warm enough so that solid CO_2 does not accumulate on the surface at either pole at any time during the year. This is equivalent to requiring that the mean daily temperature not fall below about 145°K anywhere on the planet. Since it appears, from terrestrial experience, that supplementary sources of heat (for example, radioactivity) of adequate magnitude are most unlikely, it is only through introducing, in the model, a decrease in radiative heat loss from the planet by atmospheric blanketing that such a result may reasonably be achieved. In order to retard the radiative heat loss at low temperatures and yet avoid unduly high daytime temperatures, a far-infrared absorption was introduced such that, in addition to the expected absorption of CO_2 near 15 microns, all wavelengths longer than 19 microns were blocked. This led to an effective emissivity relative to temperature which ranged from 0.1 at 140°K to 0.5 at 270°K. Although this procedure gave the desired result—a negligible quantity of CO_2 was found to condense—it led to other results that are in serious conflict with observation. Not only would the assumed absorption require the presence of substantial amounts of molecular species such as NH_3, CH_4, or H_2O in the atmosphere which have escaped detection in the visible (*20*), near-infrared (*1*, p. 351), or far-infrared (*4*) regions, but, in addition, the infrared photometry of Low (*9*) indicates that the integral brightnesses of Mars at wavelengths of 10 microns and 20 microns are similar, and consistent with the absence of any such greenhouse effect. Furthermore, the average disk temperature was found to be much higher than, and not consistent with, the radio data, as shown in Fig. 7. We are thus reluctant to accept the view that atmospheric molecular blanketing effects are present in sufficient strength to invalidate our thermal model.

In addition to blanketing by atmospheric molecular constituents, there is the possibility of blanketing by clouds of condensed H_2O or CO_2. An autumnal polar-haze cap is regularly observed (*21*), and some of the recent atmospheric models of Mars seem to suggest that solid particles of CO_2 might be commonly present in the atmosphere at all latitudes (*22*). We recognize that clouds of CO_2 or H_2O ice crystals could seriously retard thermal emission from the surface and thus reduce the accumulation of CO_2, but of course such clouds might also contribute additional frozen CO_2 to the surface by "snowfall." Theoretical analyses of various hypothetical haze and cloud conditions that might prevail in the Martian atmosphere may provide insight into the actual importance of such possibilities.

We have also examined the case in which both CO_2 and H_2O are present, with CO_2 dominant both in the atmosphere and in the polar cap. If the north polar cap is CO_2, its winter temperature of 145°K is so low that water, were it in equilibrium at this temperature, would be quite undetectable in the atmosphere. On the other hand, the *rate* at which any temporary excess of water would accumulate on the polar cap is likely to be quite low. For instance, calculations based on the CO_2 model and four additional assumptions show that the polar cap might accumulate ice only at the rate of 0.005 g cm^{-2} yr^{-1} if the mean annual content of water vapor at the equator were 10^{-3} g cm^{-2}. The four assumptions are that (i) water can be removed from the atmosphere only at a rate proportional to the water vapor content and with a time constant of 5 days, to empty a vertical column above the surface; (ii) water condenses out at this rate wherever the surface temperature falls below the local frost point; (iii) the water can be returned to the atmosphere as rapidly as the sun can evaporate it; and (iv) water vapor is transported in latitude in a "random walk" at the rate of 5 days for a 10-degree step.

Because of the precession of Mars's perihelion and rotation axis, which will periodically reverse the roles of the north and south poles as the sites of a permanent cap, we need only consider how much water would be accumulated at this rate during a fraction, $1/2\pi$, of the effective precession period of about 5×10^4 years (*23*). This would amount to about 40 g of ice per square centimeter—a quite negligible amount. We therefore conclude that it is possible to reconcile the observed presence of 10^{-3} g of water vapor per square centimeter with the presence of large amounts of frozen CO_2, including a semipermanent polar cap.

The calculation just described also showed that the amount of water vapor above each pole varied by a factor of 2 to 3 throughout the year, being about 6×10^{-4} g cm^{-2} in the winter and about 15 to 20×10^{-4} g cm^{-2} following the disappearance of the polar cap in the summer. About 20 to 30×10^{-4} g of ice per square centimeter was found to be deposited annually on the transient polar caps.

Concerning a possible means of reconciling our model with the polar cap observations previously cited, it is of interest to note that, although the amount of H_2O condensed is very small compared with the amount of CO_2, the H_2O will remain in the cap until all the CO_2 has disappeared. Furthermore, as the CO_2 evaporates, the H_2O may well become concentrated in a thin film at the top surface of the receding cap and may alter the reflective properties of the cap enough to make it appear to be composed of water ice.

If we accept the possibility that, on the average, 10^{-3} g of water vapor is present per square centimeter, we are led to some further conclusions of interest. First, if this water vapor can penetrate into the soil to a depth of at least a few meters, as should be the case for porous surface material such as is indicated by the infrared observations, it may reach a region where the temperature is perpetually below the 190°K condensation temperature at the corresponding surface-level vapor pressure of 3.7×10^{-4} millibar. The water would tend to migrate to such regions and condense as permafrost. The mean annual temperature as a function of latitude is shown in Fig. 9, together with the condensation temperatures corresponding to amounts of water vapor equal to 10^{-2}, 10^{-3}, and 10^{-4} g cm^{-2}. It may be seen that, in both hemispheres, the regions poleward of latitudes 40° to 50° will tend to trap water as permafrost.

It is difficult to estimate the quantity of water likely to be present as permafrost because we can only speculate on the depth of loose, permeable material, and on whether such loose material is saturated. Since we are here concerned with geological time scales of tens or hundreds of millions of years, it seems likely that the trapping layers are saturated with ice. However, internal heat sources may impose a downward thermal gradient which at some unknown depth will raise the temperature above the condensation temperature. It would be most surprising if the saturated trapping layer did not extend to depths of at least several tens of meters, so it seems quite possible that several hundred grams of water per square centimeter could be present in the pores of the soil.

While we cannot specify the depth to which the permafrost layer extends, we are able to estimate at what depth below the surface its top is situated. At a given latitude this level will be that at which the vapor pressure of water, averaged throughout the year, is equal to the atmospheric average vapor pressure, for then the net annual exchange of water with the atmosphere will be zero. This depth is plotted as a function of latitude in Fig. 10.

Figure 10 shows that the top of the permafrost layer should be only a few centimeters below the surface except near the boundaries of the layer, and that the layer should be some 30 centimeters nearer the surface at a given latitude in the northern hemisphere than in the southern. This depth difference, multiplied by the fractional porosity of the soil and integrated over a hemisphere, represents the amount of water potentially available for transfer between the two hemispheres in the course of the 5×10^4-year precessional cycle. We have not investigated this transfer process in detail.

Although local irregularities in the surface and subsurface conditions will introduce corresponding fluctuations in the depth to the top of the permafrost, we do not expect to find gross departures from the results obtained with the model insofar as the distribution and physical state of near-surface water are concerned. In particular, we would not expect to find large amounts of water (or ice) at any depth in the equatorial regions unless the area was extremely well isolated from the atmosphere.

The possible occurrence of liquid water is a question of some concern. An attractive idea is the possibility that soil moisture sometimes exists in the Elysium region, where relatively strong radar reflections (*24*) and persistent morning clouds (*25*) were observed during the 1965 opposition period. However, nothing in our study points toward the occurrence of liquid water at any place or time during the daily, annual, or precession cycle. The maximum temperature attained in the permafrost layer was about 200°K—far below the melting temperature of ice, (273°K). Furthermore, even if the temperature were to rise above 273°K, it is not certain that the ice would melt rather than sublime, since the total atmospheric pressure, 5 to 10 millibars, could be less than the 6.9-millibar vapor pressure of water at its triple point.

Of course, we cannot exclude the possibility that very small amounts of liquid water might form for brief periods in connection with the evaporation of the residual water ice of a polar cap after the CO_2 has disappeared if, for example, the ice frost were partially trapped in the upper few millimeters of the surface under conditions where solar heating was intense but evaporation was retarded by the limited porosity of the soil. Again, strongly deliquescent salts situated near the surface might, under some circumstances, lead to the presence of moisture in the soil.

Other Volatiles

There are no confirmed reports of other volatiles on Mars. Kiess *et al.* (*3*) have reported nitrogen oxides; more recently that subject has been discussed by Sagan, Hanst, and Young (*26*) and by Heyden (*27*). Perhaps more surprising than the possibility that nitrogen

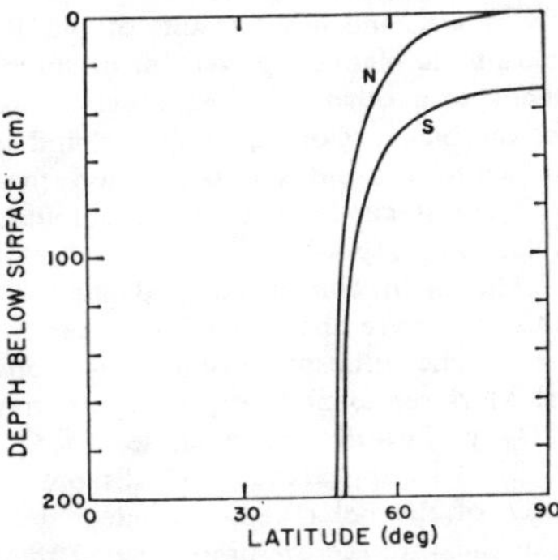

Fig. 10. Depth of top surface of H_2O permafrost, as a function of latitude. The difference in depth between the two hemispheres defines an amount of water that should be exchanged between the hemispheres or during the 5×10^4-year precessional period.

oxides are present is the apparent dearth of nitrogen itself. An upper limit of a few millibars for the partial pressure of nitrogen in the Martian atmosphere is implied by the Mariner IV occultation results. Such a low abundance would seem anomalous, on the basis of geochemical considerations, if the atmosphere of Mars was formed in a manner at all similar to the formation of the earth's atmosphere.

Carbon dioxide can be expected to be lost from Mars, by various dissociation and ionization processes, at a much higher rate than that at which nitrogen is lost. Gravitational escape will not be significant in either case if the exospheric temperatures are as low as the 550 ± 150°K estimated by Gross *et al.* (*19*) or the 85°K estimate of Johnson (*17*). Accordingly, it is difficult to understand how an enormous CO_2 enrichment relative to nitrogen could take place if both gases were present either in a primordial atmosphere or in continuing volatile emissions from the interior of Mars. One possible clue to the apparent anomaly is the suggestion that CO_2 might have been preferentially trapped upon and beneath the surface in a solid state while the entire mass of nitrogen was always in the gas phase at the surface and thus able to escape much more effectively. It is possible that on Mars, as on the moon (*28*), the presence of a low-vapor-pressure solid phase of a volatile molecular or atomic species may be of more significance in retention of the volatile than is the species' resistance to escape from the vapor phase. Another fractionation mechanism for selective retention on Mars of a heavier, condensable gas relative to a lighter, noncondensable one was discussed by Stoney (*2*) prior to 1898.

The other class of Martian volatiles about which there has been much speculation is that of organic compounds, particularly those related to possible biological activity. While our analysis is too restricted to form the basis of any discussion of the *a priori* likelihood of biogenic compounds being formed now, or having been formed in the past, on the surface of Mars, it may offer some insight into the probable redistribution of such compounds should they be present on Mars. In particular, our analysis for the evaporation and redistribution of H_2O should be roughly applicable to any volatile minor constituent. If solid (or liquid) organic compounds are present near the surface anywhere in the temperate or in the equatorial regions of the planet, they should be evaporating also, and condensing in the polar regions. If the circulation is sufficiently rapid, the mass transferred would be roughly in proportion to the ratio of the evaporation rate of the substance to the evaporation rate of water in the temperature range around 170° to 200°K. Thus it would seem that organic substances significantly more volatile than H_2O in the low-temperature range should be depleted from the soil over most of the planet and transferred to the polar areas well within a single 8000-year portion of the precessional cycle.

Summary and Conclusions

We have found that a rather simple thermal model of the Martian surface, in combination with current observations of the atmospheric composition, points strongly toward the conclusion that the polar caps of Mars consist almost entirely of frozen CO_2. This study was based upon the following principal assumptions.

1) Carbon dioxide is a major constituent of the Martian atmosphere.

2) The blanketing effect of the atmosphere is small, and due principally to the absorption band of CO_2 near 15 microns.

3) Lateral and convective heat transfer by the atmosphere is negligible.

4) The far-infrared emissivity of the Martian soil and of solid CO_2 are near unity.

5) The reflectivities of the soil and of solid CO_2 in the visible part of the

spectrum are about 0.15 and 0.65, respectively.

6) Values for soil conductivity, density, and specific heat are those characteristic of powdered minerals at low gas pressure.

7) Water is a minor constituent of the Martian atmosphere, the maximum total amount in the atmosphere being 10 to 30 $\times 10^{-4}$ g cm^{-2}.

In addition, several simplifications were made, which might have significant effects but should not alter our principal conclusions. Among these are the following.

1) Local blanketing or snowfall effects due to clouds or polar haze were ignored.

2) Dark and light areas were not differentiated in this study, although Sinton and Strong (*6*) have observed temperature differences between such areas.

3) The effects of local topography and microrelief were neglected. We believe that these must have quite significant effects at the higher latitudes, especially in connection with the evaporation of the remanent south polar cap.

4) Variation of reflectivity with angle of incidence of the sunlight was neglected.

5) Temperature dependence of soil conductivity and specific heat was ignored.

6) Effects of saturation of the soil by ice upon the thermal properties of the soil were neglected.

Although in our main investigation we used certain specific values for the various relevant parameters, we also tested the effects of moderate changes in these quantities. Specifically, the soil conductivity was varied by a factor of 3, the albedo and emissivity of the surface were changed by 15 to 20 percent, and the effects of a gross amount of atmospheric blanketing were studied, as described. Only the last of these variations had any significant effect on the model, and other results of the atmospheric blanketing were in disagreement with other physical observations of the planet. Consequently, we find it difficult to avoid the conclusion that CO_2 must condense in large amounts relative to H_2O.

The main conclusions indicated by this study are the following.

1) The atmosphere and frost caps of Mars represent a single system with CO_2 as the only active phase.

2) The appearance and disappearance of the polar caps are adequately explained on the presumption that they are composed almost entirely of solid CO_2 with perhaps an occasional thin coating of water ice.

3) If the currently reported water-vapor observations are correct, water-ice permafrost probably exists under large regions of the planet at polar and temperate latitudes.

4) The geochemically anomalous enrichment of CO_2 relative to N_2 in the present Martian atmosphere may be a result of selective trapping of CO_2 in the solid phase at and under the surface.

5) If the basic evaporation and condensation mechanisms for CO_2 and H_2O discussed in this article are correct, the possible migration of volatile organic compounds away from the warm temperate regions of the planet and their possible accumulation in the polar regions need to be carefully considered.

References and Notes

1. G. P. Kuiper, in *The Atmospheres of the Earth and Planets*, G. P. Kuiper, Ed. (Univ. of Chicago Press, Chicago, 1952).
2. G. J. Stoney, *Astrophys. J.* **7**, 25 (1898); this paper refers to an even earlier suggestion by the same author.
3. C. C. Kiess, S. Karrer, H. K. Kiess, *Publ. Astron. Soc. Pacific* **72**, 256 (1960).

3a. W. M. Sinton, in *Planets and Satellites*, vol. 3 of *The Solar System*, G. P. Kuiper, Ed. (Univ. of Chicago Press, Chicago, 1961), p. 436.

4. V. R. Stull, P. J. Wyatt, G. N. Plass, *Appl. Opt.* **3**, 243 (1964).
5. C. Leovy, *Icarus* **5**, 1 (1966).
6. W. M. Sinton and J. Strong, *Astrophys. J.* **131**, 459 (1960).
7. C. H. Mayer, T. P. McCullough, R. M. Sloanaker, *Astrophys. J.* **127**, 11 (1958); E. E. Epstein, *ibid.* **143**, 597 (1966); K. I. Kellerman, *Nature* **206**, 1034 (1965); D. S. Heeschen, *Astrophys. J.* **68**, 663 (1963).
8. J. A. Giordmane, L. E. Alsop, C. H. Townes, C. H. Mayer, *Astrophys. J.* **64**, 332 (1959); W. A. Dent, M. J. Klein, H. D. Allen, *ibid.* **142**, 1185 (1965).
9. F. J. Low, private communication (1966).
10. We neglect here the small but important variation of temperature with elevation due to (i) decrease of barometric pressure with altitude and (ii) adiabatic cooling of surface air as it flows from the edge to the middle of the polar cap.
11. A. Dollfus, in *The Atmospheres of the Earth and Planets*, G. P. Kuiper, Ed. (Univ. of Chicago Press, Chicago, 1952), p. 381.
12. L. D. Kaplan, G. Münch, H. Spinrad, *Astrophys. J.* **139**, 1 (1964).
13. A. Dollfus, in *The Origin and Evolution of Atmospheres and Oceans*, Brancazio and Cameron, Eds. (Wiley, New York, 1964).
14. V. I. Moroz, *Astron. Zh.* **41**, 350 (1964).
15. M. Spinrad, unpublished results (private communication, 1966).
16. F. J. Heyden, C. C. Kiess, W. R. Willauer, *Astrophys. J.* **143**, 595 (1966).
17. F. S. Johnson, *Science* **150**, 1445 (1965).
18. A. Kliore, D. L. Cain, G. S. Levy, V. R. Eshleman, G. Fjeldbo, F. D. Drake, *ibid.* **149**, 1243 (1965).
19. S. H. Gross, W. E. McGovern, S. I. Rasool, *ibid.* **151**, 1216 (1966).
20. T. Dunham, Jr., in *The Atmospheres of the Earth and Planets*, G. P. Kuiper, Ed. (Univ. of Chicago Press, Chicago, 1952), p. 296.
21. E. C. Slipher, in *Mars*, J. S. Hall, Ed. (Sky Publishing Corp., Cambridge, Mass.) p. 20.
22. G. Fjeldbo, W. Fjeldbo, V. R. Eshleman, *Stanford Electronics Lab. Rep. SU-SEL-66-007* (1966).
23. The period for the precession of the perihelion is about 72,000 years [see D. Brouwer and G. M. Clemence, *Planets and Satellites* (Univ. of Chicago Press, Chicago, 1961), p. 50] while that due to solar tidal precession of the rotation axis may be estimated as about 180,000 years, a value obtained by correcting the solar part of the total earth's precession, (7.5/2.4) by the appropriate factors for distance and moment of inertia. We are indebted to Professor W. M. Kaula for helpful discussion on this point.
24. R. M. Goldstein, *Science* **150**, 1715 (1965).
25. C. Capen, private communication (1965).
26. C. Sagan, P. L. Hanst, A. T. Young, *Planetary Space Sci.* **13**, 73 (1965); *ibid.*, p. 1003.
27. S. J. Heyden, *ibid.*, p. 1003.
28. K. Watson, B. C. Murray, H. Brown, *J. Geophys. Res.* **66**, 3033 (1961).

29. We are indebted to Drs. L. Davis, W. M. Kaula, Conway Leovy, R. S. Richardson, R. F. Scott, and R. P. Sharp for helpful discussions. The thoughtful criticism of Dr. W. M. Sinton is also appreciated. This research was supported in part by National Aeronautics and Space Administration grants NsG-426 and NsG 56-60.

8

Reprinted from *Science* **182**:437–443 (1973)

Polar Volatiles on Mars—Theory versus Observation

Excess solid carbon dioxide is probably present in the north residual cap.

Bruce C. Murray and Michael C. Malin

Mariner 9 produced a wealth of new information concerning processes on Mars involving solid-vapor equilibria of water and carbon dioxide (*1–3*). Interpretations of these observations in terms of theoretical models, such as that of Leighton and Murray (*4*) have been attempted. In this article we synthesize the results of the Mariner 9 mission, as they pertain to polar volatiles, and compare them with a description of the solid-vapor equilibrium relations we believe are presently active on Mars.

CO_2 Reconsidered

Leighton and Murray's (*4*) study of CO_2 and other volatiles on Mars was a simple representation of the absorption and reradiation of solar energy by solid CO_2 on Mars. In order to describe the processes taking place, an almost lunar thermal model of the martian surface was assumed, and a simplified orbit was used to facilitate computation. The model did not provide for horizontal heat transport by the atmosphere or dust, or for any interference in the polar radiation balance arising from clouds (*5*). A uniform height was assumed for the surface, and this has been contradicted by the large altitude differences discovered by ground-based radar in the late 1960's. In addition, the model range of albedos assumed was not based on actual observation. In spite of these simplifications, it was evident that the surface atmospheric pressure on Mars is limited by the polar heat balance; excess CO_2 must be accommodated in solid form at the poles rather than by an increase of CO_2 in the atmosphere.

The model predicted (i) a mean surface pressure of CO_2 similar to that actually observed by Mariner 4, about 5 millibars; (ii) a rate of recession of the seasonal polar caps comparable to that observed; (iii) a permanent north polar cap, which was also observed from the earth; and (iv) a semiannual planet-wide pressure fluctuation arising from atmospheric exchange with surface frost during the seasonal growth and disappearance of the polar caps. As a consequence, Leighton and Murray (*4*) concluded that the seasonal frost caps on Mars are composed of solid CO_2, not water-ice as had commonly been believed. The Mariner 7 spacecraft confirmed this prediction; it carried an infrared spectrometer, which directly detected characteristic absorption bands of solid CO_2 (*6*), and an infrared radiometer (*7*) which, along with the infrared spectrometer, confirmed that the brightness temperature of the frost composing the seasonal southern cap was approximately 150°K as predicted by Leighton and Murray. Further confirmation was provided by ground-based interferometric observations in 1971 (*8*).

A deficiency of the model arose from the simplified description of the orbit of Mars. The (erroneous) computation suggested a periodic variation in the average annual insolation at the two poles with a 50,000-year cycle, caused by precessions of the perihelion and spin axis. Leighton and Murray attributed the large residual cap, currently present in the north, to this supposed effect. More accurate analyses—carried out first by Kieffer (*9*) and later by ourselves and by Cuzzi (*10*)—indicate that the average annual insolation at the two poles is always identical (*11*). Inasmuch as the stability of a perennial deposit of solid CO_2 on the surface can be considered solely in terms of its annual heat balance, there is now little basis for the original Leighton-Murray explanation of north polar dominance as the site of a permanent CO_2 deposit (*12*).

Mariner 9 has increased our knowledge of the heat transport processes and circulation systems operative in the martian atmosphere. This new information suggests to Leovy (*13*) that a significant amount of heat is transported from equatorial areas to polar areas during certain seasons on Mars. Additional heat transfer also occurs during large dust storms that periodically envelop the planet. These considerations, which are based on observations, cause concern about the original Leighton-Murray model, which ignored these effects. Similarly, the polar hood, a phenomenon long observed from the earth, has been studied more completely from martian orbit in the visible, infrared, and ultraviolet wavelengths (*14*); we conclude that it may influence the oversimplified heat balance equation used in the original model.

A third possible deficiency, pointed out by Kieffer (*9*), concerns the extreme sensitivity of the equilibrium pressure and temperature to albedo values of the exposed CO_2 frost. For example, a change in albedo from .65 to .75

Dr. Murray is professor of planetary science and Mr. Malin is a graduate student in the Division of Geological and Planetary Sciences, California Institute of Technology, Pasadena 91109.

results in a change in equilibrium partial pressure of CO_2 from 13.5 to 2.3 mbar (*15, 16*). Occurrences that might, at times, change the albedo, such as dust infall on the ice surface or perhaps different textures of CO_2 frost resulting from local micrometeorological conditions, cause Kieffer to believe that the Leighton-Murray model may be inherently unstable.

Reservoir of Solid CO_2

It is most important to recognize that the only place on or within the planet where perennial solid CO_2 can now exist is on the surface near the poles, or at shallow depth (*17*) there beneath a permanent cap with a high albedo. Solid CO_2 can not survive burial beneath any low-albedo "dirt," even temporarily, since the subsurface temperature exceeds the sublimitation point of the solid and CO_2 will escape as a gas; this is the case everywhere except near the poles. In addition, any burial process (for example, eolian deposition of dust) will necessarily be slow, with individual particles warming the CO_2 around them, and sinking from solar view by subliming CO_2 into the atmosphere. Thus, a scum layer of dark (low-albedo) material may be buried beneath a topmost layer of frost, but as soon as this topmost layer is removed, the dark dust will heat up and any CO_2 beneath it will escape. Although the escape will be limited by the rate at which CO_2 can pass through the scum layer, it is unlikely that this layer could resemble a solidified, although porous medium, and the resulting rate will thus be rapid (*18*). Furthermore, McElroy (*19*) argues that the amount of CO_2 lost through atmospheric escape is probably small. There is no evidence that large amounts of CO_2 are tied up in surface materials such as carbonates (*20, 21*). Therefore, most of the CO_2 ever released from the interior of Mars is probably now in the atmosphere or included within permanent polar frost deposits. Furthermore, the discovery by Mariner 9 of extensive volcanic deposits on portions of the martian surface suggests that the total amount of CO_2 liberated to the surface probably exceeds that now present in the atmosphere (*22*). Thus, excess CO_2 in the solid form is to be expected in the polar areas.

Conversely, it would have to be considered an extraordinary accident if the total CO_2 released over the history of Mars and now available at the surface should just exactly equal that required for the formation of the observed annual caps: a small proportion less and hardly any annual caps would form at all; any larger amount would be in the form of buried solid, as is proposed here. Although the simplified model (*4*) which predicts a permanent CO_2 cap has significant deficiencies both theoretically and observationally, the seasonal caps are composed of CO_2 as predicted, excess solid CO_2 is quite likely, and a permanent deposit of solid CO_2 evidently is in equilibrium with atmospheric CO_2. Otherwise, the observed CO_2 surface pressure of about 5 mbar must be regarded as a fantastic coincidence, especially in view of the obliquity variations discovered by Ward (*23*) and the consequent hundredfold atmospheric pressure oscillations (*16*).

So, what is missing? We believe that there must be a large reservoir of solid CO_2 in gaseous equilibrium with the atmosphere (on some time scale), but buried immediately below the exposed residual water-ice cap. And we believe we have located this reservoir near the north pole.

The principal effect of such a reservoir is to average out annual and longer-term fluctuations in the polar heat balance. Heat exchange probably takes place between buried and exposed solid CO_2 through a "heat pipe" effect with gaseous CO_2 as the circulating medium, as well as by direct conduction through the water-ice cap. The temperature of the reservoir must therefore be a slowly changing function of the past polar heat balance. It is that temperature (*24*) which now regulates the pressure of the martian atmosphere. If such a reservoir exists, it might provide an answer to the difficulties concerning the stability of the exposed CO_2 cap. Short-term fluctuations in the albedo of the polar cap or in atmospheric heat transfer would be compensated by heat exchange with the large reservoir, rather than causing precipitous changes in atmospheric pressure, as might be implied by the earlier model (*4*). The characteristic response time of the reservoir to temperature variations was found to be 10^2 to 10^3 years (*16*).

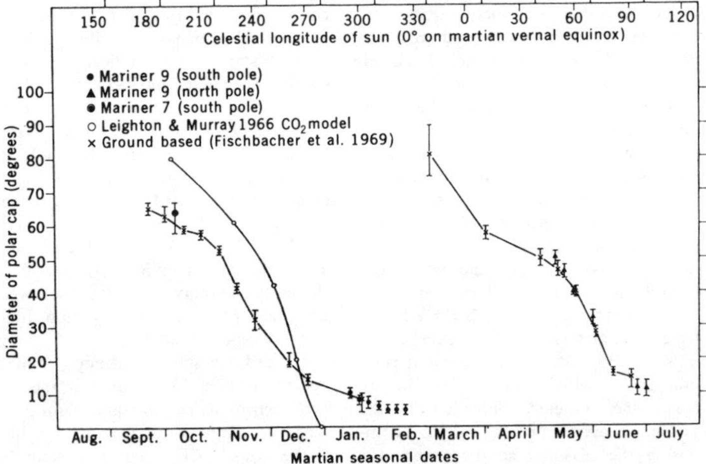

Fig. 1. Recession of the martian polar caps as a function of martian seasonal date. The error bars represent the maximum variations seen as a function of longitude. For the compilation of ground-based observations by Fischbacher *et al.* see (*25*); for Leighton and Murray see (*4*). Note the excellent agreement between the spacecraft and ground-based observations. The differences between the curves may reflect topographic control of the polar cap retreats.

Implications of Mariner 9 Observations

Thus, we are especially interested in the Mariner 9 observations for clues to the nature of the residual polar caps, and to the location and size of any subsurface solid CO_2 deposits.

The retreat of the south polar cap was monitored by Mariner 9 and the results were reported by Murray *et al.*

(*1*). As indicated in Fig. 1, the cap did not disappear entirely—contrary to the prediction of the Leighton and Murray (*4*) model of a seasonal CO_2 cap. Near the south pole a small residual cap was found which had only recently been recognized with confidence from the earth (*25*). The outline of the residual cap did not change appreciably during the peak of summer insolation in the southern hemisphere. This observation led Murray *et al.* (*1*) to postulate that the residual southern cap may be composed of water-ice rather than solid CO_2. A similar relationship in the north polar region based on observations obtained later in the mission has been described by Soderblom *et al.* (*2*). In both cases, the residual cap exhibits sharp boundaries and lies entirely within a peculiar, very smooth, layered formation which occurs only in the polar areas of Mars. Figure 2 shows a stereographic outline of the residual caps for comparison with observations of nearly a century ago. The residual caps appear to be a long-term feature of the martian landscape.

The probability that the exposed residual frosts are water-ice is reinforced by a consideration of the relative stability of solid H_2O and solid CO_2 during the maximum solar heating at the poles on Mars. The maximum surface temperatures if we consider radiative exchange only (no allowance for inward heat flow) range from 160° to 200°K for water-ice with albedo values of 0.8 to 0.6 (*26*). Inasmuch as CO_2 is the major constituent of the martian atmosphere, an exposed deposit will evaporate or condense in response to increases or decreases in the net heat balance. For example, approximately 1 meter of solid CO_2 would sublime while the residual cap is exposed during summer. Loss of water-ice, on the other hand, is limited in the higher temperature ranges (180° to 190°K) by the capacity of the cold CO_2 atmosphere to accept and transport water vapor. At lower temperatures the evaporation rate of ice itself becomes a limit. Therefore, a residual water-ice cap is much more stable than a solid CO_2 one on Mars in the summertime.

This line of inquiry suggests several other factors of interest. The temperature of the residual water-ice frost will be maintained at the condensation temperature of CO_2 (150°K) until all the CO_2 has sublimed away at the end of the seasonal recession. Only then can the water-ice begin to heat up in response to absorbed solar energy. This heating process will not be instantaneous. The rate of heating depends on the thickness of the ice, the proximity to dust and rocks, and related factors. However, if the water-ice heats up slowly enough to reach a temperature of 180° to 190°K by late in the summer, it will then begin to saturate the atmosphere immediately above it with water vapor. The polar hood phenomenon, which develops in the fall over each cap, may well involve saturated water conditions in the atmosphere above the pole (*27, 28*). Leovy (*13*) notes that an abundance of water vapor is inconsistent with simple models of atmospheric transport from equator to pole. It seems plausible to us that a residual water-ice cap at each pole could be the source for atmospheric water vapor in the fall season of each hemisphere. Similarly, the polar regions may at some seasons act as sources of atmospheric water vapor and at others as sinks. Such two-way exchanges of water between equator and pole might reduce the large amounts of water apparently lost to permanent polar deposits (*29*) and ease the difficulty encountered by Leovy.

Accordingly, the next question we wish to investigate is whether there is indeed solid CO_2 buried beneath the water-ice in the polar regions. No observational evidence of any such deposit has yet been reported. The south polar region was monitored at higher resolution and more extensively than the north. Figure 3 shows two views of a local area taken during and at the end of the martian southern summer. In the latter view, there is a considerable amount of residual frost, but also dark areas where bare ground is exposed. It is clear that the water-ice is not lying directly on top of massive CO_2 deposits in that region. A more general examination of the south polar region, as shown in Fig. 4, indicates that the white frost delineates and follows very closely the underlying terrain of the layered deposits.

The circumpolar dark bands are believed to be sloping surfaces that face the equator and separate level areas on

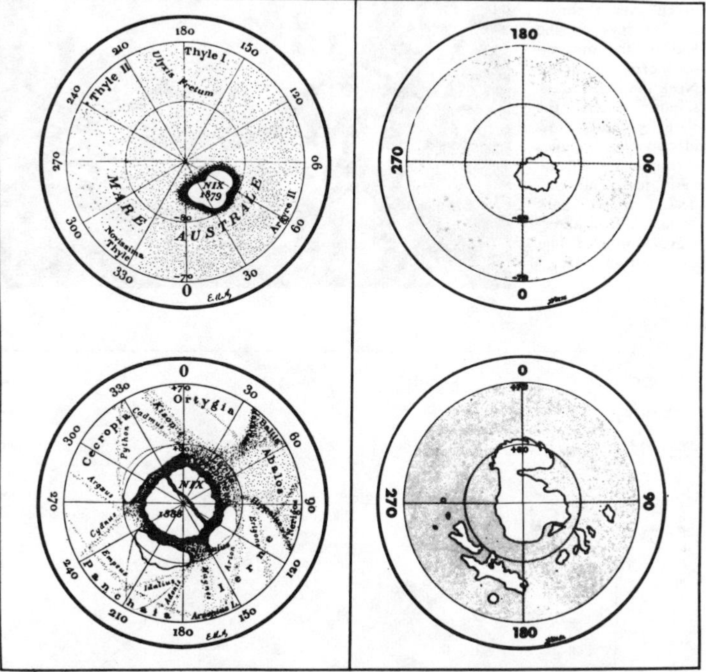

Fig. 2. Residual frost caps on Mars as seen in the 19th and 20th centuries. These stereographic projections of maps of the north and south polar caps are separated by almost 100 years. Note the latitudinal and longitudinal similarities and the close correspondence in areal extent. The 19th-century drawings by the astronomer E. M. Antoniadi, presented in Pickering (*42*), show the dates of the observations.

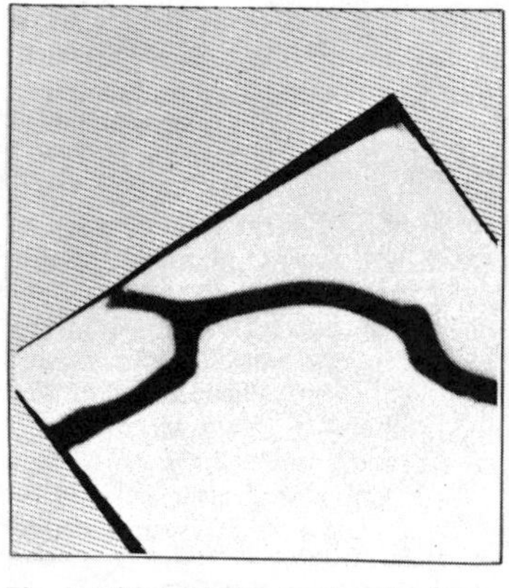

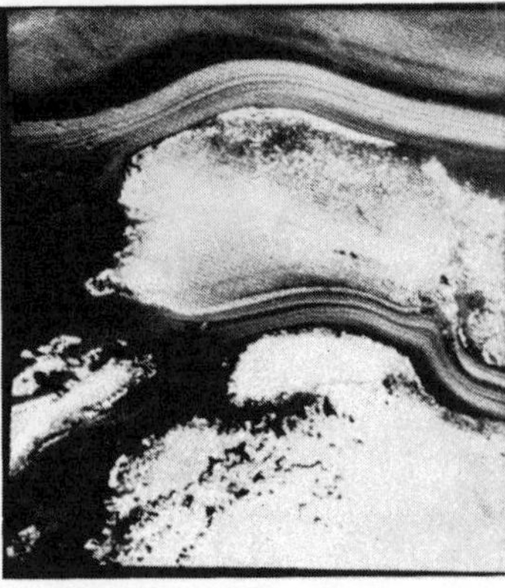

Fig. 3. High-resolution south polar frames of "the fork." The two photographs, taken 110 days apart, show a region of layered terrain and permanent polar ice near 85.4°S, 355.0°W. The dark features are bare ground. The presence of interior dark material indicates that the ice is thin enough to allow small-scale relief to show through and that it is probably water (CO_2 ice could not survive in contact with low-albedo material).

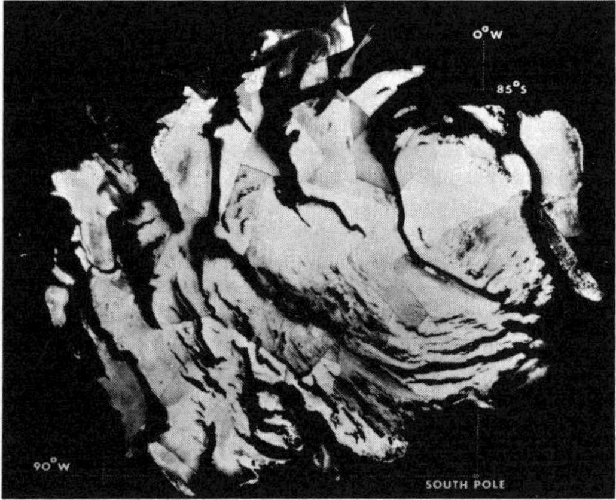

Fig. 4. Residual south polar cap of Mars, shown in a stereographic mosaic of Mariner 9 high-resolution frames taken over a period of 36 days at the end of the martian southern summer. Note the many dark regions where the bare ground (low albedo) shows through the permanent ice. The biggest, longest dark bands are equatorward slopes of layered terrain plates.

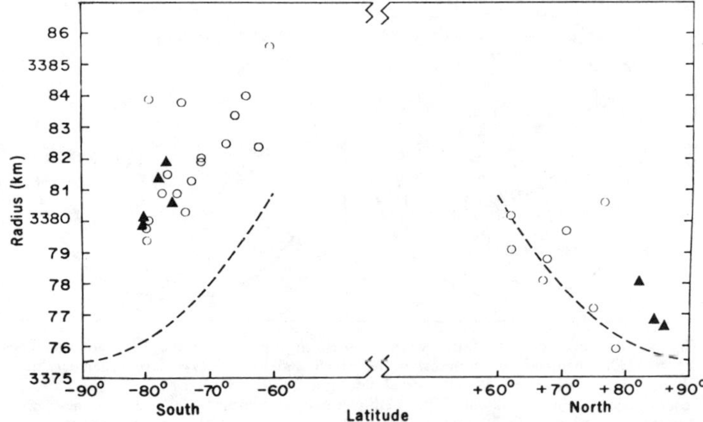

Fig. 5. Radii of Mars in polar regions, measured by the radio occultation experiment and reported by Kliore *et al.* (*31*). (Open circles) Occultation points in polar regions; (closed triangles) occultation points on layered terrains. The dashed line represents the triaxial ellipsoid which best approximates the figure of Mars. Both sets of data show that the southern points tend to be some 2 to 4 km higher than the northern points. Independent reductions of the surface pressures show the same trend, but with greater observational scatter.

which the residual frost is accumulated (*1, 30*). In 1969, Mariner 7 photographed these circumpolar features when they were covered by the seasonal CO_2 deposit. They appeared brighter then than the surrounding area because of the higher solar insolation on those slopes. For the same reason they defrosted before the Mariner 9 encounter and thus appear as dark bands in the Mariner 9 pictures.

Kliore *et al.* (*31*) report radio occultation observations which suggest to us that solid CO_2 should not be present now in the south polar regions. Figure 5 shows the planetary radii corresponding to spacecraft occultations which occurred in the polar latitudes (as viewed from the earth). It can be seen that the southern residual cap must be higher than the northern one by at least 2 kilometers. Any solid CO_2 in the south would be in contact with atmospheric CO_2 at a pressure lower by about 2 mbar than in the north (*32*). There is no reason to suppose a permanent CO_2 southern cap would be at a systematically lower temperature than the northern one. Hence, solid CO_2 deposits in the south would be out of equilibrium and would gradually be transferred to the north. A mass of CO_2 equal to that in the present atmosphere of Mars would be transferred in well under 1000 years. Thus, it seems unlikely that any significant mass of solid CO_2 could be buried beneath the south polar residual cap, and most improbable that it could be buried beneath the bare ground there because of the high subsurface temperature.

In the north polar regions, a similar appearance of the frost-coated layered terrains also places limits on the location and size of any buried CO_2 deposits over most of the residual cap. However, in one locality, indicated on Figs. 6 and 7, there is an interruption in the underlying terrain pattern sug-

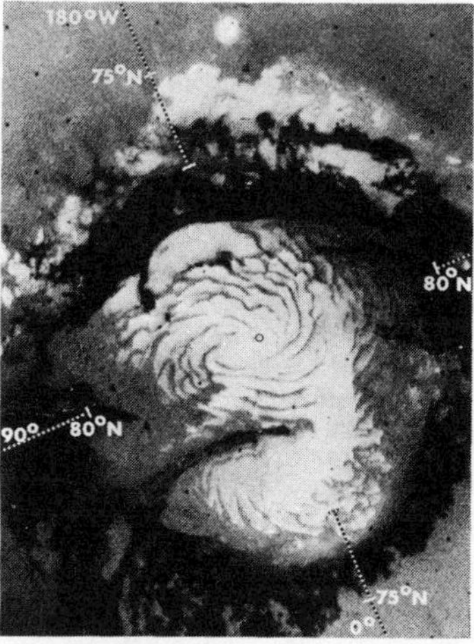

Fig. 6. Residual north polar cap of Mars. This is a stereographic mosaic of two Mariner 9 low-resolution photographs, taken 12 hours apart on 27 October 1972, on Mariner orbits 667 and 668. The three albedos are indicative of dark ground, light ground, and high-albedo ice. Except for a few outlying patches, most ice lies within two lobate forms joined at one end, above the 80th parallel. The dark arcuate features are believed to be equatorward slopes of the layered terrains. The featureless bright ice connecting the two ellipses is believed to be the area of the solid CO_2 reservoir.

gestive of a thicker deposit of frozen volatiles. We believe that this is the location of a significant accumulation of solid CO_2 buried beneath a relatively thin water-ice cover.

An important question is, how thick is the white deposit? And how can we be sure it is solid CO_2 and not water-ice? First, we think it is very unlikely that a thin deposit of CO_2 would be found on a surface sloping toward the equator. Solid CO_2 on a 5° equatorward slope experiences about the same insolation as that on a level surface 5° in latitude closer to the equator. The sharp boundary of the receding seasonal cap is evidence of the sensitivity of CO_2 to such slight insolation differences.

Water-ice is not nearly so responsive to insolation differences on a short time scale. However, the existence of the dark bands in both north and south residual caps is testimony to the preference of water for level surfaces. There is no evident reason why that pattern should change sharply in the anomalous area under consideration. Hence, we think the break in the pattern of the dark bands corresponds to the location of a thick mass of volatile.

If the white substance is thick enough to bury the local topography, just how thick must it be? The topographic relief across the dark bands in the south polar area is believed to be of the order of 300 to 500 m (*33*). If similar relief is exhibited in the north, the excess mass of frozen volatile lying on top of the layered terrain and beneath the water frost is probably at least 300 m in average thickness in order to mask the terrain configuration. On the other hand, solid CO_2 could hardly exhibit more than about 2 km in maximum relief without finding itself out of equilibrium with the atmosphere, since the same kind of redistribution which favors the north pole over the south will also limit the vertical relief of a solid CO_2 deposit. The inferred shape of the deposit is consistent with that expected for solid CO_2 responsive to slight variations in insolation (*4*) and height. Water-ice is not similarly regulated and displays a thin uniform structure in the south, for example, where it can be studied at high resolution (*34*). We cannot definitely rule out a thick deposit of water-ice, but we think it is much less likely than CO_2. A reasonable upper limit to the volume of a CO_2 deposit can be computed by using an thickness of 1 km.

We can estimate the limits on the mass of solid CO_2 contained in the thickened portion of the north residual cap as follows. The area of the thickened part, including some uncertainty, is (7 to 8.5) $\times 10^{14}$ cm^2. The thickness is 300 to 1000 m, as discussed above. Thus, the volume is (21 to 85) $\times 10^{18}$ cm^3. We use a density range of 1.0 to 1.5 g/cm^3 to allow for the possibility of some admixture of water-ice or dust buried within the CO_2 (1.0 g/cm^3) and the possibility of pure cold CO_2 (1.5 g/cm^3). Thus, the limits on the mass of CO_2 in the thickened part of the north residual cap are (21 to 127) $\times 10^{18}$ g, equivalent to 1.1 to 6.4 times the mass of CO_2 in the present atmosphere of Mars. We take a range of 2 to 5 as a reasonable estimate, recognizing that the upper limit is more sure than the lower one.

Leighton and Murray (*4*) predicted planet-wide seasonal pressure variations of the martian atmosphere of about 25 percent from peak to peak due to the transfer of solid CO_2 between the two hemispheres. So far, Mariner 9 data acquired by radio occultation and infrared and ultraviolet spectroscopy have not detected any annual pressure variation. However, the absolute accuracy of those tests may not be sufficient to rule out a small variation of the type predicted (*35*). We cannot propose any entirely satisfactory mechanism by which a solid CO_2 reservoir could stabilize the atmosphere at a constant pressure on such a short time scale. Hence, a more complicated situation (*36*) is indicated if the annual variation is really absent.

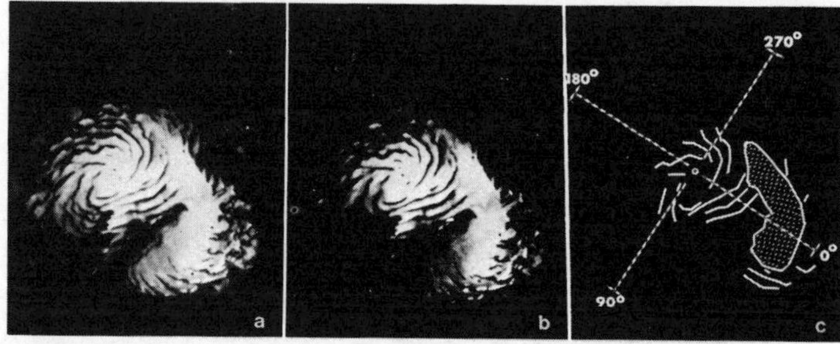

Fig. 7. The CO_2 reservoir. (a and b) Stereographic projections of the north polar caps, processed to enhance fine-scale albedo differences. Image (a) was made by isolating a range of gray levels which included light bare ground and bright ice, and then enhancing the contrast within that range. Image (b) is of the same region, with only the bright ice isolated. The faint gray markings in the center frame represent variations in brightness of less than 5 percent at high albedo, and are interpreted as photometric shading due to very slight slope differences. Thus, it is believed that the layered terrain scarps are overlain by material at least as thick as the elevation differences across the scarps. The drawing in (c) delineates the area believed to be thickened frost.

Conclusions

The residual frost caps of Mars are probably water-ice. They may be the source of the water vapor associated with seasonal polar hoods. A permanent reservoir of solid CO_2 is also probably present within the north residual cap and may comprise a mass of CO_2 some two to five times that of the present atmosphere of Mars. The martian atmospheric pressure is probably regulated by the temperature of the reservoir and not by the annual heat balance of exposed solid CO_2 (*37*). The present reservoir temperature presumably reflects a long-term average of the polar heat balance.

The question of a large permanent north polar cap is reexamined in light of the Mariner 9 data. The lower general elevation of the north polar region compared to the south and the resulting occurrence in the north of a permanent CO_2 deposit are probably responsible for the differences in size and shape of the two residual caps. The details of the processes involved are less apparent, however. It might be argued that the stability of water-ice deposits depends on both insolation and altitude. The present north and south residual caps should be symmetrically located with respect to such a hypothetical stability field. However, the offset of the south cap from the geometrical pole, the nonsymmetrical outline of the north cap, and the apparently uniform thickness of the thin, widespread water-ice all argue against control by simple solid-vapor equilibrium of water under present environmental conditions. We think that the present location of the water-ice may reflect, in part, the past location of the permanent CO_2 reservoir. The extreme stability of polar water-ice deposits increases the likelihood that past environmental conditions may be recorded there. Detailed information on elevations in the vicinity of the residual caps is needed before we can further elucidate the nature and history of the residual caps. This, along with measurements of polar infrared emission, should be given high priority in future missions to Mars.

Two conclusions follow from the limitation of the mass of solid CO_2 on Mars at present to two to five times the mass of CO_2 in the atmosphere. If all of this CO_2 was entirely sublimated into the atmosphere as a result of hypothetical astronomical or geophysical effects, the average surface pressure would increase to 15 to 30 mbar. Although such a change would have considerable significance for eolian erosion and transportation, there seems to be little possibility that a sufficiently earthlike atmosphere could result for liquid water to become an active erosional agent, as postulated by Milton (*38*). The pressure broadening required for a greenhouse effect requires at least 10 to 20 times more pressure (*39*). If liquid water was ever active in modifying the martian surface, it must have been at an earlier epoch, before the present, very stable CO_2/H_2O system developed. There can be no intermittent earthlike episodes now.

Furthermore, the present abundance of CO_2 on Mars may be an indicator of the cumulative evolution of volatiles to the surface of the planet (*40*). Thus, even the possibility of an earlier earthlike episode is dimmed. On Mars, the total CO_2 definitely outgassed has evidently been about 60 ± 20 g/cm^2. On the earth, about 70 ± 30 kg/cm^2 of CO_2 have been released to the surface (*41*). Hence, the total CO_2 devolved by Mars per unit area is about 0.1 percent of that evolved by the earth. Thus, the observational limits we place on solid CO_2 presently located under the north residual cap also may constitute considerable constraints on the total differentiation and devolatilization of the planet. If they are valid, it would seem unlikely that Mars has devolatilized at all like the earth, or ever experienced an earthlike environment on its surface.

References and Notes

1. B. C. Murray, L. A. Soderblom, J. A. Cutts, R. P. Sharp, D. J. Milton, R. B. Leighton, *Icarus* **17**, 328 (1972).
2. L. A. Soderblom, M. C. Malin, J. A. Cutts, B. C. Murray, *J. Geophys. Res.* **78**, 4197 (1973).
3. J. A. Cutts, *ibid.*, p. 4211.
4. R. B. Leighton and B. C. Murray, *Science* **153**, 136 (1966).
5. C. B. Leovy, *ibid.* **154**, 1178 (1966).
6. K. C. Herr and G. C. Pimentel, *ibid.* **166**, 496 (1969).
7. G. Neugebauer, G. Munch, S. C. Chase, Jr., H. Hatzenbeler, E. Miner, D. Schofield, *ibid.*, p. 98; G. Neugebauer, G. Munch, H. Kieffer, S. C. Chase, Jr., E. Miner, *Astrophys. J.* **76**, 719 (1971).
8. H. P. Larson and W. A. Fink, *Astrophys. J. Lett.* **171**, 91 (1971).
9. H. Kieffer, personal communication.
10. J. N. Cuzzi, thesis, California Institute of Technology (1973).
11. This ignores any slight differences in atmospheric absorption due to the somewhat different profiles of solar elevation angle versus time.
12. B. C. Murray, W. R. Ward, S. C. Yeung [*Science* **180**, 638 (1973)] have identified periodicities of 51,000 and 2,000,000 years in the maximum annual insolation which will cause north-south differences for processes that are nonlinear in insolation, such as the mobilization of water vapor.
13. C. B. Leovy, *Icarus* **18**, 120 (1973).
14. G. A. Briggs, abstract of paper presented at the third annual meeting of the American Astronomical Society, Division of Planetary Sciences (AAS-DPS), Tucson, Arizona, March 1973; V. G. Kunde, abstract of paper presented at the third annual meeting of the AAS-DPS, Tucson, Arizona, March 1973; C. A. Barth, abstract of paper presented at the third annual meeting of the AAS-DPS, Tucson, Arizona, March 1973.
15. However, such an albedo change would have to persist for 100 years or more (*16*).
16. W. R. Ward, B. C. Murray, M. C. Malin, in preparation.
17. C. Sagan [*J. Geophys. Res.* **78**, 4250 (1973)] argues that solid CO_2 will not be stable anywhere on the planet at a depth where the geothermal gradient reaches temperatures above the triple point (217°K). He estimates that depth as 2 to 3 km on Mars. However, we believe that solid CO_2 could never be buried even a few meters deep, except under permanent polar deposits of high albedo, much less reach a depth of a few kilometers.
18. Murray *et al.* (*1*) suggested the possibility that either solid CO_2 or water-ice might be included within polar sedimentary layers that are annually covered by seasonal frost. Whereas water-ice is a possible or even likely constituent of those deposits, further reflection on the stability of solid CO_2 rules out that substance as a constituent of those layers or any other exposed rock unit. For example, even an albedo as high as .50 indicates a subsurface temperature at the pole of 169°K. For an albedo of .25 the temperature will be 187°K.
19. M. B. McElroy, *Science* **175**, 443 (1972).
20. F. P. Fanale and W. A. Cannon [*Nature* **230**, 505 (1971)] have argued that small amounts of gaseous CO_2 may be trapped as adsorbed gases within the martian regolith. Later they greatly increased this estimate by the ad hoc assumption of a depth of 1 km for the regolith (*21*). Within the framework of this article we can only ignore such an ad hoc estimate, but we will consider it further as part of the dynamical history of CO_2 (*16*).
21. ———, unpublished manuscript.
22. B. C. Murray and M. C. Malin, in preparation; F. P. Fanale, personal communication.
23. W. R. Ward, *Science* **181**, 260 (1973).
24. It is possible that the temperature of the reservoir is somewhat higher than 150°K, but that the vapor pressure of CO_2 in equilibrium with it is depressed. A mechanism for this effect could be the formation of a CO_2 clathrate (hydrate) compound at the exposed interfaces of the reservoir, where water-ice is also abundant. S. L. Miller and W. D. Smythe [*Science* **170**, 531 (1970)] show that the CO_2 hydrate–CO_2 solid gas phase boundary lies some 5°K below the CO_2 solid–CO_2 gas phase boundary for martian pressures. Thus, the reservoir could conceivably exist over a larger range of temperatures, depending on the amount of clathrate present, and still maintain an equilibrium atmospheric pressure of 5 to 6 mbar.
25. G. E. Fischbacher, L. J. Martin, W. A. Baum, "Mars polar cap boundary," part A, final report to the Jet Propulsion Laboratory by the Planetary Research Center, Lowell Observatory, Flagstaff, Arizona (1969).
26. The albedo refers to a small patch of fresh frost surface. On low-resolution pictures, such as the Mariner 9 ones, such surfaces often blend in with bare ground, which leads to lower apparent albedos than actually apply to the heat balance of the frosts.
27. A. L. Lane, abstract of paper presented at the second annual meeting of the AAS-DPS, Kona, Hawaii, March 1972.
28. R. J. Curran, B. J. Conrath, V. G. Kunde, R. A. Hanel, abstract of paper presented at the third annual meeting of the AAS-DPS, Tucson, Arizona, March 1973.
29. J. A. Cutts, *J. Geophys. Res.* **78**, 4231 (1973).
30. B. C. Murray and M. C. Malin, *Science* **179**, 997 (1973).
31. A. J. Kliore, G. Fjeldbo, B. L. Seidel, M. J. Sykes, P. M. Woiceshyn, *J. Geophys. Res.* **78**, 4331 (1973).
32. This difference in height should influence the thickness and therefore the rate of recession of the seasonal CO_2 caps as well. The asymmetry between the northern and southern recessions shown in Fig. 1 between cap diameters of 50° and 15° may reflect the height differences illustrated in Fig. 5. Similarly, the correspondence between the computed and actual recessions might be improved if the prominent height differences were included.
33. K. R. Blasius, *J. Geophys. Res.* **78**, 4411 (1973).
34. There is much poorer Mariner 9 coverage of the northern residual cap than of the

southern one because of orbit, lifetime, and seasonal constraints (*2*). There is virtually no useful high-resolution coverage of the thickened area of the north polar cap.

35. Improved limits on the presence or absence of planet-wide seasonal pressure fluctuations probably can be obtained from refined analyses of the results from (i) the Mariner 9 radio occultation and ultraviolet and infrared spectroscopy experiments, and (ii) the Soviet Mars 2 and Mars 3 orbital experiments (as well as Mars 4, 5, 6, and 7 experiments now under way). Direct measurements of surface pressure from future U.S. and Soviet landers, as well as further measurements from orbit, could provide sufficiently long-term, precise pressure observations to permit a decisive determination of how the seasonal variation in the mass of the frost caps interacts with the atmosphere. This information, in turn, has important implications for transport of heat, dust, and water vapor by the atmosphere.
36. For example, pressure buffering of the seasonal CO_2 cycle in excess of that calculated by Leighton and Murray (*4*) might arise through additional semiannual heat exchange by direct conductive transfer of heat into and out of thick water-ice deposits in the residual cap. Water-ice might be a much more effective seasonal heat reservoir than solid CO_2 because of its great stability. Alternatively, Fanale and Cannon (*21*) suggest buffering by a large reservoir of adsorbed CO_2 in the martian regolith.
37. G. Briggs (personal communication) has proposed an extreme but interesting extension of our concept of a CO_2 reservoir capped with water-ice. Briggs wondered if the water-ice could effectively and indefinitely seal off the solid CO_2 from atmospheric interaction. We do not think this is likely for reasons based on long-term climatic changes on Mars. Obliquity variations occur on Mars (*23*) which will cause extreme short-term variations in the insolation received at the poles—so large, in fact, that the equilibrium pressure regime for CO_2 (solid) varies by nearly two orders of magnitude on a 50,000-year time scale. We believe that this is too rapid and too major a variation to allow isolation. Similarly, if we assume that the meteorite flux at Mars is equal to or ten times larger than the rate of the moon, we estimate impact rates at the polar regions that would produce at least one 100-m crater in 1000 years. This would effectively break any water-ice seal of the CO_2 reservoir.
38. D. Milton, *J. Geophys. Res.* **78**, 4037 (1973).
39. C. Sagan, O. B. Toon, P. J. Gierasch, *Science* **181**, 1045 (1973).
40. We exclude any hypothetical early martian episode when enormous quantities of the substance were somehow lost through escape or surface combination. Furthermore, the present earthlike abundance of CO_2 in the atmosphere of Venus argues for the persistence of CO_2 in the gaseous phase. If the maximum ad hoc estimates of Fanale and Cannon (*21*) for adsorbed CO_2 in the regolith were correct, the total CO_2 budget of Mars would be increased to approximately 2 percent of that of the earth—still a small fraction. Furthermore, this CO_2 would not be accessible to become part of the atmosphere under current conditions; the atmospheric equilibria would be virtually unchanged, and our limits on past atmospheric history would still apply.
41. A. P. Ingersoll and C. B. Leovy, *Annu. Rev. Astron. Astrophys.* **9**, 147 (1971).
42. W. H. Pickering, *Pop. Astron.* **24** (No. 4), 236 (1916).
43. We wish to acknowledge G. Biggs and J. Cutts of Jet Propulsion Laboratory, Pasadena; A. P. Ingersoll, R. B. Leighton, and R. P. Sharp of California Institute of Technology, Pasadena; and H. Kieffer of the University of California at Los Angeles for discussions, criticisms, and helpful debates concerning early drafts of this paper. S. C. Yeung, Princeton University, contributed to the early development of some of the ideas presented here. Supported in part by the National Aeronautics and Space Administration. Contribution No. 2248 of the Division of Geological and Planetary Sciences, California Institute of Technology, Pasadena 91109.

9

Reprinted from *Science* **194**:1341–1344 (1976)

MARTIAN NORTH POLE SUMMER TEMPERATURES: DIRTY WATER ICE

Hugh H. Kieffer, Stillman C. Chase, Jr., Terry Z. Martin, Ellis D. Miner, and Frank Don Palluconi

Observations of the martian south polar cap obtained by Mariner 7 conclusively demonstrated that the seasonal polar caps were composed predominantly of solid CO_2 (*1*). Whether the residual polar caps are composed of H_2O, CO_2, or a combination of them has been the subject of continued debate [for example, see (*2–4*)]. The abundance of martian surface forms attributed to fluvial erosion is a strong indication that Mars previously had a more extensive atmosphere than it does now, and permanent deposits of solid CO_2 have been invoked as stores of material that could contribute to periodic reconstitution of such an atmosphere (*5*). A persistent CO_2 polar cap would also act as a permanent cold trap and strongly influence the transport of water vapor around the planet.

A conclusive test for the absence of solid CO_2 is observation of brightness temperatures appreciably greater than 148 K, the saturation temperature of CO_2 at the martian mean total surface pressure (6.1 mbar). Carbon dioxide can be "condensed" as carbon dioxide–water clathrate ($CO_2 \cdot 6H_2O$), with an effective CO_2 density of about 0.33 g cm^{-3}, at temneratures about 5 K higher than pure CO_2. Considering both the possibility of a clathrate and that the polar surface could be at low elevation (high pressure) (*6*), temperatures above 155 K are incompatible with condensed CO_2 at the martian surface. Conclusive observations were not made by Mariner 9 (*7*).

A major objective of the Viking thermal mapping investigation has been to measure the temperature of the residual north polar cap. The infrared thermal mapper (IRTM) measures the brightness temperature at 18 to 24 μm (T_{20}) and 10 to 13 μm (T_{11}) and the total reflected solar energy (expressed as apparent albedo) simultaneously in seven 5-mrad-diameter spots with three telescopes; the fourth telescope has three detectors each at 7 and 9 μm and one in the 15-μm CO_2 band (*8*).

While the second Viking spacecraft was searching for a suitable landing area, the orbiter infrared instruments made observations which, on several revolutions, extended into the north polar region. On 31 August 1976, coverage of nearly 180° of longitude was obtained with seven accompanying pictures of a portion of the polar cap.

The measurements of 20-μm brightness temperatures over the region of the imaging coverage are shown in Fig. 1. The major thermal boundaries clearly correspond to structure in the visual brightness. The large dark regions have T_{20} near 235 K, while the main portion of

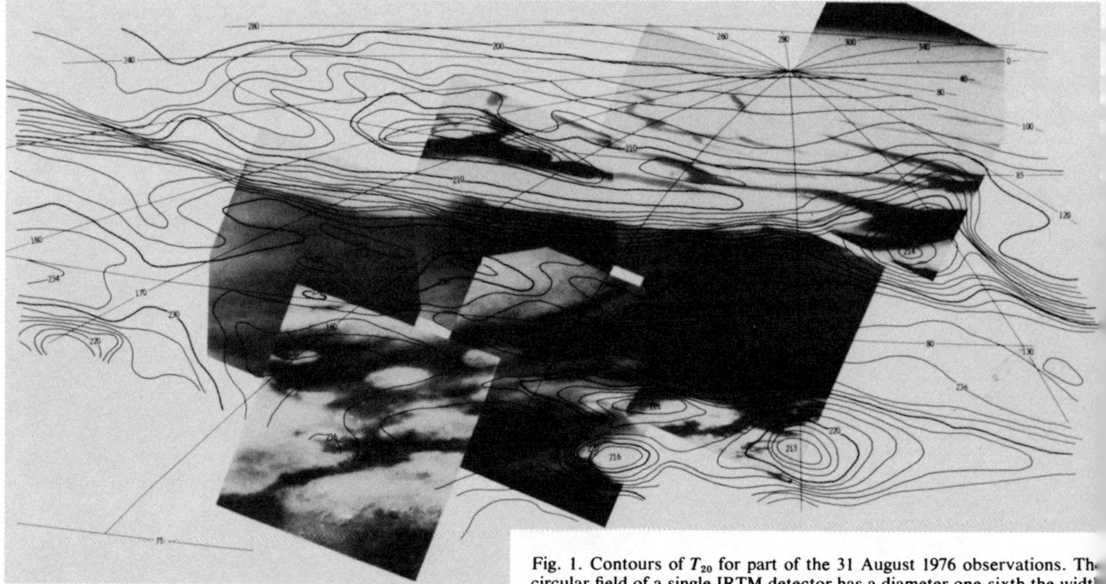

Fig. 1. Contours of T_{20} for part of the 31 August 1976 observations. Th circular field of a single IRTM detector has a diameter one-sixth the widt of a single photograph. The light areas with temperatures near or belov 220 K are water ice. The light areas in the lower left portion of the mosaic are almost entirely clouds with little actual brightness variation; th contrast between adjacent frames indicates the variation of image enhancement. At the center of the mosaic, the geometry at the time of the IRT observations was: solar incidence angle 62°, viewing angle 70°, phase angle 49°, local time 1500 hours, and range 3200 km.

the residual cap has T_{20} near 205 K. The bright areas near 77°N, 145°W are a frost-filled crater and some adjacent patches which have temperatures of 215 K or less. The warmer region at 84°N, 130°W corresponds to a dark entrant into the cap observed by Mariner 9.

Although there is certainly temperature structure below the resolution of these IRTM observations, the observed temperatures are so high that the presence of CO_2 ice can definitely be excluded. The IRTM bands are sufficiently separated in wavelength that a scene with wide temperature distribution would produce strong apparent spectral dependence. Scenes which were in fact composed of equal areas at 205 K and 240 K would have 3 K difference between T_{20} and T_{11}, approximately that observed in the regions of strong temperature contrasts. At 87°N, 160°W, where the imaging data suggest that there is little dark (warm) material in the IRTM field of view, the brightness temperatures in all bands agree at 205 K within their noise levels. This cannot be accomplished with any mixture of 150 K and 240 K material. Were this scene composed partially of CO_2 ice at 150 K, with the remainder being dark material near 240 K, the fraction of dark material required would have to vary from 29 percent for T_7 to 59 percent for T_{20}. In addition, the albedo of the CO_2 ice would have to exceed unity to fit the observed reflected sunlight. These requirements are independent of the length scale of mixing; in addition, lengths below the order of 1 m are almost certainly incompatible with the necessary temperature differences. The imaging data imply that large (scale larger than 1 km) unfrosted areas comprise at most a few percent of the IRTM field of view at this location.

The T_{11} observations of the north polar region are shown in Fig. 2. The major thermal boundary follows the outline of the bright cap area seen at the end of the Mariner 9 mission, at a season 1½ martian months earlier than the current observations. There are also several local cold areas well separated from the main polar cap. In particular, the 90-km-diameter crater Korolev, at 73°N, 196°W, which was observed near local noon, has T_{11} = 210 K in its interior, while the region immediately surrounding it is near 240 K. In the region of the irregular frost

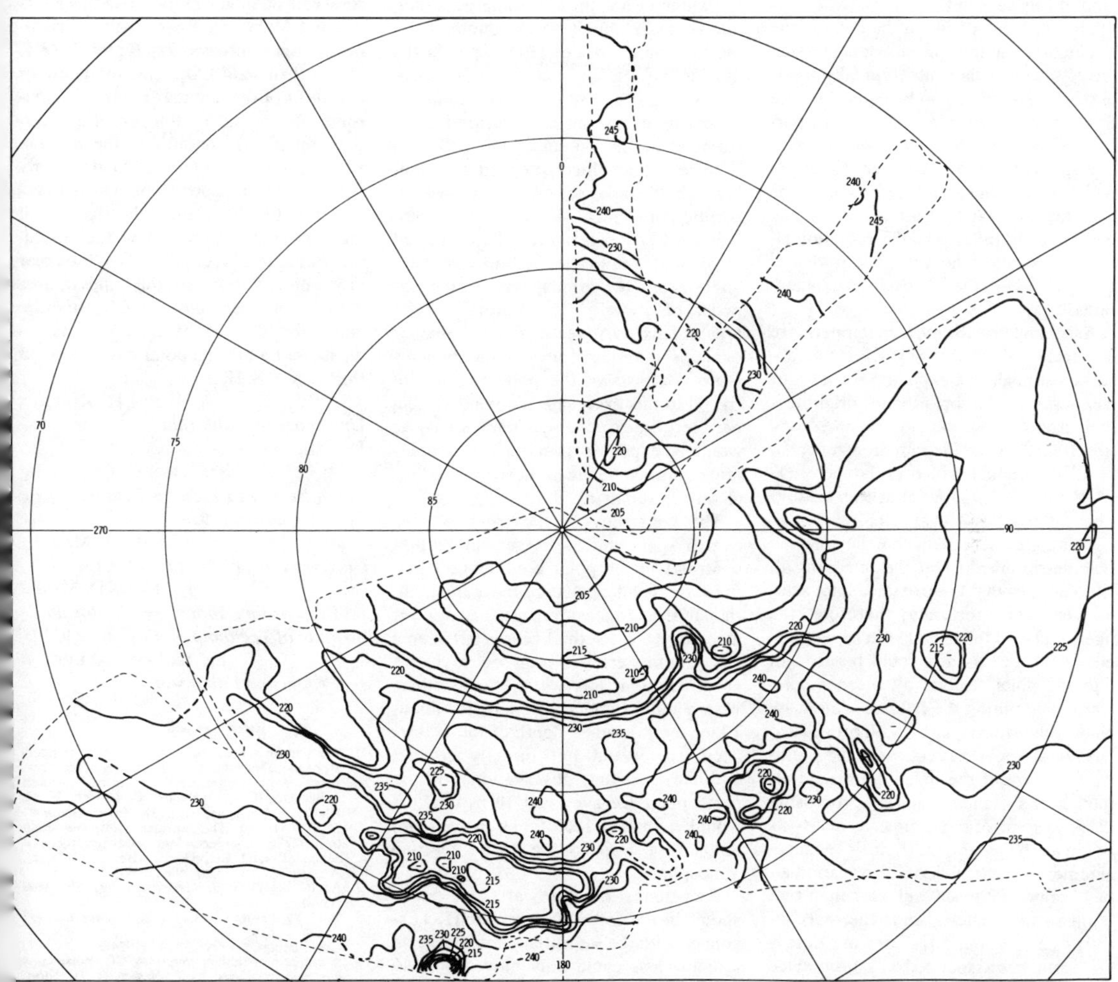

Fig. 2. Polar stereographic projection of T_{11}. The light dashed line indicates the extent of coverage. Closed contours are positive unless identified as negative (−). Near longitude 90°, the apparent decrease of T_{11} at constant latitude is probably due to the very oblique viewing. The center of each of the three sequences was at approximately 1500 hours; no correction for diurnal temperature variation has been made.

patches seen between 75°N, 180°W, and 76°N, 210°W by Mariner 9, T_{11} varies between 209 K and 220 K.

There is a strong correlation between temperature and apparent albedo A measured by the IRTM in the polar region. Nearly all points fall within $T_{20} = 240 - 136\ (A - 0.2) \pm 5$ K, with strong concentrations of data at $A = 0.24 \pm 0.05$ and $A = 0.41 \pm 0.03$; the intermediate points may largely represent observations which include appreciable fractions of both these apparent end members.

There is essentially a one-to-one correlation between areas of high albedo, areas of low temperature, and bright areas seen by Mariner 9 earlier in the summer. All of these areas must be covered with water ice. There is no evidence for any frozen CO_2 in the north polar region at this season (*9*).

The apparent albedo of the polar frosts is much lower than the albedo of clean terrestrial snow deposits (typically greater than 0.7). This could be caused by the frosts having either included dirt or partially glazed surfaces that reflect much of the sunlight in the specular direction, which is not observed by Viking. The simultaneous measurement of thermal emission and reflected solar radiation allows a test of the heat balance to distinguish between these possible explanations.

Assuming that the surface temperature is nearly in equilibrium with the absorbed sunlight, as expected for a water ice deposit, the broadband brightness temperature T and bolometric albedo A_B (the fraction of solar radiation reflected in all directions) will be related by $(T/245)^4 = (1 - A_B)$ at 80°N at this season. This yields $A_B = 0.45$ at 210 K.

This agreement between the inferred bolometric albedo and the apparent albedo implies that the frost does not scatter very anisotropically; therefore, the likely cause of the low albedo is the presence of dirt. This dirt could be brought into the polar regions by the net poleward wind during the fall and winter, by global dust storms, or by both processes.

Several bright areas occur in the southwest portion of the imaging mosaic; although the standard image processing displays some of these areas as nearly as bright as the polar cap, from IRTM measurements the areas have an apparent albedo only 0.02 to 0.05 greater than that of the surface material and a thermal contrast of only about 5 K. They are in several instances associated with ground ice areas, some are nearly circular in horizontal extent, and the largest has periodic cusps along internal bands and near its edge, characteristic of condensate clouds. The small thermal and albedo contrasts across the clouds imply that they are both low and thin; they are almost certainly water ice clouds, indicating that the lower atmosphere is saturated with H_2O.

The T_{15} measurements of the atmosphere covered latitudes from 72° to 82°N and were 168 ± 3 K throughout this region. For the slant geometry of these observations, T_{15} refers to the 0.5-mbar pressure level, about 25 km in altitude, yielding an average lapse rate over the dark regions of −2.8 K per kilometer. This is well below the adiabatic lapse rate, even ignoring a possible boundary layer, which indicates a largely stable atmosphere, although the clouds suggest some convection in the first few kilometers. The relatively high temperatures of the water ice and the atmosphere are conducive to relatively large amounts, by martian standards, of H_2O vapor in the atmosphere. High water vapor abundances were measured in simultaneous observations by the near-infrared spectrometer on the Viking orbiter (*10*).

Once a high-albedo deposit is established in the polar region, its reduced absorption of sunlight will allow it to maintain a lower temperature than unfrosted areas, and it can grow by trapping water vapor from the atmosphere at the expense of partially dehydrating warmer areas. This positive feedback behavior explains the sharp albedo and temperature contrasts of the polar region. Intrinsic to this exchange is a wind system between the frosted and dark areas, as would be expected from the 30 K temperature contrast between these two materials in the summer.

The constancy of the frost patches over 5 years and the large temperature contrasts in the polar area suggest that the water ice deposits are fairly thick, although direct measurements are not yet available (*11*). Ice thicknesses between a few centimeters, as required to extinguish the solar flux which would otherwise reach the underlying soil, and about 1 km, the presumed depth of the craters occupied, would not directly conflict with any existing observations. If the north polar ice averaged 10 m thick, it would represent 1000 times the amount amount of H_2O in the entire martian atmosphere.

Persistence of the water ice areas should be expected, as complete H_2O exchange with the saturated atmosphere on a daily basis could only move about ½ cm of ice in a martian summer. Since there is almost certainly much more H_2O in the polar caps than in the atmosphere, the polar frost areas are probably stable as long as the annual climate is not changed; it is likely that they persist over periods at least as long as half the shortest orbital precessional cycle (*12*). Because of the positive feedback discussed above, there is a potential for building very thick ice deposits. Based on the surface and atmospheric temperatures measured by the IRTM, a standard atmosphere over the dark regions would contain about 5 mg cm^{-2} of water vapor. If the outlying frost areas trap 10 percent of this water vapor on a daily basis during midsummer, they would accumulate 50 m of ice in 100,000 years.

While water ice may accumulate at the polar surface over long periods, the summertime polar environment reported here does not favor a permanent CO_2 deposit. At 5°N, 81°W, near the middle of a reservoir of solid CO_2 proposed by Murray and Malin (*3*), the measured brightness temperature was 216 K; this is 68 K higher than solid CO_2. This temperature is typical of the summer polar cap environment, and such a temperature difference for about one-fifth of the martian year represents the thermal load that any permanent CO_2 deposit would have to accommodate. It is unlikely that small areas of frozen CO_2, below the resolution of the IRTM, could exist in the polar environment observed; the temperatures would not even allow a CO_2 hydrate clathrate. The IRTM observations indicate that all of the polar condensate at this season is H_2O.

HUGH H. KIEFFER
University of California, Los Angeles 90024
STILLMAN C. CHASE, JR.
Santa Barbara Research Center, Goleta, California 93017
TERRY Z. MARTIN
University of California, Los Angeles
ELLIS D. MINER
Jet Propulsion Laboratory, California Institute of Technology, Pasadena 91103
FRANK DON PALLUCONI
Jet Propulsion Laboratory

References and Notes

1. Temperatures near 150 K, the saturation temperature of CO_2 at the mean martian surface pressure, were measured by the Mariner 7 infrared radiometer [G. Neugebauer, G. Münch, H. H. Kieffer, S. C. Chase, Jr., E. D. Miner, *Astron. J.* **76**, 719 (1971)]. The Mariner 7 infrared spectrometer gave independent spectroscopic evidence of solid CO_2 [K. C. Herr and G. C. Pimentel, *Science*, **166**, 496 (1969)].
2. A. P. Ingersoll, *J. Geophys. Res.* **79**, 3403 (1974).
3. B. C. Murray and M. C. Malin, *Science* **182**, 437 (1973).
4. G. A. Briggs, *Icarus* **23**, 167 (1974).
5. A climatic instability involving CO_2 polar caps has been proposed by C. Sagan, O. B. Toon, and P. J. Gierasch [*Science* **181**, 1045 (1973)]. Gas storage in the soil is discussed by F. P. Fanale and W. A. Cannon [*J. Geophys. Res.* **79**, 3397 (1974)].

6. The analysis of Mariner 9 radio occultation data by P. M. Woiceshyn, [*Icarus* **22**, 325 (1974)] indicates that north of 75°N the average surface pressure is 0.4 mbar greater than the martian average. A topographic map of the same region by D. Dzurisin and K. R. Blasius, [*J. Geophys. Res.* **80**, 3286 (1975)] indicates that the surface of the residual cap is on the average about 4 km higher than the unfrosted areas.
7. The Mariner 9 infrared radiometer and infrared interferometer-spectrometer did not have adequate spatial resolution to demonstrate that the frost areas of the residual north polar cap were not near 150 K. The global dust storm that occurred during the Mariner 9 mission caused considerable opacity in the atmosphere over the residual south polar cap, preventing a definite determination of the residual frost temperature.
8. The IRTM experiment is described in H. H. Kieffer, G. Neugebauer, G. Münch, S. C. Chase, Jr., E. D. Miner, *Icarus* **16**, 47 (1972); H. H. Kieffer, S. C. Chase, Jr., E. D. Miner, F. D. Palluconi, G. Münch, G. Neugebauer, T. Z. Martin, *Science* **193**, 780 (1976).
9. A buried solid CO_2 cap, which is sealed off from the atmosphere and has survived from some other climatic period, cannot be ruled out by these (and most other) observations. Although this has been suggested (*4*), a detailed theoretical treatment of its possible stability remains to be developed.
10. C. B. Farmer, D. W. Davies, D. D. La Porte, *Science* **194**, 1339 (1976).
11. On 30 September 1976, the inclination of the Viking 2 orbiter was increased to 75°. The improved viewing of the polar region by all three orbiter instruments should allow resolution of many detailed questions about these intriguing areas.
12. W. R. Ward [*J. Geophys. Res.* **79**, 3375 (1974)] has made a detailed analysis of the variation of the martian orbit and spin axis direction. The shortest period appreciably influencing the polar climate is the 175,000-year precession of the equinoxes.
13. J. Bennett has made major contributions to processing IRTM data. The assistance of P. Christensen, B. Jakosky, and A. Peterfreund, officially but totally inadequately described as data aides, is gratefully acknowledged. The imaging observations were provided by the Viking Orbiter Imaging Team, led by M. H. Carr. Financial support was provided by the NASA Viking Project Office.

18 October 1976

10

Reprinted from *Science* **194**:1329–1337 (1976)

NORTH POLAR REGION OF MARS: IMAGING RESULTS FROM VIKING 2

James A. Cutts, Karl R. Blasius, Geoffrey A. Briggs, Michael H. Carr, Ronald Greeley, and Harold Masursky

Abstract. *During October 1976, the Viking 2 orbiter acquired approximately 700 high-resolution images of the north polar region of Mars. These images confirm the existence at the north pole of extensive layered deposits largely covered over with deposits of perennial ice. An unconformity within the layered deposits suggests a complex history of climate change during their time of deposition. A pole-girdling accumulation of dunes composed of very dark materials is revealed for the first time by the Viking cameras. The entire region is devoid of fresh impact craters. Rapid rates of erosion or deposition are implied. A scenario for polar geological evolution, involving two types of climate change, is proposed.*

On 30 September 1976, Viking 2 executed an orbital plane change maneuver modifying the inclination of the Viking 2 orbit from 52° to 75°. This maneuver made the entire north polar region of Mars visible to the Viking 2 spacecraft under favorable lighting conditions through an atmosphere of widely scattered clouds. This report presents some preliminary interpretations of the geological features in the north polar region gained from study of the orbital imaging data returned to Earth and processed by 1 November 1976.

Intense interest in observing the north polar region of Mars with the Viking orbiter cameras was stimulated in 1971–72 by high-resolution (200 m) Mariner 9 observations of the south polar region (*1–5*) as well as by moderate resolution (up to 600 m) observations of the north polar region (*6*). Briefly, these observations revealed a complex mass of eroded layered deposits near the martian south pole and strongly suggested the existence of a similar terrain near the north pole. The layered polar deposits provide the first persuasive geological evidence for cyclical climatic change on a planet other than Earth. Viking 2, with an improved imaging system relative to Mariner 9 (*7*) and a capability for extensive multiframe stereoscopic and colorimetric coverage, has provided an outstanding opportunity for examining this evidence in much more detail. This opportunity began at the time of the Viking 2 orbital plane change and will end, at latest, when the region passes over the terminator during the northern autumn (the fall equinox is 4 January 1977). It is quite possible, however, that the onset of the autumn-winter high latitude haze and cloud cover known as the polar "hood" may terminate surface visibility even earlier.

Geological framework. The polar terrains revealed so far by Viking 2 images can be conveniently classified into three types: layered deposits in the central polar region, a contiguous area covered by dunes, and a cratered plains surface that appears to underlie stratigraphically both of these units. A sketch map delineating the distribution of layered terrains, dunes, and cratered plains appears in Fig. 1, which also illustrates the areal extent of the photographic coverage available for study. The materials of the perennial ice cap, identified as water ice by Viking 2 temperature measurements (*8*), occur primarily within the perimeter of the layered terrains, although isolated patches or outliers occur in physical contact with both the cratered surface and the dunes. The mosaic of long-range oblique Viking 2 images (Fig. 2) obtained before the plane change illustrates the pinwheel pattern of frost-free areas within the predominantly frost-covered layered deposits.

Characteristics of layered deposits. The layered deposits contribute more than any other feature to the geological distinctiveness and significance of the polar regions of Mars. We suspect that these layered deposits are exposed at the surface in all areas mapped in Fig. 1 where frost is absent. They are manifested on slopes by a parallel striping of the surface (Fig. 3) (*9*); however, in only a fairly small number of locations can the topographic and geological nature of the exposures be clearly discerned. In one of these areas (Fig. 3b), selected so that shading and shadowing dominates locally over brightness variations caused by albedo, a terraced slope can be recognized.

With the use of arguments presented previously for features viewed in the south polar region (*2*), a strong case can be made that erosion of a layered deposit accounts for the origin of the terraced slope. The lateral continuity of terraces and their uniformity in height suggest that the individual layers that gave rise to the terraces extend as continuous thin sheets over areas of several thousand square kilometers. Moreover, these individual sheets are not only quite constant in thickness, but the thicknesses of different sheets are rather similar to one another. The apparently unique occurrence of layered deposits in the polar region implies meteorological control over their formation. Direct deposition of dust from the atmosphere, perhaps influenced by the distribution of polar ice and modulated by climatic change, remains the most probable mechanism of accumulation. These climatic changes may have resulted from periodic, perturbation-induced changes in the orbital elements of Mars and the direction of its rotation axis (*5*).

Another distinctive attribute of the layered deposits on a planet characterized by many ancient crater-pocked surfaces is the absence of fresh impact craters. In an area of 800,000 km^2, no fresh craters can be recognized. Much of this surface is so smooth and free of other forms of topographic relief that there should be no difficulty in detecting craters down to 300 m in diameter. Similar results for the south polar region were reported from the Mariner 9 data (*1*, *2*). There are some circular structures in the size range of 2 to 8 km which may be the remnants of impact craters (Fig. 4). The production rate of impact craters this large on the planets is expected to be several orders of magnitude below the production rate of craters in the size range of 300 m to 1 km. Consequently, if these circular structures are impact craters, then the efficiency of crater obliteration in this part of Mars must vary dramatically with crater size.

Many of the frames taken by Viking merely confirm what was already known about the characteristics of layered deposits from Mariner 9 observations of the south polar region and what was strongly suspected about their characteristics in the north polar region. However, even from our preliminary analysis, we are able to report some significant new results regarding the erosional history of the layered deposits and their relationship to the formation of the present polar ice deposits.

Unconformable contact of buried landscape type. In Fig. 5, there appear several examples of unconformable contacts of the buried landscape type as evidenced by one set of terraces obliquely truncating another. The topography and stratigraphy of this relationship is illustrated by the block diagram and cross section in Fig. 6. Such relations indicate at least one major erosional interruption in the deposition of layered materials in the north polar region.

We propose that the unconformable contacts were brought about by the following series of events. First, a series of near-horizontal layers were deposited, obscuring all earlier topography and culminating in a flat or gently sloping surface. Then erosion set in, and differential erosion of the layers caused slopes to assume a terraced topographic form. The next event was another episode of deposition of layered materials which obscured the relatively small-scale terracing but otherwise conformed to the relief of the eroded slopes. Finally, another episode of erosion resulted in the formation of a set of terraces both in the most recent layered deposit and in the older deposit, perhaps by exhumation of an old buried terrace system. The two sets of terraces meet obliquely, thereby revealing the unconformable contact.

It is possible that all the unconformities of Fig. 5 were produced in the same erosional episode. If this is true, it implies a minimum of two erosional episodes in the north polar region: an earlier one associated with the unconformities and a later one responsible for the most recent development of terracing. The origin of these erosional episodes is quite uncertain.

Dune fields of the north polar region. A vast belt of dunes occupies much of the region around the permanent north polar cap (Fig. 1). Much of this belt consists of a continuous mantle of dune-forming materials (presumably sand) in which regularly spaced ridges with moderately sinuous crests are developed. Both longitudinal dunes, formed by winds directed parallel to the ridge orientation, and transverse dunes, formed by winds acting perpendicular to the ridge orientation, have been identified. In some areas (Fig. 7a), the ridge spacing and orientation are almost invariant over distances of more than 100 km; this suggests a similar consistency in the strength and direction of the wind. In other areas (Figs. 7b and 8), ridge spacing and orientation change rapidly from place to place, and the sinuosity and degree of branching and merging of the dune crests are enhanced. These characteristics indicate both greater variability in the direction of strong winds at any one place and also significant variations in the mean wind direction with location. Occasional strong winds from more than one direction are indicated by a secondary set of ridge alignments oriented obliquely to the primary ridge pattern (Fig. 7a).

On the periphery of the "sea" of transverse dunes is a transitional zone where separation and direction of ridges changes rapidly and the continuous mass of dune-forming materials breaks down into longitudinal dunes, barchans, and other approximately equant dunes whose detailed shape is not resolved (Fig. 8). Some of these dunes may be in the process of accreting to the main dune mass; others may be breaking away from it. Such a relationship is most striking within and near the mouth of Borealis Chasma (Fig. 9). A narrow stream of dune-forming materials appears to be migrating along a gently curving valley southward until it joins a large delta-shaped dune mass distinguished by highly sinuous ridge patterns. On the other

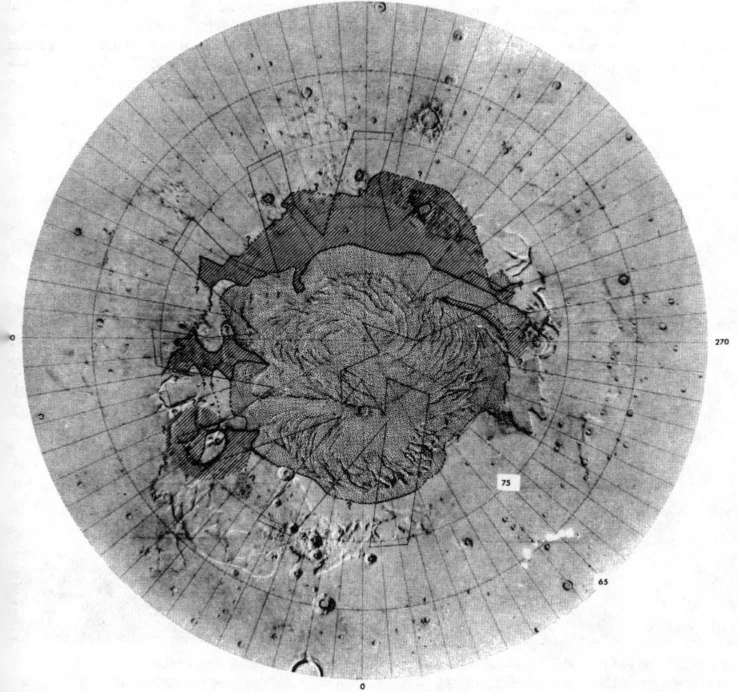

ig. 1. High-resolution photographic coverage obtained by Viking 2, up to and including data eturned and processed by 1 November 1976, is outlined by fine line. Terrains have been classied into three types: layered deposits (stipple), dunes (stripes), and cratered plains (these definiions are clarified in the text). The distribution of permanent polar ice deposits is not indicated.

side of this large reservoir of sand, the continuous pattern of dunes dissolves into a tenuous skein of linear dunes and isolated barchans. This assessment of the dynamic geological situation is preliminary; it is possible that the general pattern of migration is taking place in the opposite direction from the one contemplated above.

The north polar dune fields have a complex relationship with other topographic features and terrains. Moderate-sized craters can serve to partly anchor the dunes in a discontinuous pattern of ridges (Fig. 8a) that may only partially mantle the surface. Where the cover of dune-forming materials is continuous, buried crater rims may be expressed as subtle indentations in the otherwise smooth sinuosity of the dune crests (Fig. 8b). This can provide some indication of the depth of the dune mass if reasonable estimates of the crater rim relief can be derived. Dunes extend up and over the layered deposits in places (Fig. 10). There they clearly postdate both the deposition and erosion of layered materials.

What materials are these north polar dunes made of, from what source are the materials derived, and what accounts for their accumulation in this 80°N latitude belt as dunes of prodigious thickness and extent? A plausible model is that the source of dune-forming materials is the layered deposits. The morphology of these surfaces indicates that erosion has occurred, and it seems reasonable that a portion of the eroded material could have formed dunes. A difficulty with this idea is that the layered units are believed to have formed by the deposition of suspended dust particles. Dust, however, is much too fine to be mobilized by saltation, the transport process which is generally thought to be required for dune formation. There are ways around this objection. One could argue that dust in the polar regions has accreted to sizes large enough for saltation to operate by the process that is responsible for duricrust (*10*) formation. Alternatively, one might argue that the regimes of suspension and saltation are not so rigidly separated on the basis of particle size on Mars as they are on Earth.

Neither of these arguments appears too convincing; moreover, the circumpolar dunes display a substantial albedo contrast relative to the cratered plains and the layered deposits. Their darker appearance suggests a different composition, perhaps similar to the spectacular dark dune masses photographed elsewhere on Mars (*11*). It may be that these dune materials are derived directly from the soils of lower latitudes instead of indirectly after having first resided in the polar layered deposits. These materials may represent the missing compositional component inferred to exist from elemental deficiencies in the soils at the Viking

Fig. 2. An oblique view of the north polar cap of Mars, acquired before the Viking 2 orbital plane change, shows part of the spiral pattern of dark, frost-free bands, wherein slopes are locally steeper than the surroundings. Higher-resolution images show complex fine-scale parallel light and dark bands within these areas (Viking 2 frames 22B33 to 22B37). The region shown extends from about 80°N to 90°N (at the planet's limb). Each frame in the mosaic is 185 km wide.

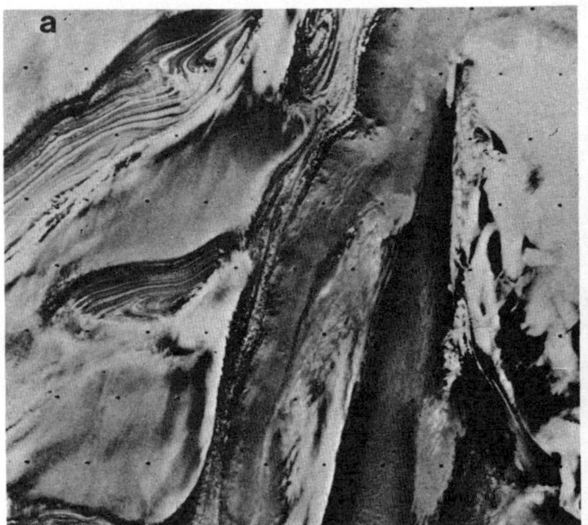

Fig. 3. Layered deposits in the north polar region of Mars. (a) Parallel banding indicative of layered deposits in the north polar region. Tonal contrasts due to albedo markings are difficult to separate from those due to differences in illumination. The topographic nature of this surface is still obscure (Viking 2 frame 59B77). The region shown is about 65 by 100 km. (b) The terraced character of some of the layered deposits is revealed in this image, in which the sun is illuminating the scene from the bottom (Viking 2 frame 57B32). The region shown measures 35 by 5[illegible] km. The frost-covered upland area at the top of the frame is separated from the dark frost-free lowland by terraced slopes, which are highlighted where they face the sun. This image has not been filtered, and different features in the scene appear in their true relative contrasts.

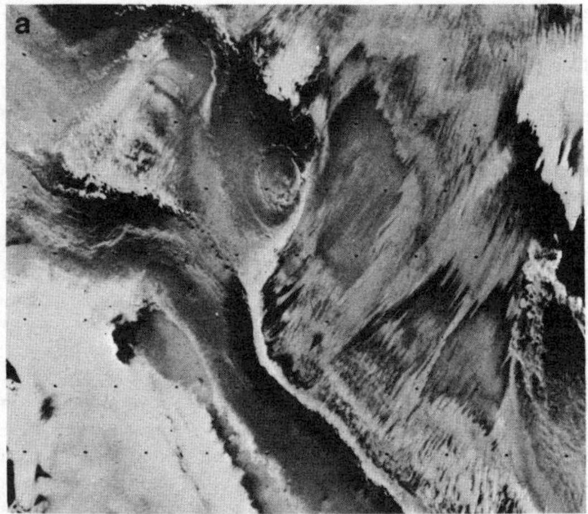

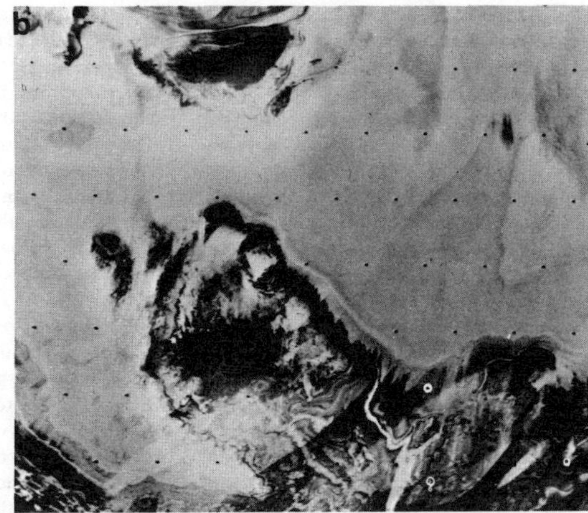

Fig. 4. These frames reveal no indication of small fresh impact craters on the layered deposits but the larger circular structures may be of impact origin. (a) The candidate impact structure is comprised of three or four distinct circular raised ridges visible just above center (Viking 2 frame 59B78). (b) The candidate impact structure is comprised of several bright markings which nearly form a closed subcircular form tangential to the edge of the ice cap (Viking 2 frame 56B79).

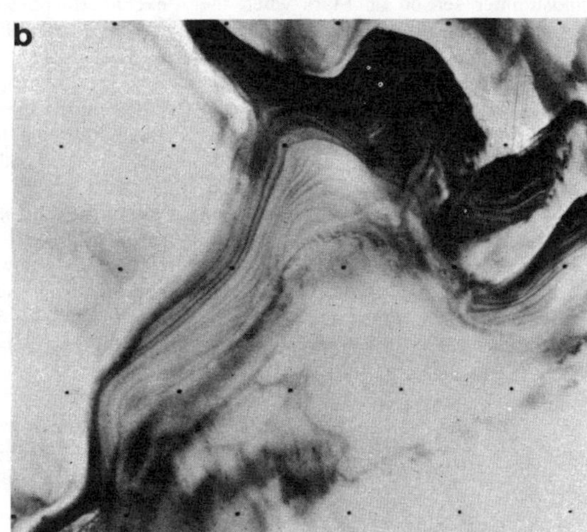

Fig. 5 (above). Unconformable contacts of the buried landscape type found within the north polar layered deposits. (a) One series of contacts is seen near the upper right center; another series, near the lower left edge of the frame. A replicating angular pattern of dark markings (see text) is located near the lower right corner. Bright searchlight patterns also appear in the frame and extend from the lower right corner toward the center of the frame. Most of the area in this frame is ice-covered (Viking 2 frame 56B84). (b) An enlarged view of the first series of unconformities. The clearest example is near the center of the frame, where the uppermost terraces, which have a rather irregular form at this point, obliquely truncate terraces at a lower level. This enlarged view has not been filtered, and different features in the scene appear in their true relative contrasts.

Fig. 6 (right). A block diagram indicating the three-dimensional structure of the topography and sedimentary layering that appears near the center of Fig. 5b. The view is similar to the one that an observer located on the surface of Mars near the apex of the deposits at top center would obtain by looking in the direction of 6 o'clock. The cross section shows the sequence of layers from which the terrain is believed to be constructed.

landing sites (*12*). Dunes on the layered deposits within the polar cap may be quite different in composition from the circumpolar dunes; they may be largely composed of particles of water ice.

Cratered plains unit. The cratered plains unit provides the substrate upon which the layered deposits and dunes have formed. Portions of this terrain surface can be observed outside the area covered by layered deposits and dunes. An analogous unit in the south polar region—the pitted plains unit (*1*)—is characterized by extensive pitting and grooving, which has been attributed to early aeolian activity (*1–3*) preceding the formation of layered deposits. No clear evidence of comparable north polar aeolian activity has been recognized in Viking 2 pictures, but the imagery of the cratered plains is somewhat affected by haze and is very limited in areal coverage (Fig. 1).

Characteristics of permanent ice. The observations of the north polar region being reported here were obtained in the midsummer season on Mars when the areocentric longitude of the sun relative to vernal equinox ranged from 130° to 140°. Earth-based observations (*13*) indicate that the polar cap should have reached its minimum size 2 or 3 months previously. Consequently, we expected to observe a deposit of perennially frozen volatiles, and the Viking 2 infrared radiometer observations (*8*) suggest that water ice and not frozen carbon dioxide is the dominant if not the sole component.

Perennial ice is observed in isolated patches on the cratered plains and dunes but covers the major part of the layered deposits (Fig. 1). On the layered deposits the ice follows the spiraling pinwheel configuration of the terrain, covering the flat areas but shunning the slopes (Fig. 2). The calculated albedo of the ice (*8*) is much lower than expected, and it appears that the albedo of ice is reduced significantly below the value for pure ice as a consequence of admixture with dust.

Relationship between perennial ice and dunes. The dune pattern in the circumpolar belt displays two distinct kinds of relationships to perennial ice. In the first, the pattern of ridges is unmodified by the presence of ice (Fig. 7a), which has apparently formed after the dune ridges assumed their present configuration. This relationship suggests that water ice is either accumulating or being redistributed in the north polar region on a time scale that is short compared to the age of the dune ridges. In fact, we cannot exclude the possibility that some of these ice deposits sublime later in the summer season.

Near other frost patches, the pattern of the dune ridges is clearly altered near and within the frost. This could imply that the perennial ice in that area preceded the most recent change in the distribution of the dunes. Better definition of the relationship of frost patches to the dunes will be possible when color images of the dune fields are acquired.

Relationship between perennial ice and layered deposits. As judged from our analysis of monoscopic imagery only, perennial ice appears to occupy primarily the flat-lying areas between sinuous and curving terraced slopes. In many instances (for example, Fig. 4b), the perimeter of the frost is parallel to the margin of the uppermost layer on the slopes. It is tempting to speculate that the present distribution of frost is controlling the accretion of material that forms the layers. If this were the case, then the terraces formed at the same time as the layers with which they are associated. However, there is evidence to contradict this model. In Fig. 5b, for example, perennial ice covers the unconformity between the most recent series of layers and an older series. Consequently, the notion that the terraces on a given slope form during an erosional episode after the deposition of the series of corresponding layers was completed is the more probable explanation of the topography.

The surface of the perennial ice is devoid of craters and any other topographic forms over the major part of the interior of the ice cap. Locally, however, there are small areas of possible dune fields, and near the margins of the ice cover there are a variety of unusual patterns in, or on, the ice. One class of phenomenon is informally known as "searchlight patterns" (Fig. 5a). These bright elongate patches are bounded by abrupt margins that sustain a linear trend for 100 km or more, maintaining a parallel or slightly diverging aspect. Wind erosion or deposition seems an obvious ex-

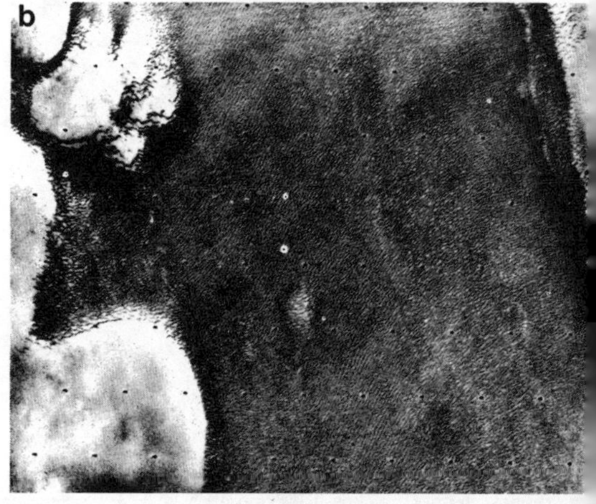

Fig. 7. Fields of transverse sand dunes from an area peripheral to the north polar ice cap. (a) Dunes here have a consistent trend (approximatel north-south) with minor sinuosity, branching, and merging. Vague circular forms are probably buried craters. Bright spots within the ridges ar ice deposits (Viking 2 frame 59B32). (b) Dunes with much more variation in direction also occur: a shorter wavelength and greater sinuosit appear in this dune field, which adjoins and in places appears to be mantled by frost deposits. Vague circular forms again are probably burie craters. The bright patches of ice near the upper left are associated with a distinct change in the dune pattern, possibly indicating that the deposit of ice preceded the development of the present dune pattern (Viking 2 frame 58B01).

planation for these features, although the exact mechanism is unknown. Linear dark markings also form angular patterns on the ice; Fig. 5a includes an example of two almost identical patterns separated by about 3 km. We have considered the possibility that lateral motion of the upper portion of a single feature caused this apparent replication, but insufficient information is known about the topography to pursue this idea in any detail. We expect that the higher-resolution pictures available after conjunction will help us elucidate the nature of features such as these.

Dynamics of the annual and perennial ice. No changes in the frost cover have been observed over the short time base of Viking observations. However, duplicate coverage has been very limited in areal extent. Model studies suggest no significant changes in frost cover in this season, and so the negative result is not surprising.

The recent elucidation of the water ice composition of the northern polar cap by Viking 2 temperature measurements now causes us to reexamine some earlier conclusions (*1*) about the annual behavior of polar frosts on Mars. A region near the edge of the permanent southern cap, known informally as "the fork," was monitored from near the beginning of the Mariner 9 mission, the first third of southern summer, to near the fall equinox (*13*). During this time the ice was observed to retreat continuously both at the edges and patchily within the ice mass. It is likely that both residual caps have the same water ice composition (*13*) and that the observations relate to the disappearance of seasonal water ice. If the water ice is laid down in the same proportion to CO_2 as the mixing ratio in the atmosphere, then about 10^{-2} g cm^{-2} would be deposited seasonally and would remain until all the seasonal CO_2 has sublimed. An upper limit for the amount of water ice that could be deposited at the residual cap would be the total water vapor in one hemisphere, assumed to average 20 precipitable micrometers: a total of 0.2 g cm^{-2} for a cap covering 1 percent of the hemispheric surface area. This seasonal accumulation of water ice is presumed to sublime gradually (since the air will rapidly saturate) during the summer. The conclusion may be reached that a very small amount of water ice is capable of providing the appearance of an unbroken ice sheet. Thus, the observation of regions within the residual caps where the cover is uniform does not necessarily imply great thickness. The thickness of water ice might be so small that it is transferred entirely from one pole to the other during the cyclic precession of the equinoxes (period, ~ 180,000 years). The southern summer solstice occurs near perihelion, so the southern cap presently experiences considerably more insolation during the summer than does the

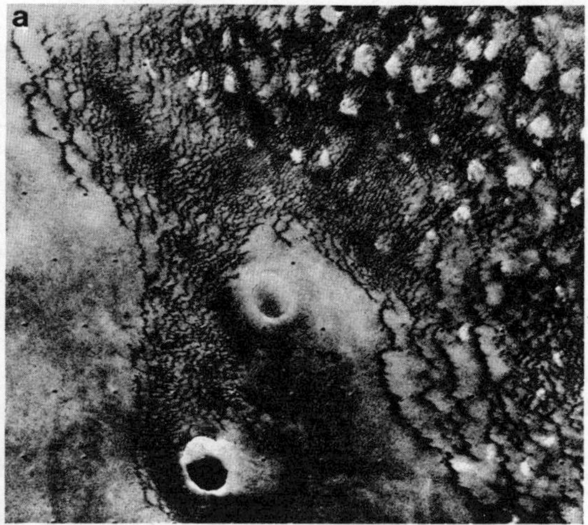

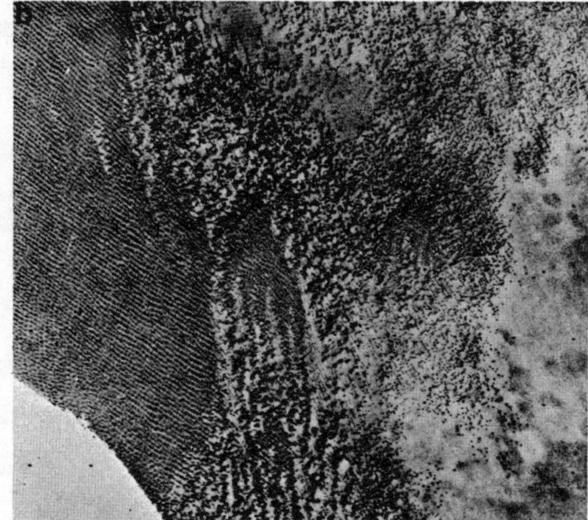

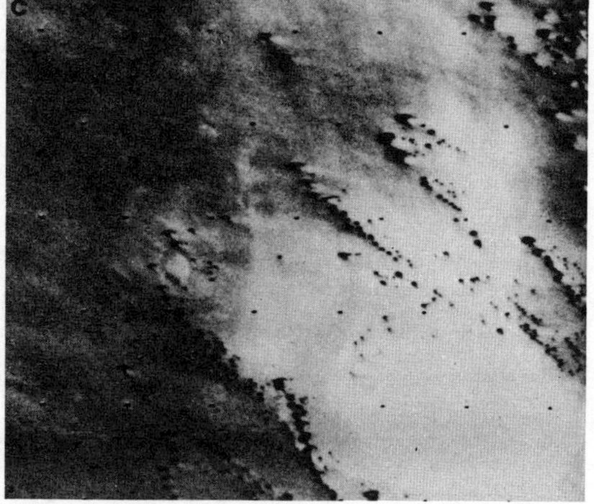

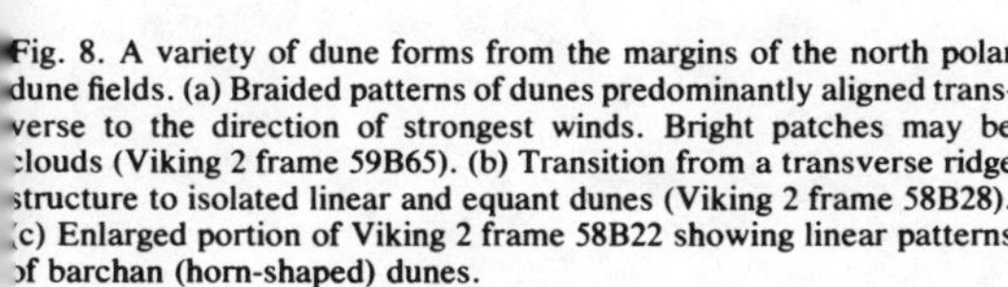
Fig. 8. A variety of dune forms from the margins of the north polar dune fields. (a) Braided patterns of dunes predominantly aligned transverse to the direction of strongest winds. Bright patches may be clouds (Viking 2 frame 59B65). (b) Transition from a transverse ridge structure to isolated linear and equant dunes (Viking 2 frame 58B28). (c) Enlarged portion of Viking 2 frame 58B22 showing linear patterns of barchan (horn-shaped) dunes.

northern cap. The relatively small size of the southern cap and its broken and patchy appearance at the end of summer may be due to the gradual transfer of water to the northern cap.

Huguenin (*14*) has discussed the photochemical weathering of surface materials by oxidation, in which atmospheric water vapor would be irreversibly lost. The rates of loss derived by Huguenin would have significant implications for the polar caps. Water released from the caps during the summer months would be used to replenish the atmospheric water lost to weathering, and the caps would thereby be diminished in thickness. Oxygen removal at a rate of 10^8 to 10^{11} atom cm^{-2} sec^{-1} (*14*) implies water vapor removal at 10^{-7} to 10^{-4} g cm^{-2} $year^{-1}$. If all this were supplied by the polar caps (areal extent, ~ 1 percent of the surface area of Mars), then the caps would be depleted at 10^{-5} to 10^{-2} g cm^{-2} $year.^{-1}$ A meter of ice could be lost in 10^4 to 10^7 years. If the caps are indeed thin enough to transfer between poles over the equinoctial cycle, then it is likely that they have a total lifetime that is very short in geological terms. In that event, the present residual caps may represent a relatively recent outgassing event (or even current outgassing at a rate only slightly greater than the weathering rate) or, less plausibly, a recent cometary impact. The Viking results are unlikely to resolve the question of ice cap thickness, however, and new remote sensing or in situ data will probably be required.

Crater retention ages in the layered deposits, dunes, and perennial ice. No fresh impact craters have been recognized on the circumpolar dune fields or the layered deposits on either ice-covered or ice-free areas. Approximately ten fresh craters with sizes large enough for detection (300 m) would be expected to form in an area this large (~ 10^6 km^2) every million years if one scales the present lunar cratering rate to Mars (*15*). These craters would be easier to recognize in some areas, the flat ice cap for ex-

Fig. 9. A dune field in Borealis Chasma. Dark dune-forming materials have apparently been transported away from the pole in a curving stream extending from the top of the frame. They are accumulated in an approximately triangular dune mass that occupies the center of the mosaic. The average trend of the sinuous ridges in the dune mass rotates in a clockwise direction through an angle of approximately 45° from the northern to the southern margin. The discontinuous dark texture on the right side arises from partial dune cover. Perennial ice is visible near the top of the frame and associated with the crater near the bottom of the frame. The bright patch near center right may be cloud (mosaic of Viking 2 frames 58B21 to 58B34).

Table 1. Outline of geological evolution of north polar region of Mars.

Stage 1	Onset of polar activity Moderate aeolian modification of the ancient volcanic terrains
Stage 2	First depositional period Layered deposits of silicate dust and possibly interbedded ice accumulate to a thickness of several kilometers
Stage 3	First erosional period Erosional attack of layered deposits results in a landscape of gently curving scarps and channels with terraced slopes
Stage 4	Second depositional period More layered deposits accumulate unconformably on top of the units formed in the first depositional period
Stage 5	Second erosional period Further erosional attack of layered deposits results in exhumation of earlier formed landscapes and reveals unconformable contacts between the deposits of the first and second depositional period. Some of the eroded material reaccumulates as a girdle of sand dunes between 75°N and 80°N
Stage 6	Recent period Ice in the permanent polar cap assumes its present form and distribution

ample, than in others such as the dunes. However, the absence of a population of impact craters indicates that the lifetime of fresh craters in this part of Mars is very much shorter than a million years. One possible explanation is that the landscape as a whole is changing at the scale of several hundred meters to a kilometer in time periods of less than a million years. An alternative possibility is that processes exist in the polar region which selectively destroy impact craters. For example, cavities in ice created by impact may "heal" at a faster rate than the surrounding landscape changes by the erosion or accretion of ice and dust.

We can estimate the deposition and erosion rates that are needed in the area of the layered deposits, for example, to remove craters at the rates implied by the estimates given above. The layered deposits occupy an area of 800,000 km^2; for this exercise we will not distinguish between the area covered by ice and the very much smaller area of frost-free ground. Using the observation that there is less than one fresh crater of 300 m or larger in an area of 800,000 km^2, one can compare this result with equilibrium populations $C_E(D)$ calculated on a variety of assumptions.

The equilibrium incremental population of craters of diameter D per 10^6 km^2 may be expressed as

$$N_E(D) = N(D) \cdot T_L(D) \qquad (1)$$

where $N(D)$ is the incremental crater production rate in craters of diameter D per 10^6 km^2 per million years, $T_L(D)$ is the crater lifetime measured in millions of years, and the cumulative equilibrium population, $C_E(D)$, is expressed as

$$C_E(D) = \int_D^\infty N(D')\, T_L(D')dD' \qquad (2)$$

If the crater lifetime is independent of diameter, then by using Eqs. 1 and 2 and the observed upper limit to the crater populations, the lifetime of craters 300 m and larger is calculated to be 0.12 million years if the lunar value for $N(D')$ is assumed.

A more realistic model for crater removal must incorporate a diameter dependence. With perhaps overly simplistic notions of crater obliteration by means of uniform removal of material from the surface or through uniform addition of material from the atmospheric suspension, the crater lifetime scales as the diameter D. Adopting incremental crater production rates proportional to D^{-4} below 1 km and D^{-3} above 1 km, one can calculate a rate of removal or addition of material to the surface based on the upper limit on crater populations. These rates are approximately 1 km per million years or 1 mm per year.

An alternative is that obliteration of craters may be caused by filling from a sheet of saltating material transported to and fro across the ice cap. There are indications of dune features on the ice cap which could imply pervasive saltation processes, perhaps involving mixtures of ice and silicate debris. In this case, crater lifetime varies as D^2 and rates of material transport are estimated to be 0.08 km^3 per kilometer of width per million years or 800 cm^3 per centimeter of width per year.

Although deposition and erosion rates calculated above are large, they are not implausibly large considering that materials have been observed to be redistributed over the surface on scales visible from orbiters in a matter of days or weeks (*16*). Perhaps, the only distinctions between polar regions and other parts of Mars where impact crater to-

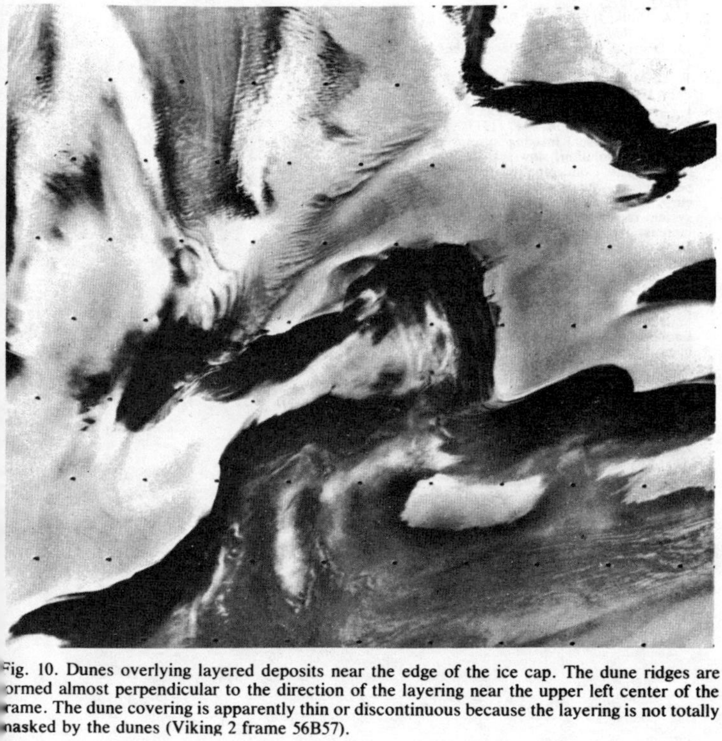

Fig. 10. Dunes overlying layered deposits near the edge of the ice cap. The dune ridges are formed almost perpendicular to the direction of the layering near the upper left center of the frame. The dune covering is apparently thin or discontinuous because the layering is not totally masked by the dunes (Viking 2 frame 56B57).

pography is more enduring are that in the polar regions the surface processes are well supplied with mobile material and are sustained for a much larger fraction of the time.

A scenario for the evolution of the north polar region. Our preliminary study of the north polar region geology leaves us with uncertainties and ambiguities concerning the events that have taken place there. Some of this confusion may well be resolved by future analyses of the imaging data. The scenario that is outlined in Table 1 is not a unique one—it may not even be the one that is most in harmony with the data—but it does offer a credible framework for the observations against which further observations and theoretical models may be tested.

Two distinct types of climate change are implied by this scenario. Climate changes of type 1 are associated with the fine-scale layering and are clearly cyclical with a relatively short and apparently regular period. They seem to involve fairly limited excursions in environmental conditions affecting, apparently, the rate of deposition. Climate changes of type 2, of which two episodes, at least, have been recognized so far, occur on a much longer time scale, perhaps two or even three orders of magnitude longer than that of type 1 changes. Type 2 climatic changes do not necessarily have a uniform duration or a regular period, and they seem to involve radical excursions in environmental conditions from a depositional regime to an erosional regime.

There is no evidence that changes in the polar climate have any connection with climatic changes postulated in connection with channel formation on Mars (*17*). The polar climatic changes indicated are probably much more subtle than the massive temperature and pressure changes required for fluvial activity in the equatorial regions. The martian channels are also reported to be extremely old (*18*) whereas erosion in the polar regions is clearly very recent. We cannot, however, exclude the possibility that accumulation of the layered deposits was contemporary with channel formation, but even if that were so, the erosion of the deposits postdated it. The layered deposits may nevertheless play an important role in the volatile history of Mars. Much of the water inferred to have been released from the planet's interior since its formation (*19*) could have been codeposited in the form of ice with dust in the polar regions.

Paradoxically, the events that are least certain in the history of the polar region are those that are closest to the present. The relationship of the layered units to the polar caps is not understood. Neither is the origin of the dune fields. The searchlight features that transect major topographic features without a change in direction are also a puzzle. We don't know whether the polar region is presently experiencing a depositional or erosional cycle. We cannot exclude the possibility that the evolution of the polar region is now controlled by a climatic regime quite distinct from those that have existed previously. If we are fortunate enough to recover higher-resolution imagery after the spacecraft passes the solar occultation period of superior conjunction, we may obtain insights into some of these tantalizing problems.

References and Notes

1. B. C. Murray, L. A. Soderblom, J. A. Cutts, R. P. Sharp, D. J. Milton, R. B. Leighton, *Icarus* **17**, 328 (1972).
2. J. A. Cutts, *J. Geophys. Res.* **78**, 4231 (1973).
3. ———, *ibid.*, p. 4211.
4. D. Dzurisin and K. R. Blasius, *ibid.* **80**, 3286 (1975).
5. B. C. Murray, W. R. Ward, S. C. Yeung, *Science* **180**, 638 (1973).
6. L. A. Soderblom, M. C. Malin, J. A. Cutts, B. C. Murray, *Icarus* **78**, 4197 (1973).
7. The Viking visual imaging system consists of two high-resolution, slow-scan television framing cameras. Conceptually similar to the Mariner camera systems used in previous Mars, Mercury, and Venus missions, the visual imaging system incorporates improvements designed to increase both spatial resolution and coverage. Each camera employs a 475-mm diffraction-limited telescope and a 37-mm-diameter vidicon, the central region of which is scanned with a raster format of 1056 lines by 1182 samples and produces a 1.54° by 1.69° field of view. The optical axes of the cameras are offset by 1.38°. Cameras are shuttered alternately, which results in contiguous swaths of images 80 km wide, with resolution better than 100 m near periapsis. Six color filters are available to restrict the image spectral bandpass to limited portions of the cameras' near-visual response characteristics. The camera systems are described in detail by J. B. Wellman, F. P. Landauer, D. D. Norris, and T. E. Thorpe (*J. Spacecr. Rockets*, in press).
8. H. H. Kieffer, S. C. Chase, Jr., T. Z. Martin, F. J. Palluconi, E. D. Miner, *Science* **194**, 1341 (1976).
9. All images published here are oriented with the north approximately toward the top of the frame and with the illumination directed from the bottom of the frame. Unless otherwise noted, the dimensions of individual frames lie in the range 60 to 100 km. Most images used have been filtered to display detail in both ice-covered and ice-free areas. Consequently, in many cases the brightness of an area in the processed image is not a reliable indicator of the presence or absence of ice cover.
10. The crustlike material observed by the Viking landers has recently been termed "duricrust," which avoids the implication of genesis that accompanied the earlier use of the term "caliche."
11. J. A. Cutts and R. S. U. Smith, *Icarus* **78**, 4139 (1973).
12. P. Toulmin III, B. C. Clark, A. K. Baird, K. Keil, H. J. Rose, Jr., *Science* **194**, 81 (1976).
13. G. A. Briggs, *Icarus* **23**, 167 (1974).
14. R. L. Huguenin, *Science* **192**, 138 (1976).
15. L. A. Soderblom, C. D. Condit, R. A. West, B. M. Herman, T. J. Kriedler, *Icarus* **22**, 239 (1974).
16. C. Sagan *et al.*, *J. Geophys. Res.* **78**, 4199 (1973).
17. C. Sagan, O. B. Toon, P. J. Geirasch, *Science* **181**, 1045 (1973).
18. M. C. Malin, *J. Geophys. Res.* **81**, 4825 (1976)
19. T. Owen and K. Biemann, *Science* **193**, 801 (1976).

20. The Viking 2 reconnaissance of the north polar region of Mars has benefitted from the dedicated efforts of large numbers of individuals. Among members of the orbiter imaging team and staff we thank especially K. Klaasen, T. E. Thorpe, R. Tyner, and J. B. Wellman for their untiring support of the polar photography effort. Financial support for the work of team members was provided by NASA through the Viking Project Office and the Office of Planetary Geology (R.G.).

Part IV

REMOTE SENSING OF THE SURFACE

Editor's Comments on Papers 11 Through 14

Until the Viking landings, earth-based telescopic spectral reflectance measurements provided one of the few available techniques for acquiring data about the chemical composition of the martian surface. While most researchers concede that the reddish coloration is produced by iron III oxides, disagreement has centered on whether limonite is a major soil constituent (Dollfus 1961; Sharanov 1961; Draper et al. 1964; Tull 1966) or merely a very thin coating on silicate grains (Van Tassel and Salisbury 1964; Younkin 1966; Salisbury and Hunt 1968, 1969). The Viking X-ray fluorescence data support the latter hypothesis (Toulmin et al. 1976; Clark et al. 1976). Lander spectral measurements are also consistent with the presence of nontronite (iron-rich montmorillonite; see Huck et al. 1977).

Another significant problem has been to establish the cause of the difference between light and dark regions, which is of interest because of the seasonal changes and once-popular vegetation hypothesis. Pollack and Sagan (Paper 11) propose that the albedo markings are created by variations in particle size and not chemistry. (Note that because of its length, only the conclusions of their paper are reprinted here.) Other support for this hypothesis comes from the relative mobility of the dark areas (Sagan et al. 1972; 1973). The alternate viewpoint is expressed by McCord, Elias, and Westphal (Paper 12) who demonstrate that the two albedo terrains possess distinctive spectral curves that imply chemical differences. These may simply represent variations in the degree of oxidation—the brighter areas being more highly oxidized (Adams and McCord 1969). Additional spectral data supporting a diversity of compositional types may be found in McCord et al (1977a, b). Probably both factors—grain size and soil chemistry—are involved. It would not be unreasonable to expect that chemical fractionation of eolian deposits according to grain size has occurred on Mars, as on Earth.

The IR spectrometer analysis of atmospheric dust from the 1971 dust storm has yielded another approach to the determination of surface composition. Hanel et al. (Paper 13) attribute the broad spectral features at 480 and 1200 cm^{-1} to dust having a SiO_2 composition of about 60 per cent, from which they conclude that "geochemical differentiation of Mars has occurred." While an even higher silica content has been inferred from the dust spectra (Aronson and Emslie 1975; Toon et al. 1977), in situ X-ray fluorescence analysis (see Paper 16, Part V) shows no more than 45 per cent SiO_2 in the soil. If atmospheric dust were predominantly composed of clay minerals, the increased SiO_2 content of the dust could be understood.

Reports of spectral features near 3μ, indicative of condensed water, have appeared in the literature (Sinton 1967; Houck et al. 1973). Pimentel, Forney, and Herr (Paper 14) discuss the distribution of bound water in the martian surface and further distinguish between ice and hydrated minerals. Another interesting finding is the latitude dependence of ice adsorption and the pronounced changes at the edge of the south polar cap that mark the condensation of ice.

The Viking Lander gas chromatograph–mass spectrometer recorded up to 1 per cent water of hydration released between 200° and 350°C in the soil samples (Biemann et al. 1976b, 1977). Possible hydrated minerals include goethite, clays, or sulfates.

11

Reprinted from *Space Sci. Rev.* **9**:289–295 (1969)

AN ANALYSIS OF MARTIAN PHOTOMETRY AND POLARIMETRY*

JAMES B. POLLACK** and CARL SAGAN**
Harvard University and Smithsonian Astrophysical Observatory, Cambridge, Mass., U.S.A.

[*Editor's Note:* In the original, material precedes this excerpt.]

5. Discussion of Results

5.1. Summary of Conclusions

We first summarize our basic conclusions and the data on which they are based. The sharp rise in reflectivity between 4500 and 6500 Å (Section 2.1.2), and the distinct negative branch at small phase angles in the polarization curve (3.4) imply independently that both bright and dark areas are composed of fine powders or of extremely porous rocks. This result is consistent with the low thermal inertia of these areas, and with the low radar reflectivity of the bright areas. The existence of Martian clouds that arise in the bright areas and have photometric properties similar to those of the bright areas suggests that the surface material is a powder rather than an extremely porous rock.

The spectra of the bright and dark areas are controlled between 3000 Å and 11000 Å mainly by their ferric oxide content. This is indicated by the low, nearly constant reflectivity between 3000 and 4500 Å, by the steep slope between 4500 and 7000 Å, and by the plateau between 7000 and 11000 Å; all fit well by goethite once a particle size is chosen that is appropriate to the absolute value of the reflectivity at one wavelength (Sections 2.1.3, 2.1.4). There appears to be no other geochemically abundant material strongly enough absorbing in the blue to reproduce this spectrum. Ferric oxide polyhydrates also match the polarization curves of bright and dark areas extremely well (Sections 3.1, 3.4.2). Because of calibration problems in normalizing Martian to solar spectra, and because the feature has a wide-band low-strength character, present observations are not sensitive enough to test the presence of ferric oxides by searching for the limonite band near 9000 Å, which arises from absorption by Fe_2O_3 (Section 2.2.1).

The large reflectivity decline between 2.4 and 3.1 μ, in both bright and dark areas, indicates the presence of significant quantities of water of hydration in both areas (Section 2.2.3). Neither Martian ice clouds nor surface carbonates seem capable of accounting for these observations. Comparisons of theoretical 3.0- to 3.8-μ spectra that allow for thermal emission with the observed spectra imply at least one molecule of water of hydration per Fe_2O_3 moiety. A small quantity of additional water may be present, either bound or adsorbed to ferric oxides or to other constituents. The peaking of the reflectivity curve near 1.4 μ also suggests the presence of water of hydration, since the water opacity begins to become appreciable at slightly longer wavelengths (Section 2.2.3). The polarimetry yields a low value of the real part of the refractive index, again implying the presence of at least one molecule of water of hydration per Fe_2O_3 (Section 3.4.2).

That ferric oxides are a *major* constituent of the bright areas, and not a thin patina of desert varnish, follows from the approximate agreement between the particle sizes implied by comparison of photometry of Mars and of pure goethite samples with the particle sizes found from the Stokes-Cunningham fallout times of observed yellow clouds, and from the bright-area thermal inertias (Section 2.1.5). The same conclusion follows from a comparison of the refractive index required to account for the polarimetry with that of a composite silicate with a goethite patina (Section 3.4.2.).

That ferric oxides are also a major constituent of the dark areas follows from the near identity of the refractive indices of bright and dark areas; this agreement derives from the very low contrast in the blue, violet, and ultraviolet (Sections 2.1.3, 2.1.4), and from the fact that the polarization curve for the dark areas can be derived from that for the bright areas merely by increasing the mean particle size (Section 3.4.4). Even during the seasonal darkening of the dark areas, their index of refraction remains almost the same as for the bright areas.

Since these considerations indicate that goethite is a principal constituent of the surface, a first approximation to mean particle radii, $\bar{a}$, on Mars can be obtained by matching laboratory and Martian reflectivities (Section 2.1.5). The bright areas are characterized by $\bar{a} \simeq 25$ μ; the dark areas outside the seasonal darkening, by $\bar{a} \simeq 100$ μ; and the dark areas during the seasonal darkening, by $\bar{a} \simeq 200$ μ. Yellow-cloud fallout times (Section 2.1.5), surface thermal inertias (Section 2.1.5), and polarimetry (Section 3.4.5) give similar results. *Both the photometry and, particularly, the polarimetry, indicate that the principal event of the wave of darkening is a change in mean particle size, by a factor* ~ 2, *with no substantial change in composition.*

Samples of pure goethite that match the observations at other wavelengths give too high a reflectivity in the region just beyond 1.1 μ. Without changing any of the other results, agreement with observation can be secured by the addition of small quantities of Fe_3O_4 or MgO, both reasonable materials by terrestrial or meteoritic analogy (Section 2.2.2). Somewhat larger amounts of these materials are required for the dark areas than for the bright areas. The presence of these materials will also decrease the detectability of the limonite band.

This model also accounts for the loss of surface contrast toward the blue in a very natural way, without any necessity for invoking a blue haze (Section 4.1). The similarity of the negative branches of the polarization curves at various wavelengths, the enhanced short-wavelength visibility of the polar caps, the ultraviolet spectra, and other evidence all imply that indeed there is no blue haze (4.1). It is suggested that seeing plays an important role in the occurrence of blue clearings, with good blue seeing, no Martian clouds, and large and contrasting dark areas enhancing the prospects for detectable blue clearings (Section 4.2).

Thus, the bright and dark areas of Mars are composed in significant part of limonite; i.e., goethite ($Fe_2O_3 \cdot H_2O$) is a major constituent, with variable amounts of bound and adsorbed water, as well as other minerals, probably present. The dark-area particles are generally larger than the bright-area particles, and the dark-area particle sizes increase during the seasonal darkening.

5.2. Objections to the Model

We next discuss various arguments that have been advanced against goethite being a principal constituent of the bright areas. We have already seen that the absence of a well-defined limonite band in Martian spectra presents no basic contradiction with the conclusions of the present paper. Coulson *et al.* (1965) have examined the angular scattering properties of samples of pulverized goethite, and have found a strong backward lobe in the scattering diagram that is not found for Mars. However, this comparison is invalid. The laboratory measurements refer to a fixed angle of incidence and varying angles of reflection, while the astronomical measurements refer to varying angles of both incidence and reflection as a function of position on the Martian disk. For Mars, the difference between angles of incidence and reflection is the same, at a given phase angle, for all points on the disk. It is only at $\Phi = 0°$ that the effect should be prominent. However, it is not a unique indicator of limonite; many materials show similar backscattering. It is significant that Coulson *et al.* find a strong backscattering lobe for all angles of incidence with little variation in reflectivity. We also note that the laboratory measurements of Coulson *et al.* are probably dependent on the geometrical properties of the sample. They smooth down the sample and, in so doing, tend to orient the normals to the grain surfaces parallel to the local surface normal. On Mars a more random distribution of orientation is expected.

An important theoretical question concerns the thermodynamic stability of goethite under Martian conditions of temperature and humidity. Adamcik (1963), and Schmalz (1959) have come to opposite conclusions on this question, chiefly because of differences in their choices of thermodynamic variables. In a more recent and careful analysis, Fish (1966) finds a substantial uncertainty as to the correct values of the thermodynamic variables, and attempts to bracket the true values. He finds that for mean diurnal surface temperatures, $\bar{T} \geqslant 250\,K$, goethite will be unstable under the low water-vapor abundances of the Martian atmosphere.

Fish concludes that goethite on Mars should be converted almost entirely to hematite.

More recently POLLACK *et al.* (1969) have estimated the reaction rates for dehydration of goethite and rehydration of hematite, as well as the equilibrium vapor pressure curve. As the rate constants decrease rapidly with decrements in temperature, the stability of goethite at the surface is determined by the peak surface temperature; as a result goethite will be unstable at this locale. However large variations in temperature occur only very close to the surface: at greater depths a low constant value of temperature pertains, which is inside the goethite stability field. Because there is a much greater volume of material in the constant temperature regime, dynamical equilibrium between dehydration at the surface and rehydration below can occur with the observed atmospheric water vapor content, despite the higher rate at the surface. Estimated dehydration times at the surface are long ($\sim 10^3$–10^4 years) and so a given grain initially at the surface should be mixed to depths at which it can be rehydrated before it has undergone much decomposition. As a result there are no tenable thermodynamic objections to goethite grains at the surface of Mars.

In addition, water may occasionally be available to provide favorable stability conditions near the surface: the frost point is reached before sunrise on Mars and in fact dawn hazes are observed. Precipitated water may lie on the Martian surface each day; water trapped in the subsurface structure as the temperature rises will greatly increase the local subsurface humidities. In this context it is interesting to note that some terrestrial goethite deposits appear in places where thermodynamic equilibrium is apparently violated (R. Siever, private communication, 1967). Clearly, much work remains to be done on the question of goethite stability.

Finally, VAN TASSEL and SALISBURY (1964) have objected to large quantities of iron oxides on Mars essentially on the grounds that (by terrestrial analogy and from cosmic abundances) large quantities of silicates should be present. They claim that the mean density of Mars is below what one would expect with $(Fe/Si) \geqslant 1$. This is true if the iron is present as the metal, as in a planetary core. But the moment of inertia of the planet and the negative results on trapped particles and magnetic fields obtained by Mariner 4 counterindicate an appreciable iron core. Iron combined with oxygen, as in limonite, has a density close to the uncompressed density of Mars. The results of the present paper refer, however, only to the surface layers of Mars; they are certainly compatible with the presence of silicates, only provided that goethite is a major constituent. However, more surface iron does seem to be present on Mars than on the earth, a conclusion discussed below.

ADAMS (1968) has recently compared his laboratory measurements with spectral reflectivity observations of Mars. To match the visual reflectivity of Mars he requires goethite, as we have. Supposing that there is no dip near 0.9 μ, but one near 1.0 μ, which he attributes to Fe^{++}, Adams estimates what mixture of goethite and Fe^{++}-containing basalt are consistent with all the observations; he achieves an upper limit of 25% goethite by weight, not inconsistent with our results. However, as pointed out above, the reality of weak features at 0.9 μ and at 1.0 μ remains unsettled. Furthermore, it seems rather arbitrary to segregate all the Fe^{++} into basalt.

In a laboratory study of the reflectivity properties of limonite samples of various sizes, SALISBURY and HUNT (1968) found that some specimens of smaller particle size exhibited a markedly lower reflectivity below 5500 Å than samples of larger particle size. This result disagrees with our prediction of increasingly lower contrasts towards shorter wavelengths, as is observed for Mars. Our discussion was based upon two assumptions: (1) that the two samples are related by a simple homology transform; i.e., that grain shape and arrangement are identical for the two samples (under such conditions, when a particle of any size is completely opaque, photons will have identical reflectivities), and (2) that, since it is very difficult to grind particles smaller then $\sim 1\ \mu$ (none of the laboratory experiments achieve particles this small), all particles are large compared with the wavelength of light. The results of Salisbury and Hunt might be attributed to a variation in grain shape with size, because of different grinding histories, and/or to a difference in intergrain arrangement, possibly because of the increasing importance of adhesion at smaller particle sizes. It would be desirable to repeat the measurements under conditions of low humidity, as apply on Mars, since the importance of adhesion may be a function of humidity. We might also point out that Hovis' measurements for the samples discussed in this paper showed no contrast reversal down to a wavelength of 5000 Å. Clearly more laboratory work is needed to understand under what conditions contrast reversal obtains.

5.3. SOME IMPLICATIONS

We now briefly explore some further implications of our results. We have found sizable quantities of bound or adsorbed water in the Martian surface layers. Strictly speaking, the photometry, polarimetry, and spectroscopy pertain only to the top millimeter or so; but when account is taken of the stirring of the dust by winds, we find that our results must apply to some greater depth. Furthermore, radar observations indicate that the bright areas are pulverized to a depth of at least 1 m (SAGAN and POLLACK, 1969); presumably, the composition does not vary over this depth. The mean atmospheric water-vapor content is $\sim 10^{-3}$ gm cm^{-2} (SCHORN *et al.*, 1967), comparable to that in a goethite specimen 10 μ thick. Thus, practically all the water on Mars is present subsurface; the amount is at least 10^2 gm cm^{-2}, and may be much larger. The water present on the surface of the earth is believed to have been altogether outgassed from the interior by volcanism and other plutonic activity (see, e.g., CAMERON and BRANCAZIO, 1964). While there may be some contribution from the infall of cometary debris on Mars, the bulk of the water there should also be endogenous. Thus, some appreciable outgassing of Mars appears to have occurred. Compared to the several-km-thick equivalent layer of water on the earth, the minimum Martian equivalent layer of meters that we have deduced implies an outgassing rate at least 0.1% that of the earth and possibly much larger. If the bulk of the N_2 outgassed in geological time is present in the contemporary terrestrial atmosphere, assuming initial conditions on the two planets to have been similar, this implies a Martian outgassing rate of > 1 mb of N_2 in geological time. Even if N_2 has never escaped from Mars, this implies no contradiction with the present atmospheric composition (SAGAN and POLLACK, 1967b). With ~10-atm CO_2 equivalent in the terrestrial sedimentary column (HUTCHINSON, 1954) and the same assumptions, a Martian outgassing rate of >100 mb CO_2 over geological time is implied. The contemporary atmosphere is unlikely to have more than ~10 mb CO_2, but surface carbonate deposits can be expected on Mars as on the earth. We also note that during the past 4.5×10^9 years, appreciable quantities of water may have been lost from Mars (depending on the history of the atmospheric structure and exospheric temperature), but not from the earth. Because of surface reaction and escape, no contradiction with the present Martian atmosphere is thereby generated, but a greater outgassing rate would be implied.

It is interesting that the iron on the Martian surface is almost entirely in the oxidized form Fe_2O_3, rather than such more reduced forms as Fe_3O_4 or FeO. By comparison, the oxidation state of terrestrial iron of the crust is close to that of magnetite, while stony meteorites generally lie between FeO and Fe (see, e.g., BIRCH *et al.*, 1942). One possible explanation for the high degree of oxidation on Mars is the photodissociation of water vapor, the escape of hydrogen to space, and the oxidation of the Martian surface by the oxygen left behind (WILDT, 1934; SAGAN, 1966). The additional outgassed water required is some three times that discussed in the previous paragraph.

The silicon content of the Martian surface material is at most comparable to that of iron. The photometric and polarimetric evidence excludes any great preponderance of silicon compounds. By contrast, the crust of the earth is composed principally of silicon and oxygen, with some aluminum and relatively little iron. The mantle seems to contain iron and silicon in approximately equal amounts; a similar statement holds for stony meteorites, while (Fe/Si) is increased in iron meteorites (BIRCH *et al.*, 1942). Thus the iron-to-silicon ratio on the Martian surface is comparable to the undifferentiated cosmic abundances.

Two possibilities have been suggested (SAGAN, 1966) to explain the higher surface (Fe/Si) on Mars than on the earth. During differentiation of the earth, a major fraction of the iron has migrated toward the core. Mars, on the other hand, has probably experienced less differentiation and has, at best, a very small core. In this case, the formation of a crust may have been inhibited and the surface composition of Mars may be then roughly characteristic of the gross composition of the planet. Alternatively, the iron content of the Martian surface is larger than that of the immediately underlying layers, permitting some substantial planetary differentiation. The iron-rich surface layers can then be attributed to meteoritic infall, Mars being much closer to the asteroid belt than is the earth. Scaling the terrestrial micrometeorite-infall rate (PARKIN and TILLES, 1967) by a factor of 10, we find an accumulation over 4.5×10^9 years of between 2 and 200 m. Because of atmospheric drag, these particles suffer negligible ablation in entering the Martian atmosphere, fall gently to the surface (cf. ÖPIK, 1958), and produce no craters. The iron-to-silicon ratio of meteorites is, as mentioned above, within the range deduced for Mars.

The Mariner 4 photographs imply that about 15% of Mars is covered by craters with diameters in excess of 9 km. Typical depth-to-diameter ratios for newly formed craters are ~0.1. From this source alone, then, an average depth of some 100 m has been pulverized. (This may be in agreement with the mean depth of cover from micrometeorites, and suggests that some substantial filling of Martian craters by micrometeorites has occurred. The craters do have a pronounced flat-bottomed filled-in appearance.) Because limonite is more friable than iron-poor silicates, it will constitute a larger fraction of the rock flour created by meteorite impact. In turn this rock flour is lifted by the winds, and will preferentially cover the surface (Rea, 1965).

The compositional similarity between bright and dark areas can be caused by the smaller planetary differentiation or by the larger micrometeoritic infall. Winds will certainly tend to homogenize the surface layers of the planet. The particle sizes of the bright-area material are such that the expected Martian winds will lift these particles and carry them fair distances. The larger dark-area particles will only saltate or bounce along the surface. Those particles that can be carried by the wind will constitute the surface layers; this provides a causal relation to explain the particle sizes deduced. The segregation of particle sizes between bright and dark areas can be understood quantitatively in terms of elevation differences. Seasonal exchange of material by winds between the bright and dark areas offers a natural explanation of the seasonal photometric and polarimetric changes, which we have seen to be due to variations in particle sizes rather than composition. These matters are the subject of a paper to be published elsewhere.

[*Editor's Note:* In the original, material follows this excerpt.]

References

[*Editor's Note:* Reference list has been abridged to include only those references cited in the text.]

Adamcik, J. A.: 1963, 'The Water-Vapor Content of the Martian Atmosphere as a Problem of Chemical Equilibrium', *Planetary Space Sci.* **11,** 355–359.

Adams, J. B.: 1968, 'Lunar and Martian Surfaces: Petrologic Significance of Absorption Bands in the Near Infrared', *Science* **159,** 1453–1455.

Birch, F., Schairer, J. F., and Spicer, H. C. (eds.): 1942, *Handbook of Physical Constants*, Waverly Press, Baltimore.

Cameron, A. G. W. and Brancazio, P. J.: 1964, *The Origin and Evolution of Atmospheres and Oceans*, John Wiley & Sons, New York.

Coulson, K. L., Bouricius, G. M. B., and Gray, E. L.: 1965, 'Effect of Surface Reflection on Radiation Emerging from the Top of a Planetary Atmosphere'. General Electric Space Sciences Laboratory Rept. R65SD64.

Fish, F. F., Jr.: 1966, 'The Stability of Goethite on Mars', *J. Geophys. Res.* **71,** 3063–3068.

Hovis, W. A., Jr.: 1965, 'Infrared Reflectivity of Iron Oxide Minerals', *Icarus* **4,** 425–430.

Hovis, W. A., Jr. and Callahan, W. R.: 1966, 'Infrared Reflectance Spectra of Igneous Rocks, Tuffs, and Red Sandstones from 0.5 to 22 μ', *Journ. Opt. Soc. Amer.* **56,** 639–643.

Hutchinson, G. E.: 1954, 'The Biochemistry of the Terrestrial Atmosphere', in *The Earth as a Planet*, vol. 2 of *The Solar System* (ed. by G. P. Kuiper), Univ. of Chicago Press, Chicago, pp. 371–433.

Öpik, E. J.: 1958, *Physics of Meteor Flight in the Atmosphere*, Interscience Publishers Inc., New York.
Öpik, E. J.: 1960, 'The Atmosphere and Haze of Mars', *J. Geophys. Res.* **65**, 3057–3063.
Parkin, D. W. and Tilles, D.: 1967, *Influx Measurements of Extraterrestrial Material.* Smithsonian Astrophys. Obs. preprint.
Pollack, J. B., Wilson, R. N., and Goles, G.: 1969, to be published.
Rea, D. G.: 1965, 'Some Comments on "The Composition of the Martian Surface"', *Icarus* **4,** 108–109.
Sagan, C.: 1966, 'Mariner IV Observations and the Possibility of Iron Oxides on the Martian Surface', *Icarus* **5,** 102–103.
Sagan, C. and Pollack, J. B.: 1967b, 'Elevation Differences on Mars'. Smithsonian Astrophys. Obs. Spec. Rep. No. 224.
Sagan, C. and Pollack, J. B.: 1969, 'Radio Evidence on the Structure and Composition of the Martian Surface (to be published); also in *J. Res. Nat. Bur. Standards* **69D** (1965), 629.
Salisbury, J. W. and Hunt, G. R.: 1968, 'Martian Surface Materials: Effect of Particle Size on Spectral Behavior', *Science* **161,** 365–366.
Schmalz, R. F.: 1959, 'A Note on the System $Fe_2O_3 - H_2O$', *J. Geophys. Res.* **64,** 575–579.
Schmidt, I.: 1959, 'Visual Problems in Observing the Planet Mars', in *Proc. of Lunar and Planetary Exploration Colloquium*, vol. I, No. 6, pp. 19–22.
Schorn, R. A., Spinrad, H., Moore, R. C., Smith, H. J., and Giver, L. P.: 1967, 'High-Dispersion Spectroscopic Observations of Mars. II: The Water-Vapor Variations', *Astrophys. J.* **147,** 743–752.
Van Tassel, R. A. and Salisbury, J. W.: 1964, 'The Composition of the Martian Surface', *Icarus* **3,** 264–269.
Wildt, R.: 1934, *Ozon und Sauerstoff in den Planeten-Atmosphären.* Veroffentlichungen Univ.-Sternwarte Göttingen, No. 38.

12

Reprinted from *Icarus* **14**:245–251 (1971)

Mars: The Spectral Albedo (0.3–2.5μ) of Small Bright and Dark Regions

THOMAS B. McCORD AND JONATHAN H. ELIAS

Planetary Astronomy Laboratory, Department of Earth and Planetary Sciences, Massachusetts Institute of Technology, Cambridge, Massachusetts 02139

AND

JAMES A. WESTPHAL

Division of Geological Sciences, California Institute of Technology, Pasadena, California 91107

Received November 9, 1970

A review of the available spectral geometric albedo measurements for Mars was presented earlier for the spectral region 0.3 to 1.1μ. A new observational study has greatly increased the store of data, especially for small Martian regions and for the infrared spectral region 1.0 to 2.5μ. Here we combine the new data with data both from the earlier review and, for the infrared spectral region, from the literature. We present a more complete picture of Martian spectral reflectivity properties than was available. This study should provide a more firm basis upon which models of Martian surface composition can be built. At visible wavelengths the Mars dark area Syrtis Major is red rather than green or grey in color; the bright area Arabia is even redder than Syrtis Major. Absorption bands, which differ between bright and dark areas, appear in the reflection curves. The 1μ absorption feature for dark areas is confirmed and more completely described. A previously unreported absorption band near 0.95μ for bright areas appears along with several absorption features in the infrared. The geometric albedo for Arabia reaches a maximum of about 0.43 at 1μ. The Bond albedo for this same area reaches a maximum of 60%. The bright area Arabia is occasionally three times brighter than the dark area Syrtis Major at red wavelengths. Published infrared reflection data available for Mars are not in complete agreement. Changes in brightness and color of Arabia are discussed which are not in agreement with traditional darkening wave theory.

The interaction of a planetary surface with incident solar radiation is one of the few phenomena which enable us to study the planet remotely. The spectral reflection properties are of particular interest in that they are affected primarily by the composition and mineralogy of the planetary surface material. In an earlier article (McCord and Adams, 1969) the available spectral reflectivity data for Mars in the spectral region 0.3–1.1μ were reviewed. These results were used to construct models for the surface composition of Mars (Adams and McCord, 1969; Plummer and Carson, 1969). An extensive observational study by McCord and Westphal (1970) of small regions of Mars (about 200km in diameter) during the 1969 apparition greatly increased and extended the available data. Several new features have appeared. These new results are incorporated with the previous data to present a new foundation for subsequent surface-model derivation.

When viewed through a telescope or from a spacecraft the Martian surface appears to be composed of three photometric provinces: bright areas, dark areas, and polar regions. In this study we ignore the polar caps for lack of data. We have chosen to discuss Arabia and Syrtis Major as representative areas for bright and dark regions, respectively. In fact there seem to be areas on Mars representing a series of brightness; these two specific areas are

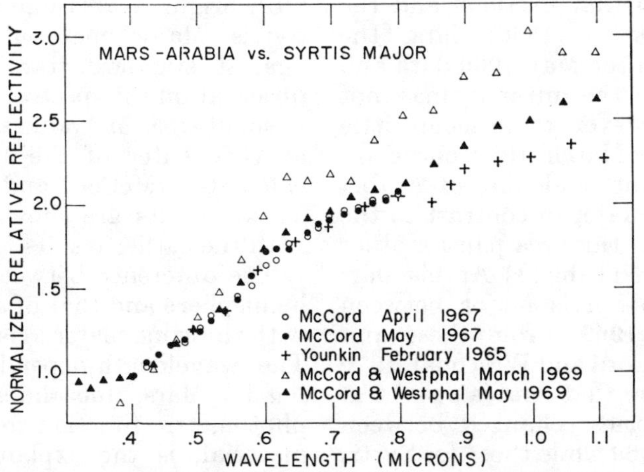

FIG. 1. The ratio of the reflectivity of Arabia to that of Syrtis Major as a function of wavelength. The ratio was scaled to unity at 0.4μ. Data presented in an earlier review (McCord and Adams, 1969) are combined with new observations (McCord and Westphal, 1970). Dates are Earth dates when observations were made.

logical end members of this series. Recent observational results (McCord and Westphal, 1970) suggest that there also exists a series of spectral reflectivity types with Arabia and Syrtis Major again representing end members.

One of the most sensitive methods of comparing the reflectivities of two sources is to compute the ratio of the reflectivities as a function of wavelength. This measurement is also easy to make accurately. The ratio data (McCord and Adams, 1969; McCord and Westphal, 1970) in the spectral region 0.3 to 1.1μ which are available for the Arabia/Syrtis Major area-pair are presented in Fig. 1. All data sets have been scaled so that a curve fitted through them will have a value of unity near 0.4μ. Table I gives the scaling factors and additional information on the observations.

As with the earlier data we see that the dark areas differ in color from the bright areas in that they are simply less red, not green or grey. The striking feature in a similar plot presented earlier (McCord and Adams, 1969) was the lack of any difference in the relative color of the two Martian areas for the observations at different times. Two new data sets have now been added. The May 1969 data differ only slightly from earlier data, except perhaps in the infrared; the March 1969 data differ more radically. Considering the chance for systematic observational error, especially in the infrared where Martian contrasts are greatest, the small Martian image size in

TABLE I

INFORMATION DESCRIBING THE OBSERVATIONAL CONDITIONS[a]

	$(I_a/I_{sm})_{0.4\mu}$	Martian season (NH)	Phase angle
February 16–17, 1965	0.98	Mid May	39°
April 8–9, 1967	1.04 ± 0.02	Early July	6°
May 12–13, 1967	0.86 ± 0.06	Late July	21°
March 13–17, 1969	1.25 ± 0.03	Late July	36°
May 25–26, 1969	1.15 ± 0.03	Early September	5°

[a] The second column is the ratio of the intensities of 0.4μ of Arabia relative to Syrtis Major.

1965 compared to that in 1969, and the scatter in Younkin's (1966) data, the difference between the May 1969 data and Younkin's data in the infrared, may not be significant. However, there seems little question that the March 1969 curve indicates a significant reddening of Arabia relative to Syrtis Major in contrast to the curves available for this area-pair for other dates. In addition (Table I) Arabia darkened relative to Syrtis Major between March and May 1969. From these and earlier results (McCord and Westphal, 1970 —Fig. 4) it appears that Arabia's spectral albedo and brightness changed between March and May 1969 while those for Syrtis Major remained constant. This is not the expected sense of change if the traditional darkening wave were responsible. Using the information in Table I it can be seen that no simple relation between the contrast at 0.4μ, the color difference, the solar phase angle, or Martian season is evident.

In the earlier review (McCord and Adams, 1969) the ratio data and data available for the geometric albedo of the entire planet were converted into spectral geometric albedos for hemispheres covered with bright Arabia-like material and with Syrtis Major material. By calibrating against standard stars during the 1969 observation the spectral geometric albedos of small areas in Arabia and in Syrtis Major at the center of the Martian disk were calculated (McCord and Westphal, 1970). These results are shown in Fig. 2 along with the earlier results.

The difference between the albedo of a hemisphere and that of a flat disk covered with the same material is obvious in Fig. 2. The wavelength-dependent limb-darkening for Mars, due almost entirely to the photometric function for Martian surface material, is the explanation. The limb-darkening wavelength dependence follows very closely the wavelength dependence of the reflectivity of Arabia (McCord and Westphal, 1970). This difference is important when constructing models for the composition of the Martian surface. The geometric albedo of the Martian bright material exceeds 0.40 (somewhat higher than has been generally considered), while the geometric albedo changes by nearly a factor of 10 between 0.4 and 0.8μ. This is truly a highly colored material.

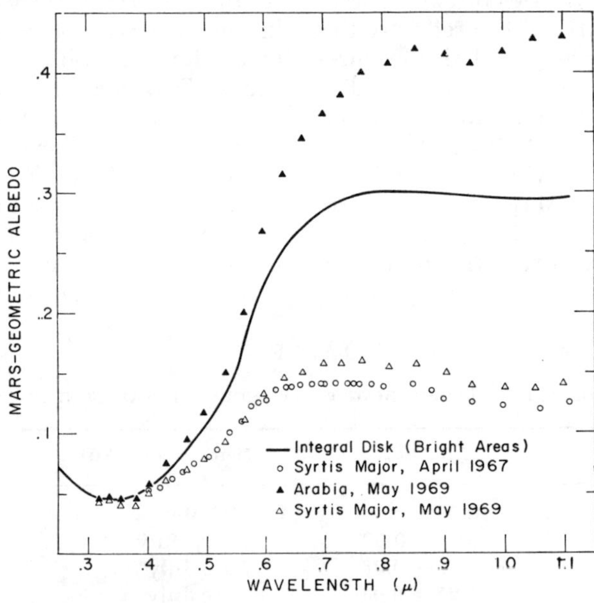

FIG. 2. The spectral geometric albedo is shown for small flat areas in Arabia and in Syrtis Major (McCord and Westphal, 1970) and for hemispheres covered with average Martian bright areas and with Syrtis Major material (McCord and Adams, 1969).

The broad absorption feature which appeared for wavelengths longward of 0.9 μ in the earlier data for Syrtis Major (McCord and Adams, 1969) is faithfully reproduced in the new data. This feature has also been confirmed by others (O'Leary and Jackel, 1970) during the 1969 apparition, although there has been some evidence to the contrary (Younkin, 1969). Also a slight depression appears near 0.8μ. The two curves for Syrtis Major agree very well, excluding the limb-darkening effect.

The new curve for Arabia contains an absorption feature between 0.9 and 1.0μ which has not been reported previously. The older data do not show it, but there is a paucity of data in this spectral region (McCord and Adams, 1969). Also it must be remembered that the older data are for average bright material, not for Arabia alone. The interpretation of this absorption feature is beyond the scope of this paper, but it is in this spectral region where absorption bands for many terrestrial minerals occur and where absorption bands occur in spectra for the Moon (McCord and Johnson, 1970) and Vesta (McCord, Adams, and Johnson, 1970).

The curves for the bright and dark regions are similar in that they both indicate a very red material. However, the absorption features which appear in the near infrared are quite different for the two areas. This difference probably indicates a different mineralogical composition and not simply a particle size difference.

The ratio of the reflectivity of Arabia to that of Syrtis Major (Fig. 3) was extended (McCord and Westphal, 1970) during the May 1969 observational period into the infrared to 2.5μ. The data presented in Fig. 3 are not scaled as in Fig. 1, but represent the true ratio between the two Martian regions. The reflectivity of Arabia increases to almost three times that of Syrtis Major between 0.3 and 1μ. Between 1.0 and 2.5μ the contrast remains about constant except for two features. The largest feature occurs between about 1.6 and 2.1μ and the other between about 1.3 and 1.4μ. These features are located in approximately the same wavelength regions as water vapor absorption but the method of data acquisition makes contamination of the data by telluric absorption unlikely (McCord and Westphal, 1970).

The ratio data beyond 1.0μ were con-

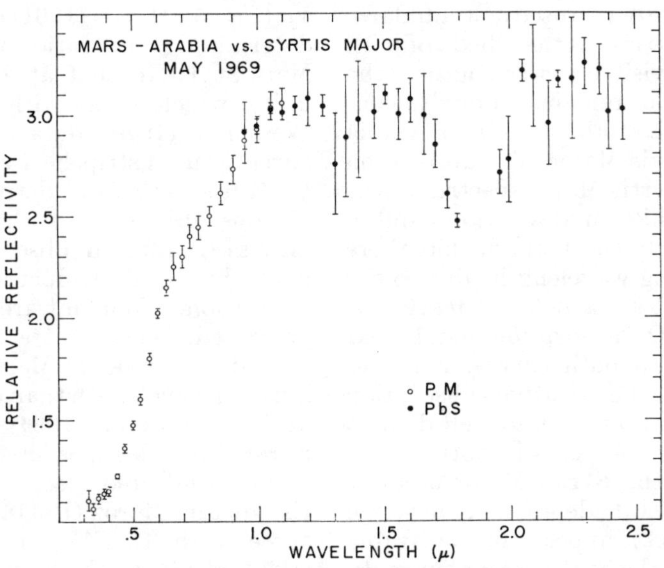

FIG. 3. The ratio of the spectral reflectivity of Arabia to that of Syrtis Major after McCord and Westphal (1970). Both photomultiplier (P.M.) and lead sulfide (PbS) detectors were used to make the observations. The ratio curve is not scaled as in Fig. 1.

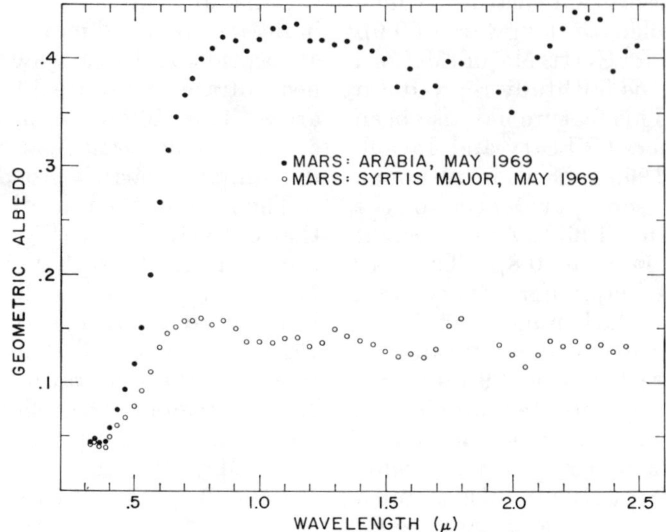

FIG. 4. The spectral geometric albedo for a small flat area in Arabia and in Syrtis Major is shown for the 0.3 to 2.5 μ spectral region as it was in late May 1969.

verted by McCord and Westphal (1970) to geometric albedos (Fig. 4) by the same procedure used to produce Fig. 2. However, there are almost no stars for which the spectral energy flux is well-known throughout the infrared spectral region studied here. Therefore, an area on the Moon, for which the spectral flux is known (McCord and Johnson, 1970), was used as a standard.

After the sharp rise in the albedo of both areas in the visible wavelengths, the geometric albedo remains roughly constant out to 2.5 μ at about 0.40 for Arabia and 0.15 for Syrtis Major. The absorption feature in the Syrtis Major spectrum near 1 μ appears less like an absorption band in the more complete spectrum in that there is no discrete long wavelength edge to the band. Nevertheless, a definite feature is present. The CO_2 absorption band near 2.0 μ is evident in both curves. Features responsible for the relative absorption apparent in Fig. 3 are not so well defined in Fig. 4. A bump appears in both curves near 1.8 μ, but the Syrtis Major bump is greater. A general depression between about 1.4 and 2.2 μ appears in the Arabia curve but not in the Syrtis Major curve. In general, several features appear in the infrared portion of the curve in Fig. 4 but the scatter in the data makes precise description impossible. This spectral region should be observed carefully during the 1971 opposition as important evidence of surface composition may be hidden in the structure in the curves.

Before the 1969 apparition there were few data available on the spectral reflectivity of Mars at wavelengths longer than 1 μ (Fig. 5). Harris (1961) quoted Kuiper as determining the ratio of reflectivity of Mars at 1 μ to that at 2.0 μ to be about unity, which agrees with the McCord and Westphal (1970) data. Tull (1966) apparently used stripchart records published by Kuiper (1952) to derive three infrared albedos. Binder and Cruikshank (1963) discussed infrared observations of Mars made by G. P. Kuiper during the 1963 opposition. Four infrared albedos were presented. These latter two data sets indicate a redder Martian surface at infrared wavelength than do the other data in Fig. 5. Moroz (1964) presented three infrared albedos measured by him as well as a near infrared spectrum determined by Esipov and Moroz (1963). The Moroz data agree with the McCord and Westphal (1970) results to the limit of their spectral coverages. Sinton (1967) discussed observations in the 2 to 4 μ region. These data are consistent with McCord and Westphal's

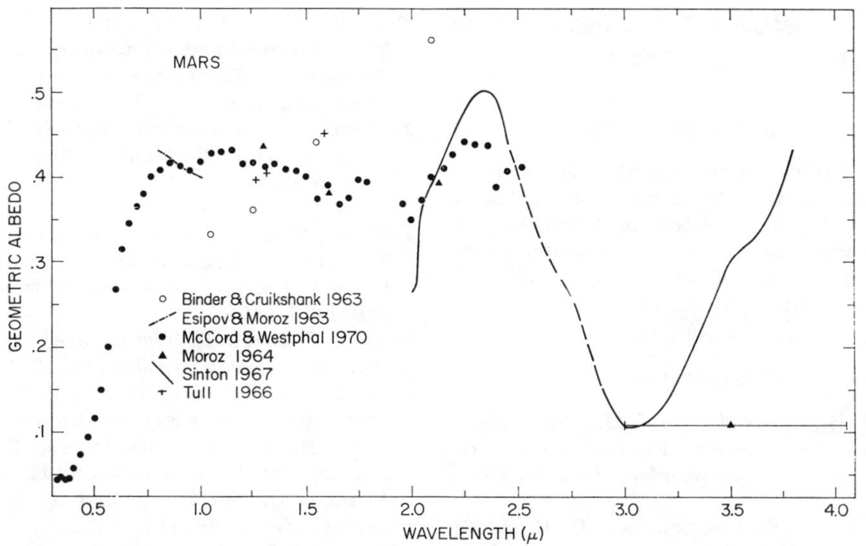

FIG. 5. The published spectral reflectivity data for Mars at infrared wavelengths is shown.

(1970) results to 2.5μ, but they disagree with Moroz's (1964) results between 3 and 4μ. Apparently Sinton experienced calibration difficulties.

There is not complete agreement among the various infrared Martian reflectivity spectra presented in Fig. 5. Binder and Cruikshank (1963) and Tull (1966) show the Martian reflectivity continuing to increase between 1.0 and 2.5μ. Kuiper (1952), Moroz (1964), and McCord and Westphal (1970) show a near constant or slightly decreasing reflectivity in the same spectral region. Sinton (1967) measured a strong depression in the Martian reflectivity near 3.0μ; Moroz (1964) suggests a uniformly low reflectivity throughout the 3.0–4.0μ spectral region. The 1.0–4.0μ spectral region clearly needs more study.

Many models for the surface composition of Mars rely on laboratory measurements of the spectral reflectivity properties of terrestrial materials. Most laboratory spectrometers used for these studies make use of an integrating sphere. Albedos measured by this method most nearly resemble the Bond albedo, defined as the percentage of light incident on a surface which is returned to space *in all directions.* The Bond albedo A of a surface is usually expressed as the product of two quantities: the geometric albedo p and the phase integral q (see, for example, Harris, 1960). The phase integral q for Mars has been discussed in detail most recently by de Vaucouleurs (1964). (See his Figs. 24 and 28.) Unfortunately data for only the integral disk of Mars is available. However, we can probably use this data to first order for a small area at the center of the Martian disk. We have chosen to use de Vaucouleurs' Case B (his Fig. 28) as most likely to represent the spectral dependence of the phase integral, as this case most nearly follows the spectral dependence of the surface reflectivity. We further assume that the integral disk values for q closely resemble the q for Martian bright areas, as most of the integral disk is composed of bright areas. Using the spectral geometric albedo for Arabia (Fig. 2) and the spectral phase integral presented by de Vaucouleurs (1964) we arrive at an estimate of the spectral Bond albedo for a flat surface covered with Arabia material.

The Bond albedo curve is quite similar in shape to the geometric albedo curve. However, the Bond albedo increases in value from about 3% at 0.35 to about 60% at 0.8μ and beyond. Thus at 0.8μ and at longer wavelengths 60% of the incident solar radiation is reflected by Arabia. Arabia is a bright surface by terrestrial soil standards. In contrast, at 0.8μ lunar upland material has a Bond albedo of about 12%.

Acknowledgments

Part of this work was done while McCord was in residence at the Department of Geological Sciences, California Institute of Technology as a visiting associate. This work was supported by NASA grants HGL-22-009-019 and NGR-22-009-473 and NSF grant GP-12774.

References

Adams, J. B., and McCord, T. B. (1969). Mars: Interpretation of spectral reflectivity of light and dark regions. *J. Geophys. Res.* **74**, 4851–4856.

Binder, A. B., and Cruikshank, D. P. (1963). Comparison of the infrared spectrum of Mars with the spectra of selected terrestrial rocks and minerals. *Commun. LPL* **2**, 37, 193–196.

Esipov, V. F., and Moroz, V. I. (1963). Spectroscopy of Mars. *Astron. Tsirk.* **262**, 1.

Kuiper, G. P. (1952). Planetary atmospheres and their origins. In "The Atmospheres of the Earth and Planets" (G. P. Kuiper, Ed.), pp. 306–405. Univ. of Chicago Press, Chicago, Ill.

McCord, T. B., and Adams, J. B. (1969). Spectral reflectivity of Mars. *Science* **163**, 1058–1060.

McCord, T. B., and Westphal, J. A. (1970). Mars: Narrowband photometry, from 0.3 to 2.5 microns, of surface regions during the 1969 apparition. *Astrophys. J.*, in press.

McCord, T. B., and Johnson, T. V. (1970). Lunar spectral reflectivity (0.30–2.50 microns) and implications for remote mineralogical analysis. *Science* **169**, 855–858.

McCord, T. B., Adams, J. B., and Johnson, T. V. (1970). Asteroid Vesta: Spectral reflectivity and compositional implications. *Science* **168**, 1445–1447.

Moroz, V. I. (1964). The infrared spectrum of Mars. *Sov. Astron.—A.J.* **8**, 273–281.

O'Leary, B. T., and Jackel, L. (1970). The 1969 opposition effect of Mars: Full disk, Syrtis Major and Arabia. *Icarus*, **13**, 437–448.

Plummer, W. T., and Carson, R. K. (1969). Mars: Is the surface colored by carbon suboxide? *Science* **166**, 1141–1142.

Sinton, W. M. (1967). On the composition of martian surface materials. *Icarus* **6**, 222–228.

Tull, R. G. (1966). The reflectivity spectrum of Mars in the near-infrared. *Icarus* **5**, 505–514.

de Vaucouleurs, G. (1964). Geometric and photometric parameters of the terrestrial planets. *Icarus* **3**, 187–235.

Younkin, R. L. (1966). A search for limonite near-infrared spectral features on Mars. *Astrophys. J.* **144**, 809–818.

Younkin, R. L. (1969). Spectrophotometry of Mars. *Trans. Amer. Geophys. U.* **50**, 230.

13

Reprinted from *Icarus* **17**:428–431 (1972)

Investigation of the Martian Environment by Infrared Spectroscopy on Mariner 9

R. Hanel, B. Conrath, W. Hovis, V. Kunde, P. Lowman, W. Maguire, J. Pearl, J. Pirraglia, C. Prabhakara, B. Schlachman, G. Levin, P. Straat, and T. Burke

[*Editor's Note:* In the original, material precedes this excerpt.]

Atmospheric Dust Composition

South-polar and midlatitude spectra recorded during the first 100 orbits show strong features whose positions and relative strengths are characteristic of silicate minerals. The broad features near 480 cm^{-1} and between 900 and 1200 cm^{-1} which appear in emission in the south-polar spectra, and in absorption over the low latitudes where the surface is warmer than the atmosphere, indicate the presence of particles suspended in the atmosphere. It has been pointed out that the mineralogical characteristics, primarily the SiO_2 content, determine the spectral position of the absorption and transmission maxima, (Lyon, 1964; Hovis and Callahan, 1966; Aronson *et al.*, 1967; Conel, 1969; Salisbury, 1970). From such data it may be possible to infer the dust compostion. However, the matter is complicated by effects of particle size and the possibility that more than one mineral group may be contributing to the recorded spectrum.

The wavenumber of maximum absorption in the interval between 900 and

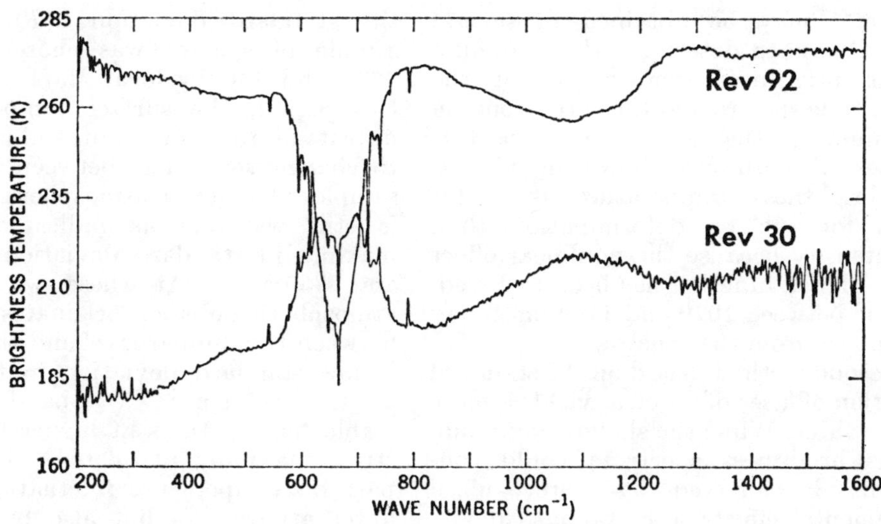

Fig. 4. Brightness temperatures calculated from spectra recorded on revolutions 30 and 92. The midlatitude spectrum shows a minimum due to silicate dust near 1085 cm^{-1}. The south polar spectrum shows a maximum at approximately the same wavenumber.

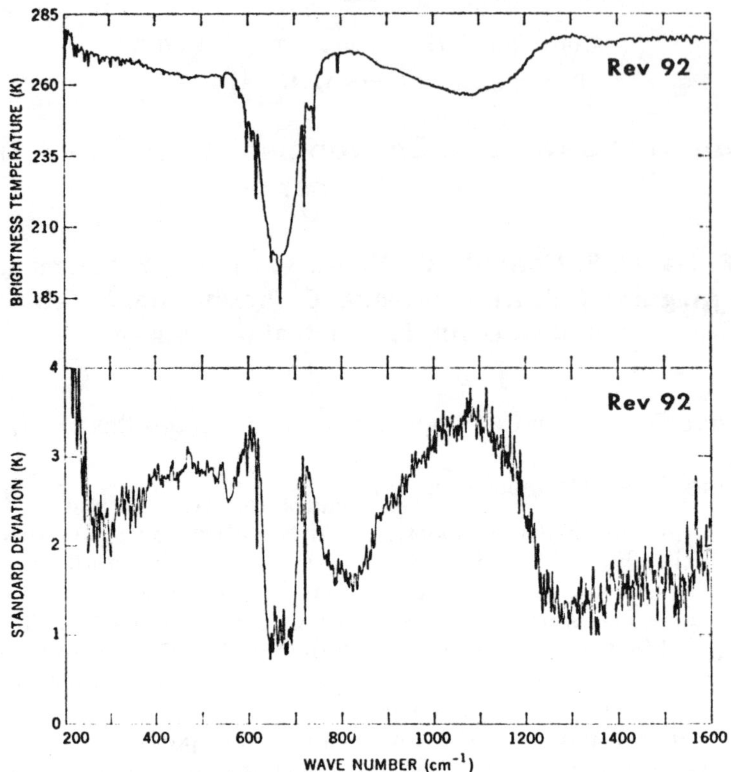

FIG. 5. (a, top) Brightness temperature average of 16 spectra recorded on revolution 92. (b, bottom) Standard deviation of the brightness temperatures within the ensemble. The maximum in the brightness temperature near 1300 cm^{-1} is caused by a transmission maximum in the dust cloud. The maximum in the standard deviation occurs at the most absorbing wavenumber of the dust.

1250 cm^{-1} may be obtained in several ways. One is to determine the minimum in the brightness temperature of the midlatitude spectra and the corresponding maximum in the south polar spectra; samples of both are shown in Fig. 4. The brightness temperature is better suited for such a determination than the intensity because the nonlinear effect of the Planck function has been removed. A value between 1070 and 1100 cm^{-1} may be deduced from the spectra.

A second method, based on the standard deviation of a set of spectra, yields a more precise value. While the shallow minimum of the brightness spectrum could conceivably be affected by unrecognized instrumental effects, the standard deviation depends much less on the absolute calibration. Shown in Fig. 5 is the same midlatitude average spectrum (Rev. 92) displayed in Fig. 4(a), together with the standard deviation $\sigma(T)$. The ensemble of spectra was chosen between 1200 and 1400 hr local Mars time. Over this period, the surface temperature is near its diurnal maximum and is expected to change very little between individual samples. The lower atmosphere, however, is still warming as indicated by the maxima in standard deviation near 600 and 730 cm^{-1}. At these wavenumbers atmospheric emission originates primarily between the 3-mbar level and the surface. A low standard deviation in the center of the 667-cm^{-1} CO_2 band indicates stable temperatures at higher levels. The strong maximum at 1090 cm^{-1} is caused in part by temperature fluctuations in the lower atmosphere but also by the local variability of the dust among spectra in the ensemble. In either case the maximum variation is expected to occur at the most absorbing wavenumber. This occurs at 1090 cm^{-1}, in agreement with the results obtained from the brightness temperatures.

It has been pointed out by Conel (1969) that the transmission maximum (Christiansen frequency) is also a good indicator of bulk chemical composition. For silicates, the Christiansen frequency shifts from 1360 cm^{-1} for quartz towards lower wavenumbers with the inclusion of other elements into the lattice structure. From the maximum of the brightness temperature in the midlatitude spectrum and from the minimum in the standard deviation (Fig. 5) a Christiansen frequency between 1280 and 1300 cm^{-1} is inferred. Again, both methods are consistent.

Although the spectral positions of the absorption maximum and minimum of the Martian dust cloud seem to be well defined, the interpretation in terms of the silica content is still difficult. The difficulties arise in part from the complexity of theoretical treatments of a cloud of scattering and absorbing particles of nonspherical shape and in part from the lack of knowledge of the complex refractive index of most minerals in the infrared. Although some effort towards a better theoretical understanding have been made by Aronson *et al.* (1967) and Conel (1969) the interpretation of the IRIS spectra in terms of the SiO_2 content must still be based on an empirical comparison with laboratory transmission spectra of mineral dust.

One extensive set of transmission measurements of mineral dust has been prepared by Lyon (1964) who imbedded mineral powders in KBr pellets. With such preparation only the absorption maxima are reliable, since the minima representing the Christiansen frequency are shifted considerably by the differences in refractive index between air or CO_2 (~1) and KBr (~1.5). Comparison of the peak absorption of the Martian dust as measured by IRIS (~1085 cm^{-1}) with the transmission spectra of Lyon indicates a reasonable agreement with dacite (68.7% SiO_2, USNM 82, Lassen Peak, CA) as well as with quartz basalt (57.25% SiO_2, USNM, Snag Lake Cinder Cone, Lassen Co., CA) but poor agreement with more basic materials.

Comparison of the Christiansen frequency measured by IRIS (~1290 cm^{-1}) with Conel's laboratory data (Conel, 1969, Fig. 5) shows good agreement with the alkali-feldspar group (SiO_2 ~65%) as well as with andesine (SiO_2 ~57%) and oligoclase (SiO_2 ~62%) of the plagioclase group, but no agreement with pure SiO_2, with micas or other minerals of lower SiO_2 content. Comparison of the IRIS-determined Christiansen frequency with a diagram kindly made available by J. Salisbury (private communication) indicates also a SiO_2 content by weight of approximately 60%.

From these comparisons the SiO_2 content of the Martian dust may be estimated as 60 ± 10%.

Geochemical Implication

Before drawing any geochemical conclusions from the chemical composition of the Martian aerosol, it is important to determine if the dust, either suspended or on the surface, is chemically representative of the Martian surface rocks, and specifically if the dust could be higher in SiO_2 than the rocks from which it was derived.

A possible parallel to the Martian situation may be provided by terrestrial wind-deposited sediments (loess), found in large areas of North America, Europe, and Asia. Pettijohn (1957, p. 378) lists chemical analyses of 5 loess samples from the United States, China, and Belgium, with the following weight percentages of SiO_2: 59.3, 61.0, 64.6, 72.8, and 74.5. These values are somewhat more siliceous than the average continental crust, which is about 60% SiO_2 according to studies by Ronov and Yaroshevsky (1969). However, it is apparent that no gross chemical fractionation occurred in these sediments.

A second aspect of the question involves possible chemical fractionation by grain size of wind-deposited sediments. As Pettijohn (1957, p. 101) shows, chemical composition of terrestrial sediments (including water-deposited ones) is strongly dependent on grain size. Fine-grained sediments tend to be richer in clay minerals (chiefly kaolinite, about 46% SiO_2), and coarse-grained ones richer in quartz (pure SiO_2); thus fine-grained sediments are generally less siliceous than coarser-grained ones. Terrestrial loess has grain sizes largely between 30 and 60 μm (Pettijohn, 1957, p. 378), compared with the average grain size of less than 5 μm for clay (Krumbein and Sloss, 1951, p. 71), a major constituent of terrestrial sediments as a group. Evidence by IRIS and other experiments (Moroz and Ksanfomaliti, 1972) indicates that the dust particle size is a few micrometers. It is therefore expected that the dust is probably somewhat less siliceous than the rocks from which it is derived.

From the above evidence, it appears that the present estimate of $60 \pm 10\%$ SiO_2 for the Martian dust composition is probably fairly representative of the mean composition of the Martian crust. This being so, it supports the conclusion that geochemical differentiation of Mars has occurred (Hanel *et al.*, 1972*a*). Comparison of various differentiated and undifferentiated bodies is shown in Table I. The conclusion that Mars is differentiated is further supported by the existence of volcanoes and lava fields discovered by the television experiment on Mariner 9 (McCauley *et al.*, 1972). Many of the lava flows, such as those in the Nix Olympica–Tharsis region, appear similar in morphology and scale to basaltic lava flows as seen on terrestrial orbital photographs (Lowman and Tiedemann, 1971).

TABLE I

COMPARATIVE SiO_2 CONTENT OF TERRESTRIAL, LUNAR, AND MARTIAN CRUSTS AND CHONDRITES

	Earth[a]		Moon		Mars	Chondrite[d]
	Continental	Oceanic	Maria[b]	Highlands[c]		
SiO_2 (wt %)	60.2	48.7	39.9	47.1	60 ± 10	46.8

[a] Wyllie (1971).
[b] Lowman (1972); average of 22 crystalline rocks from Apollo 11 and 12 sites.
[c] Average of 8 fragmental rocks from Apollo 14 site (Lowman, 1972) and 2 soil samples from Apollo 15 Appennine Front site (Apollo 15 Preliminary Science Report, 1972).
[d] Richardson chrondrite (analyst, H. Wiik), cited as typical by Wood (1968); SiO_2 given relative to stony material; makes up 34.3% (wt) of total meteorite including metallic and sulfide phases.

[*Editor's Note:* In the original, material follows this excerpt.]

REFERENCES

[*Editor's Note:* Reference list has been abridged to include only those references cited in the text.]

ARONSON, J. R., EMSLIE, A. G., ALLEN, R. V., AND McLINDEN, H. G. (1967). Studies of the middle- and far-infrared spectra of mineral surfaces for applications of remote compositional mapping of the moon and planets. *J. Geophys. Res.* **72**, 687–703.

CONEL, J. E. (1969). Infrared emissivities of silicates: Experimental results and a cloudy atmosphere model of spectral emission from condensed particulate mediums. *J. Geophys. Res.* **74**, 1614–1634.

HANEL, R. A., CONRATH, B. J., HOVIS, W. A.,

KUNDE, V. G., LOWMAN, P. D., PEARL, J. C., PRABHAKARA, C., AND SCHLACHMAN, B. (1972*a*). Infrared spectroscopy experiment on the Mariner 9 mission: Preliminary results. *Science* **175**, 305–308.

HOVIS, W. A., AND CALLAHAN, W. R. (1966). Infrared reflectance of igneous rocks, tuffs, and red sandstone from 0.5 to 22 microns. *J. Opt. Soc. Am.* **56**, 639–643.

KRUMBEIN, W. S., AND SLOSS, L. L. (1951). "Stratigraphy and Sedimentation," p. 71. W. H. Freeman Co., San Francisco, CA.

LOWMAN, P. D., JR., AND TIEDEMANN, H. A. (1971). Terrain photography from Gemeni spacecraft: Final geologic report. X-644-71-15, Goddard Space Flight Center.

LYON, R. J. P. (1964). Evaluation of infrared spectrophotometry for compositional analysis of Lunar and planetary soils. Part II: Rough and powdered surfaces. NASA Contractor Report CR-100.

MCCAULEY, J. F., CARR, M. H., CUTTS, J. A., HARTMANN, W. K., MASURSKY, H., MILTON, D. J., SHARP, R. P., AND WILHELMS, D. E. Preliminary report on the geology of Mars. *Icarus* **17**, 289–327.

MOROZ, V. I., AND KSANFOMALITI (1972). Preliminary results of the astrophysical observations of Mars from AIS Mars-3. Preprint of paper presented at URSI/IAU/COSPAR Symposium on Planetary Atmospheres and Surfaces, Madrid, Spain, 10–13 May 1972.

PETTIJOHN, F. J. (1957). "Sedimentary Rocks," Harper & Brothers, New York.

RONOV, A. B., AND YAROSHEVSKY, A. A. (1969). Chemical composition of the Earth's crust. *In* "The Earth's Crust and Upper Mantle" (P. V. Hart, Ed.). Geophysical Monograph 13, Amer. Geophys. Union, Washington.

SALISBURY, J. W., VINCENT, R. K., LOGAN, L. M., AND HUNT, G. R. (1970). Infrared emissivity of lunar surface features 2. Interpretation. *J. Geophys. Res.* **75**, 2671–2682.

14

Reprinted from *J. Geophys. Res.* **79**:1623–1634 (1974)

Evidence About Hydrate and Solid Water in the Martian Surface From the 1969 Mariner Infrared Spectrometer

GEORGE C. PIMENTEL, PAUL B. FORNEY, AND KENNETH C. HERR

Space Sciences Laboratory and Chemistry Department
University of California, Berkeley, California 94720

The distribution of mineral hydrate and solid water on Mars is central to any discussion about the presence or absence of life on that planet. In fact, it is a key factor that will guide the selection of landing sites for future landing probes. Hence continuing observational effort has been directed toward adding to our knowledge of the presence of water in condensed phases on Mars, its form, and its geographic variability.

The characteristic spectral signature of condensed phase water, including mineral hydrates, adsorbed water, and ice, is a broad absorption (half width 300–400 cm^{-1}) centered near 3 μ (e.g., 3.08 μ for ice). The breadth of this band and the shift toward longer wavelengths from the gas phase absorption (near 2.66 μ) are well known to be due to hydrogen bonding [*Pimentel and McClellan,* 1960, chap. 3]. Since water of hydration in inorganic crystals is always linked in the crystal through hydrogen bonds [*Pimentel and McClellan,* 1960, chap. 9], the spectrum of a hydrated mineral always displays intense and broad absorption that tends to fall in this same spectral range, near 3 μ (e.g., see the spectra published by *Hunt et al.* [1950], *Miller and Wilkins* [1952], *Tuddenham and Lyon* [1960], *Hovis* [1965], and *Hunt et al.* [1973]. Adsorbed water is similarly hydrogen bonded to a silicate surface [*Pimentel et al.,* 1953] and will show similar absorption. Hence these ubiquitous spectral characteristics furnish, on the one hand, a reliable indicator for the presence of bound water, including water of hydration, although without additional spectral information they have limited diagnostic specificity.

Several studies have shown that in this 3-μ spectral range Mars has a low albedo [*Moroz,* 1964; *Sinton,* 1967; *Herr and Pimentel,* 1969; *Beer et al.,* 1971; *Houck et al.,* 1973], which has uniformly been attributed to condensed phase water. *Sinton* [1967] has also found differences between light and dark areas: he measured the minimum albedo over light areas to be at 3.0 μ, a slightly shorter wavelength than what he found over dark areas, 3.1 μ. He speculated that this difference might signify compositional differences.

Because of the spectral range scanned (2–14 μ) and the excellent geographic resolution (130 km at closest approach), the Mariner 6 and 7 infrared spectra offer unique evidence about both gaseous and condensed phase water and their distributions. In fact, the 3-μ absorption of condensed phase water was immediately obvious in the Mariner Mars (MM) 6 and 7 spectra at the first data return, and it was reported in our first publication [*Herr and Pimentel,* 1969]. However, both quantitative deductions and spectral differentiation between mineral hydrate water and ice have required extensive laboratory simulation studies and comparative computer analyses of the spectra. We report here the results of these analyses relevant to the presence, distribution, and form of condensed phase water. A report about gaseous water will be published separately (J. M. McAfee, A. M. Winer, D. Horn, K. C. Herr, and G. C. Pimentel, manuscript in preparation, 1974).

EXPERIMENT

The Mariner 6 and 7 spacecraft each carried an infrared spectrometer (IRS) that recorded the infrared spectrum from 1.9 to 14.4 μ with a 10-s scan time and better than 1% spectral resolution [*Herr et al.,* 1972]. (Mariner 6 and 7 were NASA missions managed by the Jet Propulsion Laboratory (JPL) and directed by project manager H. M. Schurmeier.) Laboratory simulation spectra were recorded with a flight-qualified spectrometer identical to the Mariner 6 and 7 instruments, as is described below.

Laboratory reflection-absorption spectra. Reflection-absorption spectra of ice, solid CO_2, and various minerals were studied in the 2- to 5-μ wavelength range with the modified environmental chamber (Figure 1). The chamber walls and light baffles were blackened with 3M 101-C10 velvet black paint (emissivity ϵ = 0.94–0.96) and cooled to 77°K. This cooling practically eliminated stray light from thermal sources other than the sample itself.

The IRS was maintained at its normal operating temperature, 238°K. The samples to be studied were cooled from below by an external liquid bath that provided any desired temperature down to 77°K. The bulk sample

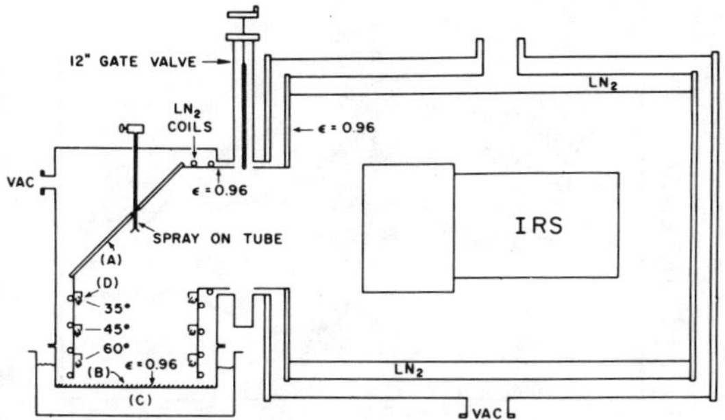

Fig. 1. Infrared spectrometer environmental test chamber. A, front surface mirror; B, sample area; C, temperature bath (77°–350°K); D, rings of tungsten lamps (glass envelopes removed).

temperature was measured with copper-constantan thermocouples covered by the sample.

When it is emplaced in the environmental chamber, the spectrometer views the sample through an optical port framed by a 12-inch gate valve and via a cooled front surface aluminum mirror. When the gate valve is closed, the sample can be deposited or changed without disturbing the vacuum environment in the instrument compartment.

Ice and solid CO_2 samples were prepared by condensing water vapor or gaseous CO_2 (Matheson, instrument grade, mass spectrometer analyzed, 99.99%) onto a 30-cm-diameter circular area coated with 3M 101-C10 velvet black paint and filling the field of view of the spectrometer. The deposition surface was held at 77°K, and the sample was deposited at a rate of about 0.7 mm/h. For mineral samples, typically 30–60 g were evenly distributed over the area. The samples, already selected according to particle size, were sifted through a Tyler standard sieve appropriate to the desired maximum grain size. Mineral sample thicknesses varied from 0.5 to 1.5 mm.

After deposition the sample compartment was evacuated to a pressure below 2×10^{-5} torr, and the inner walls and light baffles were cooled with liquid nitrogen. Then the gate valve was opened, and spectra were recorded.

Solar illumination of the samples is reasonably well simulated between 2 and 4 μ by the emission of 28-volt

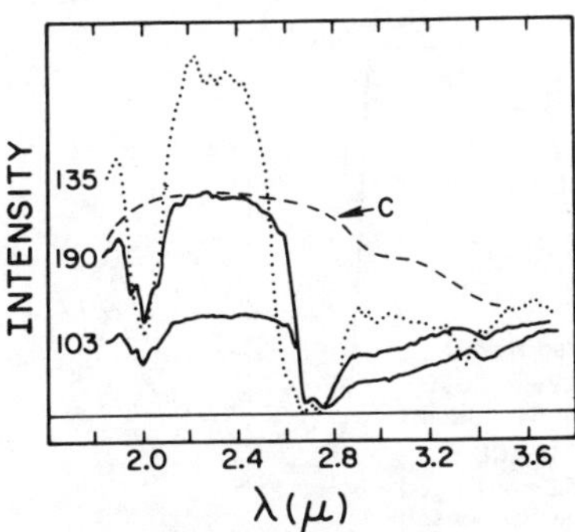

Fig. 2. Mariner 7 infrared spectra in the range 1.9–3.6 μ (normalized by the factor (cos IN)$^{-1}$). Spectrum 7-135: polar cap, 69°S, 356°E, IN = 60°. Spectrum 7-190: Hellas, 41°S, 63°E, IN = 66°. Spectrum 7-103: Meridiani Sinus, 7°S, 1°E, IN = 2°. The C represents the celestial sphere (see text). All intensities in this and subsequent figures are plotted on a linear scale.

Chicago miniature CM 7 tungsten lamps (glass envelopes removed). Three rings of these lamps are provided (radius 16.2 cm), and each ring holds 48 equally spaced lamps. The three rings were placed at different heights above the sample (5.7, 10.2, and 21.6 cm) to give a variety of illumination angles. The spectrometer's field of view sees at normal incidence the part of the sample contained between two concentric circles, the outer circle (radius 12.5 cm) being defined by the acceptance angle of the primary mirror and the inner circle (radius 5.1 cm) by the blocking of the secondary mirror. For a particular ring of lamps the collected light comes from a range of illumination angles. The contribution of a given angle depends upon the amount of area and the intensity of its illumination at that angle. For each lamp the intensity depends inversely upon the square of the distance from that lamp ρ and directly upon the sine of the incidence angle θ. If the contribution is weighted by $\sin \theta/\rho^2$, the light collected from one of the rings can be characterized by its average angle of illumination, near 35° for the lowest ring, 45° for the second, and 60° for the highest.

RESULTS

The Martian and laboratory spectral data will be presented first and then the systematics that are to be found in them. The significance of these systematics will be considered in the subsequent section.

Martian spectra. Figure 2 shows three MM7 spectra, 7-135 recorded over the southern polar cap, 7-190 recorded over Hellas (at 2.2 μ this is the brightest region observed except for polar cap spectra), and 7-103 recorded over Meridiani Sinus (at 2.2 μ the darkest region observed). The instrumental stray light background has been removed by using a deep-space spectrum (7-2) as the baseline. Each spectrum has also been divided by cos IN (IN is the angle of solar illumination measured from the normal), so that the vertical displacements in the figure show relative brightness B (i.e., relative albedo) if the Lambert law is reasonably followed. If B is measured at 2.2 μ and then normalized by dividing by the polar cap brightness (Mariner 7 spectrum 7-135), all of the nonpolar Martian values fall between 0.64 and 0.28.

Also included in Figure 2 is the solar spectrum reflected off an unknown part of the spacecraft as recorded by the Mariner 6 instrument 11 days after encounter during a search of the celestial sphere (normalized to spectrum 7-190 at 2.2 μ). Since these celestial sphere spectra show no evidence of pronounced absorptions by the spacecraft surfaces, they furnish a useful I_0 line. This I_0 line confirms the Martian origin of a broad absorption in the region 2.85–3.3 μ, as is characteristic of water in the condensed phase. The band does not have a well-defined minimum, and no systematic shifts in this wavelength could be discerned. This absorption can be more readily distinguished in an I/I_0 plot (based on the solar background I_0 curve) (Figure 3). Two MM7 spectra are shown: 7-94, recorded over a relatively bright region, Thymiamata, and 7-100, recorded 1 min later over a dark area in Meridiani Sinus. For spectrum 7-100 the contour of the CO_2 band at 2.70 μ (3700 cm^{-1}) has been sketched in (dashed line). The shaded area represents the additional absorption not due to CO_2 that is attributable to condensed phase water. The figure shows clearly that the extent of this absorption varies from one geographic locale to another.

This difference seen in Figure 3 encourages the hope that the Mariner 6 and 7 spectra can give information about the geographic variability of condensed phase water. To validate such interpretations, it is important to establish the photometric reproducibility of the IRS instruments. The best opportunity to do this is provided by the MM6 spectrum 6-193, since its field of view is centered only 34 km south of that of MM7 7-100 and its phase angle is only 7° different. In a plot like that of Figure 3 the 6-193 spectrum is indistinguishable from 7-100 within the noise level shown. This close match shows that photometric accuracy near 3 μ is much better than the variability of observed albedos. Interpolating to the intersection point of the two tracks, −3.4°S, 358.9°E, gives 2.2-μ brightness measurements from the two spectrometers that differ by only 1.7% and relative brightness measurements at 3.1 μ (i.e., $I(3.1\ \mu)/I(2.2\ \mu)$) that differ by only 5.4%. These differences (which are possibly caused by phase angle difference) are gratifyingly small when they are compared with the brightness variability of more than a factor of 2 and changes of relative brightness at 3.1 μ as large as the change displayed between MM7 spectra 7-94 and 7-100 (which differ by a factor of 1.5). There is no doubt that experimental uncertainty (a few percent) is much smaller than the variability of apparent albedo in the 3-μ spectral region.

Laboratory spectra. The infrared spectra of quite a range of mineral samples have been recorded and will be reported separately (K. C. Herr, P. B. Forney, and G. C. Pimentel, manuscript in preparation, 1974). These samples include the igneous rock materials rhyolite, obsidian, andesite, gabbro, four basalts of different ratios of plagioclase feldspars to pyroxenes, and the minerals pyroxene, olivine, albite, quartz, sandstone, goethite, siderite, and calcite. Figure 4 extracts from this comprehensive study three samples whose spectra most resemble those of Mars near 9 μ where silicate minerals absorb. At 2.2 μ none of the spectra show evidence of absorption, despite the differing albedo. In contrast, the figure shows that all three samples absorb throughout the region 2.7–3.3 μ (3700–3000 cm^{-1}). This is the ubiquitous absorption that is displayed by almost all terrestrial silicates due to water of hydration. The depth of the band increases as hydrate content increases. (We have also found that the depth of this band depends upon particle size. For example, for red basalt for particle sizes below 250 μ the depth decreases as particle size decreases, undoubtedly owing to increased surface reflections (K. C. Herr, P. B. Forney, and G. C. Pimentel, manuscript in preparation, 1974).) If the intense CO_2 absorption near 2.7 μ (the shaded area in Figure 4) that obscures part of the hydrate band is taken into account, the red basalt seems to most nearly match the Martian counterpart absorption in this region as well as in others. Hence we have used red basalt as substrate in our studies of solid ice and solid CO_2 discussed below.

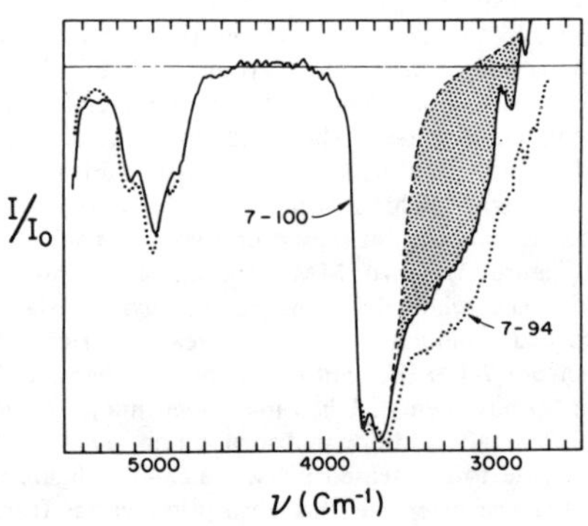

Fig. 3. Ratio spectra using the celestial sphere I_0 curve. Dotted line represents spectrum 7-94 (Thymiamata, 9°N, 353°E); solid line represents spectrum 7-100 (Meridiani Sinus, 3°S, 359°E).

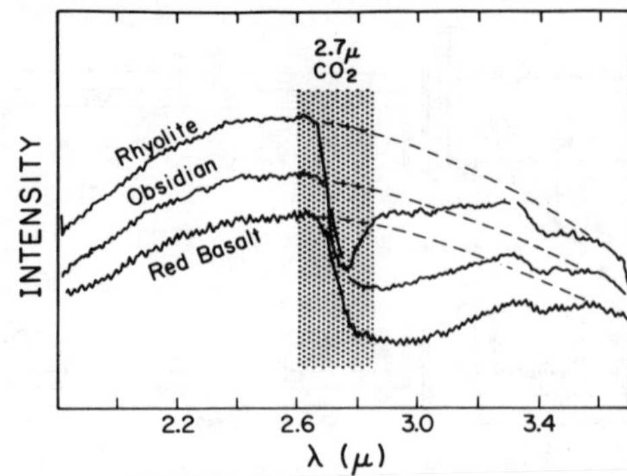

Fig. 4. Infrared spectra of three terrestrial silicates (particle size 40–160 μ, illumination angle 60°). The ordinates are displaced for clarity.

Figure 5 shows, in the dotted spectrum, the spectrum of the red basalt shown in Figure 4. The dashed spectrum shows how the spectrum changes when 3.2 mm of solid carbon dioxide is deposited on this mineral. The brightness generally rises except where CO_2 absorbs (at 2.1, 2.3, 2.75, 3.0, and 3.3 μ). These absorptions furnish the characteristic signature of finely divided solid CO_2, as was discussed earlier [*Herr and Pimentel,* 1969; *Kieffer,* 1970*a*]. The brightness increase caused by the CO_2, a factor of about 1.5 at 2.2 and 3.1 μ, is large and similar to the about doubled brightness displayed at the edge of the Martian polar cap at these two wavelengths (see also *Kieffer* [1970*b*]).

The solid curve (Figure 5) shows the spectrum obtained after a known volume of water was evaporated to give a very

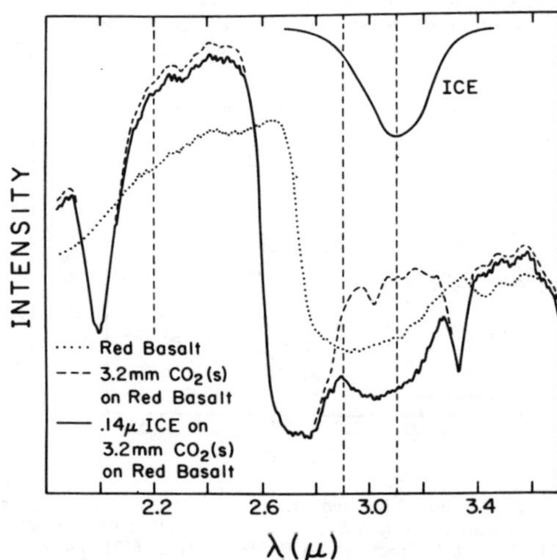

Fig. 5. The infrared spectra of solid CO_2 and ice, deposited successively on red basalt. Illumination angle is 60°.

TABLE 1*a*. Mariner 6 Albedos at 2.2, 2.9, and 3.1 μ

Spectrum	Location	*IN*, deg	Solar Hour Angle	Surface Temperature,* °K	*B*(2.2 μ)†	R_1§	R_2¶
			Candor				
153	6.6°N, 286.2°E	24.9	10.6	279	36.4	0.241	0.867
154	3.7°N, 291.2°E	19.2	11.0	281	34.4	0.224	0.861
155	1.7°N, 294.8°E	15.1	11.2	284	35.1	0.213	0.844
			Juventae Fons				
157	-1.3°S, 300.4°E	9.0	11.6	283	32.5	0.207	0.846
158	-2.5°S, 302.7°E	6.5	11.8	282	31.0	0.203	0.836
159	-3.6°S, 305.0°E	4.6	11.9	286	31.2	0.215	0.814
160	-4.6°S, 307.0°E	3.6	12.1	284	22.0	0.263	0.850
			Aurorae Sinus				
161	-5.6°S, 309.0°E	3.9	12.2	289	28.5	0.203	0.872
163	-7.2°S, 312.7°E	6.7	12.5	290	23.9	0.226	0.875
164	-8.0°S, 314.4°E	8.4	12.6	290	23.5	0.211	0.854
165	-8.7°S, 316.1°E	10.1	12.7	290	21.2	0.247	0.882
166	-9.4°S, 317.7°E	11.8	12.8	294	18.5	0.244	0.926
167	-10.0°S, 319.3°E	13.5	12.9	292	17.2	0.287	0.882
169	-11.2°S, 322.4°E	16.8	13.1	291	19.4	0.262	0.884
170	-11.7°S, 323.9°E	18.4	13.2	289	22.3	0.234	0.876
171	-12.2°S, 325.4°E	19.9	13.3	294	18.7	0.248	0.847
172	-12.7°S, 326.9°E	21.5	13.4	296	18.2	0.265	0.908
			Eos				
173	-13.1°S, 328.3°E	23.0	13.5	293	17.3	0.287	0.886
			(Slew)				
176	2.7°N, 338.1°E	34.3	14.3	283	27.0	0.239	0.849
177	2.1°N, 339.3°E	35.4	14.4	286	26.4	0.212	0.896
			Margaritifer Sinus				
178	1.7°N, 340.4°E	36.3	14.4	277	24.4	0.277	0.855
179	1.1°N, 341.6°E	37.4	14.5	275	33.7	0.233	0.800
181	0.2°N, 344.1°E	39.6	14.7	273	38.3	0.222	0.789
			Aram				
182	-0.3°S, 345.3°E	40.8	14.8	273	38.6	0.216	0.814
183	-0.7°S, 346.6°E	42.0	14.9	275	37.5	0.213	0.821
184	-1.1°S, 347.8°E	43.1	14.9	276	31.8	0.233	0.812
185	-1.4°S, 349.0°E	44.3	15.0	277	28.4	0.240	0.886
187	-2.1°S, 351.4°E	46.7	15.2	282	25.7	0.241	0.888
188	-2.3°S, 352.7°E	47.9	15.3	282	25.4	0.239	0.905
189	-2.6°S, 353.9°E	49.1	15.4	283	21.4	0.236	0.987
			Meridiani Sinus				
190	-2.8°S, 355.1°E	50.3	15.4	282	19.9	0.261	0.886
191	-3.0°S, 356.3°E	51.5	15.5	281	22.1	0.249	0.871
193	-3.4°S, 358.7°E	53.9	15.7	276	21.1	0.294	0.858
194	-3.6°S, 359.9°E	55.1	15.8	283	21.3	0.266	0.883
195	-4.2°S, 359.3°E	54.4	15.7	287	23.8	0.422	0.850
			(Slew)				
199	-8.6°S, 309.4°E	4.9	12.5	289	19.6	0.277	0.944
			Aurorae Sinus				
200	-10.7°S, 314.6°E	10.4	12.7	284	19.3	0.330	0.865
201	-12.1°S, 318.5°E	14.5	13.0	287	21.8	0.290	0.920
			Eos				
202	-13.3°S, 321.9°E	17.9	13.2	289	23.2	0.270	0.924
203	-14.2°S, 324.8°E	21.0	13.4	290	20.6	0.304	0.901
			Pyrrhae Regio				
205	-15.7°S, 330.0°E	26.3	13.8	285	20.8	0.321	0.889
206	-16.3°S, 332.4°E	28.7	14.0	290	20.5	0.320	0.905
207	-16.8°S, 334.6°E	30.9	14.1	291	20.1	0.337	0.862
208	-17.2°S, 336.8°E	33.1	14.3	289	19.8	0.356	0.878
209	-17.6°S, 338.8°E	35.1	14.4	287	21.4	0.317	0.909
211	-18.3°S, 342.7°E	39.0	14.7	284	22.2	0.311	0.898
212	-18.2°S, 344.7°E	40.8	14.8	281	21.3	0.337	0.912
			(Slew)				
213	-16.2°S, 346.8°E	42.7	14.9	280	24.0	0.322	0.880
214	-16.5°S, 348.6°E	44.4	15.1	276	25.5	0.316	0.866
215	-16.7°S, 350.3°E	46.1	15.2	273	28.7	0.299	0.876
217	-17.0°S, 353.6°E	49.4	15.4	271	34.5	0.263	0.838
218	-17.1°S, 355.3°E	51.0	15.5	268	32.1	0.284	0.841
219	-17.2°S, 356.9°E	52.5	15.6	266	35.2	0.273	0.858

TABLE 1a. (continued)

Spectrum	Location	IN, deg	Solar Hour Angle	Surface Temperature,* °K	B(2.2 μ)†	R_1§	R_2¶
220	-17.3°S, 358.4°E	54.1	15.7	265	34.8	0.264	0.824
			Deucalionis Regio				
221	-17.3°S, 0.3°E	55.9	15.9	263	36.8	0.267	0.834
223	-17.3°S, 3.3°E	58.9	16.1	261	35.5	0.277	0.857
224	-17.3°S, 4.8°E	60.3	16.2	258	35.8	0.283	0.838
225	-17.3°S, 6.3°E	61.8	16.3	257	35.7	0.265	0.886
226	-17.2°S, 7.7°E	63.2	16.4	255	35.2	0.286	0.875
227	-17.1°S, 9.2°E	64.6	16.5	252	36.4	0.278	0.832
229	-16.9°S, 12.0°E	67.4	16.7	251	34.2	0.282	0.901
230	-16.8°S, 13.4°E	68.8	16.8	249	33.7	0.287	0.866
231	-16.6°S, 14.8°E	70.1	16.9	247	34.1	0.269	0.914
232	-16.4°S, 16.2°E	71.5	17.0	242	34.0	0.286	0.913
233	-16.3°S, 17.5°E	72.9	17.1	242	31.7	0.290	0.854
235	-15.8°S, 20.2°E	75.6	17.3	239	33.5	0.294	0.816
236	-15.6°S, 21.6°E	76.9	17.4	239	32.3	0.280	0.914
237	-15.4°S, 22.9°E	78.3	17.5	237	33.2	0.265	0.818
238	-15.1°S, 24.2°E	79.6	17.6	237	31.7	0.254	0.982
239	-14.8°S, 25.6°E	80.9	17.7	235	29.7	0.244	1.246
241	-14.2°S, 28.2°E	83.6	17.8	233	28.6	0.270	0.887
242	-13.9°S, 29.5°E	84.9	17.9	231	26.5	0.236	1.327
243	-13.6°S, 30.8°E	86.3	18.0	231	26.5	0.209	1.676
244	-13.2°S, 32.1°E	87.6	18.1	229	35.4	0.210	1.423
245	-12.8°S, 33.4°E	88.9	18.2	230	49.3	0.185	1.687
			(Terminator)				

Terminator: 18.26 solar hour angle, 34.4° longitude.
*See discussion of Table 1 in text for information (K. C. Herr, P. B. Forney, D. K. Stone, and G. C. Pimentel, manuscript in preparation, 1974).
†$I(2.20\ \mu)/\cos IN$.
§$I(3.10\ \mu)/I(2.20\ \mu)$.
¶$I(2.90\ \mu)/I(3.10\ \mu)$.

thin layer of ice (0.14 μ) deposited evenly over this CO_2-covered basalt sample (which was held at 77°K). This deposition procedure gives finely divided ice and would probably be classified by *Kieffer* [1970*b*] as low density. The brightness at 2.2 μ is almost unaffected, but there is a striking increase in absorption near 3 μ. The ratio of the last two spectra (solid/dashed) closely matches the spectrum of solid ice, whose band profile is shown in Figure 5, as measured in separate experiments. The ice is characterized by maximum absorption at 3.10 μ (3230 cm^{-1}), and when it is deposited on basalt alone, on solid CO_2, or on stainless steel, it displays less absorption at 2.90 μ (3450 cm^{-1}) than at the band center, 3.10 μ. Absorption at 3.10 μ approaches 100% for ice layers exceeding 0.5 μ in thickness.

Figure 5 shows that both ice and water of hydration in terrestrial minerals absorb near 3.10 μ but that ice absorbs relatively less at 2.90 μ than hydrate water. Consequently, an increase in the amount of condensed phase water at a given locale would cause a reduction in the apparent albedo at 3.10 μ (relative to, say, that at 2.2 μ). The possibility that ice might contribute to this increased absorption can be assessed by examining the relative intensities at 2.90 and 3.10 μ, since this ratio is less than unity for all mineral hydrates that we have studied and exceeds unity for ice.

With these key frequencies in mind we list in Table 1 all of the MM6 and 7 spectra recorded in the 1.9- to 3.6-μ spectral region. For each spectrum we list latitude, longitude, solar hour angle, surface temperature, angle of solar illumination *IN*, the brightness at 2.20 μ divided by cos *IN* ($B(2.2\ \mu) = I(2.20\ \mu)/\cos IN$), the brightness ratio $R_1 = I(3.10\ \mu)/I(2.20\ \mu)$, and the brightness ratio $R_2 = I(2.90\ \mu)/I(3.10\ \mu)$. (Estimates of surface temperature have been made from the MM6 and 7 IRS data on the basis of absolute brightness at three wavelengths, 5.45, 7.77, and 12.12 μ. The results are in good agreement and in general accord with the MM6 and 7 radiometer results [*Neugebauer et al.*, 1971]. The IRS temperature data will be reported elsewhere (K. C. Herr, P. B. Forney, D. K. Stone, and G. C. Pimentel, manuscript in preparation, 1974). Only the 5.45-μ temperature estimate is quoted here since it is measured at about the desired geographic locale and for both the MM6 and the MM7 instrument. This temperature is that measured during the specified spectrum, though, of course, 9.2 s after the beginning of the scan.) The R_1 ratios fall in the range 0.43–0.15 (to be compared with the celestial sphere ratio 0.70). Increase in the total amount of condensed phase water is signaled by a decrease in R_1, although not unambiguously so, since this ratio can also be affected by particle size [*Hovis*, 1965; K. C. Herr, P. B. Forney, and G. C. Pimentel, manuscript in preparation, 1974]. The presence of ice would be indicated by a rise in R_2.

SYSTEMATIC CORRELATIONS

Careful examination of the data in Table 1 reveals a series of systematic correlations involving $B(2.2\ \mu)$, R_1, and R_2 with various parameters: local temperature, latitude, and proximity to the polar cap. Fortunately, there are sufficient data to examine these relationships more or less independently.

Brightness and temperature. Before the 3-μ behavior manifested in R_1 and R_2 is examined, it is fruitful to consider

TABLE 1*b*. Mariner 7 Albedos at 2.2, 2.9, and 3.1 μ

Spectrum	Location	*IN*, deg	Solar Hour Angle	Surface Temperature,* °K	$B(2.2\ \mu)$†	R_1§	R_2¶
Thymiamata							
92	14.9°N, 350.1°E	26.2	11.3	265	31.8	0.231	0.839
93	11.7°N, 351.6°E	22.7	11.4	272	34.4	0.225	0.871
94	9.0°N, 352.8°E	19.7	11.5	271	34.7	0.206	0.883
95	6.5°N, 353.9°E	17.0	11.6		36.1	0.218	0.855
97	2.4°N, 355.9°E	12.4	11.7	278	33.1	0.220	0.872
98	0.5°N, 356.8°E	10.3	11.8	287	27.7	0.252	0.845
99	-1.2°S, 357.7°E	8.4	11.8	295	22.2	0.291	0.871
100	-2.9°S, 358.6°E	6.5	11.9	297	22.5	0.265	0.861
101	-4.5°S, 359.4°E	4.9	12.0		19.7	0.291	0.883
Meridiani Sinus							
103	-7.4°S, 1.1°E	2.3	12.1	298	17.0	0.295	0.888
104	-9.0°S, 2.1°E	2.4	12.1	299	16.5	0.314	0.852
105	-10.1°S, 2.7°E	3.1	12.2	298	18.1	0.300	0.896
106	-11.6°S, 3.7°E	4.7	12.3	298	18.6	0.315	0.854
Deucalionis Regio							
107	-14.6°S, 4.1°E	6.9	12.3		21.4	0.284	0.852
(Slew)							
109	-41.5°S, 343.7°E	35.1	10.9	272	28.7	0.280	0.817
Argyre I							
110	-44.9°S, 323.3°E	47.3	9.5	250	29.0	0.304	0.809
111	-46.0°S, 317.3°E	51.4	9.1	241	26.7	0.276	0.955
112	-47.2°S, 318.6°E	51.2	9.2	237	26.2	0.286	0.914
113	-48.3°S, 320.0°E	51.1	9.3		24.3	0.288	0.934
115	-50.6°S, 322.6°E	51.1	9.5	237	26.7	0.247	0.957
116	-51.6°S, 323.9°E	51.2	9.6	238	27.1	0.256	0.913
117	-52.7°S, 325.3°E	51.3	9.7	239	27.9	0.255	0.909
Mare Oceanidum							
118	-53.7°S, 326.6°E	51.5	9.8	240	26.2	0.268	0.875
119	-54.7°S, 327.9°E	51.7	9.9		27.8	0.237	0.886
121	-56.7°S, 330.5°E	52.3	10.1	245	27.6	0.227	0.938
122	-57.7°S, 331.9°E	52.7	10.1	244	26.4	0.223	0.998
Mare Australe							
123	-58.6°S, 333.3°E	53.1	10.2	239	28.1	0.215	0.903
124	-59.6°S, 334.7°E	53.5	10.3	241	26.9	0.211	1.006
125	-60.5°S, 336.1°E	53.9	10.4		31.1	0.203	0.959
Polar Cap Edge							
127	-62.3°S, 339.1°E	54.9	10.7		47.0	0.164	1.210
128	-63.2°S, 340.7°E	55.5	10.7		50.1	0.192	1.287
South Polar Cap							
129	-64.1°S, 342.3°E	56.0	10.9		56.4	0.174	1.312
130	-65.0°S, 344.0°E	56.6	11.0		62.6	0.189	1.248
131	-65.8°S, 345.7°E	57.2	11.1		59.8	0.278	1.140
133	-67.5°S, 349.4°E	58.5	11.4		62.3	0.247	1.098
134	-68.3°S, 351.4°E	59.2	11.5		63.1	0.261	1.137
135	-69.1°S, 353.5°E	59.9	11.6		61.1	0.259	1.167
136	-69.9°S, 355.7°E	60.6	11.8		63.7	0.262	1.140
137	-70.7°S, 358.1°E	61.4	12.0		60.4	0.276	1.117
139	-72.2°S, 3.2°E	62.9	12.4		58.5	0.240	1.084
140	-72.9°S, 6.0°E	63.7	12.6		58.6	0.253	1.087
141	-73.6°S, 9.1°E	64.6	12.7		62.0	0.238	1.135
142	-74.2°S, 12.3°E	65.4	13.0		62.2	0.227	1.130
143	-74.8°S, 15.8°E	66.3	13.2		61.3	0.204	1.100
145	-75.9°S, 23.5°E	68.1	13.7		61.2	0.151	1.097
146	-76.4°S, 27.8°E	69.0	14.0		54.1	0.184	1.086
147	-76.8°S, 32.3°E	69.9	14.3		58.0	0.179	1.126
148	-77.2°S, 37.2°E	70.9	14.7		53.3	0.252	1.153
149	-77.4°S, 42.2°E	71.9	15.0		52.6	0.231	1.104
151	-77.8°S, 53.0°E	73.9	15.8		48.4	0.236	1.107
152	-77.8°S, 58.6°E	74.9	16.1		52.9	0.235	1.131
153	-77.7°S, 64.2°E	76.0	16.5		53.7	0.188	1.044
(Slew)							
159	-7.7°S, 355.7°E	2.4	(*11.9*)		(*11*)	(*0.174*)	(*0.586*)
Deucalionis Regio							
160	-14.9°S, 0.5°E	6.4	12.2	284	23.4	0.330	0.897
161	-18.8°S, 3.5°E	11.2	12.5		31.2	0.304	0.880

TABLE 1*b*. (continued)

Spectrum	Location		*IN*, deg	Solar Hour Angle	Surface Temperature,* °K	*B*(2.2 μ)†	R_1§	R_2¶
				Pandora Fretum				
163	-24.3°S,	8.2°E	18.2	12.8	286	30.2	0.298	0.898
164	-26.4°S,	10.3°E	21.1	12.9	285	30.5	0.297	0.950
165	-28.3°S,	12.3°E	23.8	13.1	283	32.5	0.303	0.915
166	-30.0°S,	14.3°E	26.2	13.2	283	34.6	0.288	0.870
167	-31.6°S,	16.2°E	28.5	13.3		32.4	0.298	0.890
				Noachis				
169	-34.3°S,	20.0°E	32.7	13.6	283	29.9	0.305	0.902
170	-35.6°S,	21.9°E	34.7	13.7	280	31.5	0.286	0.891
171	-36.7°S,	23.8°E	36.6	13.8	280	31.8	0.290	0.863
172	-37.7°S,	25.6°E	38.4	14.0	279	31.9	0.286	0.886
173	-38.7°S,	27.5°E	40.2	14.1		30.3	0.305	0.875
175	-40.4°S,	31.3°E	43.6	14.4	273	32.3	0.280	0.880
176	-41.2°S,	33.2°E	45.3	14.5	275	30.2	0.301	0.907
177	-41.9°S,	35.1°E	46.9	14.6	273	27.0	0.293	0.902
178	-42.6°S,	37.0°E	48.4	14.7	271	25.9	0.297	0.907
179	-43.2°S,	39.0°E	50.0	14.9		25.1	0.297	0.901
				Hellespontus				
181	-44.2°S,	42.8°E	53.0	15.1	265	22.0	0.322	0.906
182	-44.6°S,	44.8°E	54.5	15.3	263	24.0	0.333	0.953
183	-44.7°S,	47.0°E	56.0	15.5	260	23.8	0.323	0.928
				(Slew)				
184	-40.3°S,	51.4°E	57.6	15.6	255	34.5	0.295	0.926
				Hellas				
185	-40.6°S,	53.2°E	59.0	15.9		34.6	0.293	0.944
187	-41.0°S,	57.2°E	62.1	16.2	251	35.6	0.287	0.981
188	-41.1°S,	58.6°E	63.2	16.3	248	36.5	0.296	0.923
189	-41.2°S,	60.8°E	64.8	16.4	244	38.2	0.281	0.949
190	-41.2°S,	62.5°E	66.1	16.6	242	39.0	0.300	0.891
191	-41.2°S,	64.3°E	67.5	16.7		37.9	0.297	0.939
193	-41.1°S,	67.8°E	70.2	16.9	239	37.6	0.327	0.882
194	-40.9°S,	69.5°E	71.5	17.0	237	37.9	0.316	0.915
195	-40.7°S,	71.2°E	72.8	17.2	235	36.3	0.319	0.956
196	-40.5°S,	72.9°E	74.1	17.3	234	38.8	0.311	0.910
197	-40.2°S,	74.6°E	75.4	17.4		36.6	0.334	1.007
199	-39.6°S,	77.9°E	78.0	17.6	231	37.3	0.341	1.008
200	-39.2°S,	79.5°E	79.3	17.7	230	39.2	0.359	0.917
201	-38.8°S,	81.1°E	80.6	17.8	228	38.5	0.364	0.930
202	-38.4°S,	82.7°E	81.9	17.9	227	37.6	0.364	0.947
203	-37.9°S,	84.2°E	83.2	18.0		37.5	0.427	0.898
205	-36.8°S,	87.3°E	85.8	18.3	224	40.0	0.395	1.041
206	-36.2°S,	88.7°E	87.1	18.4	223	45.6	0.404	0.949
207	-35.6°S,	90.2°E	88.4	18.5	223	74.1	0.356	1.020
208	-34.9°S,	91.6°E	89.7	18.6	221	316	0.252	1.241
				(Terminator)				
209	-34.2°S,	93.0°E	91.0	18.7			0.343	1.362
211	-32.7°S,	95.8°E	93.6	18.9	220		0.253	1.195
212	-31.9°S,	97.1°E	94.9	19.0	219		0.418	1.171
213	-31.0°S,	98.4°E	96.3	19.1	219		0.387	1.089
214	-30.2°S,	99.7°E	97.6	19.2	218		0.674	0.893

Entries in italics are data taken when the instrument was crossing the limb.

*See discussion of Table 1 in text for information (K. C. Herr, P. B. Forney, D. K. Stone, and G. C. Pimentel, manuscript in preparation, 1974).

†*I*(2.20 μ)/cos *IN*.

§*I*(3.10 μ)/*I*(2.20 μ).

¶*I*(2.90 μ)/*I*(3.10 μ).

the systematic interdependence of brightness at 2.2 μ and local surface temperature. Since the surface temperature displays both a strong diurnal variation and a strong latitude dependence [*Neugebauer et al.*, 1971; K. C. Herr, P. B. Forney, D. K. Stone, and G. C. Pimentel, manuscript in preparation, 1974], it is necessary to isolate data over restricted ranges of solar hour angle and latitude. The optimum solar hour angle is between 11.5 and 15.0, within which the surface temperature reaches a maximum and changes rather slowly. Figure 6 shows a plot of the available MM6 data within this hour angle range and for near-equatorial latitudes between 3°N and 14°S. There are 29 data points shown, 15 from the first track ending in Aurorae Sinus and, after the first slew, nine from the track ending in Meridiani Sinus. Then, after the second slew, there are five more observations in Aurorae Sinus. Each measured brightness is plotted versus the average of the two temperatures measured before and after that scan, as is indicated by the error bars. The general trend shows clearly that dark areas on Mars are considerably warmer than adjacent

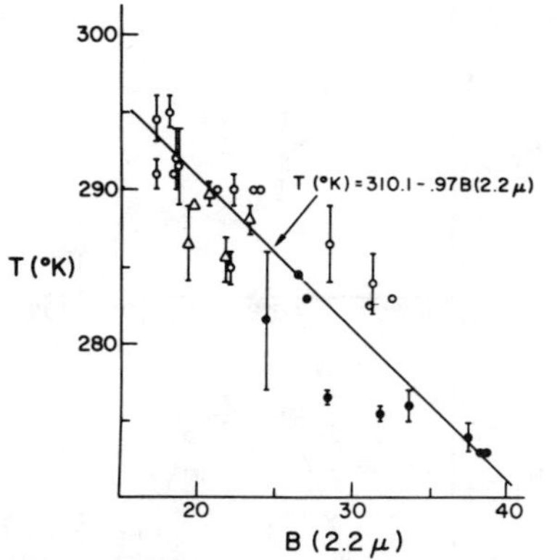

Fig. 6. Surface temperature versus albedo at 2.2 μ, $B(2.2\ \mu)$. Latitude 3°N–14°S, solar hour angle 11.5–15.0. Mariner 6 spectra 6-157 to 6-173 (open circles), 6-176 to 6-185 (solid circles), and 6-199 to 6-203 (triangles).

bright areas. At latitudes near the equator and just after noon (i.e., at the warmest time of day) doubling the albedo lowers the temperature by about 19°. This correlation is most vividly evident in the MM6 spectra 6-176 to 6-191, for which the field of view passed successively over the dark region Margaritifer Sinus, then over the very bright region Aram, and finally over Meridiani Sinus, the darkest area that we observed. The lower plot in Figure 7 shows how the 2.2-μ albedo displays these well-known features. In the upper plot the solid curve shows the experimental temperatures, again plotted against hour angle. Then the dotted curve shows the temperatures calculated from the 2.2-μ brightness by using the straight-line relationship defined by the points in Figure 6. The agreement is remarkably good until the late-afternoon diurnal temperature drop begins to be evident (at times later than 15.0).

Brightness and $R_1 = I(3.1\ \mu)/I(2.2\ \mu)$. There has been much discussion and speculation about the likelihood that compositional and/or particle size variations contribute to the striking albedo changes that define the prominent markings on Mars [e.g., *Adams and McCord*, 1969; *Binder and Jones*, 1972; *McCord*, 1969; *Salisbury and Hunt*, 1969]. If such variations exist, they might be reflected in the variability of the hydrate absorption band near 3 μ, as was suggested by *Sinton* [1967]. To investigate this possibility and to eliminate any latitude dependence, we have examined separately the data for each of two 10° latitude ranges, 0°–10°S (26 data) and 10°S–20°S (43 data). In each of these latitude ranges the data are sufficiently numerous to provide a reasonable statistical picture of the possible correlations between brightness $B(2.2\ \mu)$ and the two brightness ratios R_1 and R_2. Even so, the scatter of the data is such that it is beneficial to examine averages.

Figure 8 shows the average 3.1-μ relative brightness R_1 plotted versus the 2.2-μ brightness B. Each point averages over all data within a 5-unit brightness range, beginning with 15.1–20.0. The error bars represent $\sigma_n = \sigma_1/(n^{1/2})$, where σ_1 is the standard deviation of a single datum and n is the number of data averaged. The magnitude of σ_1 was uniformly about 10% of R_1. In the 0°–10°S latitude range there is a significant (30%) drop in R_1 as brightness increases. This drop is much less evident at 10°S–20°S (about 10–15%), and no significant trend could be observed at 40°S–50°S, the other latitude range in which the data are sufficiently numerous to provide a test. The observations north of the equator, in the range 10°N–0°, though few, are in accord with the 0°–10°S curve.

When R_2 is examined in a similar manner, there is only a slight trend discernible and only in the 0°–10°S latitude range. Here R_2 seems to decrease systematically from an average of about 0.90 in a dark region ($B \approx 19$) to about 0.82 in a light region ($B \approx 38$). This modest change is almost masked by the scatter of the data.

Latitude and R_1, R_2. To investigate the dependence of R_1 and R_2 on latitude, it is desirable to consider only data from a limited brightness range. We have selected the 2.2-μ brightness interval $25.0 \leq B \leq 35$. Table 2 summarizes the averages of all such data, grouped in latitude intervals of 10° (20°N–10°N, 10°N–0°, etc.). Average hour angles and surface temperatures are given in the table, and they indicate that the trends in R_1 and R_2 cannot be ascribed to systematic variation in either variable.

Figure 9 shows that both R_1 and R_2 depend upon latitude.

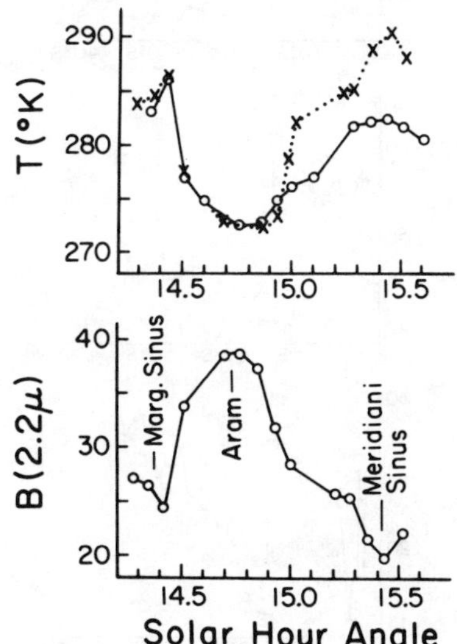

Fig. 7. Surface temperatures calculated from albedo at 2.2 μ; spectra 6-176 to 6-191. Crosses, calculated temperatures; circles, experimental temperatures.

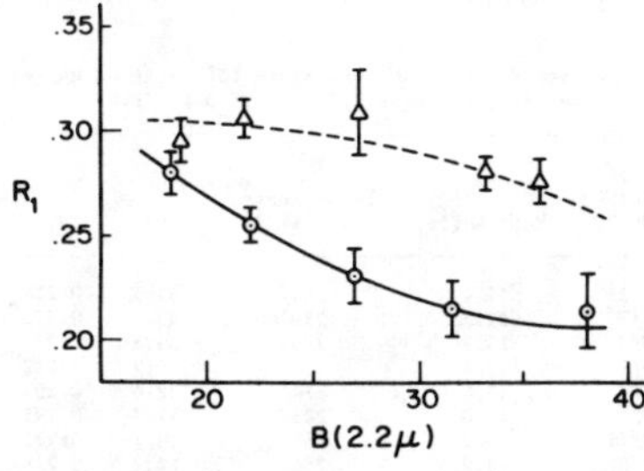

Fig. 8. Plot of $R_1 = I(3.1\ \mu)/I(2.2\ \mu)$ versus $B(2.2\ \mu)$. Circles represent latitudes of 0°–10°S; triangles, 10°S–20°S.

Again, the error bars shown are $\sigma_n = \sigma_1/(n^{1/2})$ values. Apparently, R_2 tends to increase as latitude becomes more southerly. In contrast, R_1 seems to be high in the mid-southerly latitudes, 10°S–50°S, and low both near the equator (and north) and far south beyond 50°S.

Polar cap and R_1, R_2. The MM7 data provide an excellent view of the behavior of R_1 and R_2 in the vicinity of the polar cap edge (the so-called 'collar') and onto the polar cap. There are 15 observations north of the polar cap in the latitude range 41°S–60°S with slowly varying morning hour angle (from 9.1 to 10.4) and approximately constant brightness (in the range 24.3–29.0). The polar cap edge is reached at 60.5°S, and there are 24 more observations over the cap, extending to 78°S and to a late-afternoon hour angle of 16.5.

The values of B, R_1, and R_2 are shown in Figure 10 plotted against latitude. The uppermost plot of the 2.2-μ brightness shows a precipitous rise beginning at 60.5°S and finally leveling off at 65°S. Then the brightness remains reasonably constant to 76°S, after which it wanes somewhat. The 5° latitude range 60°S–65°S defines the edge of the polar cap, and our temperature measurements show that the rising brightness can be explained by increasing ground coverage by solid CO_2 (K. C. Herr, P. B. Forney, and G. C. Pimentel, manuscript in preparation, 1974). The decreasing brightness for the seven most southerly measurements (76°S–78°S) is also of special interest. In the spectra for this region there is almost no trace of the 2.3-, 3.0-, and 3.3-μ features of solid CO_2. Yet there is no doubt that the polar cap fills the IRS field of view. Some possible explanations of the loss of these features have been listed by *Herr and Pimentel* [1969].

The 3.1-μ relative brightness R_1 decreases steadily as the latitude changes from 40°S to 60°S, whereas R_2 shows a tendency to rise. At the polar cap edge both R_1 and R_2 show discontinuities: R_1 drops and remains low to latitude 65°, and R_2 rises to the highest value observed anywhere on the planet (1.3) in this same interval. Thereafter R_2 remains above 1.1 throughout the polar cap observations. The value of R_1 rises and remains reasonably stable in the range 0.24–0.28 until about 75°S and beyond, the polar cap region where the brightness begins to decrease and where the 'forbidden' spectral features in the CO_2 spectrum are lost.

Plainly, the three curves in Figure 10 correlate in defining a 5° latitude range at the edge of the polar cap that is unique. This range might well constitute a collar to the polar cap. In this region the brightness suggests partial ground coverage, R_1 suggests unusually large absorption due to condensed phase water, and R_2 suggests that this absorption is due to ice.

Further evidence about the cause of the R_1 and R_2 excursions is provided by ratio spectra in which one Mars spectrum is used as an I_0 curve for another spectrum from the collar range. The best opportunity is offered by spectra 7-129, recorded at latitude 64.1°S, and 7-131, recorded 20 s later and farther south on the polar cap at latitude 65.8°S. These two spectra differ in brightness by only 6%, so that surface coverage is about the same, but they display extreme contrasts in R_1 and R_2. The lower curve in Figure 11 shows the ratio of

TABLE 2. Average Values of R_1 and R_2 in 10° Latitude Ranges at Fixed Brightness, $B(2.2\ \mu)$ = 25.0 to 35.0

Latitude	No. of Data (MM6/7)	Hour Angle	Temperature, °K	B	R_1	R_2
13.3°N	0/2	11.4	269	33.1	0.228	0.855
2.9°N	6/3	12.7	279	30.9	0.223	0.860
-2.5°S	8/0	13.5	280	29.3	0.223	0.857
-16.6°S	16/1	16.2	256	32.7	0.282	0.873
-27.3°S	0/4	13.0	284	32.0	0.297	0.908
-35.7°S	0/6	13.8	281	31.3	0.295	0.884
-42.7°S	0/11	13.0	261	29.1	0.291	0.897
-55.1°S	0/9	9.9	240	27.2	0.238	0.932
-60.5°S	0/1	10.4		31.1	0.203	0.959

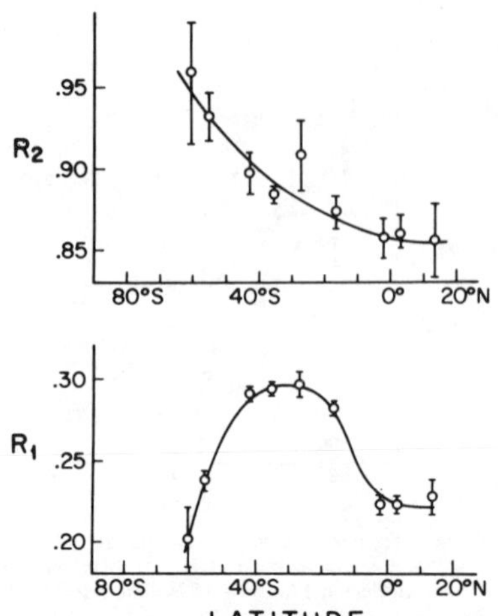

Fig. 9. Variation of R_1 and R_2 with latitude (B(2.2 μ) between 25.0 and 35.0).

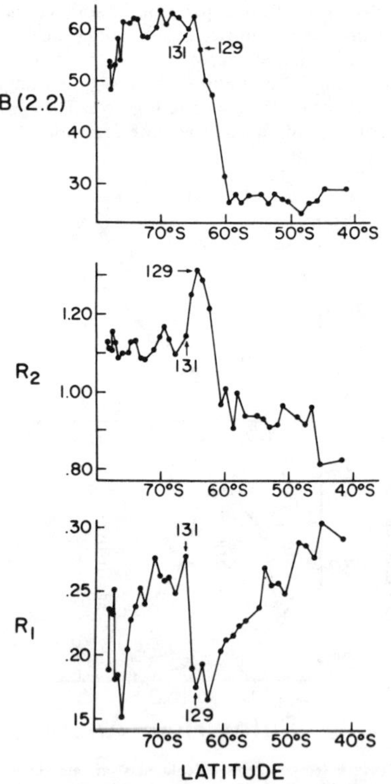

Fig. 10. Behavior of B(2.2 μ), R_1, and R_2 at the polar cap.

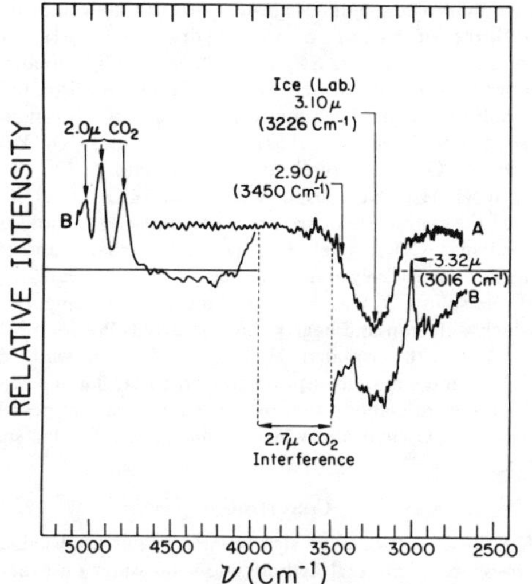

Fig. 11. Ratio spectra at the edge of the polar cap. Spectrum A, laboratory reference spectra, where I is 0.08-μ ice on red basalt and I_0 is red basalt. Spectrum B, polar cap spectra, where I is spectrum 7-129 and I_0 is spectrum 7-131.

these two spectra. They have been scaled to equal brightness over the 2.2-μ region, this scaling producing clearly recognizable and characteristic signatures of gaseous CO_2, the triplet near 2 μ, and of solid CO_2, the sharp spike at 3.3 μ. These features are inverted, and it is thus shown that absorption in spectrum 7-131 exceeds that in 7-129 in each case. The ratio spectrum also includes a broad minimum centered at 3.1 μ. Since this feature is not inverted, it indicates heavier absorption in spectrum 7-129. This feature closely resembles the signature of ice, as is shown by the upper curve, which is a ratio of two laboratory spectra: the spectrum of 0.08-μ ice on red basalt divided by the spectrum of that same sample of red basalt. The correspondence of the two ratio spectra is obvious. We see that Mars spectrum 7-129 seems to differ from Mars spectrum 7-131 by the superposition of the ice spectrum. It is significant that the expected partial pressure of atmospheric water vapor, 10^{-3}–10^{-4} torr (J. M. McAfee, A. M. Winer, D. Horn, K. C. Herr, and G. C. Pimentel, manuscript in preparation, 1974) would permit condensation at surface temperatures between 190° and 200°K. This range is encompassed by our measured surface temperature for spectrum 7-124, 241°K (at latitude 59.6°S), and the condensation temperature of solid CO_2 at the ambient atmospheric pressure, near 147°K. In fact, our measured temperature (which is, presumably, an effective temperature that averages covered and uncovered regions) passes through 194°K for spectrum 7-129, apparently consistent with the ice interpretation (K. C. Herr, P. B. Forney, D. K. Stone, and G. C. Pimentel, manuscript in preparation, 1974).

DISCUSSION

The data reported here were recorded on July 31 and August 5, 1969. In the southern hemisphere, where most of the observations were made, these dates corresponded to an early Martian spring. The subsolar point latitudes were 8.1°S and 9.3°S for the MM6 and MM7 encounters, respectively.

The comparisons in Figure 5 leave little room for doubt that the 2.85- to 3.3-μ absorption band can be reproduced by terrestrial minerals. The breadth of this absorption characteristically associates it with water of hydration and/or ice. Figure 3 shows that there are significant changes in this absorption in different locales. Clues to the causes of these changes are presumably contained in the systematic correlations displayed in Figures 6–10. These correlations add considerably to our knowledge of the variability of surface hydration on Mars, and they invite interpretation. They will be considered again in the same sequence that was presented in the preceding section.

Brightness and temperature. The fact that Martian surface temperatures qualitatively anticorrelate with visual albedo has been deduced by *Neugebauer et al.* [1971] as they attempted to model their recorded thermal data. This relationship can be made quantitative with the IRS data on the basis of the brightness at 2.2 μ (see Figure 6). The implications of these measured albedos for thermal modeling of the surface minerals will be explored in a separate report (K. C. Herr, P. B. Forney, D. K. Stone, and G. C. Pimentel, manuscript in preparation, 1974). The immediate relevance here is that surface temperature could well be an operative factor in the degree of surface hydration. Hence investigations of the dependence of the 3-μ absorption upon locale and hour angle are more revealing when comparisons are confined to a limited brightness range. Coincidentally, such a limitation groups measurements of similar phase angles.

Brightness and R_1. It is interesting that only in the most northerly observations (above 10°) is there a significant dependence of R_1 upon brightness B (2.2 μ) (Figure 8). In assessing this fact it is perhaps relevant that the most striking Martian albedo contrasts seem to be confined to latitudes more northerly than 20°S. Thus in Table 1 there are 82 observations south of 20°S, and only four of these display a brightness below 25.0. In contrast, there are 92 observations north of 20°S, and 40 of these are darker than 25.0. Perhaps both the R_1 dependence upon B and the more pronounced light/dark contrasts near the equator are caused by compositional and/or particle size variability that is more or less confined to this equatorial belt. Since R_1 (as well as B) depends upon the 2.2-μ intensity, it is useful to note that no Martian spectra showed evidence of an absorption band in this spectral region except over the polar cap.

Whatever the explanation of the geographic localization, we conclude that near the equator the relative brightness at 3.1 μ decreases as 2.2-μ brightness increases. Since brightness increase is accompanied by temperature decrease (Figure 6), the 0°–10°S R_1 correlation might be attributed to changing surface temperature. If so, lowering the surface temperature lowers R_1, as would be expected if hydration increases. However, in view of the modest temperature decrease (~20°), the observed 30% decrease in R_1 is probably at least in part due to composition and/or particle size differences.

It is appropriate to comment upon the impact of our data on the contention of *Houck et al.* [1973] that the hydrate absorption is centered at 2.8 μ and on *Sinton*'s [1967] conclusion that for light areas on Mars there is an albedo minimum near 3.0 μ and that this minimum shifts to longer wavelengths, near 3.1 μ, over dark areas. The discrepancy between the Mariner 6 and 7 results and those of Houck et al. is obviously understandable in terms of the high scatter of their data. Their deduced minimum albedo, 2.8 μ, is within the range of intense absorption by CO_2 near 2.7 μ (Figure 2), which absorption band is not even distinguishable in their spectral

Figure 5. Turning to the discrepancy with Sinton's minimum albedo (at 3.0–3.1 μ), we note that in the present results, at all latitudes except those over the polar cap and at or beyond the terminator, R_2 was always found to be below unity. It is thus manifested that over most of the planet there is no minimum albedo in the 3-μ spectral region. Thus our data tend to contradict Sinton's conclusion.

Also deserving note is the observation just mentioned that, for both MM6 and MM7, R_2 seemed to rise above unity just at the terminator. This rise could signify ice condensation at nightfall, either on parts of the planetary surface or in the atmosphere. Unfortunately, at dusk the signal is very small in this spectral range, so that the confidence level in this conclusion is not high.

Latitude and R_1, R_2. Table 2 shows that the dependence of R_1 upon latitude (Figure 9) cannot be ascribed to a systematic variation of either hour angle or surface temperature. The latter conclusion is strengthened by the observed latitude dependence of R_2 (Figure 9) since near the equator R_2 varies only slightly with brightness, hence with temperature.

The monotonic increase of R_2 as the IRS field of view moves south, coupled with the falling value of R_1, suggests that absorption due to ice might be occurring below 40°S. Figure 5 shows that a film of ice a few hundredths of a micron thick would account for the small trend observed. Furthermore, such an interpretation of the R_2 behavior would not necessarily imply that the ice is on the planetary surface; ice crystals in the lower atmosphere, as has been proposed by *Fanale and Cannon* [1971], or, of course, in the upper atmosphere, as well, would have the same effect on R_1 and R_2.

This explanation of the R_2-trend leaves unexplained the drop in R_1 at northerly latitudes. We are investigating the possibility that this is due to a phase angle effect, since the observations at the mid-latitudes 15°S–45°S were made at large phase angles (~83°), whereas lower angles were involved in the other regions (35°–55°).

The polar cap and R_1, R_2. Certainly the most spectacular and significant changes in R_1 and R_2 are those recorded at the edge of the polar cap. The 2.2-μ brightness shows that at the time of these late-morning observations (hour angle 10.4–11.0) the surface is only partially covered between latitudes 60°S and 65°S. Just in this 5° latitude range R_2 is peaked at 1.3, and R_1 drops to the lowest values recorded on the planet (Figure 10). These correlating changes are just those that we have shown to be associated with ice formation (Figure 5). That ice is a likely explanation is attested by the ratio spectrum shown in Figure 11 and by our thermal measurements, which record average temperatures that reach the ice condensation temperature at about 63°S. This latter evidence also favors the likelihood that this ice is on the planetary surface, not in the atmosphere. Taken all together, the observations indicate that there is ice on parts of the surface at the edge of the waning polar cap.

The value of R_2 remains high (above unity) over the polar cap beyond this collar region, although R_1 returns to an intermediate value (near 0.25). This might signify that there is still condensed phase H_2O absorption but somehow differently placed. For example, in the collar the ice might be on the exposed ground, whereas farther south the ice might be under or on solid CO_2. Laboratory experiments have shown that a layer of ice can be detected in the 3-μ spectral region underneath as much as a centimeter of solid CO_2. One possibility that can be put aside is that this broad 3-μ band is due to molecular H_2O suspended at high dilution in solid CO_2 (rather than aggregated in clusters of H_2O molecules or in crystallites of carbon dioxide hydrate, $CO_2 \cdot 6H_2O$, as suggested by *Miller and Smythe* [1970]). There is a wealth of experience with the infrared spectra of molecules isolated in inert solid matrices, including the spectrum of H_2O in solid nitrogen [*Van Thiel et al.*, 1957] and H_2O in solid CO_2 (Y. M. Huang and G. C. Pimentel, unpublished data, 1970). In the latter work H_2O was suspended in solid CO_2 at 20°K to provide the characteristic spectrum of isolated H_2O molecules trapped in a CO_2 crystal. The spectrum is sharp, and it is located in an entirely different spectral region (near 2.7 μ). Then the solid CO_2 was warmed to discover the temperature at which diffusion and aggregation occurred. We found that near 125°K the isolated H_2O molecular spectrum disappeared, and a new absorption appeared near 3 μ in a broad band closely resembling that of unannealed ice (but possibly due to the CO_2 hydrate, which would have a similar spectrum).

CONCLUSIONS

The present work adds significantly to our knowledge of the presence of mineral hydrate and solid water on Mars. It provides the first opportunity to investigate the distribution of such water and its dependence upon other planetary variables.

The data show that the R_1 intensity ratio $I(3.1\ \mu)/I(2.2\ \mu)$ depends systematically upon brightness (at least near the equator) and upon latitude. Such evidence signifies some sort of compositional and/or particle size variability that might include variability of the extent and nature of hydration. (It is to be noted that the mineral spectra presented by *Hovis* [1965] show striking dependence upon particle size at 2.0 μ but that they change very little near 3 μ. It is not clear, however, what this implies for R_1 because the hydrated samples that Hovis studied are so heavily hydrated that they are almost completely absorbing for all particle sizes (implying $R_1 \approx 0$). We have also investigated albedo as a function of particle size, with particular attention to basalts. These results will be reported separately (K. C. Herr, P. B. Forney, and G. C. Pimentel, manuscript in preparation, 1974).) If we attribute the systematic changes of R_1 to increased hydration, hydrate water content is largest in very bright areas near the equator and north of it and at latitudes more southerly than 50°S.

The R_2 intensity ratio $I(2.9)/I(3.1)$ has more specific diagnostic value: a high value is characteristic of ice, for which R_2 exceeds unity. Because an ice layer only a few hundredths of a micron thick is readily detected, R_2 is a very sensitive indicator. Thus the present data put to rest the classical speculation that light areas on Mars might generally owe their brightness to frost formation, whether at high altitudes, as suggested by *Slipher* [1962] and *Tombaugh* [1966], or at low altitudes, as suggested by *Sagan and Pollack* [1966] and *O'Leary and Rea* [1967].

In the MM6 and 7 observations this R_2 ratio rises steadily at the more southerly latitudes, and it is accompanied by a drop in R_1. The same effect is noted at the terminator, although less definitely. These changes could be due to ice thinly covering a small fraction of the planetary surface in particularly cold spots, possibly on partially shaded slopes, as was suggested by *Balsamo and Salisbury* [1973]. At southerly latitudes the fraction so covered seems to increase as the polar cap edge is approached. However, the most distinctive occurrence of this characteristic R_1, R_2 pattern is found at the edge of the polar cap, from 60°S–65°S, where ratio spectra closely resemble the ice spectrum as well. This is strong evidence that ice is formed on the planetary surface at the edge of the polar cap.

Acknowledgments. We gratefully acknowledge research support from the National Aeronautics and Space Administration that permitted the development of the infrared spectrometer and to the Jet Propulsion Laboratory, which supported the fabrication and use of the flight instrumentation. Space Sciences Laboratory Report, series 14, issue 60, August 1, 1973.

REFERENCES

Adams, J. B., and T. B. McCord, Mars: Interpretation of spectral reflectivity of light and dark regions of Mars, *J. Geophys. Res., 74,* 4851–4856, 1969.

Balsamo, S. R., and J. W. Salisbury, Slope angle and frost formation on Mars, *Icarus, 18,* 156–163, 1973.

Beer, R., R. H. Norton, and S. V. Martonchik, Astronomical infrared spectroscopy with a Connes-type interferometer, 2, Mars, 2500–3500 cm^{-1}, *Icarus, 15,* 1–10, 1971.

Binder, A. B., and J. C. Jones, Spectrophotometric studies of the photometric function, composition, and distribution of the surface materials of Mars, *J. Geophys. Res., 77,* 3005–3020, 1972.

Fanale, F. P., and W. A. Cannon, Adsorption on the Martian regolith, *Nature, 230,* 502–503, 1971.

Herr, K. C., and G. C. Pimentel, Infrared absorptions near three microns recorded over the polar cap of Mars, *Science, 166,* 496–499, 1969.

Herr, K. C., P. B. Forney, and G. C. Pimentel, Mariner Mars 1969 infrared spectrometer, *Appl. Opt., 11,* 493–501, 1972.

Houck, J., J. B. Pollack, C. Sagan, D. Schaack, and J. Decker, High altitude infrared spectroscopic evidence for bound water on Mars, *Icarus, 18,* 470–480, 1973.

Hovis, W. A., Jr., Infrared reflectivity of iron oxide minerals, *Icarus, 4,* 425–430, 1965.

Hunt, G. R., L. M. Logan, and J. W. Salisbury, Mars, components of infrared spectra and the composition of the dust cloud, *Icarus, 18,* 459–469, 1973.

Hunt, J. M., M. P. Wisherd, and L. C. Bonham, Infrared absorption spectra of minerals and other inorganic compounds, *Anal. Chem., 22,* 1478–1497, 1950.

Kieffer, H., Interpretation of the Martian polar cap spectra, *J. Geophys. Res., 75,* 510–514, 1970*a*.

Kieffer, H., Spectral reflectance of CO_2 and H_2O frosts, *J. Geophys. Res., 75,* 501–509, 1970*b*.

McCord, T. B., Comparison of the reflectivity and color of bright and dark regions on the surface of Mars, *Astrophys. J., 156,* 79–86, 1969.

Miller, F. A., and C. H. Wilkins, Infrared spectra and characteristic frequencies of inorganic ions, their use in qualitative analysis, *Anal. Chem., 24,* 1253–1294, 1952.

Miller, S. L., and W. D. Smythe, Carbon dioxide clathrate in the Martian ice cap, *Science, 170,* 531–533, 1970.

Moroz, V. I., The infrared spectrum of Mars (1.1 to 4.1μ) (in Russian), *Astron. Zh., 41,* 350–361, 1964. (*Sov. Astron. AJ, 8,* 273–281, 1964.)

Neugebauer, G., G. Münch, H. Kieffer, S. C. Chase, Jr., and E. Miner, Mariner 1969 infrared radiometer results: Temperatures and thermal properties of the Martian surface, *Astron. J., 76,* 719–728, 1971.

O'Leary, B. T., and D. G. Rea, Mars: Influence of topography on formation of temporary bright patches, *Science, 155,* 317–319, 1967.

Pimentel, G. C., and A. L. McClellan, *The Hydrogen Bond,* pp. 67–141, 255–295, W. H. Freeman, San Francisco, Calif., 1960.

Pimentel, G. C., C. W. Garland, and G. Jura, Infrared spectra of heavy water adsorbed on silica gel, *J. Amer. Chem. Soc., 75,* 803–805, 1953.

Sagan, C., and J. B. Pollack, Elevation differences on Mars, *J. Geophys. Res., 73,* 1373–1388, 1966.

Salisbury, J. W., and G. R. Hunt, Compositional implications of the spectral behavior of the Martian surface, *Nature, 222,* 132–136, 1969.

Sinton, W. M., On the composition of Martian surface materials, *Icarus, 6,* 222–228, 1967.

Slipher, E. C., *A Photographic History of Mars,* chap. 3, Lowell Observatory, Flagstaff, Ariz., 1962.

Tombaugh, C. W., Evidence that the dark areas on Mars are elevated mountain ranges, *Nature, 209,* 1338, 1966.

Tuddenham, W. M., and R. J. P. Lyon, Infrared techniques in the identification and measurement of minerals, *Anal. Chem., 32,* 1630–1634, 1960.

Van Thiel, M., E. D. Becker, and G. C. Pimentel, Infrared studies of hydrogen bonding of water by the matrix isolation technique, *J. Chem. Phys., 27,* 486–490, 1957.

Editor's Comments on Paper 15

15 **SURKOV et al.**
Preliminary Results of Investigations of Gamma Radiation from Mars from Mars 5 Observations

The surface abundances and distribution of natural radioactive elements (U,Th,K^{40}) yield significant evidence concerning the extent of crustal differentiation, since the large ions of these elements concentrate in the early stages of partial melting. Crystallization of the upwelling magma results in near-surface enrichment of these elements. The level of surface radioactivity, coupled with other data, such as heat flow values, can provide information about the distribution of heat sources in the interior and place constraints on the thermal evolution of the planet.

A useful techniqure for measuring surface radioactivity in planetary exploration has been orbital gamma-ray spectrometry. Because gamma rays are absorbed by the atmosphere, the method is limited to planets with a very thin or no atmosphere. Gamma-ray spectrometry has been successfully applied to the geochemical mapping of the Moon (Metzger et al. 1973). The western lunar maria were found to be significantly more radioactive than the eastern maria. These geographical variations probably reflect chemical inhomogeneities in the surface brought about by igneous differentiation. Gamma-ray spectrometer analyses have been carried out on the surface of Venus by Soviet unmanned spaceprobes (Vinogradov 1973; Surkov et al. 1976). Rock types inferred from levels of radioactivity range from granitic (Venera 8) to basaltic (Venera 9 and 10), thereby indicating a well-differentiated planet.

Yu. A. Surkov and others (Paper 15) present initial results from the Soviet Mars 5 gamma-ray spectrometer experiment. Measurements were obtained from an average altitude of 2300 km over a large area including Thaumasia, Mare Erythraeum, and Margaritifer Sinus, which covers a variety of geologic units from sparsely to heavily cratered plains and cratered uplands (see Paper 6, Part II). The observed gamma-ray energy spectra from Mars 5 appear to lie between the extremes represented by terrestrial mafic and sialic rocks, although Surkov et al. conclude that ". . . for a vast region of Mars the uranium, thorium and potassium content corresponds to terrestrial igneous rocks of basic composition." Further analysis of the data led to a K/U ratio close to 10^4, with K and U contents similar to terrestrial mafic rocks (Surkov et al. 1977a). (Actual values are K = 3000 ppm, U = 1 ppm, Th = 5 ppm; from Surkov et al. 1977b, cited by Anders et al. 1977). The Viking X-ray fluorescence experiment and volcanic morphologies tend to support the Soviet conclusion (see Paper 16, Part V and Paper 24, Part VII).

15

Reprinted from *Space Research* COSPAR **14**:993–1000 (1976)

PRELIMINARY RESULTS OF INVESTIGATIONS OF GAMMA RADIATION FROM MARS FROM MARS 5 OBSERVATIONS

Yu. A. Surkov, L. P. Moskalyova, F. F. Kirnozov, V. P. Kharyukova, O. S. Manvelyan and O. P. Shcheglov

V. I. Vernadsky Institute of Geochemistry and Analytical Chemistry, Moscow, USSR

The paper presents a description and a preliminary analysis of the data obtained by the gamma-ray spectrometer aboard the Mars 5 automatic interplanetary station. The aim of the experiment was to determine the content of natural radioactive isotopes and the main elements in the rocks forming the Martian surface.

Gamma-ray spectra are given on the basis of measurements on the route from earth to Mars and when the probe was in the vicinity of Mars. Components due to nuclear interactions of cosmic rays with the atmosphere and rocks of Mars are singled out in the gamma-ray spectra. The content of natural radioactive isotopes in the rocks of the area under investigation is evaluated.

1. Introduction

During the past few years the intensity and the spectral composition of gamma-radiation from rocks forming the surface layers of the moon and the planets have been measured aboard space vehicles. The main purpose of such experiments is to determine the chemical composition and the rock type.

The possibility of determining the rock type according to the characteristic gamma radiation emitted was first pointed out in 1967 [1]. This idea was subsequently developed in a number of theoretical papers: [2, 3] deal with some processes leading to the emission of gamma-ray quanta at the lunar surface; in [4] it is shown that measurements of gamma radiation near the Martian surface would facilitate identification of the types of Martian rocks.

The first measurements of gamma-ray spectra over the lunar surface showed that such studies were promising [5]. The gamma-ray spectrometers installed on the Venera 8 (Venus 8) probe and on Apollos 15 and 16 made it possible to determine the type of the rocks forming the Venusian and lunar surfaces and the content of natural radioactive elements in them [6, 7]. Here the results of the first such experiment to study the gamma radiation from Mars are considered.

2. The Experiment

The Mars 5 automatic interplanetary station was launched in February 1974. The equipment on board included a gamma-ray spectrometer with a NaI(Tl) crystal, 63 mm × 63 mm in size, placed in a scintillating plastic container. Signals caused by gamma-ray quanta and charged particles are separated by the pulse height analyser. The energy range was from 0.6 to 8 MeV. The number of channels is 256; the channel capacity is 2^{16} pulses. The resolution at the 0.661 MeV line of ^{137}Cs is 11%.

The spectrometer was calibrated over a wide energy range of gamma radiation by means of standard sources. Regularities in the gain coefficient variation and the zero drift of the energy scale were established.

3. Gamma Radiation Measurements from Mars 5

In accordance with the programme of the probe's operation a series of measurements was performed en route from earth to Mars at distances of 61.2, 85.4 and 93.8 million kilometres from the earth. One of measured spectra is shown in Fig. 1; this background spectrum represents a continuum distribution with

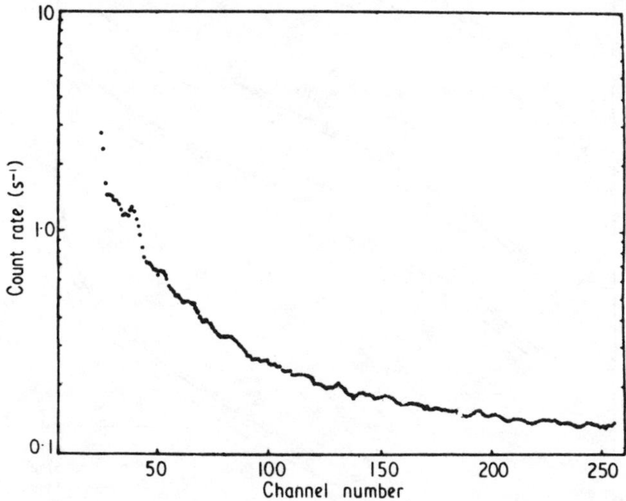

Fig. 1. Background gamma-radiation spectrum observed in interplanetary space.

weakly pronounced peaks. This gamma radiation is mainly due to nuclear interactions between cosmic rays and the material of the interplanetary station and the detector. Also, the ^{40}K radioisotope, contained in small amount in the material of the detector and the spacecraft, also makes a contribution to the background gamma radiation.

Several gamma-radiation measurements were carried out when the spacecraft was nearest Mars. The altitude when measurements commenced was 2600 km, the minimum altitude was 1760 km. Measurements ceased at a height of 4190 km; the average height above the planet was about 2300 km. The exposure time during each measurement was 60 minutes.

During measurements near the planet the detector recorded gamma-ray quanta coming from some effective area which is determined by the instrument's spatial resolution. In order to determine the size of this area S the fluxes of gamma-ray quanta from a spherical layer on the planet uniformly distributed with radioactive material were calculated for altitudes ranging from 100 to 10000 km. Fig. 2 shows the fractional intensities calculated for various energies of gamma-ray quanta as a function of the planetary effective area S for several values of the heights H. The intensity is given as a percentage of the total intensity determined by the detector's solid angle.

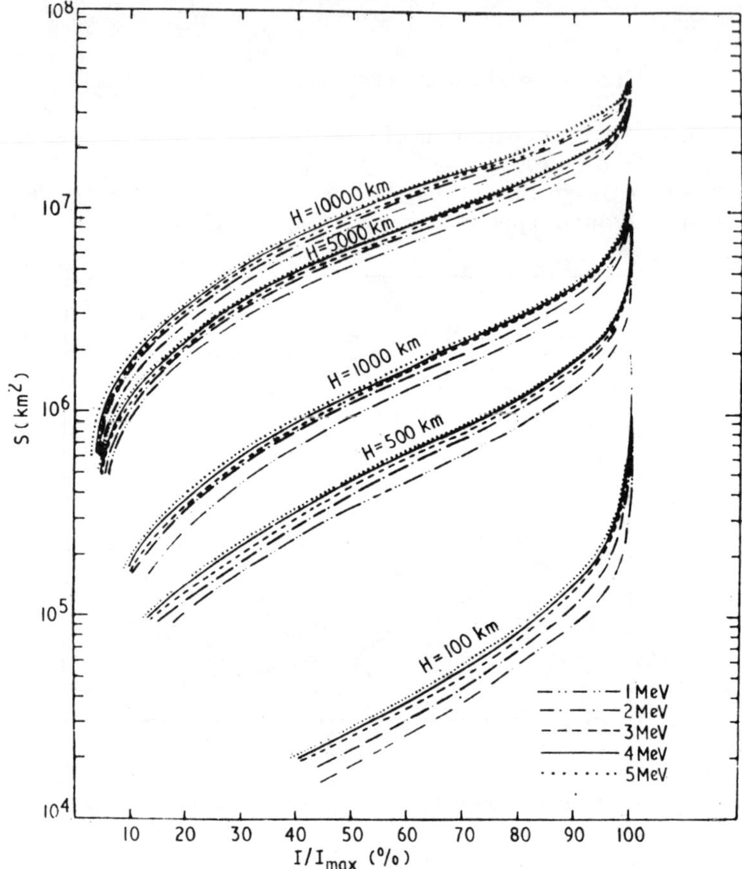

Fig. 2. Intensity of gamma radiation expected, as a function of the height of the spacecraft and the radiating area of the Martian surface S. (Intensity is expressed as a percentage of maximum intensity over the entire viewing angle of the apparatus, I/I_{max}.) — ·· — ·· — 1 MeV; — · — · — 2 MeV; ----- 3 MeV; ——— 4 MeV; ······ 5 MeV.

Fig. 3 shows the radius of the effective area as a function of height H for three energies. If we mean by the planet's effective area a surface section, the intensity of radiation from which is 90% (or 50%) of the total intensity, then at a height of 2000 km and an energy of 3 MeV the radius of the effective area R_{eff} is about 1600 km (or $\sim$ 900 km).

The map of Mars (Fig. 4) shows the probe's trajectory over which the gamma-ray spectrometer operated and the effective area of the planetary surface from which gamma radiation was registered. The external boundary of the hatched band determines the region corresponding to the effective area from which 90% of the gamma radiation is collected, while the internal boundary is for 50%. It is evident from the figure that in all Mars 5 measurements the detector "scanned" the areas from 25° N to 50° S and from 130° W to 320° E, i.e. mainly the Thaumasia, Argyre and Coprat areas, the Phoenix Lake, the Sheba and Pearl Bays. The influence of the surface relief on the recorded intensity can be regarded as insignificant.

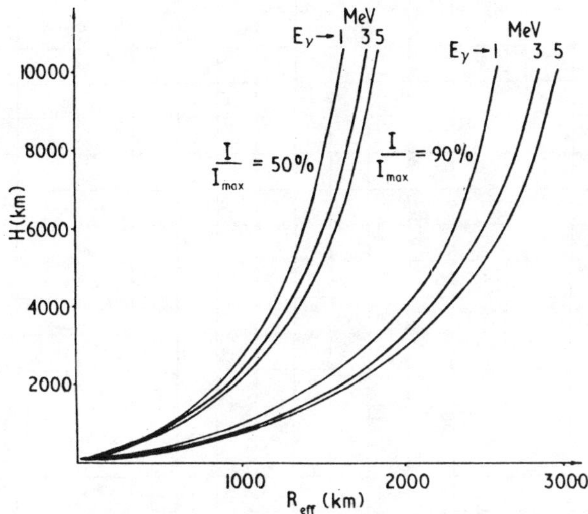

Fig. 3. Radius of effective area as a function of the orbit height.

Fig. 5 shows some Martian gamma radiation spectra measured from the orbit of Mars 5. The integral intensity of the gamma radiation recorded here is on the average 109 ± 6 pulses per second from 1 to 8 MeV. In order to determine the Martian gamma radiation the background radiation of 41 ± 5 pulses per second for the same energy range has to be subtracted.

A preliminary analysis of the experimental data was performed to find out the contribution to the gamma radiation of natural uranium, thorium and potassium radioisotopes carrying important geochemical information. Fig. 6, curve 1, shows a spectrum of Martian gamma radiation. Weakly pronounced peaks correspond to gamma radiation with energies 1.4—1.5; 1.8—1.9; 2.5—2.6; 3.5—3.7; 4.4—4.5; 6.0—6.15 and 6.8—6.9 MeV. Gamma-ray quanta of such energies are due both to the decay of potassium, thorium and uranium (1.46, 2.62 and 1.76 MeV respectively) and to nuclear interactions of cosmic rays and secondary nucleons (mainly neutrons) with elements whose presence can be expected in the Martian atmosphere and rock: carbon (4.43 MeV), oxygen (3.68, 6.13 and 6.93 MeV), magnesium (1.83 and 2.62 MeV), aluminium (3.54 and 6.8 MeV) and iron (1.82 and 6.02 MeV).

From the data we have obtained during the experimental simulation on accelerators and high balloons the contributions of the gamma radiation of the atmosphere and the rock were determined [4, 8, 9]. Curve 2 in Fig. 6 represents a simulated spectrum of gamma radiation induced by cosmic rays in the atmosphere

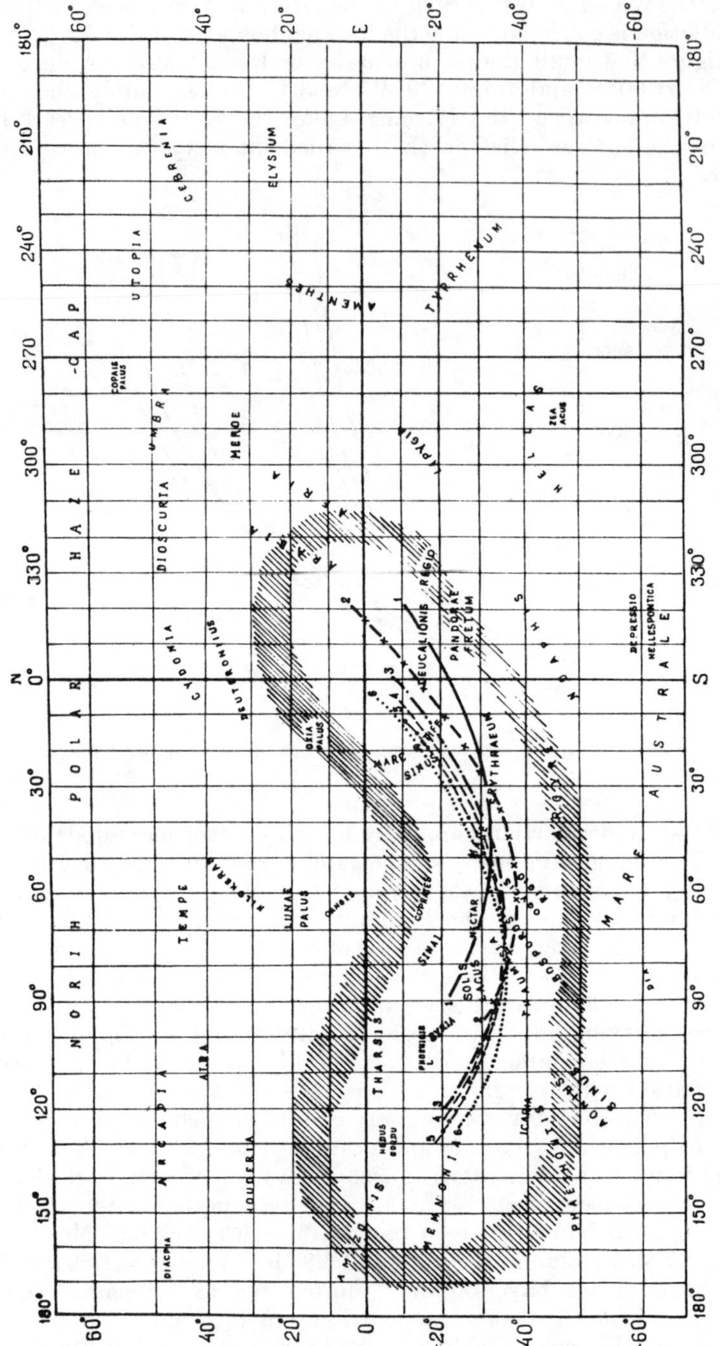

Fig. 4. Map of Mars with projections of the probe trajectory.

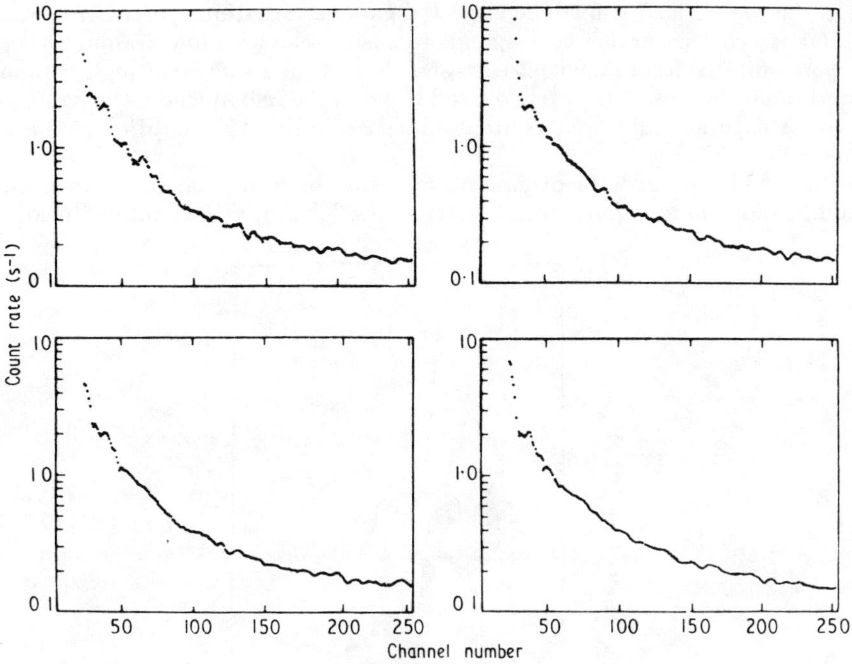

Fig. 5. Gamma-radiation spectra of the Martian surface measured from the orbit of Mars 5.

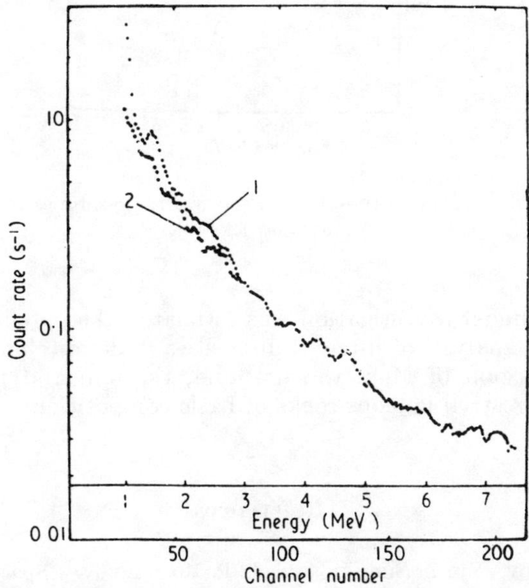

Fig. 6. Gamma-radiation spectrum measured from the orbit of Mars 5 with the background subtracted (1) and the spectrum of gamma-radiation induced by cosmic rays (2).

and in the rock. Fig. 7 shows the expected gamma radiation spectra of terrestrial and Martian rocks: curve 1 corresponds to basic rocks with low uranium, thorium and potassium content; curve 2 corresponds to acid rocks with high content of natural radioisotopes. The hatched band 3 shows the region where the gamma-ray spectra of natural radioisotopes are situated according to the different measurements.

Using calibration spectra of gamma radiation of terrestrial rocks with known uranium, thorium and potassium contents the level of their content in Martian

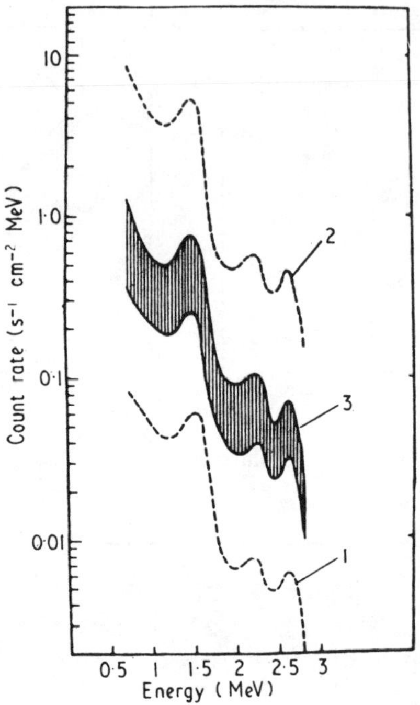

Fig. 7. Intensity levels of gamma-radiation of natural radioisotopes in terrestrial (1, 2) and Martian rocks (3).

rocks of the area under investigation was estimated. The preliminary evaluation based only on the analysis of integral intensities in definite energy ranges shows that for a vast region of Mars the uranium, thorium and potassium content corresponds to terrestrial igneous rocks of basic composition.

References

[1] A. P. Vinogradov, Yu. A. Surkov and L. P. Moskalyova, Space Research VII, 71 (1967).
[2] T. W. Armstrong and R. G. Alsmiller, Astronaut. Acta, **14**, 143 (1969).
[3] T. W. Armstrong, J. Geophys. Res. **77**, 524 (1972).

[4] Yu. A. Surkov, L. P. Moskalyova and V. P. Kharyukova, Space Research XI, 181 (1971).
[5] A. P. Vinogradov, Yu. A. Surkov et al., Kosm. Issled. **5**, 374 (1967).
[6] A. P. Vinogradov, Yu. A. Surkov, F. F. Kirnozov and V. N. Glazov, Geokhimia, No. 1, p. 3 (1973).
[7] R. C. Reedy, J. R. Arnold and J. I. Trombka, J. Geophys. Res. **78**, 5847 (1973).
[8] Yu. A. Surkov, L. P. Moskalyova et al., Appl. Nucl. Spectrosc. **3**, 246 (1972).
[9] Yu. A. Surkov and L. P. Moskalyova, Abstracts of paper at Nat. Conf. on Cosmic Rays, Tashkent 1968. USSR Academy of Sciences Press, Moscow 1968.

Part V

GEOCHEMISTRY—IN SITU ANALYSES

Editor's Comments on Paper 16

16 **BAIRD et al.**
Mineralogic and Petrologic Implications of Viking Geochemical Results from Mars: Interim Report

The chemical composition of the martian surface has been determined by X-ray fluorescence measurements at the two Viking landing sites (Clark et al., 1976, 1977). A disadvantage of the X-ray fluorescence method is that mineral composition can only be inferred indirectly. Baird et al. interpret (Paper 16) the chemical data in terms of probable mineralogy and petrology. The chemical composition does not correspond to any single igneous rock type and probably represents a mixture (Clark et al. 1977). Baird et al. find that nontronite (iron-rich montmorillonite) is a likely component, which could have formed by UV radiation (Huguenin 1974), or subglacial volcanism—as palagonite (Toulmin et al. 1977). Montmorillonite had been proposed as a constituent of atmospheric dust (Hunt et al. 1973), although other interpretations also exist (Aronson and Emslie 1975; Toon et al. 1977).

The mafic character of the martian soil together with the volcanic morphology supports the view that basaltic rocks are widespread over the martian surface (Paper 23, Part VII; also Carr et al., 1977b). Major exposures of sialic rocks are considered unlikely and therefore the degree of chemical differentiation on Mars appears lower than on Earth.

The chemical analyses resolve an old dispute over the nature of the surface materials. Limonite, a once popular choice, cannot be a major soil component, neither can iron oxide grain coatings exceed 0.25μ in thickness; otherwise X-ray absorption effects would cause gross overestimates in light element abundances.

16

Reprinted from *Science* **194**:1288–1293 (1976)

MINERALOGIC AND PETROLOGIC IMPLICATIONS OF VIKING GEOCHEMICAL RESULTS FROM MARS: INTERIM REPORT

A. K. Baird, Priestley Toulmin, III, Benton C. Clark, Harry J. Rose, Jr., Klaus Keil, Ralph P. Christian, James L. Gooding

Abstract. *Chemical results from four samples of martian fines delivered to Viking landers 1 and 2 are remarkably similar in that they all have high iron; moderate magnesium, calcium, and sulfur; low aluminum; and apparently very low alkalies and trace elements. This composition is best interpreted as representing the weathering products of mafic igneous rocks. A mineralogic model, derived from computer mixing studies and laboratory analog preparations, suggests that Mars fines could be an intimate mixture of about 80 percent iron-rich clay, about 10 percent magnesium sulfate (kieserite?), about 5 percent carbonate (calcite), and about 5 percent iron oxides (hematite, magnetite, maghemite, goethite?). The mafic nature of the present fines (distributed globally) and their probable source rocks seems to preclude large-scale planetary differentiation of a terrestrial nature.*

Both Viking landers (VL) are in relatively low, basinal areas of Mars (VL1 at Chryse Planitia and VL2 at Utopia Planitia) characterized by abundant red-colored fine material and scattered blocks of generally angular rocks. At the site of VL1 there is more geologic diversity than at the VL2 site: pitted and relatively unpitted rocks, both light and dark colored; abundant dunelike deposits; and outcrops of apparent bed rock, crudely layered and perhaps intruded by a dike (*1*). The VL2 site, by contrast, is a flat plain more densely covered with angular blocks of uniformly dark-colored, pitted, fine-grained rocks. At both sites, but more prominently at VL2, the uppermost surface of fine material appears to be cemented into a crust 1 to 2 cm thick, which seems to range widely in mechanical strength. As of this report, four samples (three at VL1 and one at

VL2) have been delivered to the x-ray fluorescence spectrometers (XRFS) aboard the landers. We believe all four are martian fines, ranging in physical character from a loose, powdery consistency to more or less indurated or cemented duricrust. Deliveries of pebble-sized fractions of rock material were attempted during the second and third sequences on both VL1 and VL2; an attempt was also made to deliver coarse material residual in the sampler after an acquisition for the VL2 biology investigation. On VL2 no coarse samples were received, while on VL1, the chemical similarities to previously analyzed fines and the low bulk density of the material, suggest delivery of cemented fines only. Further attempts will be made to acquire the geochemically important rock material, but this report is confined to a discussion of the possible mineralogic and petrologic implications that can be derived from the analyses of martian fines from the two Viking landing sites, approximately 6500 km apart.

Samples of Mars surface materials were delivered to the Viking 1 XRFS on sols (*2*) 8, 34, and 40 and to the Viking 2 XRFS on sols 29 and 30 (Table 1). Photographs of the sample sites have been published (*3, 4*).

As with the other Viking lander soil analysis experiments (biology and molecular analysis), samples for the XRFS instruments were obtained from sites located approximately 1.5 to 2.5 m from the spacecraft and from maximum depths of 6 cm below the surface. Two different modes of surface sampler operation, one for fines and the other for pebble-sized grains, were used in acquiring samples. In the "fines" mode, the surface sampler sifts the bulk sample through its 2-mm sieve directly into the XRFS funnel. In the "pebble" mode, the surface sampler was first positioned away from the lander and the inverted collector head was vibrated for 60 to 90 seconds in order to sieve fine particles (< 2 mm) from the bulk sample. The coarse fraction retained in the collector head was then delivered to the XRFS funnel through a 12-mm screen built into the funnel. Simulations performed with the Viking science test lander were used to verify the effectiveness of this technique in removing the fines from mixtures of pebbles and silt-sized rock powder.

At the VL1 site, the first sample (designated S1) was obtained from a drift deposit of windblown material composed predominantly of silt-sized particles (*3*). Two successive acquisitions (the nominal sampling sequence for the XRFS) were taken from the headwall of a trench excavated approximately 3 hours earlier, during sample collection for the biology and molecular analysis experiments. It is likely that S1 was composed largely of material from below the interface of the martian surface and atmosphere.

The second and third VL1 samples, S2 and S3, were obtained in the "pebble" mode from an area of surface concentration of pebble-sized material (*3, 5*). Sample S2 was obtained from alongside the near end of a trench that had been excavated on sol 31 during sampling for the molecular analysis experiment. The S2 acquisitions uncovered several small rocks or soil clods, the average size of which indicates that sampling may have included material from a depth of 4 cm. The S3 sample was obtained 6 sols later by two acquisitions in which the surface sampler penetrated the headwall of the adjacent sol 31 trench.

The first XRFS sample (U1) at the second Viking landing site was obtained (sol 29) from a surface exposure of duricrust that had been sampled on sol 21 for the molecular analysis experiment. The sample probably included material from the immediate surface and from as deep as 4 cm below the surface. The sample was delivered to the XRFS by fines-mode operation of the surface sampler.

The XRFS instruments on Viking require long, repeated analyses of each sample in order to accumulate adequate data for quantitative results (*6*). Before enough data can be pooled for quantitative calculations, however, it is possible to treat the received spectra as qualitative "signatures" of rock or mineral classes by visual comparison with a library of standard spectra. This library of over 1000 spectra of terrestrial and meteoritic materials was accumulated before the landed operations with the use of flightlike x-ray spectrometers (*7*). Although the library includes spectra from a diverse array of minerals and rocks, it is strongly weighted toward potential Mars analogs, such as basaltic volcanic products, soils derived from basaltic volcanics, various evaporite-rich soils, clays and clay soils, samples from

Table 1. Description of Mars surface samples analyzed by the Viking landers 1 and 2 XRFS instruments. Abbreviation: Acq, acquisition.

Samples	Local lander time (sol/hr:min)	Nature of material	Depth (cm)*	Grain size (mm)
VL1, S1		Drift deposit of aeolian sediment	4 to 6	Fines ≤ 2
Acq 1	8/10:46			
Acq 2	8/11:33			
VL1, S2		Surface concentration of pebble-sized fragments; possibly an aeolian lag deposit	0 to 4	2 to 12
Acq 1	34/10:19			
Acq 2	34/11:19			
VL1, S3				
Acq 1	40/10:39	Same deposit as for S2	2 to 5	2 to 12
Acq 2	40/12:19			
VL2, U1				
Acq 1	29/13:42	Duricrust	0 to 4	Fines ≤ 2, partly cemented
Acq 2	30/10:42			

*Sampling depth below the surface. †The dominant grain size of the delivered sample, inferred from imagery and the known mode of surface sampler operation. The silt size for S1 was derived from an estimate by the Viking physical properties team (*3*).

Table 2. Chemical composition of martian samples (as oxides).

Oxide	Composition (percent by weight)					
	S1	S3	U1	Difference S3 minus S1	Difference U1 minus S1	Estimated error*
SiO_2	44.7	43.9	42.8	−0.8	−1.9	± 5.3
Al_2O_3	5.7	5.5	†	−0.3	†	± 1.7
Fe_2O_3‡	18.2	18.7	20.3	+0.4	+2.1	±2.9
MgO	8.3	8.6	†	+0.3	†	± 4.2
CaO	5.6	5.6	5.0	0.0	−0.6	± 1.1
K_2O	0.1	0.1	0.0	0.0	−0.1	±0.1
TiO_2	0.8	0.8	1.0	0.0	+0.2	± 0.3
SO_3	7.7	9.5	6.5	+1.7	−1.2	± 1.3
Cl	0.7	0.9	0.6	+0.2	−0.1	±0.3
Total	91.8	93.6	†			± 7.8

*Applies specifically to S1 but also generally representative of S3 and U1. †Value not yet available. ‡Total Fe as Fe_2O_3.

Antarctic dry valleys, and dune sands.

Three library spectra (proportional counters 1, 2, and 4) per sample cover the light elements (Z = 12–13), the intermediate elements (Z = 14–22) and the heavier elements ($Z \geq 24$). Signature comparisons are particularly sensitive for the latter two groups. In several instances spectra from a particular terrestrial sample proved very similar to first martian results for one or two detectors, but in no case did spectra from all three detectors compare closely with those from the terrestrial sample. However, these signature comparisons yielded clear evidence that the Mars results were indicative of (i) a relatively mafic, iron-rich and aluminum-poor igneous rock, or the degradation products of such a rock type and (ii) at least a binary mixture of materials, such as mafic igneous products and a sulfur compound. These preliminary observations have been described (*6*).

After 22 sols, sufficient spectral data had been accumulated on S1 to permit calculations of numerical values for the elements. Interim conclusions can now be drawn on the quantitative analysis of S1, but these may be subject to change as further improvements are made in our knowledge of the concentrations of the lightest elements (*8*). Our discussion applies directly to the analysis of S1, although the similarities among all samples suggest a common petrogenesis. Table 2 lists the composition of S1, S3, and U1 (as oxides, following petrologic convention) with the oxide differences from S1 for S3 and U1.

Overview. Relatively high values of CaO, MgO, and Fe_2O_3 (and low K_2O and Al_2O_3) require that the sample be mafic or mafic-derived. Oxide summation to 92 to 93 percent precludes the presence of very substantial amounts of carbonates, nitrates, or other low atomic number ($Z \leq 11$) elements, such as Na (*9*). Results from gas chromatograph–mass spectrometry (GCMS) analyses (*10*) demonstrate the presence of H_2O, although the total amount is uncertain for several reasons, including the short duration (30 seconds) of heating to a maximum temperature of 500°C. Relatively high S in our results implies the presence of sulfates, but other S compounds cannot be excluded (*8*). The low content of Al_2O_3 and moderately high SiO_2 place strict constraints on any postulated igneous origin.

The admixture problem. Iron is present in more than one compound. Lander color imagery shows that some Fe must be in a highly oxidized form giving the prominent red-to-orange coloration to surface fines (*5*). We have concluded, however, that this oxide coating must be thin or discontinuous (or both) because low-energy x-ray emissions from light elements are detected (*8*). Fe must also be present in a magnetic fraction amounting to at least 3 percent by mass, probably as magnetite or maghemite (*11*). We conclude that the bulk of the Fe is probably present in a silicate phase or phases. Abundant SO_3, if present as sulfate or sulfates, may represent further admixture with a principally silicate sample. Mg, Ca, and possibly Fe and Na are candidate cations; Clark *et al.* (*8*) present arguments suggesting the presence of magnesium sulfate as a cementing agent in the fines. It seems reasonable to suppose that some of the balance unaccounted for in our analysis represents CO_2 bound as carbonate, possibly $CaCO_3$, and Na as NaCl.

Computer modeling. Analytical results have been modeled in three forms: (i) computer searches of best-fit matches to chemical analyses of thousands of samples of terrestrial, lunar, and meteoritic materials; (ii) normative mineral calculations of martian results; and (iii) mixture calculations to give best-fit proportions of selected ingredient rocks and minerals.

Computer searches yield rank-ordered matches of chemical analyses to martian results. By reducing the "fit" requirements, larger and larger lists of "matches" can be generated (with progressively poorer agreement). In general, the best fits include mafic and ultramafic rocks and minerals, some Fe-rich soils and montmorillonitic clays, lunar materials, and carbonaceous meteoritic materials. High Fe, low Al, and low trace element concentrations appear to constrain these fits. Most analyses in the search library are of well-defined mineral phases or single rocks; soils, however, are included. Some matches are computer-selected

Table 3. Normative mineral calculations, by a modification of the standard CIPW procedure.

Mineral	1*	2†	3‡
Quartz (SiO_2)	27	39	21
Feldspar	19	1	20
Or ($KAlSi_3O_8$)	(5)	(100)	(5)
An ($CaAl_2SiO_8$)	(95)		(95)
Pyroxene	30	12	47
Wo ($CaSiO_3$)	(16)		(14)
En ($MgSiO_3$)	(84)	(100)	(53)
Fo ($FeSiO_3$)			(33)
Hematite (Fe_2O_2)	22	15	
Magnetite (Fe_3O_4)		3	2
Ilmenite ($FeTiO_3$)		2	2
Sphene ($CaTiSiO_5$)	2		
Corundum (Al_2O_3)		6	
Kieserite ($MgSO_4$)		13	
Calcite ($CaCO_3$)		9	
Troilite (FeS)			8

*Norm of S1 analysis, assuming all Fe as Fe_2O_3 neglecting SO_3. †Modified norm of S1, assuming all SO_3 as $MgSO_4$, all CaO as $CaCO_3$, 10 percent of Fe as FeO. ‡Modified norm of hypothetical precursor composition for S1, assuming all S as FeS, 90 percent of oxide Fe as FeO.

Table 4. Chemical compositions of computer-modeled mixtures compared to S1.

Item	Composition (percent by weight)			
	Mixture 1	Mixture 2	Mixture 3	S1
		Oxide		
SiO_2	55.1	46.0	43.6	44.7
Al_2O_3	8.3	8.0	6.9	5.7
Fe_2O_3	19.5	19.0	18.4	18.2
MgO	10.1	9.6	9.0	8.3
CaO	2.4	2.0	5.6	5.6
K_2O	0.0	0.0	0.0	0.1
TiO_2	0.0	0.0	0.9	0.8
SO_3	0.0	9.4	7.3	7.7
		Mineral		
Nontronite	51	52	47	
Montmorillonite	19	21	17	
Saponite	30	13	15	
Kieserite		16	13	
Calcite			7	
Rutile			1	

because of particular features of the Mars analysis (for example, abundant S and low trace elements), and the remaining elemental correlations are poor. Other matches are close for all major elements except one or two. This is especially true for correlations with Fe-bearing montmorillonite clays, if one excludes the S in the Mars analysis. The search results reinforce our belief that the Mars sample should be considered a mixture.

The apparent compositional uniformity of the martian fines from such widely separated localities gives rise to a number of questions and interpretations. Does the bulk chemical composition (with suitable allowances for oxidation, carbonation, and hydration) of the fines represent the overall composition of the primary bedrock materials exposed over a similarly planet-scale area? One way to study this problem is to examine calculated norms to see whether they are compatible with plausible models for the composition of surface rocks on Mars. The detailed results of the norm calculations depend to a considerable degree on assumptions as to the roles of S, Na, and CO_2, and to the degree of Fe oxidation postulated for the primary rocks; but, in every case, the result is significantly quartz-normative and pyroxene-rich (Table 3).

The norm in the first column (Table 3) does not include the S in the analysis, and therefore represents only the silicate or oxide fraction of the material. All the Fe is assumed to be trivalent, in accordance with the apparently highly oxidized character of the surface materials. In the second column (Table 3) we show a norm modified by the assumptions (i) that all the S is present as $MgSO_4$ (or a hydrate such as kieserite, $MgSO_4 \cdot H_2O$), (ii) that all the CaO is present in calcite, and (iii) that a fraction (arbitrarily 10 percent) of the Fe is divalent. The resulting normative assemblage contains a very large proportion of free silica, which results from the removal of the bases CaO and MgO from a silicate fraction already low in bases relative to SiO_2. This depends in part on the plausibility of calculating all CaO as $CaCO_3$. We note that the sum of oxides (and Cl) determined for S1 is about 92 percent; the 8 percent discrepancy from 100 percent could be accounted for by 0.4 percent Na for NaCl, 4.4 percent CO_2 in $CaCO_3$, and about 3.2 percent H_2O as structural water in smectite, in general (although perhaps coincidental) agreement with the analog composition to be discussed below.

The third column (Table 3) shows the norm of a composition derived from S1 on the assumption that 90 percent of the Fe is divalent, and all the S is present as FeS. This would represent the protolith from which S1 might have been derived if the weathering process had been isochemical except with respect to oxidation (and hydration?). This seems somewhat unlikely, as it would be difficult to derive an igneous rock of this composition on a planetary scale by straightforward processes of igneous differentiation. One might imagine that fractional crystallization or partial melting of a parent material extremely low in alkalies and very low in alumina would yield such a product, but most models of planetary accretion would predict that Mars as a whole, as compared to Earth, was enriched in the volatile alkali elements compared. Furthermore, if a large part of the surface is underlain by siliceous, differentiated rocks, the large-scale igneous differentiation required would seem to entail a fairly advanced state of planetary differentiation, that is, a concentration of sialic material near the surface. In fact, the abundances of Ar isotopes in the atmosphere (*12*) suggest extremely limited outgassing from the body of the planet and a quite modest K content in that (presumably shallow) portion that has outgassed. Both inferences are consistent with a poorly differentiated planet. Furthermore, igneous processes on the moon, where alkalies are known to be deficient (at least relative to terrestrial experience), have not led to products like the quartz-anorthite-pyroxene aggregates (Table 3, column 3) that would correspond to the fines of Mars if they had resulted from "isochemical" weathering.

Table 5. Chemical composition (as oxides) of martian analog 6040 compared to range of Mars preferred values.

Oxide	Analog 6040	S1 range*
SiO_2	40.8	39.4 to 50.0
Al_2O_3	6.8	4.0 to 7.4
Fe_2O_3	15.5	15.3 to 21.1
MgO	5.1	4.1 to 12.5
CaO	7.4	4.5 to 6.7
Na_2O	0.5	
K_2O	0.15	0.0 to 0.2
TiO_2	0.84	0.5 to 1.1
CO_2	3.6	
SO_3	7.0	6.4 to 9.0
Cl	0.6	0.4 to 1.0
Total	88.3	

*From Table 2, S1 value ± estimated error.

Computer mixture modeling with the compositions of a number of mineralogic and petrologic ingredients can yield close chemical fits to analyses of the Mars surface. For example, the progressive improvement in fit, beginning with three end-members of the montmorillonite family (montmorillonite, nontronite, and saponite), obtained by adding kieserite ($MgSO_4 \cdot H_2O$), calcite ($CaCO_3$), and rutile (TiO_2) is shown in Table 4. An approximately equal (50 : 50) mixture of type 1 carbonaceous chondrite and average tholeiitic basalt yields a chemical composition which agrees, within error limits, with the S1 results for all oxides including SO_3, except for somewhat higher Al_2O_3. Other mixture models can be made with, for example, basaltic rock compositions with admixed montmorillonite family minerals, plus sulfates. The compositions of these mixes are, however, dominated by mineral assemblages that are characteristic of mafic igneous rocks. It appears that computer mixing of a number of generally mafic or montmorillonitic compositions (or both), plus sulfates and chlorides, will give fairly close fits to the Mars results.

Laboratory analog modeling. On the basis of the results of computer mixtures, an analog chemical material has been developed from nontronite (59.1 percent by weight), montmorillonite (21.7 percent), kieserite (11.8 percent), calcite (5.9 percent), halite (1 percent), and titanium dioxide (0.5 percent). The chemical composition of this mix is shown in Table 5 and spectra for proportional counters (PC) Nos. 1, 2, and 4 (measured in a flightlike laboratory spectrometer) are compared with Mars results in Fig. 1. It must be emphasized that we regard this mix as a *chemical*, but not necessarily a mineralogical, analog to the Mars fines.

Developing a chemical analog has emphasized some problems with respect to mineralogical interpretation of the Mars results. Among iron-rich smectites used as candidate components, Fe_2O_3 is usually in excess of 25 percent by weight and silica about 40 percent. Dilutions to reduce the iron commonly produce unrealistically low silica values compared to Mars results. The principal constituent of the derived mix is a nontronitic montmorillonite (*13*). The addition of bentonitic montmorillonite (*14*) provides the necessary alumina and serves to increase the silica content of the mix. Kieserite, calcite, and halite were selected to provide the magnesium, sulfur, calcium, and chlorine, but other compounds obviously could have been used to achieve the same result (for example, dolomite, gypsum, and antarcticite). The nontronitic montmorillonite contains insufficient titanium; this was supplemented with TiO_2 reagent powder. Terrestrial clay samples contain variable amounts of rutile, ilme-

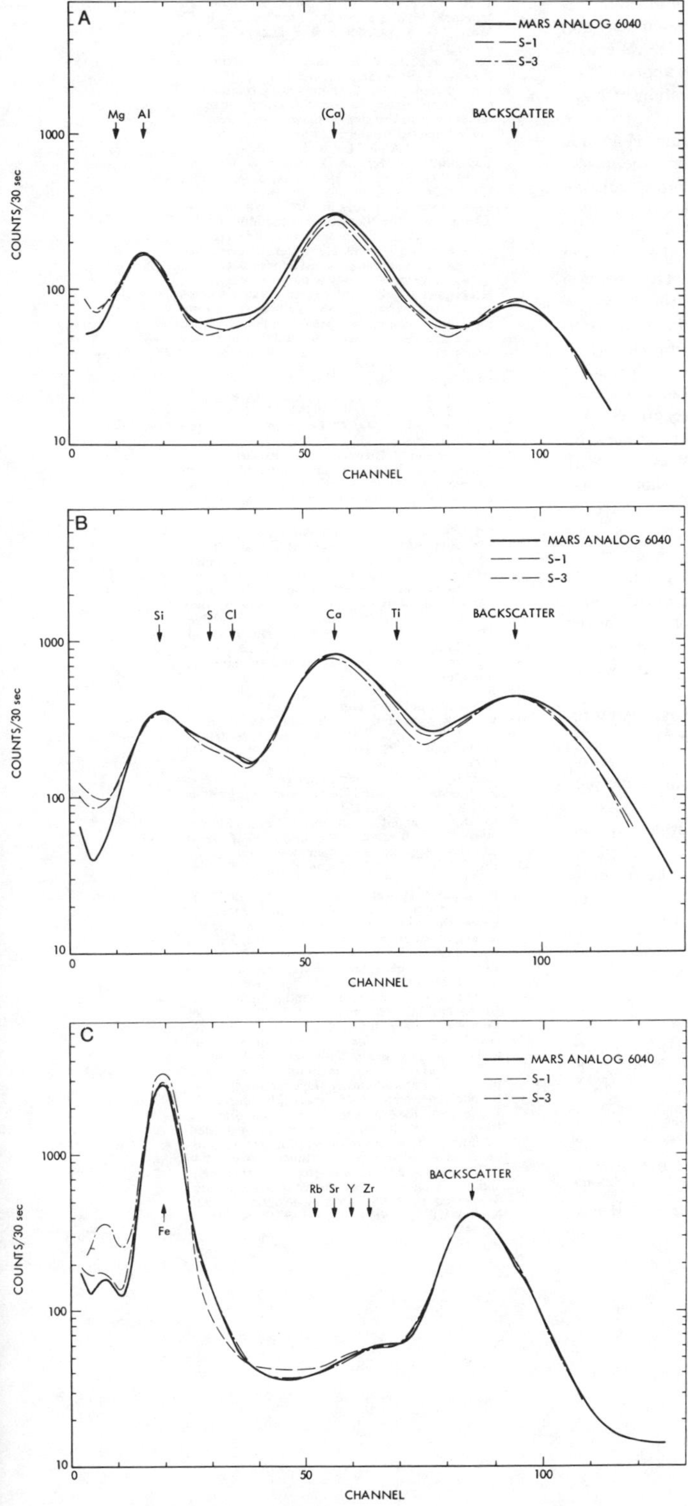

Fig. 1. Spectra of Mars analog material (our ref. 6040) compared to pooled spectra of S1 and S3 from Chryse Planitia. (A) The PC1 spectra for Mg and Al determinations (Ca is determined by PC2); (B) PC2 spectra for Si through Ti determinations; and (C) PC4 spectra for Fe and trace element determinations. Composition of analog is given in Table 5.

nite, anatase, or leucoxene as minor constituents.

A mineralogic-petrologic model. It seems likely that the Mars fines represent not the composition of primary rocks, but the products of some sort of chemical weathering process (*15*). In fact, as we have noted, a very large fraction of the overall major-element composition can be matched quite closely by a mixture of iron-rich clays characteristic of the terrestrial weathering and alteration of mafic igneous rocks. The color, albedo, and texture of certain fresh-appearing rock surfaces at the Viking landing sites (especially at VL1) are consistent with a mafic or ultramafic composition (*16*). While the present mineralogy of the fines obviously also includes "free" iron oxides, very likely sulfates, and probably carbonates and chlorides as well, the uniform composition at the two landing sites suggests that all the phases in the fines may be very intimately associated, quite possibly as extremely fine intergrowths (so that particle-to-particle compositional variation is slight). We might speculate that oxidation-dehydration of an iron-rich clay could lead to destabilization and destruction of the clay mineral structure, giving rise to just such a very fine grained aggregate. As a specific example, oxidation of a ferroan saponite could yield a mixture of nontronite, montmorillonite, and free iron oxides, with excess Mg and Ca (both "structural" and "exchangeable") available for fixation as sulfate (*17*). Ion exchange and absorption characteristics of the material could be significantly different from montmorillonite clays in a terrestrial environment. Derivation of the (relatively) less oxidized clay is even more speculative, but could have resulted from weathering processes in an ancient, more humid climate, or by direct interaction between mafic magma and extensive bodies of ice at or near the surface.

Genetic implications of mineralogic model. On Earth the formation of ferroan montmorillonites is associated with a ferromagnesian parent rock interacting with abundant water. Parent materials include basalt, basaltic glasses and palagonites (*18*), diabase (*19*), and ultramafic plutonic rocks (*19*). In the oceans, basaltic extrusions associated with ridge systems interact with oxygenated seawater to form nontronitic clays (*20*). Ferroan clays are a common, if not abundant, result of hydrothermal alteration of mafic rocks (*17*).

Prior to the Viking mission, Huguenin (*21*) advanced theoretical arguments and experimental evidence for photochemical surface reactions with the potential for forming clay minerals from pyrogenic silicates in an environment similar to that on the surface of Mars. His reaction mechanism requires only the very low partial pressure of O_2 and H_2O known to be present in the present atmosphere of Mars. So far as we are aware, however, no clay minerals have yet been produced in the laboratory by such reactions. Furthermore, photochemical processes on mineral surfaces may be effective mostly on the very fine dust grains that are carried to high levels in the atmosphere, in which case some other agency for prior comminution must be sought. The sequential combination of argillization at or near the surface, and photochemical desiccation or oxidation (or both) of the resulting dust in the high atmosphere is one hypothesis consistent with these conditions.

Our mineralogic model has, therefore, two major requirements for petrogenesis on Mars. (i) There must be exposed on the surface of Mars abundant mafic parent materials for clay formation and, conversely, there cannot be large, exposed source areas providing alkali-rich granitic materials. (ii) There must be (or must have been) sufficient water or ice on Mars to interact with mafic rock material to form ferroan clays.

References and Notes

1. T. A. Mutch, A. B. Binder, F. O. Huck, E. C. Levinthal, S. Liebes, Jr., E. C. Morris, W. R. Patterson, J. B. Pollack, C. Sagan, G. R. Taylor, *Science* **193**, 791 (1976).
2. The martian day (24 hours, 35 minutes, 39 seconds) has been designated a sol; sols are numbered consecutively from the day of landing of each lander.
3. R. W. Shorthill, H. J. Moore, II, R. F. Scott, R. E. Hutton, S. Liebes, Jr., C. R. Spitzer, *Science* **194**, 91 (1976).
4. R. W. Shorthill, H. J. Moore, II, R. E. Hutton, R. F. Scott, *ibid.*, p. 1309.
5. T. A. Mutch, R. E. Arvidson, A. B. Binder, F. O. Huck, E. C. Levinthal, S. Liebes, Jr., E. C. Morris, D. Nummedal, J. B. Pollack, C. Sagan, *ibid.*, p. 87.
6. P. Toulmin III, B. C. Clark, A. K. Baird, K. Keil, H. J. Rose, Jr., *Science* **194**, 81 (1976).
7. For efficiency, the data in most ground-based tests were taken via a multichannel analyzer rather than in the single-channel, step-scanning mode imposed by the flight unit.
8. B. C. Clark, A. K. Baird, H. J. Rose, Jr., P. Toulmin III, K. Keil, A. J. Castro, W. C. Kelliher, C. D. Rowe, P. H. Evans, *Science* **194**, 1283 (1976).
9. The Viking XRFS does not detect Na. We have assumed that the Na content of the Mars fines is very low, because (i) the analytical deficit (difference between sum of constituents determined and 100 percent) can be satisfactorily accounted for by other constituents with no, or very little, Na required; and (ii) the other alkali elements (K, Rb) are extremely low in all the Mars samples. Other assumptions are possible with regard to the analytical deficit; the model(s) proposed herein are based on our stated assumptions.
10. K. Biemann, J. Oro, P. Toulmin III, L. E. Orgel, A. O. Nier, D. M. Anderson, P. G. Simmonds, D. Flory, A. V. Diaz, D. R. Rushneck, J. A. Biller, *Science* **194**, 72 (1976).
11. R. B. Hargraves, D. W. Collinson, C. R. Spitzer, *ibid.*, p. 84.
12. T. Owen and K. Biemann, *ibid.* **193**, 801 (1976).
13. This material is an alteration product of a hedenbergite pyroxenite; we thank D. M. Morton, U.S. Geological Survey, for supplying the clay sample.
14. Bighorn bentonite, a commercial product.
15. The suggestion of nonisochemical weathering obviously raises the question of mass balance. Where and what are the compositional complements to the weathering products? This question can be addressed quantitatively only when compositional data on fresh rocks are available. Thus far we have not been successful in obtaining a sample of rocks from either Viking site, but this remains a high-priority goal for the extended mission.
16. R. E. Arvidson, personal communication.
17. J. Hower, personal communication.
18. C. S. Ross and E. V. Shannon, *Am. Ceram. Soc. J.* **9**, 77 (1926); H. M. Koster, *Beitr. Mineral. Petrogr.* **7**, 71 (1960); E. Bolter, *ibid.* **8**, 111 (1961).
19. C. S. Ross and S. B. Hendricks, *U.S. Geol. Surv. Prof. Pap. 205-B* (U.S. Government Printing Office, Washington, D.C., 1945).
20. E. Bonatti and O. Joensuu, *Science* **154**, 643 (1966); J. L. Bischoff, *Clays and Clay Minerals* **20**, 217 (1972); personal communication.
21. R. L. Huguenin, *J. Geophys. Res.* **79**, 3895 (1974).

22. We thank J. Hower (NSF) and J. Bischoff (U.S. Geological Survey) for sharing their extensive knowledge on terrestrial clay formation; the following members of the U.S. Geological Survey: R. R. Larson, J. M. Hammarstrom, C. S. Zen, J. Lindsay, P. Hearn, and J. Jensen for analytical aid; D. McIntyre, Pomona College, for a literature search of iron-bearing montmorillonites; N. Brosnahan, E. Hartman, and J. MacKay, of Pomona College for technical assistance; N. Nickle, of JPL, for assisting in the organization and operation of our x-ray laboratory; J. Gliozzi and the surface sampler team for their efforts in planning for and acquiring Mars surface samples for our experiment; R. Hargraves, E. Morris, A. Binder, and R. Arvidson for reviewing drafts of this report; R. R. Moore for providing needed calibration data through operation and maintenance of flightlike x-ray instruments. We thank S. Dwornik and G. J. Wasserburg for their help and encouragement with the x-ray experiments. Supported by NASA (Viking Program) grants NAS 1-11855, NAS 1-11858, L-9717, and NAS 1-9000.

Part VI

EROSION AND WEATHERING

Editor's Comments on Papers 17 and 18

At a time when astronomers were still attributing the seasonal color changes on Mars to vegetation, D. B. McLaughlin was outlining his "volcanic-aeolian hypothesis" in a series of papers. He was motivated by the application of geological principles to the interpretation of martian surface features, up to then exclusively within the province of astronomy. He proposed (Paper 17) that the seasonal changes in dark markings resulted from winds blowing across a thin veneer of dark volcanic ash. Ash, rather than lava, was favored because of its mobility, and because less water would be evolved with the small amounts of ash needed than with lava. (Actually, ash rather than lava is associated with release of volatiles.) The scarcity of water on Mars had, in fact, been G. Kuiper's chief objection to McLaughlin's theory (Kuiper 1955). But by 1957, Kuiper was more willing to consider wind scouring as an alternate explanation of the seasonal changes.

McLaughlin was remarkably farsighted in recognizing eolian activity and volcanism as major forces shaping the martian surface. On the other hand, current opinion explains the variable bright and dark markings in terms of differences in particle size or chemical composition of wind-blown *dust* rather than volcanic ash (Pollack and Sagan 1967 and Paper 18, Part VI; Sagan et al. 1971, 1972, 1973; Cutts 1971; Greeley et al. 1974; Arvidson 1972, 1974). While the Mariner 9 and Viking missions revealed abundant evidence for volcanism, most of the products of effusive activity probably consist of lava flows, although some pyroclastic materials may be locally present (West 1974).

The eolian hypothesis for the seasonal "wave of darkening" has been further developed by C. Sagan and J. B. Pollack (Paper 18). Although the paper is of historical interest today, many of the conclusions are still valid. Eolian processes on Mars have been analyzed in a number of more recent articles (Sagan et al. 1971, 1972, 1973; Gierasch and Sagan 1971; Hess 1973; Sagan and Bagnold 1975; Pollack et al. 1976; Iverson et al. 1976).

According to the Sagan-Pollack model, deposition of a thin layer of fine dust will brighten an area; wind scouring of the fines will darken it, thus explaining the "regeneration" of dark areas. Seasonal changes are caused by shifting wind patterns. Dust storms occur more frequently during southern hemisphere summers, when Mars is at perihelion (Capen 1971; Golitsyn 1973). The model emphasizes particle-size variation as the cause of seasonal darkening. However, spectral data suggest that chemical variations may also play a role (McCord et al. 1971, 1977a, b; Adams and McCord 1969).

It has been recognized that the dark markings are more variable than the light ones (Sagan et al. 1972, 1973). The Viking Orbiter pictures confirm this observation (Veverka et al. 1977). Comparison of Mariner 9 with Viking Orbiter pictures show, on the whole (with a few documented exceptions), that the distribution of light streaks has changed relatively little in the four years between 1972 and 1976. Veverka et al. (1977) conclude that the spacecraft data do not show evidence for the telescopically observed seasonal "wave of darkening." On the other hand, the spacecraft TV cameras have not monitored the martian surface continuously over several seasonal cycles.

17

Reprinted from *Astron. Soc. Pac. Pub.* **68**:211–218 (1956)

THE VOLCANIC-AEOLIAN HYPOTHESIS OF MARTIAN FEATURES

DEAN B. MCLAUGHLIN
Observatory of the University of Michigan

INTRODUCTION

In a recent issue of these *Publications*, Kuiper has discussed the nature of the Martian canals and dark areas.[1] I can agree with most of his ideas concerning the true forms of the canals, but with minor reservations that need not be detailed here. Concerning the more extensive maria, however, I believe that his basic assumptions are incorrect. His conclusion that volcanism is negligible appears to me premature, in view of the observational facts and the meteorological and geological principles that point to the presence of volcanoes.

Discussion of these questions could easily develop into an unprofitable and undesirable controversy. We still lack perfectly direct and decisive evidence; we can talk only of probabilities, not of conclusive proofs. But a close opposition of Mars is imminent and discussion *now* may influence the planning of observational programs.

It is my claim that much reliable observational data can be rationally interpreted in terms of wind-deposited ash from volcanoes that can be "pinpointed" on the Martian map.[2] This does not mean that an alternative hypothesis may not later give an equally consistent explanation. But, to date, no other adequate suggestion has been made, and known geologic processes contain no inkling of a possible alternative. No veteran observer of Mars seems to have given much thought, from the geologic point of view, to the specific facts of the oriented, banded pattern, the pointed bays, and the conspicuous nonseasonal changes of certain markings.

VOLCANISM VERSUS VEGETATION?

The vegetation hypothesis and the volcanic hypothesis are not mutually exclusive. After all, we do have both vegetation and volcanism on the earth! However, it does appear possible to account for the seasonal changes without vegetation,[3] and these seasonal changes, in the last analysis, are the sole evidence that vegetation is present. It may be, for example, that vegetation can grow only in areas that are subject to frequent falls of fresh ash and the moisture that would accompany it.[4]

My volcanic-aeolian hypothesis was developed expressly to account for the *forms* of the markings, which the vegetation hypothesis never attempted to do. It was only after the forms had been accounted for that I attempted to apply the mechanism to the seasonal changes. Perhaps my hypothesis is valid *only* for explaining the forms and must be supplemented by the growth of plants upon the fresh ash to explain the seasonal changes. But it does account for the forms—and the growth of vegetation, taken by itself, does not.

THE NATURE OF THE DARK MARKINGS

It is important to clarify a statement concerning the banded forms of the maria that may be misinterpreted by some who are not intimately acquainted with the observations of Mars. On page 280, Kuiper states:[1] "The basis for . . . [McLaughlin's] suggestion is the fishbone type of arrangement, symmetric to the equator, *indicated by his drawings*." The italics are mine. A reader who had not critically studied many drawings and photographs could judge this to mean that my sketches were the only series of observations in which such a pattern was depicted. This is, of course, not correct. The banding is plainly shown on numerous drawings by Antoniadi[5] and many other observers.[6] It is present on all the maps of Mars that have been published since 1864. And more to the point, it is conspicuous on numerous photographs, for example the series by Barnard,[7] those by Slipher,[8] and those by Lyot.[9]

My own observations merely introduced me to the problem. Before developing the hypothesis of wind deposition and volcanism, I had examined a great number of drawings, photographs, and maps by observers whose experience with Mars greatly exceeded mine. These fully confirmed the banding that I had seen. Since the systematically oriented and banded pattern exists, it demands a physical explanation. This I have attempted to supply.

Kuiper states: "The [dark] markings are similar to those of the flooded regions on the moon, like Mare Nubium . . ." No evidence is presented in support of this statement, and I believe it is proved incorrect by the facts. The dust whipped up in the frequent dust storms must settle somewhere, on the maria as well as on the deserts. After millions of years the outlines of any original lava-flooded areas would have been entirely concealed and topographic distinctions between them and their surroundings obliterated. Therefore, even if radioactive melting and flooding did occur in Mars's remote past, the *present* features would be largely a pattern of wind-deposited material, unless lava flows have continued right up to recent times.

If the pattern now visible on Mars is due to lava flooding of recent date, then it must be due to volcanism on a scale vastly exceeding anything required by my ash hypothesis. Much water vapor must have been evolved along with the lava, and Mars should have not only observable water vapor in its atmosphere, but probably standing pools on the surface. Thus, the scarcity of water is even more difficult to explain than in the case of my ash hypothesis. A very thin layer of ash will account for all the observed color contrasts, and the accompanying water would be correspondingly small in amount.

Kuiper presents a drawing of the region of Dawes' forked bay as he saw it, with the remark that it does not support the hooked appearance across the equator. The term "hooked," as I used it, did not refer to pointed forms, but to the abrupt bending of the markings. His drawing shows the well-known hooked (bent) form that has been recorded by every observer since Dawes. But the eastern point is very blunt and the western point, which is also blunt, has two trails leading northward to small dark spots. Presumably he considers that the bluntness of the points is opposed to an origin of the dark material from so small a source as a volcano. Actually, we could expect to see very sharp points *only* during or soon after an eruption. As soon as desert dust begins to be drifted onto the edges of the dark deposit, points may be blunted and boundaries modified. Perhaps it is overinterpretation to point out that the small dark spots with dark "canals" leading to the western prong of the bay *could be* regarded as volcanoes with long banners of ash trailing to leeward!

Kuiper remarks that Antoniadi's map does not give "convincing support" for my hypothesis. This has me puzzled, because it was from Antoniadi's map more than from any other one source that I developed my hypothesis in the first place. Nor do I see the markings of which he says: "When prominent north-south features . . . are traced across the equator, the westernmost extremity is just as likely to lie at 20° N as on the equator itself." Probably we could agree that the Syrtis Major bends prematurely at about 10° N. The majority of features to which he refers must represent combinations of less conspicuous markings than those I considered, and not parts of the smooth flow pattern. Probably each interested reader should study the maps for himself and draw his own conclusions.

Large departures from a mathematically exact pattern are to be expected. Air masses in a planet's atmosphere cannot behave like projectiles moving in a vacuum. The wind systems of the earth show great local deviations from the ideal or schematic pattern, yet we do not argue from these deviations that the Coriolis effect does not exist or that the physical basis of meteorology is incorrect!

I know of no evidence clearly opposed to the suggestion that the present arrangement of Martian maria is mainly explicable as a pattern of wind-deposited dust, regardless of the mechanism of origin of the dust. Once the wind pattern is recognized, point sources of dark material become a practical necessity. Volcanoes are not only the most obvious interpretation, but the only adequate one I have been able to find.

VOLCANISM AND WATER

Kuiper correctly points out the most serious problem raised by the suggestion of volcanism. From the scarcity of water he concludes that volcanism is negligible. I agree that the dryness of Mars is surprising, but is lack of volcanism the only possible conclusion? It appears to me equally legitimate to turn the problem around and ask: since there is such strong evidence for volcanism, can we make our theories fit the observed scarcity of water? There are several ways of removing from the atmosphere the water that volcanoes may inject into it. Some could condense on ash particles in the air and become incorporated in the surface deposits, where it would combine chemically with minerals of the ash. Probably most of it would freeze into ice spicules and remain long in the high levels of the atmosphere, not showing as vapor. If there is vegetation (which I still admit as a possibility) it may take up practically all moisture that is not otherwise accounted for. I do not entirely follow Kuiper's argument for the extremely slow loss of water by photolysis, but even if it is correct, it does not appear to eliminate volcanism, in view of other possible ways in which water may disappear. The evidence and the reasoning by which the idea of volcanism was reached are so straightforward that I can hardly discount them in favor of a circuitous and hypothetical argument.

Kuiper's objections may be slightly reduced (though not by the several orders of magnitude he mentions) by my downward revision of the estimate of Martian volcanism. My first impression, based on the large areal extent of the maria, was that activity must be on a scale greatly exceeding that on the earth. Some months before Kuiper's paper appeared, I had already concluded that Martian volcanism was merely comparable to that on the earth.[10] This revision depended in part upon a recognition that only very small annual increments to the maria were necessary, but it was based chiefly upon the rarity of large changes that could be interpreted as due to great explosive eruptions. Thus, in a little less than two centuries, there have been only six records of really great temporary markings that required (according to my estimate) ash falls as great as, or in excess of, one cubic kilometer. Most of these are discussed elsewhere,[11] and will merely be listed here. They are: Cyclops, about 1780; Hydaspis Sinus, about 1860; Phasis, shortly before 1877; the distortion of Solis Lacus in 1926; the band across Noachis in 1928; and the new marking that appeared east of Thoth in 1954.[12] The first four of these furnish powerful arguments in support of volcanism, and the other two are simply noncommittal.

SOME SUGGESTED OBSERVATIONS

In a recent note I suggested that water vapor might be detected by spectrographic observations limited to the maria and especially to the pointed bays.[13] A search should be made for volcanic cones at the points of bays and in the larger oases. However, terrestrial experience indicates that the dimensions of a cone may not indicate the eruptive capacity of the volcano. An eruption cloud should be readily observable, but the chance of catching a volcano in the act of violent eruption is very small. Still, these observations appear distinctly worth trying.

In 1954 Pettit and Richardson observed a high-level (blue) cloud that showed remarkably little motion during a month.[14] They indicate that, according to E. C. Slipher, it "had an intermittent existence, forming in the afternoon and disappearing at night." Certain strong knots fell near oases. Is it absurd to suggest that this may have been a cloud formed by condensation of steam above or near active (not necessarily violently erupting) volcanoes? This suggestion is supported by the occurrence, in 1941, of a small and dense cloud whose position overlapped their knot "a."[15] Observations of this kind are of the greatest interest. If clouds form by preference in certain local spots, there must be a physical reason for such behavior. Correlations between the positions of clouds and those of oases or vertices of bays have never been investigated. Do such correlations exist, or do they not?

Obviously much work needs to be done on the comparison of reflectivity of the Martian surface with that of terrestrial rocks and dusts, in all accessible wavelengths.

CONCLUSION

The arguments for and against volcanism on Mars can be summarized as follows. On the one hand we have the objection that the scarcity of water is incompatible with appreciable volcanism. We cannot know how much weight to assign to this argument until all the processes of removal of moisture from the atmosphere have been carefully evaluated, and until we have a more reliable estimate of the amount, or the limiting amount, of water in the atmosphere. A few objections have been based upon some departures of the banded pattern from the mathematical ideal and upon what appears to be a demonstrably incorrect assumption as to the origin of the maria. I can attach no appreciable weight to these latter arguments, for reasons already given in detail.

Against these objections we have the remarkable consistency of the forms of the maria. They are elongated in directions that agree with the trade winds. They exhibit a curving "flow pattern" that crosses the equator. The direction of curvature is consistent with the Coriolis effect. The maria are mainly south of the equator, as they must be if the strongest winds are associated with the maximum insolation, at perihelion. The maria have tributary and terminal bays that are funnel-shaped and pointed (whether sharply or bluntly). These satisfy the requirements of local sources of dark dust to be carried by the winds. Several temporary markings that have appeared suddenly had the same funnel form and their curvature agreed with the Coriolis effect. Thus, the volcanic-aeolian hypothesis gives a highly satisfactory account, in terms of known physical mechanisms, of a large body of facts *for which no other explanation has ever been offered,* to the best of my knowledge. The hypothesis was arrived at only after consideration of all known major geologic processes, and elimination of those that are impossible or negligible under Martian conditions.

The only circumstance that makes the Martian surface an astronomical problem at all is the fact that it is the surface of *another* planet and not that of our own! For that reason the gathering of data must be done by astronomical methods. But the pure astronomer cannot be self-sufficient. Astronomical methods and habits of thought have been of little help in solving problems of the earth's surface features. These are not the province of astronomy, but of geology. Similarly, interpretation of the surface of Mars is a geological (and meteorological) problem, requiring the application of geological methods of thinking, modified to suit Martian conditions. This does not imply that a geologist's interpretation will surely be correct. It does mean that *without* the application of geological thinking, there is scant hope of ever solving the problem.

I am indebted to Dr. Kuiper for making available a draft of his paper prior to publication, and to Dr. Lawrence Aller for comments on some aspects of the problem. This carries no implication concerning Dr. Aller's views.

[1] G. P. Kuiper, *Pub. A.S.P.*, **67**, 271, 1955.

[2] D. B. McLaughlin, *Bull. Geol. Soc. Amer.*, **65**, 715, 1954; *Obs.*, **74**, 166, 1954; *Pub. A.S.P.*, **66**, 161, 1954.

[3] D. B. McLaughlin, *Pub. A.S.P.*, **66**, 226, 1954.

[4] D. B. McLaughlin, *A.J.*, **60**, 268, 1955.

[5] E. M. Antoniadi, *La Planète Mars* (Paris: Librairie Scientifique Hermann et Cie., 1930), plates 6–10.

[6] For example, A. Dollfus, *l'Astr.*, **67**, 85, 1953, plates 7 and 8.

[7] G. P. Kuiper and M. R. Calvert, *Ap. J.*, **105**, 215, 1947.

[8] E. C. Slipher, *The Telescope*, **7**, 101, 1940. See also F. L. Whipple, *Earth, Moon and Planets* (Philadelphia: The Blakiston Company, 1941), pp. 210 and 218.

[9] B. Lyot (G. P. Kuiper, ed.), *Ap. J.*, **101**, 255, 1945, plate 24; H. Camichel, *Bull. Astr.*, **18**, 175, 1954, plate 6.

[10] D. B. McLaughlin, *Bull. Geol. Soc. Amer.*, **66**, 769, 1955; *A.J.*, **60**, 270, 1955; *Mars, 1954* (Report of International Mars Committee), p. 22, 1955.

[11] D. B. McLaughlin, *A.J.*, **60**, 261, 1955.

[12] E. Pettit and R. S. Richardson, *Pub. A.S.P.*, **67**, 64, 1955; see also E. C. Slipher's photographs in *Sky and Tel.*, **14**, 360, 1955.

[13] D. B. McLaughlin, *Pub. A.S.P.*, **68**, 71, 1956.

[14] E. Pettit and R. S. Richardson, *Pub. A.S.P.*, **67**, 65, 1955.

[15] O. H. Truman, personal communication. See also *Sky and Tel.*, **15**, 311, 1956.

18

Reprinted from *Nature* **223**:791–794 (1969)

Windblown Dust on Mars

by

CARL SAGAN
JAMES B. POLLACK
Laboratory for Planetary Studies, Cornell University, Center for Radiophysics and Space Research, Ithaca, New York 14850

The wave of darkening in the Martian springtime, which some say may have a biological explanation, can be explained in terms of windblown dust.

In the Martian springtime there is a sequential enhancement of contrast between bright and dark areas, called the wave of darkening, which statistically occurs first at higher latitudes although there are marked variations in behaviour for individual areas[1-3]. The biological explanation of the darkening is that organisms inhabiting the dark areas grow and reproduce during the Martian spring[4]. An alternative hypothesis is that the seasonal darkening of the dark areas is due to the movement of dust particles to the bright areas[2,5,6]. Here we shall develop the windblown dust model, starting from the assumptions of major elevation differences on Mars, the presence of fine dust, and of winds adequate to move the dust. All three assumptions have some measure of observational support.

We begin by estimating the dependence on particle size and atmospheric density, ρ, of the minimum or

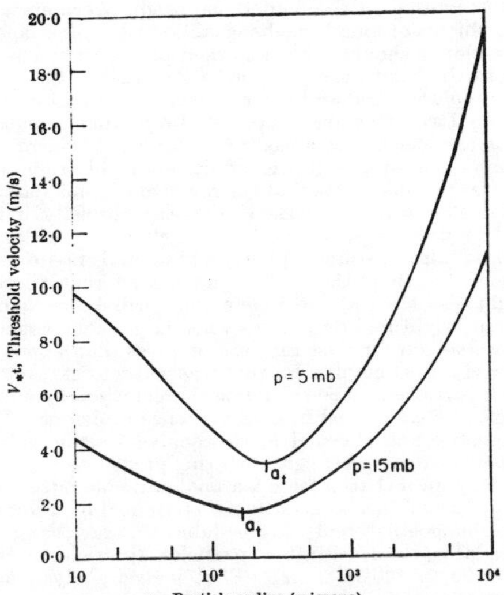

Fig. 1. Threshold velocity for dust mobility on Mars as a function of particle radius, for two choices of total surface pressure.

"threshold" stress, S_t, that must be exerted by surface winds to set grains in motion. Instead of S_t, it is usual to introduce V_{*t} defined by $S_t = \rho V^2_{*t}$. Following the classic procedure outlined by Bagnold[7], Ryan[8] calculated V_{*t} for Martian conditions, but for higher surface pressures than are currently indicated. Fig. 1 shows a recalculation of V_{*t} as a function of particle radius for pressures believed to be representative of highland and lowland areas on Mars[9]. An important feature of the curves is the minimum threshold velocity near 200 microns radius. Usually all particles with V_{*t} below the local meteorological value of V_* will be set into motion. For any grain motion to occur, V_* must, evidently, exceed $(V_{*t})_{min}$, the lowest value of V_{*t} for any particle size.

We see from Fig. 1 that, for all particle sizes, V_{*t} is larger in the highlands than in the lowlands. The density dependence of $(V_{*t})_{min}$ is particularly interesting. A comparison of the two curves of Fig. 1 with each other and with Ryan's calculations at high pressures shows that $(V_{*t})_{min} \propto \rho^{-2/3}$. This can also be derived analytically[9]. It is highly unlikely that the dependence of V_* on ρ will exactly offset the dependence of $(V_{*t})_{min}$ on ρ, so the existence of sizable elevation differences on Mars implies that dust is more easily raised at some locales than at others. If some of the dust can be carried sizable distances, so that there is effective exchange of material between highlands and lowlands, particle size segregation and a consequent contrast difference will be established; the smaller particles will tend to be brighter (see, for example, ref. 22). We can thus understand the existence of bright and dark areas on Mars without invoking differences in chemical composition.

Rather large values of the wind velocity are required to produce dust storms on Mars. If we assume that conditions of neutral stability hold throughout the frictional boundary layer, then[9] we find that wind velocities in excess of 300 km h^{-1} above the boundary layer are required to initiate grain movement for a surface pressure of 10 mbar. But the lower atmosphere of Mars is highly unstable during the day[10,11], so milder winds than indicated are required to start dust movement[10]. Expected uncertainties[9] in $(V_{*t})_{min}$ are factors of two or three. Also theoretical calculations indicate average wind speeds of about 100 km h^{-1}, again with an uncertainty of a factor of two or three, and gust velocities may be considerably larger[11-13]. It is interesting that transverse velocities $\sim$ 100 km h^{-1} or higher of large dust clouds on Mars have been reported by de Vaucouleurs (ref. 14, and a symposium on the surface of Mars, New York, 1967). Thus large scale winds may indeed be capable of producing Martian dust storms, but V_* will not then greatly exceed $(V_{*t})_{min}$. We assume that, characteristically, $V_* \simeq 1{\cdot}5\ (V_{*t})_{min}$ during a dust storm.

Owen (quoted in ref. 15) and Neubauer[10] have drawn attention to the part that dust devils may play in raising dust on Mars. Typically, these small scale disturbances are generated when the surface temperature greatly exceeds the mean air temperature, as occurs on Mars[11,16]. Contrary to Neubauer's claim, large scale winds do not induce high vortex velocities within a dust devil. Indeed, intense large scale winds inhibit the formation of dust devils[17,18]. The high velocities attained inside dust devils (on the Earth, speeds of 100 km h^{-1} have been measured) are due to angular momentum conservation as air spirals from exterior to interior. It is not clear that dust devils will be important on Mars—the large scale wind velocities may always be too great for them to form. And if the air is too unstable, as it may be on Mars, spontaneous convection currents may arise in many locales, preventing the formation of dust devils. If dust devils are frequent on Mars, V_* may greatly exceed $(V_{*t})_{min}$ and very large vertical velocities may occur within them.

There are three types of transport[7] once motion has begun: suspension, saltation and creep. Once small particles are lifted off the ground, eddy currents are able to carry them to considerable heights; in the absence of further turbulent support such particles fall out according to the Stokes–Cunningham equation and, until they reach the surface, are described as suspended. Particles of intermediate size are lifted by winds to only modest heights, along modified ballistic trajectories, and quickly return to the ground; this motion is called saltation. Finally, larger particles may be moved, not by the wind itself, but by momentum exchange with saltating grains; these particles never leave the ground but "creep" slowly in the direction of the wind. Saltating particles may also exchange momentum with smaller grains, lifting them into suspension. Only those dust grains put into suspension can be exchanged between major topographical areas, in timescales of the order of a year or less.

Suspension occurs for particles with radii less than some critical value, a_s, as gravity is a volume force and aerodynamic lift a surface force. When the vertical winds are predominantly upward, as within a dust devil, we expect that a_s can be obtained from an exact equality between the vertical wind velocity and the terminal velocity. For dust devils, the vertical wind velocities are comparable to the horizontal wind velocities. When we use a logarithmic velocity profile to relate V_* to the velocities near the top of a particle's trajectory[8], we find for all values of V_* that a_s significantly exceeds the largest particles that can be set into motion. Thus a dust devil initially puts all particles that it moves into suspension. This result should be insensitive to the choice of velocity profile.

To estimate a_s for Mars in the case of motion caused by large scale winds, we take the ratio of the terminal velocity to the critical vertical velocity near apapsis to be constant and assume that the latter scales as V_*. Thus $a_s \propto (V_*/g)^{1/2}$, where Bagnold's laboratory[19] results are used to define the proportionality constant. When $V_* = 1{\cdot}5\ (V_{*t})_{min}$, this implies $a_s \simeq 125$ microns at the 10 mbar pressure level on Mars, and only slightly different results at other pressure levels.

This value for a_s provides a lower bound on the grain sizes participating in saltation. An upper bound is given by the condition $V_{*t} = V_*$. For large scale winds, subject to the constraint $V_* = 1{\cdot}5\ (V_{*t})_{min}$, an upper bound of about 550 microns at the 10 mbar level is obtained on the size of saltating particles. No saltating particles are expected in dust devils. If V_* greatly exceeds $(V_{*t})_{min}$ for dust devils, then particles ranging in size from microns to centimetres could be put into suspension. Creep occurs only when saltating particles are present, and for particles with a maximum size about six times that of the saltating grain[7].

How far will grains put into suspension on Mars travel before returning to the ground? As a first approximation we assume that particles initially lifted to half a scale height, or 5 km, remain airborne for a time determined by their terminal velocities. Their lateral motion then equals their fallout time multiplied by the mean windspeed, here taken to be 50 km h^{-1}. We find 5, 75, and 500 micron particles travel 6,500 km, 50 km and 1 km respectively. Thermal convection leads to sizable vertical wind velocities throughout the daytime troposphere, velocities comparable with the terminal velocity of a 300 micron particle[20]. No such winds are predicted for night-time. Particles with 100 micron radii will therefore travel greater distances than calculated, while much smaller particles remain unaffected, because their lifetimes are longer than 12 h; particles much larger are unaffected by thermal convection. We infer that all particles put into suspension by large scale winds will travel distances comparable with those between bright and dark areas, while only those particles with radii less than 200 or 300 microns, put into suspension by dust devils, will travel far from the dust devil. In both cases, locales characterized by frequent dust storms will be denuded of their smaller particles, and so will appear dark.

We now attempt to correlate dust storm frequency with elevation. We have already found that $(V_{*t})_{min} \propto \rho^{-2/3}$. We now inquire into the ρ-dependence of V_*, assuming, for convenience, $V_* \propto \rho^{-n}$. To first order, for dust devils, $n=0$; dark and bright areas on Mars have very similar peak surface temperatures[20a] and hence similar temperature discontinuities between surface and air. In this case, dust is less easily raised in highlands, and dark areas should be lowlands. For large scale winds, the equation of mass continuity implies $n = 1$, and therefore that dark areas are highlands. While the analysis seems to imply that dark areas are highlands if dust is raised by large scale winds and lowlands if by dust devils, more detailed analyses are certainly warranted. For example, because windward sides of elevations are less sheltered than leeward sides, there may be a tendency for windward sides to be darker[9].

In any season we expect an equilibrium between the mass flux of small particles exchanged between bright and dark areas such that the relative concentration of small particles (and hence contrast differences) varies inversely with the relative frequency of dust storms in the two locales. Because of seasonal variations in the wind patterns and in temperature discontinuities between surface and atmosphere, the absolute and relative frequencies of dust storms should also vary seasonally, leading to a darkening cycle with the seasons. The time of greatest relative dust storm frequency (although not necessarily of greatest absolute frequency) will be the time of maximum darkening. To deal with relative frequencies we need a crude knowledge of the V_* distribution function. Because we lack this for dust devils, we can make no predictions on the time of maximum darkening for grains raised by dust devils. For large scale winds, theoretical calculations show velocities to be about a factor of three larger in winter than in summer[9,13]. We assume a Gaussian distribution of wind velocities, and expect $(V_{*t})_{min}$ during summer to correspond to a position on the high velocity tail of the velocity distribution function. On this basis, sample calculations[9] indicate that the relative frequency of dust storms and hence the seasonal darkening should reach a maximum during the beginning of the summer, in agreement with observations.

We now compare our windblown dust model with observations.

(1) According to the model, the bright areas are composed chiefly of small, exchangeable dust grains, and so yellow clouds should have a marked photometric resemblance to the bright areas. This is the case[21]. (2) Photometric, polarimetric and thermal inertia data all indicate that the dark areas are composed of a powdered material just as are the bright areas[20a,22]. (3) It is evident that the deposition of fine dust cloud material will brighten an area. Our model shows that the removal of small particles by a dust storm can darken an area, especially if it is already a dark area, or desert region close to a dark area. There is almost direct photographic and photometric evidence for this[21,23]. (4) We have seen that particles of radii less than about 100 microns will be exchanged between bright and dark areas when large scale winds are the mechanism for dust raising; and less than about 250 microns for dust devils. Bright areas will consist primarily of such particles and so should be characterized by a mean particle size calculated to be some tens of microns. Both photometric and thermal inertia analyses again indicate particle sizes in agreement with this prediction[22]. Similarly, we expect that outside seasonal darkening, the mean particle size of dark areas will be determined by a mixture of exchangeable and suspendable, non-exchangeable suspendable, and saltating particles (particle sizes of about 100 to 400 microns). Photometric analysis again indicates such sizes outside of and during darkening, with the exact numerical results in better agreement with the predictions for large scale winds than for dust devils[22]. Particle sizes inferred from thermal inertias for several dark areas are in agreement with the photometric results[20a]. (5) There will be an upper bound to the contrast that dark areas can achieve; it occurs when all the exchangeable particles are removed. Such a phenomenon is indicated in a histogram presented by Focas[23] of the maximum degree of darkening for the centres of various dark areas. This diagram indicates a lower bound of 0·43 on the ratio of the brightness of the dark areas at 5800 Å to that of reference bright areas. Because the two intervals closest to 0·43 have the largest populations, we infer that this bound is real and not the result of unfavourable statistics. This limit corresponds[9] to an average particle size of about 300–350 microns, in agreement with our predictions. (6) The model emphasizes particle size modulation rather than chemical change as the cause of seasonal darkening. Analysis of the polarimetric changes that occur during darkening supports this view[22]. (7) If a bright area initially contains a significant number of non-exchangeable particles, then it can be expected to brighten at the same time that adjacent dark areas darken; it will receive the smaller particles lifted off the dark areas. This prediction is unique to the windblown dust model of seasonal changes, and would be difficult to understand in a biological model. We expect that the southern hemisphere bright areas, where there are many adjacent dark areas, would be more likely to show this effect. Fig. 2 shows confirmatory photographic evidence. The Hellas–Eridania area, at top left, has a lower albedo than the bright area to its right and bottom. The photograph at right displays the seasonal darkening. Here, Hellas has an albedo comparable with the reference bright areas. Photographs from the Lowell Observatory Documentation Center confirm these results, and there are similar photographs for the 1954, 1956 and 1958 oppositions[24]. Elsewhere[9] we have argued that the particular instances of brightening cited are not the result of the presence of either clouds or frost deposit.

The windblown dust model can also explain the rapid regeneration of dark areas temporarily covered by bright material; winds would quickly return the fine powder back into suspension[25]. Secular changes in the configuration of bright and dark areas can be understood in terms of shifting wind patterns; the areas most susceptible to these irregular modulations seem to have the smallest extents and apparently the shallowest slopes[25]. Certain

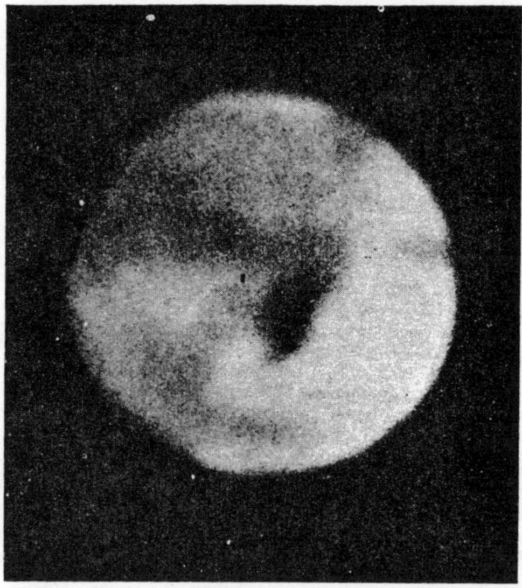

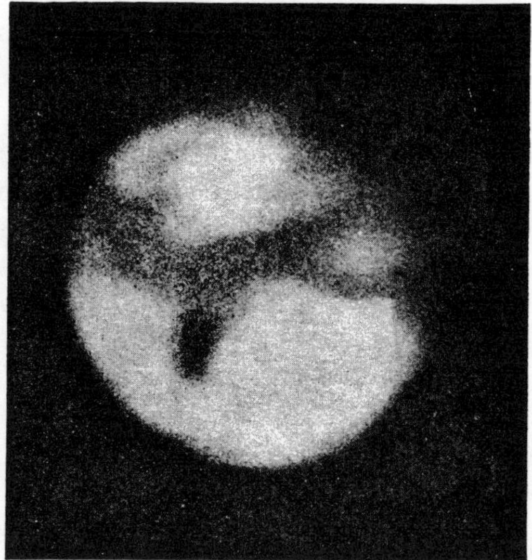

Fig. 2. Seasonal brightening of the Hellas–Eridania region. The photograph at left was taken on July 3, 1907 (Martian date, April 7); the photograph on the right was taken on October 4, 1909 (Martian date, July 1). After Slipher, 1961.

properties of the Martian canals can also be understood within the context of this theory[26]. The success of wind-blown dust models does not, of course, argue against life on Mars.

This research was supported by grants from the US National Aeronautics and Space Administration.

Received February 24, 1969.

1 Pollack, J. B., Greenberg, E. H., and Sagan, C., *Planet. Space Sci.*, **15**, 817 (1967).
2 Kuiper, G. P., *Astrophys. J.*, **125**, 307 (1957).
3 Dollfus, A., *Ann. d'Astrophys.*, Suppl. No. 4 (1957).
4 Sagan, C., *Biology and the Exploration of Mars* (edit. by Pittendrigh, C. S., Vishniac, W., and Pearman, J. P. T.) (National Academy of Sciences, Washington, 1966).
5 Sharonov, V. V., *Priroda Planet*, 507 (Moscow, 1958).
6 Rea, D. G., *Nature*, **201**, 1014 (1964).
7 Bagnold, R. A., *The Physics of Blown Sand and Desert Dunes* (Methuen, London, 1941).
8 Ryan, J. A., *J. Geophys. Res.*, **69**, 3759 (1964).
9 Sagan, C., and Pollack, J. B., *Smithson. Astrophys. Obs. Spec. Rep.* 255 (1967).
10 Neubauer, F. M., *J. Geophys. Res.*, **71**, 2419 (1966).
11 Gierasch, P., and Goody, R. M., *Planet. Space Sci.*, **15**, 1465 (1967).
12 Leovy, C., *Icarus*, **5**, 1 (1966).
13 Wen Tang, *Planetary Meteorology, GCA Corp. Rep. No. NASw*-1227 (1967).
14 de Vaucouleurs, G., *The Physics of the Planet Mars* (Faber and Faber, London, 1954).
15 Kuiper, G. P., *Comm. Lunar Planet. Lab., Univ. Arizona*, **2**, 79 (1964).
16 Sinton, W. M., and Strong, J., *Astrophys. J.*, **131**, 459 (1960).
17 Ives, R. L., *Bull. Amer. Meteorol. Soc.*, **28**, 168 (1947).
18 Sinclair, P. C., *Mon. Weather Rev.*, **22**, 8 (1964).
19 Bagnold, R. A., *The Physics of Blown Sand and Desert Dunes*, 76 (Methuen, London, 1941).
20 Gierasch, P., and Goody, R. M., *Planet. Space Sci.*, **16**, 615 (1968).
20a Morrison, D., Sagan, C., and Pollack, J. B., *Icarus* (in the press).
21 Slipher, E. C., *The Photographic Story of Mars* (edit. by Hall, J. S.) (Sky Publ. Corp., Cambridge, Mass., and Northland Press, Flagstaff, Arizona, 1962).
22 Pollack, J. B., and Sagan, C., *Smithson. Astrophys. Obs. Spec. Rep.* 258 (1967); *Space Sci. Rev.* (in the press).
23 Focas, J. H., *Ann. d'Astrophys.*, **24**, 309 (1961).
24 Dollfus, A., *Ann. d'Astrophys.*, **28**, 722 (1965).
25 Pollack, J. B., and Sagan, C., *Icarus*, **6**, 434 (1967).
26 Sagan, C., and Pollack, J. B., *Nature*, **212**, 117 (1966).

Editor's Comments on Papers 19, 20, and 21

19 BAKER and MILTON
Erosion by Catastrophic Floods on Mars and Earth

20 WEIHAUPT
Possible Origin and Probable Discharges of Meandering Channels on the Planet Mars

21 MASURSKY et al.
Classification and Time of Formation of Martian Channels Based on Viking Data

V. R. Baker and D. J. Milton (Paper 19) have compiled data that indicate erosion of the large martian outflow channels during catastrophic floods, like those that created the Channeled Scablands of eastern Washington state and the similar, but smaller, features in Iceland. J. H. Bretz, in the 1920s, had deduced that the Channeled Scablands were formed when an ice-dammed lake suddenly burst, about 18,000 years ago, and drained within days or a few weeks. However his hypothesis was not widely believed until 1969 (Bretz 1969). Baker and Milton, in their paper, compare the morphologies of the martian channels with the Channeled Scablands. All of the characteristic Scabland forms were observed on Mars, except for giant current ripples, which were probably below the resolution limits of the Mariner 9 TV cameras. Viking orbital imagery strengthens the hypothesis (Carr et al. 1976; see also Paper 21).

D. J. Milton (1974) proposed that decomposition of ice-CO_2 clathrate could release enormous volumes of water rapidly from buried permafrost. (See Paper 14, Part IV, for evidence concerning the probable existence of a substantial permafrost layer.) The clathrate compound, $CO_2 \cdot 6H_2O$, is stable where pressures exceed 10 bars and temperatures are above 0°C. Reducing the load pressure (by opening crevasses, for example) would dissociate the clathrate and water would quickly drain away, thus leading to the collapse of the overlying rock or sediment. However, S. J. Peale et al. (1975) point out that liquid water would already be stable under the conditions Milton outlines. Furthermore, the clathrates require almost twice as much heat to melt per gram as does ordinary ice. Mutch et al. (1976) also point out that the amount of clathrate is limited by the atmospheric concentration of water vapor. Thus the role of clathrates on Mars appears relatively minor.

In order to gain a better understanding of how martian channels develop, J. C. Weihaupt (Paper 20) has selected Nirgal Vallis in Mare Erythraeum for more detailed analysis. The calculation of discharge assumes that the terrestrial relation between meander wavelength and mean discharge of rivers also applies to Mars. However, as can be seen in Table 1, comparison of the morphology

Table 1 Comparison of geometry of terrestrial rivers, lunar sinuous rilles, and martian channels.

	Meander length/ radius curvature		*Meander length/ width*		*Radius of curvature/width*
Rivers*	4.7		10.9		3.1
Sinuous Rilles**	3.9		2.51		.614
Nirgal Vallis*	3.78	3.32 group a	1.72	2.59 A	.489
		4.38 group b		1.42 B	

*Data from Weihaupt (Paper 20).
**Data from VG.

shows that the meandering portion of Nirgal Vallis resembles lunar sinuous rilles more closely than rivers on Earth (Sharp and Malin 1975; Milton 1973). On the other hand, Nirgal Vallis has up to third-order branched tributaries, which follow a relationship similar to the "law of stream orders" on Earth:

Terrestrial rivers (average of four, Vermont)
$$\text{Log No} = 2.24 - 0.607\theta$$
$$\text{Nirgal Vallis Log No} = 2.62 - 0.862\theta$$

These facts suggest both volcanic and fluvial processes may have formed the channel (Sharp and Malin 1975).

An important objective is to establish the age of the martian channels. The existence of liquid water on Mars has been linked to a period of denser atmosphere, which could have resulted either from outgassing accompanying Tharsis volcanism or from climatic oscillations related to variations in orbital parameters (Ward 1973; 1974; Murray et al. 1973), although recent evidence suggests that obliquity changes may be insufficient to generate a dense enough atmosphere.

Determination of the *relative* age of a planetary surface from crater densities is independent of uncertainties in impact fluxes and can form the basis for a stratigraphic succession, as done successfully for the moon (Mutch 1972). H. Masursky et al. (Paper 21) establish by crater counts on Viking orbital photos that the channels are ancient, but span a wider range of time than inferred from the lower resolution Mariner 9 imagery (Malin 1976). In most cases, the channel floors are distinctly younger than the terrain into which they are incised. However, most of the channels are older than Arsia Mons (the southernmost of the Tharsis volcanoes). Whereas the Soderblom et al. (Paper 27, Part IX) crater chronology gives an age of ~ 0.5 billion years for Arsia Mons, the more accurate data of Neukum and Wise (Paper 28, Part IX) indicate an age of ~ 3.0 billion years for the volcano. Malin (1976) found that the large channels *predate* the sparsely cratered plains and are about as old as the Lunae Planum unit (heavily cratered plains unit of Carr et al. in Paper 23, Part VII). On the other hand, the data presented in Paper 21 shows that most of the channels are *younger* than the Lunae Planum unit (equivalent to the Vedra and Bahram Valles surfaces). This discrepancy in relative ages could have arisen from including prechannel impact-lithified craters that were exhumed by the flood scour, in crater counts on the lower resolution Mariner 9 imagery.

19

Reprinted from *Icarus* **23**:27–41 (1974)

Erosion by Catastrophic Floods on Mars and Earth

VICTOR R. BAKER

Department of Geological Sciences, University of Texas at Austin, Austin, Texas 78712

AND

DANIEL J. MILTON

U.S. Geological Survey, Menlo Park, California 94025

Received March 18, 1974; revised May 6, 1974

The large Martian channels, especially Kasei, Ares, Tiu, Simud, and Mangala Valles, show morphologic features strikingly similar to those of the Channeled Scabland of eastern Washington, produced by the catastrophic breakout floods of Pleistocene Lake Missoula. Features in the overall pattern include the great size, regional anastomosis, and low sinuosity of the channels. Erosional features are streamlined hills, longitudinal grooves, inner channel cataracts, scour upstream of flow obstacles, and perhaps marginal cataracts and butte and basin topography. Depositional features are bar complexes in expanding reaches and perhaps pendant bars and alcove bars. Scabland erosion takes place in exceedingly deep, swift floodwater acting on closely jointed bedrock as a hydrodynamic consequence of secondary flow phenomena, including various forms of macroturbulent votices and flow separations. If the analogy to the Channeled Scabland is correct, floods involving water discharges of millions of cubic meters per second and peak flow velocities of tens of meters per second, but perhaps lasting no more than a few days, have occurred on Mars.

Introduction

The channel-like landforms on Mars revealed by the 1971 Mariner 9 Orbiter are diverse in character and presumably are equally diverse in origin. Those Martian channels that resemble the sinuous rilles of the Moon may have been produced by thermal incision of lava flows (Carr, 1974), the steep-headed valleys of the fretted terrain may have been produced by headward sapping by ground-ice decay (Sharp, 1973) or artesian processes (Milton, 1973), and so on. This paper is concerned with a distinct class of Martian channels characterized generally by breadths of several kilometers or tens of kilometers, slightly sinuous courses, and locally anastomosing patterns. The largest are Ares, Tiu, and Simud Valles (Chryse Channels[1]); the best high-resolution photographic coverage is of Mangala Vallis (Amazonis Channel) and Kasei Vallis (Lunae Palus Channel). These channels are of particular importance because they have been adduced as evidence for the former presence of liquid water on the Martian surface (McCauley *et al.*, 1972; Milton, 1973; Sagan *et al.*, 1973), which has profound implications for Martian history.

We believe that these channels were created by catastrophic floods. Comparable floods occurred on Earth during the Pleistocene Epoch, as documented in the Channeled Scabland of eastern Washington produced by breakouts of ice-dammed Lake Missoula (Bretz, 1923, 1932, 1969; Baker, 1973, 1974), in the Snake River Plain of Idaho, flooded by the overflow of Lake Bonneville (Malde, 1968), and as are inferred in the scabland terrain of Wright Valley, Antarctica (Smith, 1965; Warren, 1965). Table I lists the characteristic features produced by these floods

[1] Martian topographic features were assigned names by the International Astronomical Union in late 1973. Earlier papers used informal names based on pre-Mariner areography.

with examples from the Channeled Scabland and probable equivalents in the Martian channels. Our identification on the Mariner photography is made with confidence for some features; for less distinctive features, our identification is correspondingly tentative. Finally, identification of some critically diagnostic but small features may (or may not) await recognition during future exploration of Mars.

Erosional Forms and Processes

The catastrophic-flood hypothesis proposed for the Channeled Scabland by

TABLE I

CHARACTERISTIC EROSIONAL AND DEPOSITIONAL FEATURES OF SCABLAND CHANNELS AND PROBABLE MARTIAN EQUIVALENTS AS INTERPRETED FROM MARINER 9 ORBITER IMAGERY

Feature	Scabland examples		Martian examples	
	Location[a]	Scale	Location[b]	Scale
Overall Regime:				
Large-scale bedrock channels requiring erosion to depths of hundreds of meters	Grand Coulee and Cheney–Palouse Scabland	See Table II	Mangala Vallis, Kasei Vallis, Ares Vallis, Tiu Vallis and Simud Vallis	See Table II
Low sinuosity	Grand Coulee and Cheney–Palouse Scabland	See Table II	Mangala Vallis, Kasei Vallis, Ares Vallis, Tiu Vallis and Simud Vallis	See Table II
Regional anastomosis of bedrock channels	Entire Channeled Scabland	300 × 300 km	Ares, Tiu and Simud Valles	1800 × 1200 km
Erosional Features:				
Streamlined Hills	Cheney–Palouse Scabland	Smaller ones, 0.5 × 2 km; larger 3 × 6 km	Ares Vallis (1° N, 17° W) Elysium Planitia (31° N, 229° W)	10 km in length 5 km in length
Longitudinal Grooves	Lenore Canyon and Hartline Basin	Spacing 100–500 m	Kasei Vallis	Spacing about 500 m–2 km
Inner Channel Cataracts	Dry Falls	6 km wide, 120 m high	Kasei Vallis	10–20 km wide
Erosionally exhumed circular structures	Crab Creek Scabland near Odessa (sagflowout structures)	80–250 m	Kasei Vallis (craters)	About 6 km
Tributary alcoves to main channel (marginal cataracts)	Grand Coulee	2 km wide, 250 m high	Kasei Vallis? (middle reach) Mangala Vallis?	5–10 km wide
Scour upstream from flow obstacle	Dry Coulee and High Hill Anticline (Park Lake 7.5′ Quadrangle)	500 m wide, 10 km long	Kasei Vallis	4 km wide, 10 km long
Etching of regional joint patterns	Cheney–Palouse Scabland (Starbuck 15′ Quadrangle)	Linear fractures, 10 km long over an area 25 × 20 km	Kasei Vallis	Fractures 10–20 km long in an area approximately 20 × 20 km
Butte and basin topography	All scoured basalt channels	Buttes 100 m wide and 30 m high	Mangala Vallis?	May be below the resolution capability of the Mariner imaging system
Depositional Features:				
Expansion bar complexes	Quincy Basin	500 km²	Mangala Vallis	1000 km²
Pendant bars	Wilson Creek Area (Wilson Creek 15′ Quadrangle)	4 km long, 1 km wide	Ares Vallis? (1° N, 17° W)	10 km long, 2 km wide
Eddy bars	Grand Coulee	10 km long, 2 km wide	Perhaps Kasei Vallis (middle reach) but difficult to distinguish from terraces	20 × 50 km
Giant current ripples	Most scabland channels especially Cheney–Palouse Scabland tract	Spacing, 30–130 m; Height 1–7 m	Below the resolution capability of the Mariner imaging system.	

[a] Named features are shown in Fig. 1, located by latitude and longitude in Table II, or found on the indicated U.S. Geological Survey topographic map.
[b] Named Valles are located by latitude and longitude in Table II.

J Harlen Bretz (1923) initially found little favor and has won general acceptance only after 50 yr of contention with a variety of hypotheses thought to be more in accord with uniformitarian principles. The principal objection to Bretz' hypothesis was the lack of any evident source for the immense quantities of water required, an objection not answered until the volume and rate of emptying the 640-m-deep ice-dammed Lake Missoula were shown to be adequate (Pardee, 1942). The hypothesis of the occurrence of floods on Mars is likewise based on morphologic evidence, and its acceptability should not depend on that of specific hypotheses for the cause of floods (Milton, 1974). It is interesting that some of the alternative explanations for scabland features have reappeared in the brief history of discussion of Martian channels: incision of alcoves by spring sapping (Snake River Plain–Stearns, 1936, reinterpreted by Malde, 1968, as subfluvial cataracts; Mars–Sharp, 1973; Milton, 1973); salt weathering and eolian removal of debris (Wright Valley–Selby and Wilson, 1971; Mars–Malin, 1973); erosion during drainage of a lake (Channeled Scabland–Allison, 1933; Mangala Vallis–Milton, 1973); lava channels (Channeled Scabland–W. C. Alden in 1927 as recounted by Bretz, Smith, and Neff, 1956; Mars–Carr, 1974).

The Channeled Scabland region (Fig. 1) coincides with the part of the Columbia Plateau that was subjected to periodic catastrophic flooding during the late Pleistocene. Bretz (1923) introduced the term "scabland" to refer to the chaotically eroded tracts of bare basalt that occupy the floors of channels cut through the thick loess cover of the plateau. These channels

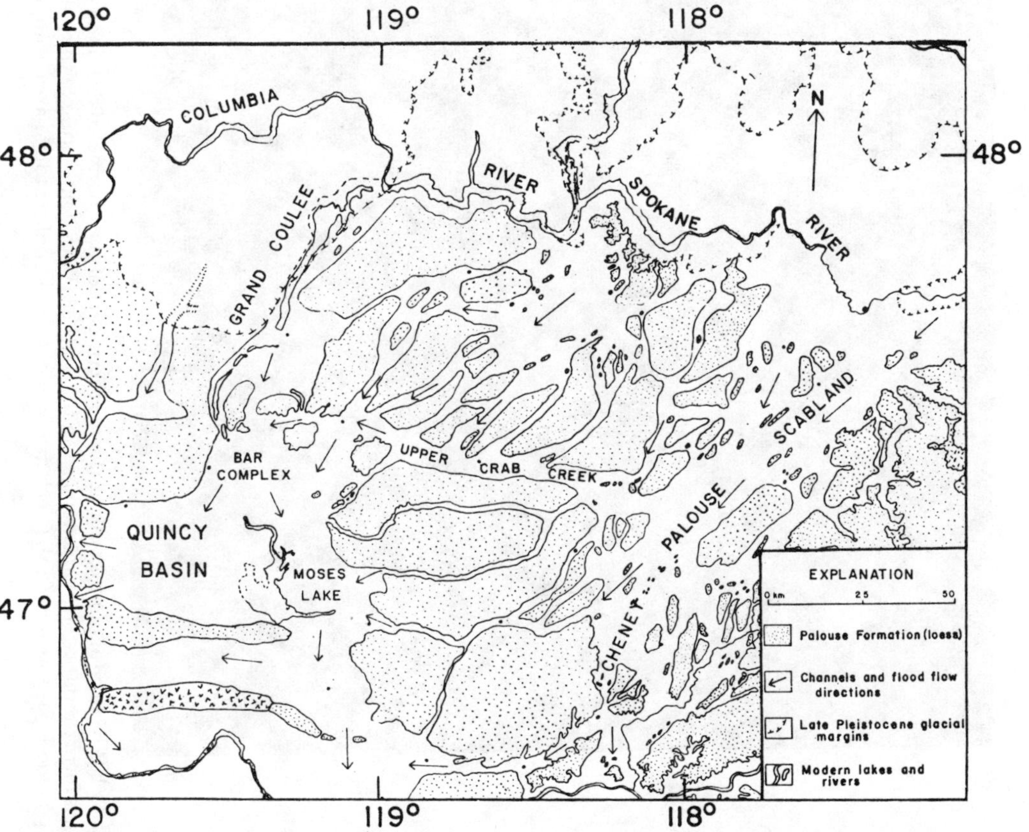

FIG. 1. Map of the Channeled Scabland in eastern Washington showing the distribution of channels and the remaining loess cover.

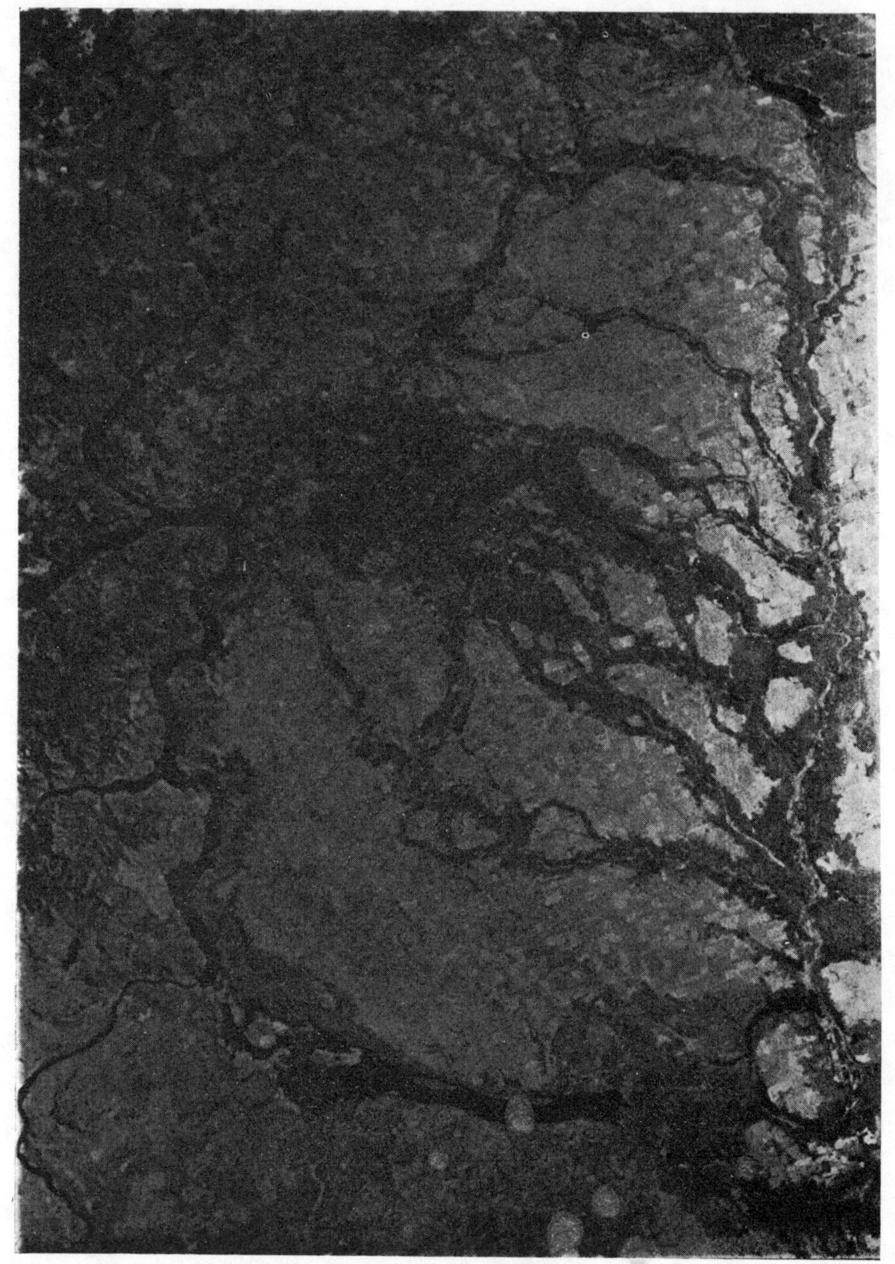

FIG. 2. ERTS orbital photograph of the northern part of the Channeled Scabland (composite of multispectral imagery, July 26, 1972). Scabland channels form the dark-toned anastomosis in the southern part, contrasting to the patchwork of wheat farms on the lighter Palouse loess. Colombia and Spokane Rivers at top, Crab Creek at bottom, Grand Coulee at left; area is approximately 90 × 120 km.

are remarkably conspicuous, even when viewed from orbital altitudes (Fig. 2).

A variety of features serve as high-water marks of the last major scabland flood, now thought to have occurred about 18 000 yr ago, and allow reconstruction of high-water surfaces and flow depths. Combining these with the channel geometry, Baker (1973) utilized hydraulic engineering procedures to calculate discharges and mean flow velocities. Additional paleohydraulic information is provided by the numerous giant current ripples that occur throughout the region. From such evidence it was shown that discharges as great as $21.3 \times 10^6 \, m^3/sec$ were conveyed through the Channeled Scabland. Because this discharge was sustained by $2.0 \times 10^{12} \, m^3$ of lake volume, flooding at the peak rate would have persisted less than a day. Even with gradually waning flows, flooding probably lasted only a week or two. Most of the erosion, however, probably occurred in the day or two of sustained high discharge. The amazingly rapid outflow of Lake Missoula is in accord with what is known of historical glacial breakout floods (Clague and Mathews, 1973). Water flowed 100–200 m deep down the steep regional dip slope of the Columbia Plateau (6–8 m/km) so that preflood valleys were reduced to mere channel-bottom roughness elements. Accordingly, the erosive process in the Scabland was subfluvial channel scour. Subfluvial erosion has been documented experimentally and field examples suggest that the process may be much more important than formerly believed (Schumm and Shepherd, 1973). It was remarkably effective in the Scabland; during the brief duration of the flood, channels were incised in bedrock as deep as 200 m. The Scabland channels differ fundamentally from normal valleys of comparable size that are the product of many thousands of years of valley incision. Although rare narrow gorges have been cut almost

TABLE II

COMPARATIVE GEOMETRY OF SELECTED MARTIAN CHANNELS WITH MISSOULA FLOOD CHANNELS ON THE COLUMBIA PLATEAU, EASTERN WASHINGTON. VALUES ARE ONLY APPROXIMATE, BUT THEY ALLOW A RELATIVE COMPARISON OF THE TWO REGIONS

Regional name	Approximate location Latitude	Longitude	Channel width[a] (km)	Channel length (km)	Channel depth (m)	Sinuosity[b]
Kasei Vallis (northern channel)	26°N	65°W	10	450	Indeterminate	1.15
Kasei Vallis (southern channel)	22°N	66°W	20	600	2500	1.06
Mangala Vallis	6°S	151°W	6	350	Indeterminate	1.05
Ares Vallis	5°N	20°W	40	1500	Indeterminate	1.04
Grand Coulee				80	250	1.07
Upper Coulee	47° 50′N	119° 8′W	8			
Hartline Basin	47° 35′N	119° 20′W	20			
Cheney–Palouse Scabland Tract				120	100	1.05
Lamont	47° 12′N	117° 53′W	15			
Benge	46° 53′N	118° 5′W	40			
Upper Crab Creek	47° 20′N	118° 30′W	6	100	100	1.01

[a] Individual channel width at the approximate location rather than total width of a channel complex.

[b] Calculated as the length of the channel center line divided by length of a straight line joining the end points of the measured reach.

wholly by fluvial action, normally in river valleys material is transported to the valley bottom by slope processes and then removed by the river, especially during floods. For example, the Grand Canyon of Arizona is a valley in which the channel of the Colorado River now (or at any point in past time) occupies only a small fraction. A subfluvially scoured channel, unlike a river valley, can have locally reversed gradients and closed depressions, since it is only necessary that the upper surface of water have a continuously downward gradient from head to mouth.

The mechanical aspects of devastating floods are poorly understood (Scheidegger, 1973). Nevertheless, the distinctive scabland erosional forms seem to be a hydrodynamic consequence of exceedingly swift, deep flood water acting on closely jointed bedrock (Baker, 1974). Scour was most pronounced in the constricted reaches of the western Scabland, where flow depths of 60–120 m and water surface gradients of 12 m/km indicate peak flood flow velocities as high as 30 m/sec. Under these conditions, secondary flow phenomena, including various forms of vortices and flow separations termed "macroturbulence" (Matthes, 1947), produced intense pressure and velocity gradients that plucked fragments of columnar basalt from the irregular channel boundaries. The plucking of the columnar colonnade and the undermining of the more resistant massive entabulature of the basalt flows produced butte and basin topography as the characteristic channel-bed scour feature.

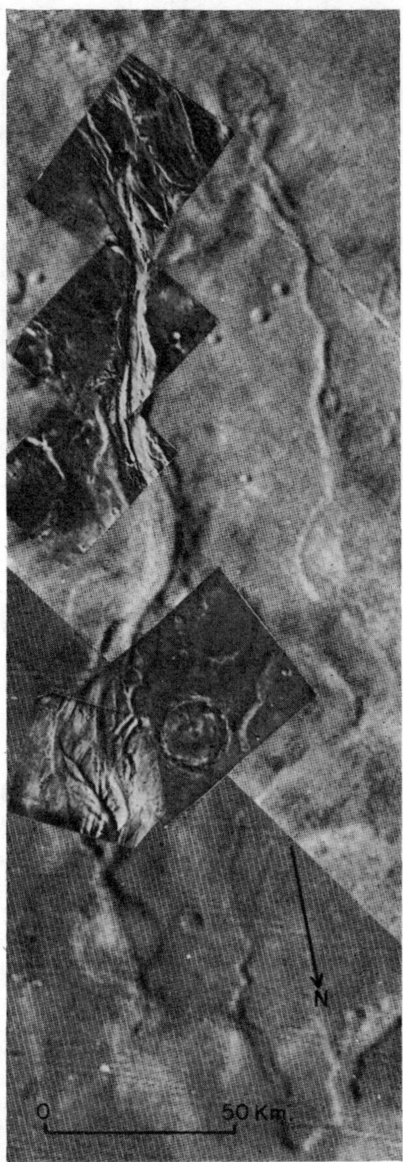

Fig. 3. Mangala Vallis. The almost straight single channel in middle section appears to be structurally controlled. Small features on the adjacent plain suggest butte and basin topography and notches in the bank may be marginal cataracts, which would indicate overbank flow. Downstream, an expansion bar complex occupies the widened reach of the channel. Composite of Mariner 9 wide- and narrow-angle frames. Narrow-angle frames in a larger format may be seen in Schumm (1974).

Channels on Mars

The Martian channels are larger than their terrestrial counterparts (Table II). The evidence for catastrophic floods is not sufficient to indicate whether they excavated the Martian channels (corresponding to the Missoula flood and the major Scabland channels) or only modified preexisting valleys (corresponding to the Bonneville flood and the Snake River Plain). Analytic photogrammetry (Blasius, 1973) is possible in a small segment of Kasei Vallis for which stereoscopic coverage is available. Approximate height measurements (kindly furnished by K. R. Blasius) indicate a total relief of more than 5km (Fig. 5). The upper 2km is occupied by a scalloped cliff that shows no sign of fluvial erosion and may be part of an older valley wall preserved above the high-water mark. Mangala Vallis (Fig. 3) and Ares Vallis, on the other hand, show evidence of overbank flow. If the downcutting occurred during a flood or floods, it is not necessary that the water depth ever equalled the difference in elevation of the high water mark and the low point of the channel floor. Nevertheless, the width and depth of the apparently flooded sections of the Martian channels are enormous.

The general pattern on both Earth and Mars is one of large channels eroded in bedrock, forming a regionally anastomosing complex with individual channels that have a low sinuosity (Table II). Normal rivers transporting a fine-grained load usually have a single meandering channel with a sinuosity greater than 1.5. This pattern represents an adjustment of a graded longitudinal profile over a period of years to continuous transport of water and sediment (Mackin, 1948). Catastrophic flood channels, in contrast, follow established preflood topography. Anastomosis occurs in the Channeled Scabland because preflood valleys did not have the capacity to convey the Missoula flood discharges without spilling over preflood divides into adjacent valleys. This would be expected wherever preflood topography is regionally smooth, as on Mars in the complex of channels of which Mangala Vallis is the largest. Individual channels may be controlled by relatively straight geologic structures, such as High Hill anticline in the Channeled Scabland or the fault scarp that may bound the west side of the middle reach of Mangala Vallis (Fig. 3) (Milton, 1973). Aside from this, the high shear stress gradient that occurs on the inside of meander curves in extreme floods may lead to channel straightening and partly account for the relatively straight character of scabland channels.

Erosional features of catastrophic floods vary greatly with the nature of the material eroded. Lithologies on Mars can only be inferred from the topography. In Kasei Vallis there appears to be a resistant

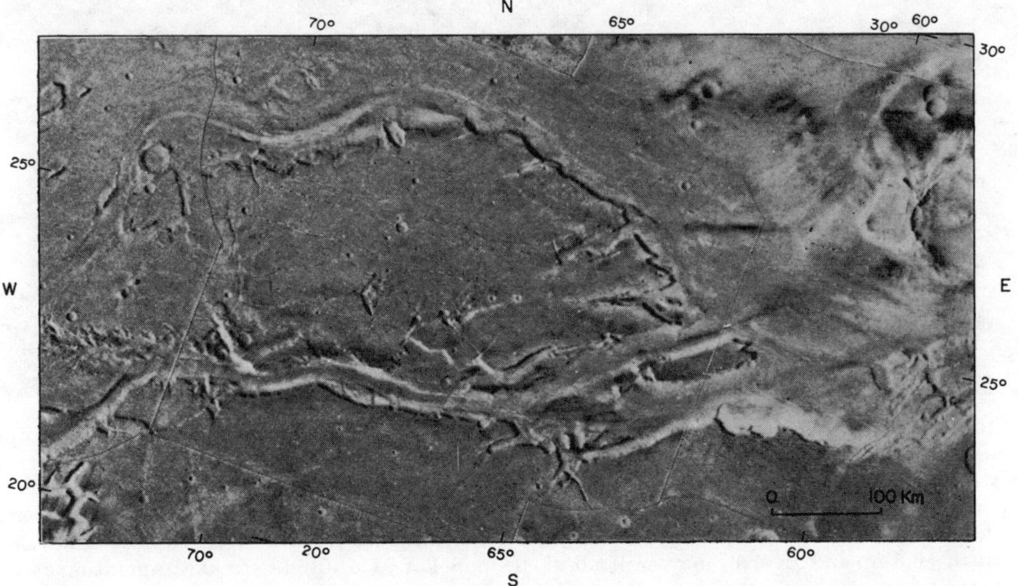

Fig. 4. Kasei Vallis (bottom) and unnamed channel (top), apparently formed or modified by floods flowing from west to east. Part of U.S. Geological Survey rectified mosaic of quadrangle MC-10, prepared from Mariner 9 wide-angle frames. Scale 1:1 000 000.

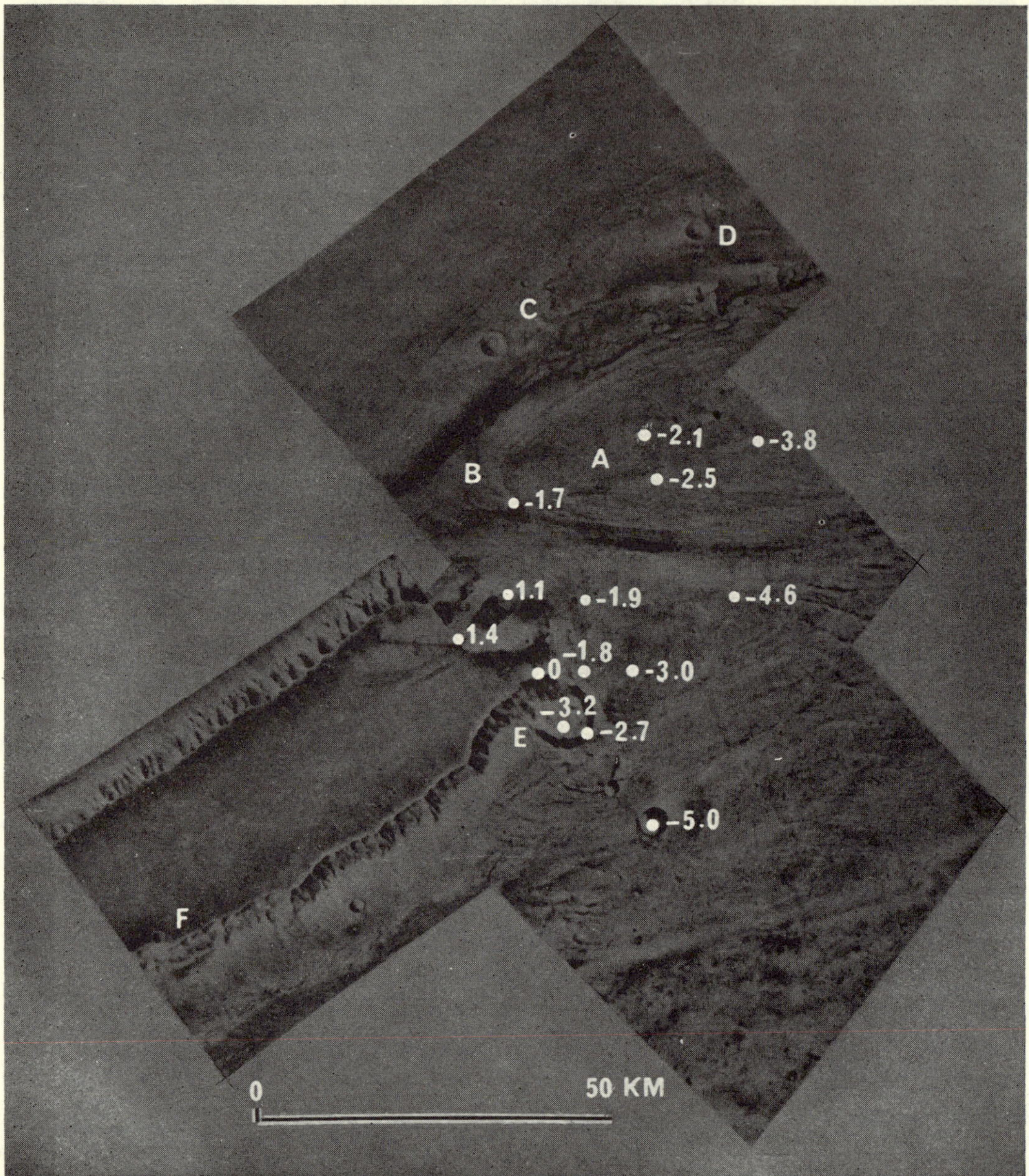

FIG. 5. Part of lower Kasei Vallis near 62°W, 25°N, with landforms indicating catastrophic flood erosion. Features discussed in the text include: (A) streamlined hill; (B) a horseshoe cataract that has worked around the head of the hill; (C) a complex of horseshoe cataracts; (D) a crater apparently controlling the site of a lower cataract; (E) a partly exhumed crater; and (F) stratification in a talus-free section of cliff. Spot elevations (in km from an arbitrary datum) are subject to individual errors of ±0.3 km and an error in overall tilt of the model of ±1.5°. Light streaks originating at some craters and promontories indicate recent winds blowing westward. Rectified mosaic of Mariner 9 narrow-angle frame 10277404, 10277474, 12866133, and 12866203.

basal unit overlain by a weaker bedded unit (note stratification in the cliff at F, Fig. 5). This is capped by a thinner resistant unit, probably basalt from the evidence of structural features in nearby areas that resemble those of the lunar maria.

The characteristic landform of the loess eroded by the Missoula flood is the streamlined hill with a blunt rounded upstream prow and a pointed downstream end, eroded as islands or entirely subfluvially to a form offering minimum resistance to rapidly flowing water (Fig. 6). Streamlined hills are found in several Martian channels, but it is not always possible to determine whether they are eroded country material or bars of sediment deposited in the channel. A large streamlined hill in the mouth of Kasei Vallis (Figs. 4, 5) is clearly eroded bedrock since transverse lineaments exposed on the channel floor and on the adjacent plateaus also extend across the hill. The material composing the hill appears to correspond to the bedded unit in the island to the west.

Characteristic forms of erosion in basalt in the Channeled Scabland are longitudinal grooves, subfluvial cataracts, and butte and basin topography. Grooves (Fig. 7) were apparently produced by powerful roller vortices (also called secondary circulation cells) that develop with their filaments parallel to the flow direction. Allen (1971) described the morphologic effects of these vortices from laboratory flume experiments and noted (Allen, 1970) that the very regular spacing of longitudinal vortices is a function of velocity and depth of flow. Flume studies in simulated bedrock (Shepherd, 1972; Shepherd and Schumm, 1974) indicate that longitudinal

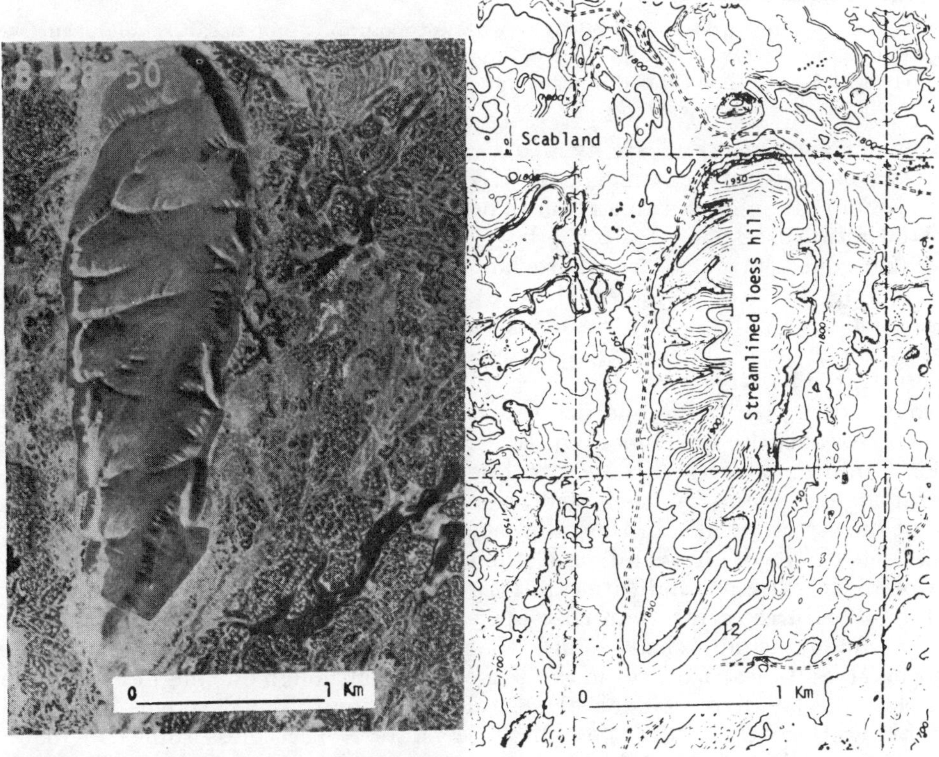

FIG. 6. Loess hill near Macall, Washington streamlined by flood erosion. A small cataract, heading an inner channel, has worked its way around the blunt upstream end of the hill (on map). Longitudinal grooves and butte-and-basin topography can be seen in the marginal scablands. Water depths and velocities averaged 12 m/sec for depths of 30–40 m in this area during the flood maximum. Map contour interval is 10 ft (~3 m).

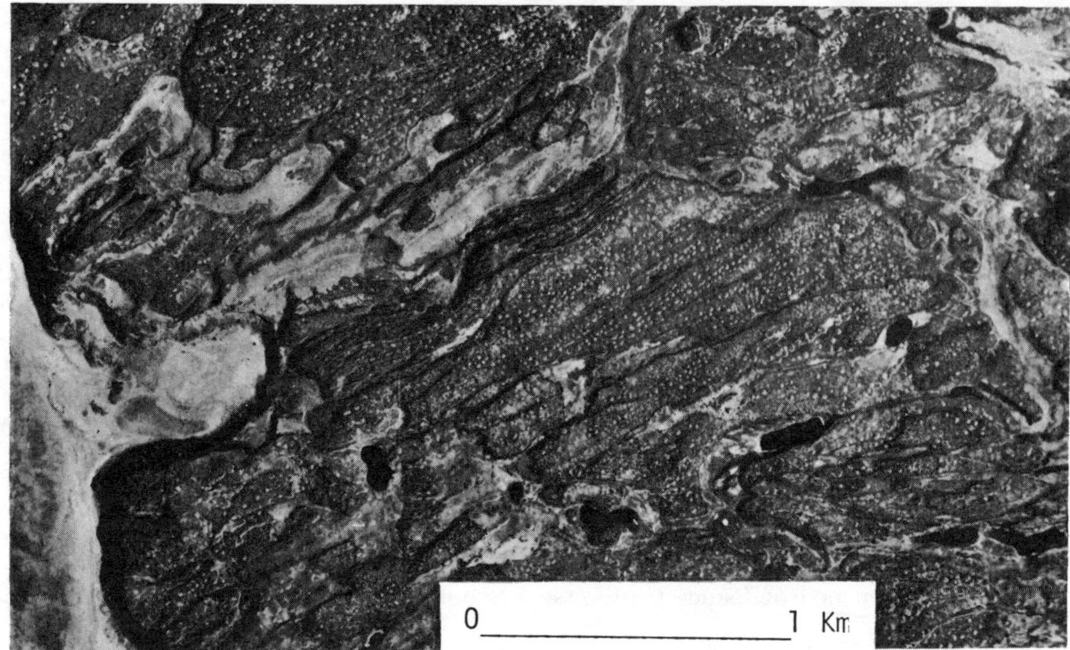

FIG. 7. Longitudinal grooves and butte-and-basin topography on a basalt scabland surface east of Jasper Canyon in the Lower Grand Coulee. Small mounds of eolian silt in the swales emphasize the linear pattern. A horseshoe-shaped inner channel cataract is at left.

grooves develop early and are eventually replaced by a deep inner channel headed by a nickpoint that migrates upstream. The scabland form corresponding to the latter is the abandoned cataract, largest of which is Dry Falls in Grand Coulee (Fig. 8). These developed subfluvially rather than by the plunge pool undercutting classically illustrated by Niagara Falls.

Grooves on a giant scale and dry cataracts are excellently developed in Kasei Vallis. Conformity of the grooves to the general channel geometry, their deflection around the streamlined hill, their absence in the lee of the high island, and their relation to the cataracts indicate that they are fluvial features rather than the superficially similar wind-etched joints that occur on Mars. Linear features in the same area clearly related to regional structural patterns rather than to the channel geometry could have been etched by either fluvial or eolian action, although their large size suggests the former. Dry cataracts in Kasei Vallis have the same scalloped rims composed of horseshoe-shaped segments as those in Grand Coulee (Fig. 5C). An upper horseshoe cataract (Fig. 5B) appears to have worked around the head of the streamlined hill to a point at which it would have been opposed to the general flow direction. Analogous situations are found in the Scabland, for example, at Dry Coulee around High Hill anticline or, on a smaller scale, the unnamed coulee around the streamlined loess hill (Fig. 6).

Butte and basin topography in the Channeled Scabland is a product particularly of the plucking of columnar basalt to form potholes and basins. Butte and basin topography cannot be identified with certainty in the Mariner photographs. Typical scabland buttes and basins are of a size that would be near B-frame resolution, although one might expect them, like the longitudinal grooves, to be larger on Mars. Nevertheless, some irregular Martian terrain may be analogous, for example on the marginal channels of Mangala Vallis (Fig. 3). The ability of deep, high-velocity flow of catastrophic floods to lift debris from channel floors and hence form basins

FIG. 8. Dry Falls cataract, 120 m high, with upper Grand Coulee in background. Longitudinal grooves are visible just upstream from the cataract.

may be evinced in some closed depressions on the floor of Martian channels. The larger craters in Kasei Vallis appear to predate the channel cutting and to have been buried and exhumed (note the large crater only partly emergent from beneath the high island (Fig. 5E) and the crater that appears to have controlled the outline of the inner channel (Fig. 5D). Although it is evident that material has been moved out of many closed depressions elsewhere on Mars where no traces of floods are seen, these craters may have been scoured out by floodwaters. An intriguingly similar landform in the Channeled Scablands is produced by the sag flowouts of the Columbia River Basalt (McKee and Stradling, 1970). Although their structural origin is very different from a crater, they owe their morphologic expression to stripping of overlaying basalt flows by Missoula flooding (Fig. 9).

Tributary alcoves to the main Scabland channels were produced by headward erosion of marginal cataracts (Bretz, 1932). Such an origin is possible for alcoves in the middle reach of Kasei Vallis (Fig. 4), but this would imply an overbank flow on the adjacent plateau that is not otherwise indicated. More probable examples are found in Mangala Vallis (Fig. 3) where erosional features on the adjacent plateau indicate overbank flooding.

Depositional Features

In environments where the velocity of flow is lowered, bedload is deposited to form subfluvial bars. Expansion bars form downstream from constrictions, as in Quincy Basin downstream from the lower Grand Coulee. The bar complex occupies $500 km^2$ and is composed of streamlined bar forms truncated by scour channels (Fig. 10). The basin fill has 60 m of relief and includes channels with reverse down-

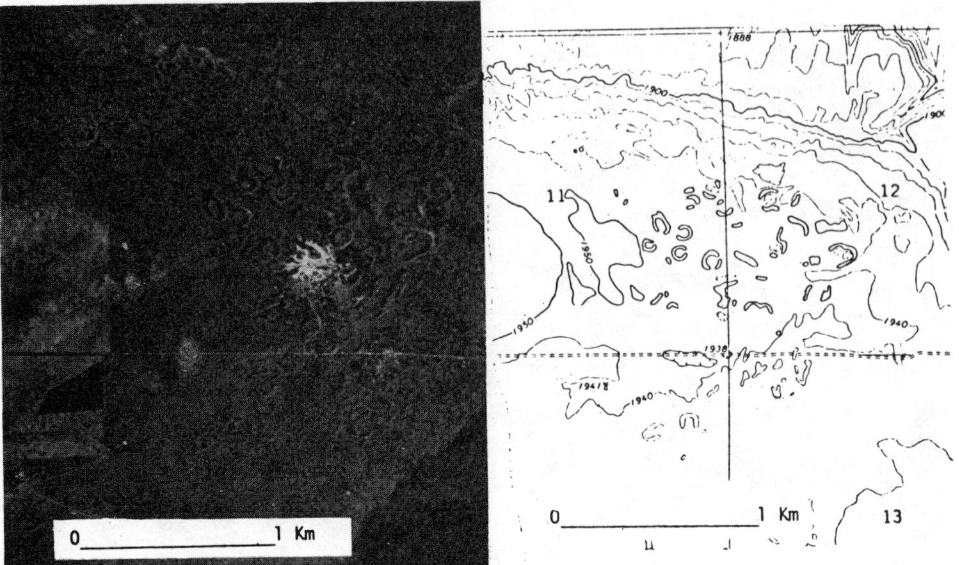

FIG. 9 Sag flowout structures near Karakul Hills in the Cheney–Palouse scabland tract. The ringlike forms are dikes that formed when lava escaped through tension joints surrounding sags on the cooled flow crust. Hydraulic plucking by the Missoula floods caused the rings to stand in relief, exhumed from overlying flat layers of basalt.

FIG. 10. View south (downstream) of a part of the Quincy Basin expansion bar comple Moses Lake (left) occupies one of several sinuous channels that truncate the streamlined ba Compare to Mangala Vallis (Fig. 3). (Photograph by David A Rahm.)

stream gradients and many closed depressions. Bretz and others (1956) interpreted the bar complex as a gravel fill of one flood incised by several later floods. Baker (1973) reinterpreted it as mainly the product of a single flood with deposition at peak flow and incision during the waning stages. An important factor in the development of the Quincy Basin bar complex is the presence of flow constrictions both upstream and downstream. The upper constriction allowed an influx of high-velocity water carrying much sediment while the lower constriction provided hydraulic damming of the basin during the rising phase of the flood. Deposition occurred as the flood-wave filled the basin. Subsequently, the network of scour channels developed as the basin drained, probably while the inflow decreased.

Forms very similar to those of the Quincy Basin are found on Mars in Mangala Vallis in the reach where it widens to 30km between 5-km narrows up- and downstream. Channels truncate and cut what appear to have once been streamlined bar forms as much as 10km long. The channels do not all grade uniformly downstream but some appear to rise in a downstream direction or contain closed depressions. Although one of us has compared this reach to a braided section of an aggrading terrestrial river (Milton, 1973), we now believe that an expansion bar complex provides a more satisfactory terrestrial analog. This accords with the location only in the wide reach rather than along the entire course, as would be expected if it was formed by an aggrading river, and the overall appearance of an erosional as much as depositional pattern.

Pendant bars, longitudinal bars, and eddy bars form in terrestrial flood channels downstream from bedrock projections, on the margins of channels, and in alcoves or in the mouths of tributary valleys, respectively. In the Mariner photograph it is difficult to distinguish these from streamlined hills or terraces eroded from bedrock, but some of the midchannel features of Ares Vallis may be pendant bars and the features occupying broad reentrants on the margin of middle Kasei Vallis (Fig. 4) and particularly in the constricted reach of Mangala Vallis (Fig. 3) may be eddy bars.

Trains of giant current ripples were not discovered until late in the investigation of the Scabland but have become generally accepted as the clearest proof of the catastrophic flood hypothesis. In form and spacing they resemble transverse sand dunes (and were so interpreted when first noted on aerial photographs), but they contain foreset bedded basalt fragments as large as 1.5m. Their chords range from 20 to 130m and their heights from 0.5m to 7m. This is below the resolution of the Mariner or Viking orbital imaging systems. A prime Viking landing site, near the mouths of Ares and Tiu Valles, could well be in a giant-ripple terrain, however. Although ripple trains in the Scabland are far less conspicuous from a fixed viewpoint on the ground than from the air, the possibility of their presence should be kept in mind during interpretation of the landscape viewed by the television camera on the Viking Lander.

Conclusions

Many of the individual features described in this paper could reasonably be explained by alternative hypotheses. For example, eolian phenomena in extremely arid deserts produce remarkably streamlined erosional forms and lineations. Again, some of the broader features of the channel systems might be taken to indicate fluvial action without extreme flooding. But the hypothesis of catastrophic flooding appears uniquely able to account for the entire assemblage of features in the major channels of Mars. Several points should be significant in seeking a cause for the floods.

1. Flooding was a local phenomenon. The water was concentrated in a small number of discrete channels generally flowing northward from the equatorial cratered terrain to the lower plains.

2. Vast quantities of water were supplied so rapidly as to provide peak discharges of millions of cubic meters per second. Even the dense terrestrial atmosphere cannot provide water fast enough to yield

such discharges from comparable-sized catchment areas. On Earth, only dam bursts or subglacial volcanic eruptions have yielded flows capable of significant macroturbulent erosion.

3. Water need only have persisted briefly on the Martian surface. Macroturbulent scour by Missoula floods was accomplished in a matter of at most a day or two.

4. When the flooding occurred, whether it was coeval in the separate channels, and whether in individual channels it was a single or a repeated event, are questions still to be answered.

Normal continuous processes of erosion and deposition act with so much greater intensity on Earth than on Mars that traces of catastrophes are soon erased. The Channeled Scabland of the Columbia Plateau preserves the record of events perhaps rare in the history of the Earth. The dominant geomorphic processes operating in this area throughout the Pleistocene have been catastrophic floods and episodic eolian activity (loess deposition) in a region of arid climate and basalt bedrock. Perhaps no terrestrial landscape finds a closer analogue on Mars than this.

Acknowledgments

Preliminary versions of this paper were read by W. D. Manley, Jr., and R. G. Sheperd, C. A. Hodges and M. H. Carr. Work by D. J. Milton was supported by Planetology Programs, NASA Headquarters, under contract W-13204, and Mariner Mars 1971 Project, Jet Propulsion Laboratory, California Institute of Technology, under contract WO-8122. Financial aid to V. R. Baker was provided by National Science Foundation grant GA-21478 and by The Geological Foundation of The University of Texas at Austin.

References

Allen, J. R. L. (1970). "Physical Processes of Sedimentation." American Elsevier, New York.

Allen, J. R. L. (1971). Bed forms due to mass transfer in turbulent flows: a kaleidoscope of phenomena. *J. Fluid Mech.* **49**, 49–63.

Allison, I. S. (1933). New version of the Spokane flood. *Geol. Soc. Amer. Bull.* **44**, 675–722.

Baker, V. R. (1973). Paleohydrology and sedimentology of Lake Missoula flooding in eastern Washington. *Geological Society of America Special Paper 144*, 79 pp.

Baker, V. R. (1974). Erosional forms and processes for the catastrophic Pleistocene Missoula floods in eastern, Washington. *In* "Fluvial Geomorphology" (Marie Morisawa, Ed.), Proceedings 4th Annual Geomorphology Symposium, Binghampton, N.Y. Publication of Geomorphology, State University, New York, Binghampton, N.Y.

Blasius, K. R. (1973). A study of Martian topography by analytic photogrammetry. *J. Geophys. Res.* **78**, 4411–4423.

Bretz, J H. (1923). The channeled scablands of the Columbia Plateau. *J. Geol.* **31**, 617–649.

Bretz, J H. (1932). The Grand Coulee. *American Geographical Society Special Publication 15.*

Bretz, J H. (1969). The Lake Missoula floods and the channeled scabland. *J. Geol.* **77**, 505–543.

Bretz, J H., Smith, H. T. U., and Neff, G. E. (1956). Channeled scablands of Washington; new data and interpretations. *Geol. Soc. Amer. Bull.* **67**, 957–1049.

Carr, M. H. (1974). The role of lava erosion in the formation of lunar rilles and Martian channels. *Icarus* **22**, 1–23.

Clague, J. J., and Mathews, W. H. (1973). The magnitude of jökulhlaups. *J. Glaciol.* **12**, 501–504.

Mackin, J. H. (1948). Concept of the graded river. *Geol. Soc. Amer. Bull.* **59**, 463–512.

Malin, M. C. (1973). Salt weathering on Mars (unpublished manuscript).

Malde, H. E. (1968). The catastrophic late Pleistocene Bonneville flood in the Snake River Plain, Idaho. *U.S. Geological Survey Professional Paper 596*, 52 pp.

Matthes, G. H. (1947). Macroturbulence in natural stream flow. *Amer. Geophys. Union Trans.* **28**, 255–262.

McCauley, J. F., Carr, M. H., Cutts, J. A., Hartmann, W. K., Masursky, H., Milton D. J., Sharp, R. P., and Wilhelms, D. E (1972). Preliminary Mariner 9 report on the geology of Mars. *Icarus* **17**, 289–327.

McKee, B., and Stradling, D. (1970). The sag flowout: A newly described volcanic structure. *Geol. Soc. Amer. Bull.* **81**, 2035–2044.

Milton, D. J. (1973). Water and processes of degradation in the Martian landscape. *J. Geophys. Res.* **78**, 4037–4047.

Milton, D. J. (1974). Carbon dioxide hydrate and floods on Mars. *Science* **183**, 654–656.

Pardee, J. T. (1942). Unusual currents in glacial Lake Missoula, Montana. *Geol. Soc. Amer. Bull.* **53**, 1569–1600.

Sagan, C., Toon, O. B., and Gierasch, P. J. (1973). Climatic change on Mars. *Science* **181**, 1045–1049.

Scheidegger, A. E. (1973). Hydrogeomorphology. *J. Hydrol.* **20**, 193–215.

Schumm, S. A. (1974). Structural origin of large Martian channels. *Icarus* **22**, 371–384.

Schumm, S. A., and Shepherd, R. G. (1973). Valley floor morphology: evidence of subglacial erosion? *Area* (*Inst. British Geog.*) **5**, 5–9.

Selby, M. J., and Wilson, A. T. (1971). The origin of the labyrinth, Wright Valley, Antarctica. *Geol. Soc. Amer. Bull.* **82**, 471–476.

Sharp, R. P. (1973). Mars: fretted and chaotic terrains, *J. Geophys. Res.* **78**, 4073–4083.

Shepherd, R. G. (1972). A model study of river incision. Unpublished M.S. thesis, Colorado State Univ., Fort Collins, Colorado, 135 pp.

Shepherd, R. G., and Schumm, S. A. (1974). Experimental study of river incision. *Geol. Soc. Amer. Bull.* **85**, 257–268.

Smith, H. T. U. (1965). Anomalous erosional topography in Victoria Land, Antarctica. *Science* **148**, 941–942.

Stearns, H. T. (1936). Origin of the large springs and their alcoves along the Snake River in southern Idaho. *J. Geol.* **44**, 429–450.

Warren, C. R. (1965). Wright Valley: Conjectural volcanoes. *Science* **149**, 658

20

Reprinted from *J. Geophys. Res.* **79**:2073–2076 (1974)

Possible Origin and Probable Discharges of Meandering Channels on the Planet Mars

J. G. Weihaupt

Department of Geology, School of Science, Indiana University–Purdue University Indianapolis, Indiana 46205

Our current knowledge of the geologic and geomorphic processes operative on the planet Mars is fragmentary at best. However, the recent 'flyby' trajectories of instrumented spacecraft have provided us with a new and exciting glimpse of the apparently geologically active planet. The Mariner 9 spacecraft took more than 7000 photographs of the Martian surface between November 13, 1971, and October 27, 1972 [*Hammond*, 1973]. During nearly 700 orbits of Mars, Mariner 9 obtained data on the thermal character of the surface, the Martian gravity field, the atmospheric surface pressure, the atmospheric composition, the atmospheric circulation, and the tectonics, volcanics, and geomorphic features of the red planet.

On the basis of the photographic evidence obtained by Mariner 9 a geologic map of Mars that outlines four classes of terrain has been constructed [*Carr et al.*, 1973]. These four classes are (1) primitive cratered terrain, (2) sparsely cratered eolian plains, (3) circular radially symmetric volcanic constructs such as shield volcanoes, domes, and craters, and (4) tectonic erosional units such as chaotic and channel deposits. One of the most striking results of the Mariner project has been the discovery of a variety of channels, most of which seem to occur in the fourth terrain class.

Masursky [1973] has identified four classes of channels that appear to be of various ages. The first class of channels is characterized by multiple braiding. Examples of this class can be seen on the border of the flat, low Chryse area of Mars. The second class of channels is characterized by sinuous channels. Some of these resemble intermittent terrestrial stream channels, and others show primary and secondary tributaries and dendritic patterns and seem to lack water sources on the planetary surface. The latter type, found on the high-level plateau surface in the Mare Erythraeum, Memnonia, and Rasena regions, may have been supplied at a previous time by rainwater [*Masursky*, 1973].

A third channel class, found in the ancient cratered terrain, consists of complex networks of tiny coalescent channels on the sides of craters. These channels also have a fluvial appearance and may have formed from the collection of rainwaters [*Masursky*, 1973]. The fourth channel class, associated with volcanic areas, exhibits channels that begin on the flanks of volcanic craters and become less well defined downslope, in contrast to fluviatile channels [*Masursky*, 1973].

Preliminary Discussion of Origin

Although the origin of these channels is not yet known, some (like those in the Mare Erythraeum region) appear to be geologically recent and appear not to have resulted from a single catastrophic event [*Hammond*, 1973]. Figure 1 is a photograph of one such channel taken by Mariner 9. The immediate alternative causes of channels like the one in Figure 1, which suggest themselves, are that channels of this type are either lava channels or running water channels. *Cameron* [1964] has suggested that certain channels (rilles) on the moon may have been caused by nuées ardentes (fluidized suspensions of volcanic ash in volcanic gas); other less likely alternatives include ammonia, hydrocarbons, or even carbon dioxide [*Milton*, 1973].

The difficulty of accepting the Mare Erythraeum type channels as lava flow channels is that they are 'all quite distinct from the lava channels seen on the moon and also on Mars' [*Hammond*, 1973]. Similarly, *Milton* [1973] points out that unroofed lava channels, like those found on the moon, do not have dendritic patterns like the pattern shown in Figure 1. The difficulty of assigning a running water origin to the channels is that water cannot exist in liquid form on the Martian surface at the temperatures and pressures that now exist there. It would either evaporate or freeze. However, *Sagan et al.* [1973] believe that climatic changes could have liberated sufficient water to permit the formation of channels by liquid water. Since evaporation of the Martian polar caps would raise the Martian atmospheric pressure to 1 bar (comparable to that on earth) and would permit liquid water to exist in a stable phase at daytime temperatures, the possibility of rainfall in Martian equatorial latitudes exists. This idea is consistent with the suggestion by *Murray et al.* [1972] that the orbit of the planet changes from a shape of near circularity to one of more elliptical character within a period of about 2 million years. Such changes in the shape of the orbit of Mars could bring about periodic climatic change that would give rise to the evaporation of the polar caps. This explanation is also consistent with the apparent geomorphic recency of some of the Martian channels.

Dimensions and Relationships of Channel Variables

The discovery of the Martian channels has therefore generated a controversy over whether they are the result of flowing water or the result of some other process. This report examines the pattern and the dimensions of the channe

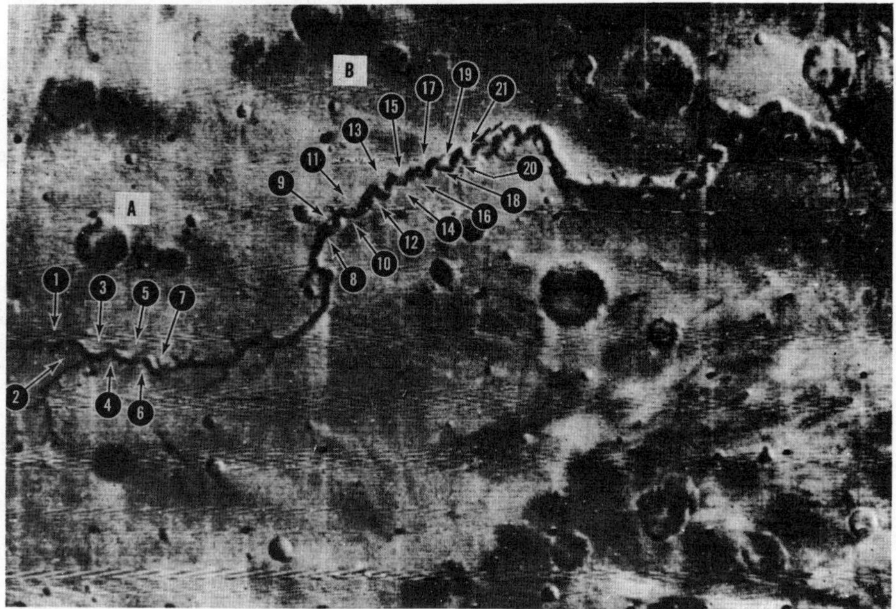

Fig. 1. Mariner 9 photograph of a sinuous channel photographed from an elevation of 1666 km in the Mare Erythraeum region of Mars. The channel is roughly 400 km long. The center of the photograph is at 29°00'S, 40°00'W. The group A segment contains meander bends 1–7, and the group B segment contains meander bends 8–21 [*Milton*, 1973, p. 4042].

meanders shown in Figure 1 in an effort to discover whether the dimensional channel variables are consistent with those expected for meandering rivers on earth.

It has been established reasonably well that stable naturally meandering rivers exhibit consistent and almost linear relationships between the wavelengths, channel widths, and radii of curvature of their meanders [*Leopold et al.*, 1964]. Each of these three variables can be measured on the photographs obtained by Mariner 9. Furthermore, these relationships appear to hold regardless of the size of a meandering river [*Leopold and Wolman*, 1960], so that very large meandering rivers reflect the same dimensional relationships among their meanders as those found for medium and small rivers. Because of these dimensional relationships it seems logical to attempt to apply the equations that represent these relationships to the Martian channel meanders.

Examination of Figure 1 reveals that two segments of the Martian channel exhibit meanders that can be analyzed morphometrically. The first segment (labeled A in Figure 1) consists of 7 meander bends, and the second segment (labeled B in Figure 1) consists of 14 meander bends, the total being 21. For each meander bend the wavelength L, the radius of curvature r_m, and the channel width w were determined. The wavelengths of the Martian channel meanders for this sample range from about 5.0 km to 16.5 km and average 9.7 km. The radii of curvature values range from 0.6 km to 2.45 km and average 1.28 km, and the channel widths range from 1.75 km to 4.0 km and average 2.8 km.

The relationship between meander wavelength and radius of curvature for natural meandering rivers on earth is found [*Leopold et al.*, 1964] to be

$$L = 4.7r_m^{0.98} \tag{1}$$

If the exponent in this equation is assumed to be unity, the equation for the sample of Martian meanders is found to be

$$L = 3.78r_m \tag{2}$$

This is in fairly good agreement with (1). However, a reexamination of Figure 1 shows that the meanders of the group A segment are somewhat different from the meanders of group B. Group A meanders have somewhat longer wavelengths, somewhat larger radii of curvature, and slightly narrower channels than group B meanders. In contrast to (2), the equation for group A meanders is

$$L = 3.32r_m \tag{3}$$

still in fairly good agreement with the relationship between wavelength and radius of curvature that is found for meandering rivers on earth. Derivation of the comparable equation for group B meanders gives

$$L = 4.38r_m \tag{4}$$

in even better agreement with (1).

Similarly, the relationship between wavelength and channel width for earth meanders [*Leopold et al.*, 1964] can be expressed as

$$L = 10.9w^{1.01} \tag{5}$$

Once again, on the assumption that the exponent of the equation is unity, the equation that describes the sample of 21 Martian meander bends is found to be

$$L = 1.72w \quad (6)$$

Interestingly, this equation reveals that the mean channel width of the Martian meanders is an order of magnitude larger than is characteristic of earth meanders. Just as the equations for wavelength and radius of curvature relationships of group A and group B were derived, the comparable equations for wavelength and channel width relationships were derived. These equations are

$$L = 2.59w \quad (7)$$

for group A meanders and

$$L = 1.42w \quad (8)$$

for group B meanders. Equations (7) and (8) reflect the fact that the group A segment of the Martian channel is slightly narrower than the group B segment.

Leopold and Wolman [1960] in their investigation of 50 meandering rivers on earth found that the ratio r_m/w is remarkably consistent from one meandering river to the next. Their investigation reveals a mean r_m/w value of 3.1 for all the rivers studied, two thirds of the rivers having values between 1.5 and 4.3. Subsequent studies by other investigators have confirmed the relative constancy of this ratio (J. G. Weihaupt, unpublished manuscript, 1973). Calculations of the r_m/w values for the sample of 21 Martian meander bends give a range of values from 0.213 to 1.225, with a mean value of 0.489. These unusually low values reflect the unusually great width of the meandering channels of Mars.

ESTIMATES OF DISCHARGE

G. H. Dury in his investigations of underfit rivers has been one of the more recent investigators to again call attention to the meandering valleys that exist in many places on earth [*Dury*, 1964]. These valleys apparently represent the relatively large channels of ancient rivers. In many cases the valley widths are 20 times the widths of smaller streams that now trickle over their floodplains. The much larger ancient channels are thought to represent a period of increased precipitation associated with glacial times [*Dury*, 1965]. Climatic change has apparently reduced the amount of water available from the atmosphere, this reduction leaving smaller streams whose channel cross-sectional areas are as much as 300 times smaller than the cross-sectional areas of the ancient rivers and whose discharges are as much as 60 times less than the discharges of the ancient rivers.

If the Martian channels represent ancient rivers from a time when the climate of Mars was different, then some estimate of the amount of water required to produce or maintain the channels can be made. The relationship between meander wavelength and mean annual discharge for meandering rivers on earth can be expressed as

$$L = 168Q^{0.46} \quad (9)$$

where Q is the mean annual discharge [*Allen*, 1970]. Equation (9) may be rewritten as

$$Q = (L/168)^{2.17} \quad (10)$$

in order to calculate the expected mean annual discharge of an earth river comparable to that of the Martian channel shown in Figure 1. On the basis of (10), the discharge of a river similar to that represented by the Martian channel would be approximately 2700 m^3/s (95,000 ft^3/s). This figure is surprisingly small in comparison with the mean annual discharge of the Mississippi River (16,000 m^3/s, or 560,000 ft^3/s), which has a channel only a fraction the size of the Martian channel.

On the other hand, in view of the disparity between the wavelengths and the channel widths of the Martian meanders, it may be more appropriate to attempt to estimate the discharge of the Martian channel on the basis of its width. *Inglis* [1941] has related channel width to discharge, using the following equation:

$$w = 4.88Q^{0.5} \quad (11)$$

For estimating discharge, the equation may be rewritten as

$$Q = (w/4.88)^2 \quad (12)$$

On this basis the discharge of an earth river with a width comparable to that of the Martian channel is likely to be about 100,000 m^3/s (3,600,000 ft^3/s). This figure is considerably greater than that for the Mississippi River but somewhat less than the roughly 225,000 m^3/s (8,000,000 ft^3/s) mean annual discharge of the Amazon River.

DISCUSSION AND INTERPRETATION

Although these results do not demonstrate that the meandering channel of Mars was formed from running water, they tend to substantiate the validity for Mars of the quantitative relationship between meander wavelength and radius of curvature. This validity is suggested by the fairly good agreement between equations (1), (2), (3), and (4). In contrast, the lack of agreement between the wavelengths and the channel widths on the one hand and between the radii of curvature and the channel widths on the other cannot be explained by contemporary earth meander theory. Four possible explanations suggest themselves. The first possible explanation is that different rates of discharge at different times may have resulted in a channel whose characteristics cannot therefore reflect a single consistent flow regime. A second possible explanation is that the equations that relate channel variables so well for natural meanders on earth simply do not apply in their present form in the weaker gravitational field of Mars. A third possibility is that the Mare Erythraeum type channels were not formed by running water but by some other fluid or mechanism. A fourth possible explanation for the disparity between Martian channel widths and the other channel dimensions is that two or more processes may be involved, so that some of the original channel or meander dimensions have been altered by secondary processes.

The first explanation, i.e., different rates of discharge may have produced the Mare Erythraeum channel, can be examined on the basis of our current knowledge of terrestrial channels. *Leopold et al.* [1964] point out that although the discharge of terrestrial streams varies seasonally and secularly, the meander length is a function of the dominant discharge. This implies that a meandering channel is not likely to reflect more than one discharge in its meandering geometry. For this reason and because no evidence is yet available for multiple discharges that might produce meander dimensions incompatible with the dimensions given by the equations referenced above, the variable discharge hypothesis is provisionally rejected.

The second explanation, i.e., the equations used above do not apply for Martian channels because of the weaker gravitational field, can be examined on the basis of the g (gravity term) in the appropriate equations. The relationships between the channel variables expressed in (1) and (5) are ultimately functions of the discharge Q as expressed in (9), (10), (11), and (12). Discharge in turn is a function only of the cross-sectional area of a channel and the mean velocity of flow in the channel. By examining the Manning equation [*Chow*, 1964], which is an expression of mean velocity of flow, i.e., $v = 1.49(R^{2/3}s^{1/2}/n)$, where v is the mean flow velocity, R is the hydraulic radius, s is the slope, and n is the Manning roughness coefficient, it appears that no gravitational term g is needed to express this relationship. It seems unlikely therefore that the weaker gravitational field of Mars can account for the dimensional inconsistencies seen in the Mare Erythraeum channel (Figure 1).

The third explanation, i.e., the Mare Erythraeum type channel, was formed by a fluid other than water, has been examined by *Milton* [1973]. Furthermore, it should be recognized that in such a case, one should consider the possibility that water may have played a role at one stage in the development of such channels and some other fluid may have played a role at another stage in its development. If such a possibility is entertained, it is possible that the wavelengths and the radii of curvature of the meanders could owe their origin to one kind of fluid and the channel widths could owe their origin to another kind of fluid. Milton has pointed out that the upstream portion of such a channel has a dendritic pattern and tributaries that are characteristic of entrenched arroyo systems on earth. This implies a water origin. However, the downstream portion of such a channel has irregular windings, a large depth/width ratio, and interlocking tight bends, which are features typical of lunar sinuous rilles of unroofed lava tubes. Because of the internal coherence 'of the entire channel system,' however, *Milton* [1973] regards the dual-origin hypothesis as being unlikely. Furthermore, the dendritic pattern exhibited by the upstream portion of channels such as this is common in terrestrial river systems and quite uncommon in lava flow channels. For these reasons a water origin is considered to be more probable than a lava or combined lava/water origin.

The fourth explanation, i.e., two or more processes may be involved, so that some of the original channel dimensions have been modified, is considered to be a more promising explanation of the disparity between the Martian channel widths and other channel dimensions. If it is assumed that the width of the Mare Erythraeum channel was once smaller and has since been modified, then the relationships between wavelength, radius of curvature, discharge, and width on Mars will be more nearly comparable to the same relationships found for terrestrial meandering channels. That is, the only dimension that must be changed to get such agreement is width, and it is then necessary to identify a process or processes that could accomplish the channel widening that this argument suggests.

Milton [1973] points out that the controlling factor in the excavation of the side canyons of the Grand Canyon on earth is artesian sapping, that is, undermining of the head walls by springs. He suggests therefore that artesian sapping may have caused cliff retreat of the walls of Martian channels such as that in the Mare Erythraeum region. Similarly, *Sharp* [1973] has suggested that dry sapping may occur in Martian canyons in such a way that cliffs are undermined by sublimation of ground ice without production of a liquid phase. In either case, *Milton* [1973] believes that some mechanism of cliff recession is necessary to explain the many wide valleys that characterize the Martian landscape.

If Milton and Sharp are correct in their assessment of slope retreat, it appears that a mechanism for channel widening may exist on Mars. And if the Martian channels of the Mare Erythraeum type were once narrower than they are today, it appears that we can expect good agreement between the dimensional relationships found for meandering channels on earth and those that exist for these meandering channels on Mars.

If the Mare Erythraeum meandering channel was narrower when the meanders were formed, our use of (12) to calculate the discharge of the channel is likely to have produced an exaggerated discharge figure (100,000 m³/s, or 3,600,000 ft³/s). It is concluded therefore that if the channel in Figure 1 originated from flowing water, its minimum probable discharge was 2700 m³/s (95,000 ft³/s) and its actual discharge was perhaps somewhat greater.

A corollary of course is that if the meandering channels of Mars exhibit dimensional relationships comparable with those found for meandering rivers on earth, it is logical to suggest that the Mare Erythraeum type channels of Mars were produced by running water in a manner not unlike that operating on earth. The source of the water still remains unresolved and may have been a function of a climatic change that increased the precipitation regime or produced liquid water by melting snow, ice, or permafrost on or in the Martian surface.

REFERENCES

Allen, J. R. L., *Physical Processes of Sedimentation*, p. 130, American Elsevier, New York, 1970.

Cameron, W. S., An interpretation of Schröter's valley and other lunar sinuous rilles, *J. Geophys. Res.*, *69*, 2423, 1964.

Carr, M. H., H. Masursky, and R. S. Saunders, A generalized geologic map of Mars, *J. Geophys. Res.*, *78*, 4031, 1973.

Chow, V. T., *Handbook of Applied Hydrology*, p. 7-24, McGraw-Hill, New York, 1964.

Dury, G. H., Subsurface exploration and chronology of underfit streams, *U. S. Geol. Surv. Prof. Pap. 452-B*, 1–56, 1964.

Dury, G. H., Theoretical implications of underfit streams, *U.S. Geol. Surv. Prof. Pap. 452-C*, 1–43, 1965.

Hammond, A. L., The new Mars: Volcanism, water, and a debate over its history, *Science*, *179*, 463, 1973.

Inglis, C. C., *Res. Publ. 4*, p. 7, Cent. Irrig. and Hydrodyn. Res. Sta., Delhi and Calcutta, India, 1941.

Leopold, L. B., and M. G. Wolman, River meanders, *Geol. Soc. Amer. Bull.*, *71*, 770–774, 1960.

Leopold, L. B., M. G. Wolman, and J. P. Miller, *Fluvial Processes in Geomorphology*, pp. 296–297, W. H. Freeman, San Francisco, Calif., 1964.

Masursky, H., An overview of geological results from Mariner 9, *J. Geophys. Res.*, *78*, 4009, 1973.

Milton, O. J., Water and processes of degradation in the Martian landscape, *J. Geophys. Res.*, *78*, 4037, 1973.

Murray, B. C., W. R. Ward, and S. C. Yeung, Periodic insolation variations on Mars, *Science*, *180*, 638, 1973.

Sagan, C., O. Toon, and P. Gierasch, Climatic change on Mars, *Science*, *181*, 1045, 1973.

Sharp, R. P., Mars: Troughed terrain, *J. Geophys. Res.*, *78*, 4063 1973.

(Received June 14, 1973;
revised February 15, 1974.)

21

Reprinted from *J. Geophys. Res.* **82**:4016–4038 (1977)

Classification and Time of Formation of Martian Channels Based on Viking Data

HAROLD MASURSKY, J. M. BOYCE, A. L. DIAL, G. G. SCHABER AND M. E. STROBELL

U.S. Geological Survey, Flagstaff, Arizona 86001

Fluviatile and volcanic Martian channels, first discovered on Mariner 9 pictures, have been reexamined by using Viking orbital photography. The superior discrimination of the Viking photographs, resulting from clearer atmospheric conditions and an improved camera system, has permitted us to map additional channels and to estimate their relative ages, using a technique based on crater counting. Broad channels like the Ares and Tiu/Simud Valles are situated along the margin of the southern highlands near Chryse Planitia, the landing site of Viking 1. They originate in areas of collapsed terrain that may have been formed when subsurface water-ice (permafrost) was melted by geothermal heat from deep-seated volcanic centers. When permafrost melting reached an abrupt topographic slope, the interstitially stored meltwater 'lakes' were breached suddenly, releasing the great floods that modified the channels. The volume of material involved in the collapsed terrain is large enough to furnish the water calculated to have filled the broad channels. Conditions are reviewed for persistence of liquid water on Mars under present and more favorable pressures and temperatures. Sinuous channels of intermediate size, like the Ma'adim and Hrad Valles and other shorter, stubby channels, have multiple tributaries; in the limited coverage available, they appear to result from 'spring sapping,' with the underground permafrost meltwater emerging in box canyons at their heads. The widespread distribution of this type of channel makes their origin by local geothermal heating less likely; climatic warming may be required to explain their formation. The final fluviatile type, dendritic channel networks, has the widest areal distribution and appears to have been formed during at least two episodes. The filamentous channels in their source areas (often the rims of craters) seem to resemble terrestrial river systems; rainfall would seem to be required to form these features. All these channel types debouch onto lowland plains or crater floors, where they disappear in short distances; these abrupt terminations may have resulted from percolation and/or evaporation. Simple and complex lava channels are common; they originate at volcanic centers and are usually morphologically distinct from the aqueous channels. Three types of lava channels are recognized. The wide variation in crater densities implies varying channel ages. Water must have flowed on the Martian surface at many different times in the past, although this would be possible only with great difficulty under the present Martian thermal conditions. Based on a crater flux curve derived by Soderblom et al. (1974) the fluviatile channel ages vary from 3.5 to 0.5 Gy. Lava channel ages range from 3.5 Gy to an age too young to date by the crater counting technique (perhaps 200 m.y.). Methods for dating the channels and volcanic episodes are still insufficiently developed to determine whether episodes of volcanic heating and climatic change are coincident. It is possible that the large floods and volcanic eruptions might trigger a short 'interglacial' interval. Alternatively, the floods may be related to episodic volcanic activity, and the dendritic channels to rainfall that was associated with independent interglacial climatic episodes resulting from variations in solar output or other causes.

INTRODUCTION

Possible fluviatile channels on Mars were first recognized from Mariner 9 photographs (H. Masursky, in the paper by *Driscoll* [1972]). Channel descriptions, classification, and possible genesis, based on Mariner imaging, have been reported by many authors [*Masursky*, 1973; *Milton*, 1973; *Baker and Milton*, 1974; *Hartmann*, 1974; *Schumm*, 1974; *Sharp and Malin*, 1975; *Malin*, 1976; *Pieri*, 1976; *Masursky*, 1976; *Nummedal*, 1976; *Baker*, 1977]. This paper reconsiders the classification, genesis, and age relations of several channel systems based on Viking photographs taken from July 1976 to February 1977 (Figure 1 and Table 1).

Because it is difficult to retain water on the Martian surface under the present climatic conditions, many observers are still convinced that the channels photographed were eroded by the wind [*Cutts et al.*, 1976; *Cutts*, 1977; D. Anderson, personal communication, 1976] rather than by flowing water. Three types of terrain features indicate that the channels were carved by liquid flow: (1) Martian channels having many tributaries form integrated systems (Figure 2*a*), like the integrated terrestrial stream systems first noted by *Playfair* [1802] (see also *Hack* [1956] and *Leopold and Maddock* [1953]); (2) the configuration of the complex anastomosing channels (Figure 2*b*) and the shapes of islands within the channels (Figure 2*c*) are unlike wind-eroded shapes (Figure 2*d*) [*Milton*, 1973; *Baker and Milton*, 1974; *Nummedal*, 1976; *Carr et al.*, 1976]; and (3) features typical of fluid erosion are seen within the Ares channel near its shoreline, whereas the terrain outside (Figure 2*c*) is unmodified.

Martian wind-eroded forms occur at differing elevations, such as near the base of Olympus Mons (Figure 2*d*) and near the base and summit of Arsia Mons, but fluvial erosional forms appear to be controlled by elevation and local topography.

AGE DETERMINATION METHODOLOGY

The density of impact craters superposed on a surface is generally assumed to be proportional to the relative age of the surface. Viking photographs were used to make counts of craters occurring within Martian channels and on adjacent surfaces in order to obtain relative ages for the channels. Craters larger than ~0.3 mm in diameter were counted; on 1500-km photographs, 0.3 mm equals ~0.4 km. In most cases the counts were done by the same observer to ensure internal consistency. Obvious secondary craters and endogenic craters

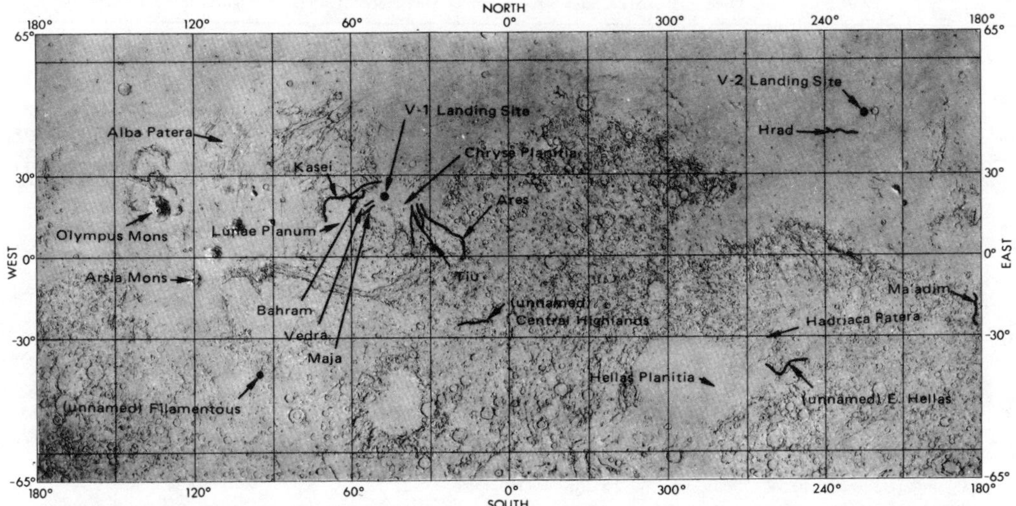

Fig. 1. Index map showing location of Martian fluviatile and volcanic channels described in this report. All names are approved by the International Astronomical Union. The following list gives the figures associated with each channel or other feature: Ares, Figures 2*b*, 2*c*, 5*c*, 5*d*, and 7*c*; Bahram, Figure 4*a*; Hrad, Figure 8*c*; Kasei, Figure 6*c*; Ma'adim, Figure 3*c*; Maja, Figures 6*a*, 7*a*, and 7*b*; Tiu, Figure 8*a*; Vedra, Figures 2*a* and 4*c*; (unnamed) East Hellas, Figure 9*a*; (unnamed) southern highlands, Figures 10*a* and 10*b*; (unnamed) 'filamentous,' Figure 9*c*; Alba Patera, Figures 10*c*, 11*d*, and 12*a*; Arsia Mons, Figure 11*b*; Chryse Planitia, Figures 2*b*, 2*c*, 3*a*, 6*a*, 6*c*, 7*a*, and 7*b*; central highlands, Figures 3*a*, 9*a*, 9*c*, 10*a*, and 10*b*; Hadriaca Patera, Figure 12*c*; Lunae Planum, Figures 4*a* and 4*c*; and Olympus Mons, Figure 2*d*.

were excluded from the counts. The crater count data were used to obtain cumulative crater size-frequency curves, normalized to an area of 1 km², for each channel or surface area counted. All crater curves have been plotted against an idealized average lunar curve [*Neukum et al.*, 1974]. The crater size-frequency distribution curve for the Viking 1 landing site is also included for comparison (see Figures 3*a* and 3*b*). Error bars are standard error ($\pm(N/A)^{1/2}$, where N is the cumulative number of craters in a class interval/unit area A).

The crater size-frequency distribution curves for the channels can then be compared by noting their relative displacements; areas having the lowest crater density determinations are assumed to be the youngest in age [*Masursky and Crabill*, 1976*a*, *b*]. This assumption has been tested by earlier workers [*Shoemaker*, 1962; *Shoemaker and Hackman*, 1962; *Baldwin*, 1964] and found to be valid. Early workers also discovered that the form and shape of the lunar curves for areas of widely differing locations and ages are similar and that the Martian curves derived from Mariner 9 data are also similar. These curves usually have regular slopes of −2° to −4°. However, some of our Martian curves derived from Viking data show unusually low slopes or anomalous bends in the slope. We infer that a surface, such as that shown in Figure 3*c* (Ma'adim Valles), has been modified by continuous erosion or deposition by either wind or water because the surface curve has a lower than normal slope (Figure 3*d*). A bend in the curve, connecting two parallel but offset segments, is usually interpreted to mean that a particular size range of craters has been destroyed or buried without affecting craters of other sizes [*Hartmann*, 1973, 1977]. This situation occurs when lava flows or other types of mantles are thick enough to obliterate the larger craters [*Neukum and Horn*, 1976]. We interpret the surface cut by Bahram Valles (Figure 4*a*) to be resurfaced because of the obvious bend seen in the Bahram surface curve (Figure 4*b*). Photographic resolution is another factor that can affect the shape of the curve, causing an apparent reduction in crater density at small diameters, so that the curve acquires a progressively lower slope or 'rollover.' Resolution of photograph 22A76 of Vedra Valles (Figure 4*c*) is shown to be ~300 m by the curve rollover at that position (Figure 4*d*). Figure 5*a* shows the relation between the effective photographic resolution, as determined by the rollover position on the crater curve, and changes in photographic resolution with variations in distance between the orbital camera and the surface photographed by the camera (the 'range'). Figure 5*b* is the highest-resolution picture of the surface of Mars yet taken. Abundant craters down to the limit of resolution are clearly shown.

It is conceivable that these variations in curve slope and shape could all result from changes in the impact flux, but the data do not seem to support such an hypothesis. No regular variation in slope angle or slope discontinuity has been observed on a planet-wide basis. The observed variations in crater density distribution curves are therefore interpreted as reflecting variations in the age and erosional state of the terrain rather than changes in the impact flux. The relative ages of channels and surfaces show clearly in the several crater density plots discussed.

Several attempts have been made to correlate Martian flux curves with lunar curves and radiometric ages of returned lunar samples [*Soderblom et al.*, 1974; *Chapman*, 1976; *Neukum and Wise*, 1976; *Hartmann*, 1977] so that absolute ages for the Martian events can be determined. The shape and position of both the lunar and the Martian flux curves, from which absolute ages are ascertained, are still subjects of intense de-

TABLE 1. Data Used to Prepare Crater Size-Frequency Distribution Curves for Figures in Text

Figure	Feature	Area Counted, km²	Number of Craters Counted		Median Number of 1-km Craters per 1 km²	Range, km	Coordinates		Viking Photo Numbers
			Per 500 m	Per 1 km			Latitude	Longitude	
2*a*	Vedra Valles, other dendritic channels					1630–1640	18°–19°N	54°–56°W	Mosaic: 46A53–46A60, 47A52, 47A54, 47A56, and 47A58
2*b*	Ares channel, NE of A-1 landing site					1920–1950	24°N	31°W	Mosaic: 3A13 and 3A14
2*c*	Ares channel, NE of A-1 landing site					1590	20°–22°N	30°–32°W	Mosaic: 4A48–4A54
2*d*	Surface, NE of Olympus Mons					3140	25°N	132°W	48B15
3*a*	V-1 landing site					1655	22°N	48°W	20A71
3*b*	V-1 landing site surface curve	11,000	120	20	2.1×10^{-3}	1650–1810	22.5°N	48.0°W	20A71, 20A46, 20A48, 20A50, 20A52, 20A69, 20A71, and 20A73
3*c*	Ma'adim Vallis					7320	18°S	182°W	88A69
3*d*	Ma'adim surface curve	69,000	0	200	$2.9 \times 10^{+1}$	7320	18°S	182°W	88A69
4*a*	Bahram Vallis					1750–1780	22°N	55°–57°W	Mosaic: 22A33, 22A34, 22A36, 22A38, and 22A40
4*b*	Bahram surface curve	10,000	200	40	1.2×10^{-2}	1750–1780	22°N	55°–57°W	22A34, 22A36, 22A38, and 22A40
	Bahram channel curve	680	6	2	2.9×10^{-3}	1750–1780	22°N	55°–57°W	22A34, 22A36, 22A38, and 22A40
4*c*	Vedra Valles					1710	20°N	55°W	22A76
4*d*	Vedra surface curve	9,200	130	35	3.7×10^{-3}	1680–1690	19°–24°N	52°–54°W	22A56, 22A76, 22A77, and 22A78
	Vedra channel curve	400	2	0	6.0×10^{-4}	1680–1690	19°–24°N	52°–54°W	22A56, 22A76, 22A77, and 22A78
5*a*	High-resolution at Alba Patera curve	80	2	1		317	39°N	107°W	268A22
	A-1 landing site	2,700	32	8	3.3×10^{-3}	1920–1950	23°–24°N	31°W	03A13, 03A14, and 4A25
	B-3 landing site	19,900	154	29	1.4×10^{-3}	3540	47°N	231°W	09B10
5*b*	High resolution at Alba Patera photo					317	39°N	107°W	268A22
5*c*	Hydaspis Chaos					2700	3°N	27°W	83A37
5*d*	Tiu Valles					2710	40°N	27°W	83A38
6*a*	Maja Vallis					1659	19°N	51°W	20A56
6*b*	Maja Vallis surface	11,000	120	15	2.0×10^{-3}	1659–1700	50°–52°N	19°–20°W	20A55, 20A56, 20A58, 44A43, and 44A44
6*c*	Kasei Vallis mouth					2080	25°N	50°W	20A26
6*d*	Kasei Vallis surface at mouth	8,130	155	18	2.2×10^{-3}	2080	25°N	50°W	20A26 and 20A28
7*a*	Maja Vallis mouth					1660	21.3°N	47.8°W	20A62
7*b*	Maja Vallis in Chryse Planitia					1820	22.2°N	44.0°W	20A40
7*c*	Ares Vallis					2700–3050	7°–12°N	21°–27°W	Mosaic: 83A18, 83A19, 83A20, 83A21, 83A47, 83A48, 83A49, and 83A50
7*d*	Ares surface curve	31,500	230	100	6.0×10^{-1}	2700–3060	8°–10°N	24°–25°W	83A18, 83A19, 83A48, 83A49, and 83A50
	Ares channel curve	4,000	35	6	1.5×10^{-3}	2700–3060	8°–10°N	24°–25°W	83A18, 83A19, 83A48, 83A49, and 83A50
8*a*	Tiu Valles					2700–2710	4°N	27°W	Mosaic: 83A06, 83A08, 83A0[illegible], 83A37, and 83A38
8*b*	Tiu surface curve	28,100	185	65	2.0×10^{-1}	2710–3050	1°–6°N	16°–29°W	Mosaic: 83A02, 83A04, 83A06, 83A08, 83A10, 83A34, 83A36, 83A37, 83A38, and 83A39
	Tiu channel curve	7,000	22	2	2.8×10^{-4}	2710–3050	1°–6°N	16°–29°W	Mosaic: 83A02, 83A04, 83A06, 83A08, 83A10, 83A34, 83A36, 83A37, 83A38, and 83A39
8*c*	Hrad Valles					3960	42°N	226°W	9B54
8*d*	Hrad surface curve	13,600	40	16	1.0×10^{-2}	3960–4060	41°–44°N	222°–239°W	9B22, 9B43, 9B45, 9B47, 9B48, 9B50, 9B52, 9B54, and 9B56
	Hrad channel curve	1,640	21	3	1.5×10^{-4}	3960–4060	41°–44°N	222°–239°W	9B22, 9B43, 9B45, 9B47, 9B48, 9B50, 9B52, 9B54, and 9B56
9*a*	East Hellas unnamed channel					8660–8840	39°–48°S	260°–270°W	Mosaic: 97A48–97A54 and 97A60–97A68
9*b*	East Hellas surface curve	500,000	0	670	2.4×10^{-2}	8660–8840	39°–48°S	260°–270°W	97A48–97A54 and 97A60–97A68

TABLE 1. (continued)

Figure	Feature	Area Counted, km^2	Number of Craters Counted: Per 500 m	Number of Craters Counted: Per 1 km	Median Number of 1-km Craters per 1 km^2	Range, km	Coordinates: Latitude	Coordinates: Longitude	Viking Photo Numbers
	East Hellas channel curve	22,800	0	6	2.6×10^{-4}	8660–8840	39°–48°S	260°–270°W	97A48–97A54 and 97A60–97A68
9c	Unnamed filamentous channel					8890	43°S	94°W	63A09 (partial frame)
9d	Unnamed filamentous channel surface	40,000	0	62	6.0×10^{-1}	8890	43°S	94°W	63A09 (partial frame)
10a	Unnamed, central highlands					8550	27°S	14°W	84A47
10b	Unnamed, central highlands					8590	24°S	9°W	84A43
10c	Alba Patera dendritic channels					1780	44°N	104°W	04B57
10d	Alba (dendritic) surface curve	2,700	0	3	1.1×10^{-3}	1780	44°N	104°W	04B57
11b	Arsia Mons flank					6950	12°S	120°W	52A04 (partial frame)
11c	Arsia flank surface curve	7,400	0	3	2.7×10^{-4}	6530	9°S	121°W	39A09
11d	West Alba Patera					4200	42°N	118°W	07B88
12a	Alba Patera summit					4120–4220	43°–44°N	108°–111°W	Mosaic: 7B60 and 7B62
12b	Alba summit surface curve	5,500	0	123	2.2×10^{-3}	4120–4220	43°–44°N	108°–111°W	Mosaic: 7B60 and 7B62
12c	Hadriaca Patera					8950–8975	28°–34°S	265°–274°W	Mosaic: 97A40–97A42
12d	Hadriaca surface curve	64,000	0	110	4.6×10^{-1}	8957	30°S	269°W	97A42

bate. We have used the *Soderblom et al.* [1974] calibration curve, rather than the *Neukum and Wise* [1976] curve, to obtain the estimates of absolute ages for Martian surfaces that we quote here. We favor the *Soderblom et al.* [1974] curve because it is consistent with the data recently presented by *Shoemaker* [1977] for asteroid populations; his data show that the impact flux at Mars is higher by a factor of ~2, in relative production of craters 4–10 km in diameter, than the lunar flux. Based on the *Soderblom et al.* [1974] curve the estimated ages of the Martian fluviatile channels vary from 3.5 Gy to 0.5 Gy; lava channels vary from 3.5 to an age too young to determine by crater counts (less than 200 m.y.). These estimates were made by extrapolating the curves to the diameter range used by *Soderblom et al.* [1974]. These age estimates may change following additional analysis of the Viking orbiter imaging data.

DESCRIPTION OF CHANNELS AND IMPLICATIONS FOR CHANNEL FORMATION

Some of the broad channels, such as the Ares and Tiu/Simud valles (Figure 1), rise in the highlands area south of Chryse Planitia, the landing site of Viking 1. These channels head in areas of collapsed terrain, like Hydaspis Chaos (Figure 5*c*), and then flow into the Chryse lowland. The spacing and size of the collapsed areas resemble those of volcanic centers in the Tharsis plateau and the Elysium region. A possible mechanism for the release of water is subterranean emplacement of local volcanic complexes which melt the overlying permafrost layer. When the outward spreading front of meltwater (from ice wedges and interstitial ice) intersects a cliff face, the underground 'lake' rapidly drains over the lowered ice dam in the subsurface; as the water emerges from the base of the chaotic terrain, it modifies preexisting fault troughs or other topographic lows and erodes channel floors (Figure 5*d*).

Other broad channels, like the Maja and Kasei Valles (Figure 1), have less obvious areas of collapsed terrain in their source areas (on Lunae Planum) than do the broad channels that rise in the highlands south of Chryse Planitia. Abundant evidence for permafrost activity does occur, however, at Lunae Planum and elsewhere on the planet [*Carr and Schaber*, 1977]. Photographs of the Maja and Kasei Valles near their mouths (Figures 6*a* and 6*c*) show crater densities that are similar to those seen at the Viking 1 landing site (Figure 3*a*), implying similar ages for the three areas; curves derived from crater counts of the three areas (Figures 3*b*, 6*b*, and 6*d*) also imply similar ages.

Viking temperature and water vapor measurements support the theory that abundant water is present and probably was also present on Mars in the past [*Kieffer*, 1976; *Kieffer et al.*, 1976*a*; *Farmer et al.*, 1976]. These data show that the residual polar ice caps are water-ice and that the atmosphere is saturated with water vapor during the Martian night.

When the proposed channel flow reaches the lowland plain, it spreads widely and forms complex braided networks. In the region directly west of the Viking 1 landing site the floodwater apparently was dammed behind marelike ridges until it overflowed low points and cut gaps in the ridge crests (Figure 7*a*). Evidence of stream erosion occurs far out into the Chryse basin, where grooves have been cut on the downslope sides of mare ridges (Figure 7*b*). Evidence of stream erosion is seen as far as 350 km from the canyon mouths of the Ares, Tiu/Simud, and Maja Valles. Evidence of ponded water is not visible in photographs of the basin, nor do shorelines, beach ridges, or bars appear to be present; photographs of terrestrial

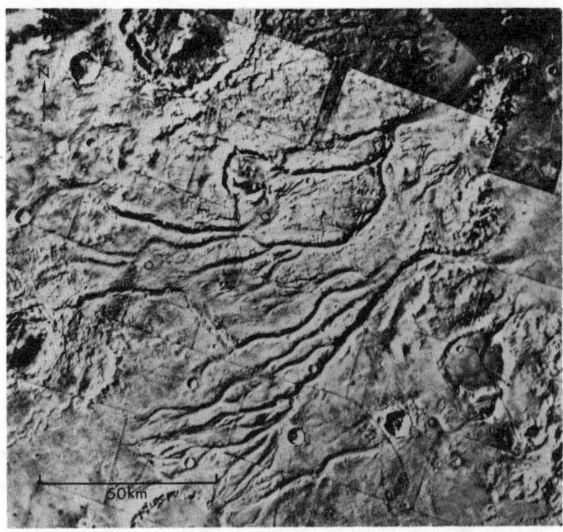

Fig. 2*a*. Photomosaic of two dendritic channel systems that drain Lunae Planum west of Chryse Planitia. Vedra Valles is the northern channel system shown. The integrated system and accordant tributary junctions are similar to terrestrial drainage systems. The tributaries start at points, not in collapsed terrain or box canyons. Collected rainfall seems necessary to erode this type of channel system.

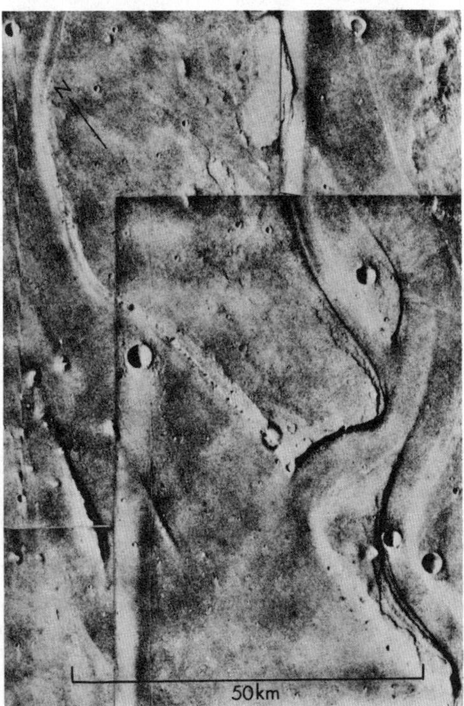

Fig. 2*b*. Photomosaic showing anastomosing channels at the north end of Ares Vallis in Chryse Planitia near the originally designated A-1 landing site. The streamlined forms and intersecting channels are characteristic of water erosion. Discrete layers, possibly basaltic lava flows, show along the edges of the islands. A number of small impact craters are seen that formed after the channel was eroded. The islands are about 100 m high.

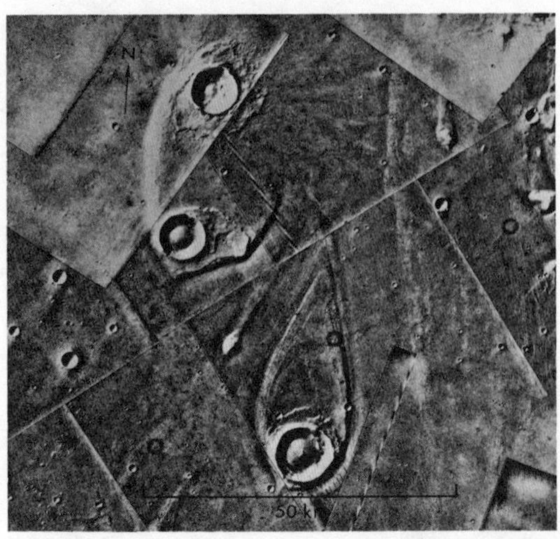

Fig. 2*c*. Photomosaic of part of Ares Valles in Chryse Planitia. The islands have impact craters at the south (upstream) ends and long streamlined forms extending downstream. In the central island a prow points upstream—a feature commonly observed in terrestrial streams that have rapid turbulent flow. The rock layers, probably basaltic lava flows, show clearly where they have been eroded by the flowing water. Ejecta blankets from some impact craters have been partially cut away by the channel, indicating that the craters preceded the channel. The ejecta blanket at the north end of the northern island lies across the channel, showing that it was later. Smaller obstacles at the south end of the picture have deep moats at the upstream end, confirming the turbulent flow. In the north east corner is a shoreline about 100 m high. The terrain to the west has surface grooves cut by the water, while the terrain to the east is unmodified. The slope of the channel floor, determined photogrammetrically, is 10 m/km to the north.

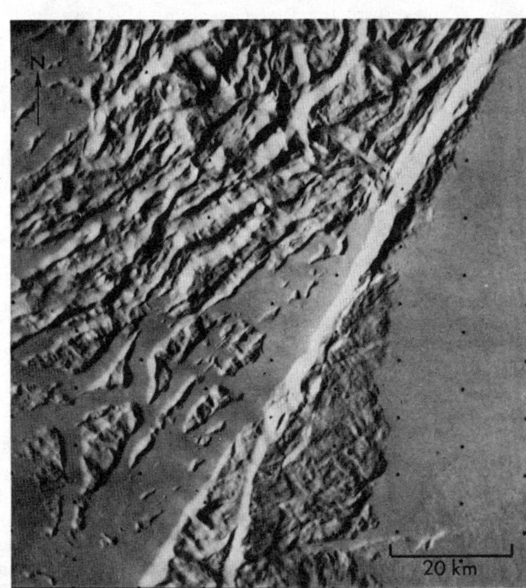

Fig. 2*d*. Photograph (48B15) of part of the aureole around the base of Olympus Mons 600 km north of the volcanic summit. The picture is about 100 km wide. The ridges probably are yardangs—positive features formed by wind erosion of old lava flows. The ridges are shaped like inverted boat hulls as are terrestrial yardangs; *McCauley* [1973] first pointed out the morphologic similarity of the Martian and terrestrial features. The shapes are distinct from the shapes produced by flowing water in the Mars channels.

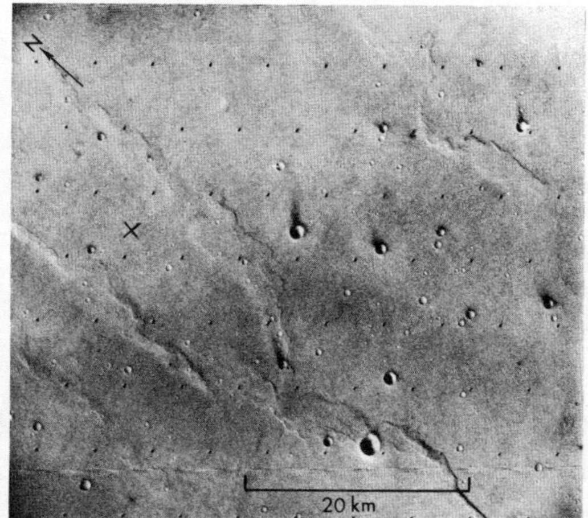

Fig. 3*a*. Photograph (20A71) of the Viking 1 landing site (marked by a cross) in Chryse Planitia. The ridges resemble mare ridges on the moon that most commonly form in basaltic lava flows. Also visible are many impact craters ranging up to 1.5 km in diameter and crater streaks which are characteristic of wind erosion and deposition. Wind dunes and drifts are shown in the lander pictures. Grooves that probably were formed by water are visible 60 km southwest of the largest crater near the bottom of the picture.

Fig. 3*c*. Oblique photograph (88A69) of Ma'adim Vallis. This sinuous intermediate size valley south of Elysium Planitia has many short, stubby tributaries that join the main channel. The channel cuts across an ancient crater in the middle of the picture; two sizable young craters lie in the channel.

dry lakes, taken at comparable resolution, show such features clearly. Either the Martian water disappeared by percolation, evaporation, or freezing before it ponded in the Martian playas, or these features are covered by younger deposits of either volcanic or eolian origin. However, only a small portion of the water calculated to flow through the channels could percolate into the subsurface [*Kieffer*, 1976; *Kieffer et al.*, 1976*b*, *c*; D. Davies, personal communication, 1977] because the permafrost layer that lies at a shallow depth would prevent deep percolation; the remainder of the water would have to freeze and then sublimate. However, there is no geomorphic evidence in the photographs taken to date for either ponded or frozen water within or adjacent to the Martian channels.

Crater counts show that the heavily cratered Martian terrain is quite ancient, whereas the less cratered channel floors are younger. Figure 7*c* shows Ares Vallis where it dissects part of the ancient cratered highlands; Figure 7*d* shows relative age relations of Ares Vallis and the surface it cuts. The Tiu/Simud

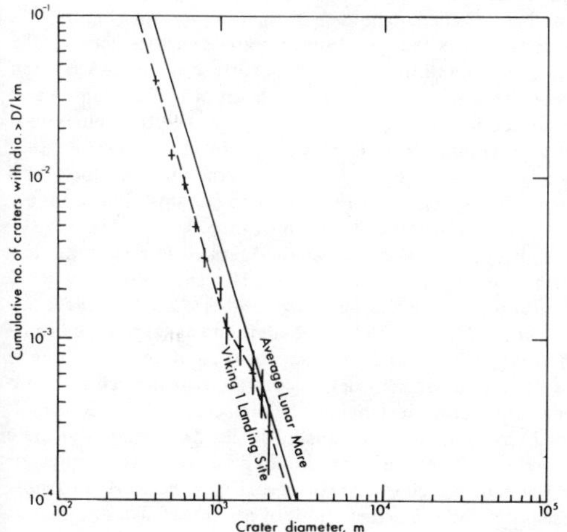

Fig. 3*b*. Crater size-frequency distribution curve of the Viking 1 landing site. It follows the lunar curve with one offset, possibly indicating a modifying event partway through the development of the surface.

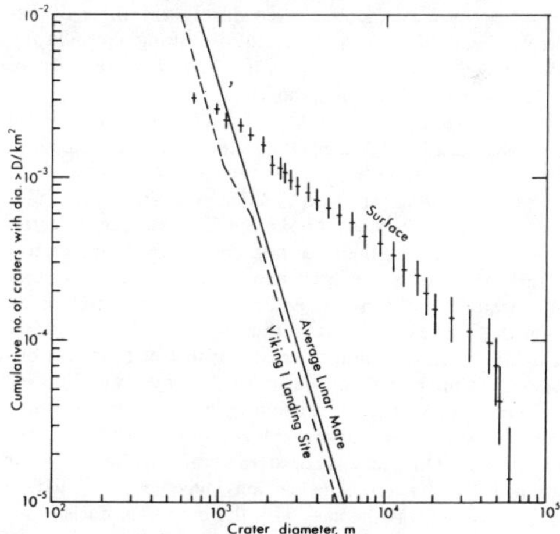

Fig. 3*d*. Crater size-frequency distribution curves for Ma'adim Vallis and the adjacent terrain. The upland is heavily cratered; the low slope on the crater curve may indicate continual degradation of craters smaller than 14 km in diameter. The obliquity of the photograph may also have biased the data for the curves.

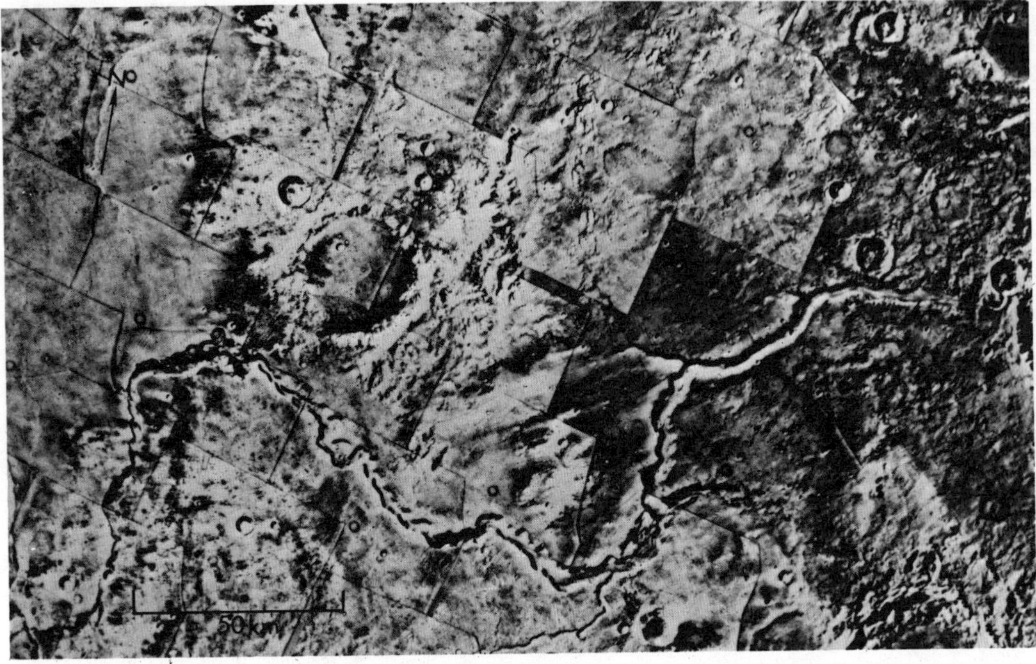

Fig. 4*a*. Viking mosaic of Bahram Vallis due west of the Viking 1 landing site. The sinuous, intermediate size channel is cut into Lunae Planum.

channel, shown in Figure 8*a*, and the crater curve for this channel (Figure 8*b*) show a lower density of impact craters than do the Ares or Kasei channels; Tiu is therefore thought to be younger. These crater counts indicate that volcanic or geothermal heating may have occurred during a number of distinct episodes. Additional high-resolution Viking photography of Martian channels will allow us to establish better relative age relations for the broad channels.

Sinuous channels like the Ma'adim (Figure 3*c*), Bahram (Figure 4*a*), and Hrad Valles (Figure 8*c*) are intermediate in size and have multiple tributaries; they are recognized as far north as 44°N and as far south as 45°S. Sinuosities and braids on the channel floors closely resemble features seen in terrestrial stream channels; their courses are smoothly integrated, and their tributaries are accordant. Their tributaries commonly head in box canyons and resemble terrestrial desert streams that are fed by springs rather than by collected surface precipitation; some tributaries are dendritic (see following discussion). We infer that the Martian box canyons are formed by 'spring sapping,' that is, erosion by water escaping from a permeable layer, with subsequent collapse of the overlying resistant rock. On Mars, such streams are probably fed by escaping ice meltwater. The ice may have been melted by volcanic heat, as was the case with the broad channels; however, their wide areal distribution suggests that an episode of climatic warming is a more likely cause for their formation.

Many short, stubby channels that head in box canyons occur either along the continental margin, where the ancient cratered highlands meet the northern lowlands, or along the rim of Valles Marineris. These streams closely resemble the tributaries of the intermediate size sinuous valleys and probably have a similar origin.

The floors of sinuous channels are wide enough to allow valid crater counts to be made of the younger channel floors, as well as of the unchanneled adjacent terrain (Figure 8*c*). Figure 8*d* shows the variation in crater densities between the Hrad channel and the surrounding surface; a similar variation between an unnamed channel northeast of Hellas Planitia and the surface it cuts (Figure 9*a*) is seen in the crater curves for that area (Figure 9*b*). The variation in crater densities that exists between counts made of the intermediate channels is as wide as that which exists for the broad streams; they probably were formed over a similar span of time.

Dendritic channel networks on Mars closely resemble terrestrial stream patterns (Figures 9*c*, 10*a*, and 10*b*). They occur ubiquitously in the southern highlands [*Masursky*, 1973; *Sagan et al.*, 1973; *Pieri*, 1976] and are found as far north as 40°N and as far south as 45°S. The networks commonly head on the outside rims of ancient degraded craters in the central highlands, originating just below the rim crest. No trace of collapsed terrain or box canyons in the headwater areas of these channels is seen at the resolution of pictures obtained thus far; the filamentous character of the stream network probably persists beyond the present limit of resolution. These networks resemble terrestrial stream networks that are the result of collected rainfall. The channels do not appear to be deeply incised, and, in the ancient cratered terrain, they dissect the outer rims of the craters, as do the intermediate channels. The

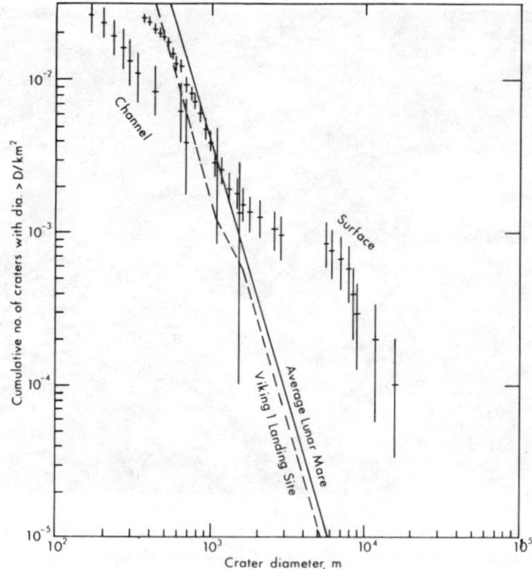

Fig. 4*b*. Crater size-frequency distribution curves for Bahram Vallis and the prechannel surface. The crater curve for the older surface has a sharp inflection indicating an erosional or depositional interruption in the crater distribution. The surface curve lies well above the average lunar mare curve, indicating greater antiquity. The channel crater curve lies along the Viking 1 landing site curve as do the curves for the Maja and Kasei valles; all of these channels emerge from Lunae Planum and flow across the western part of Chryse Planitia toward the Viking 1 landing site.

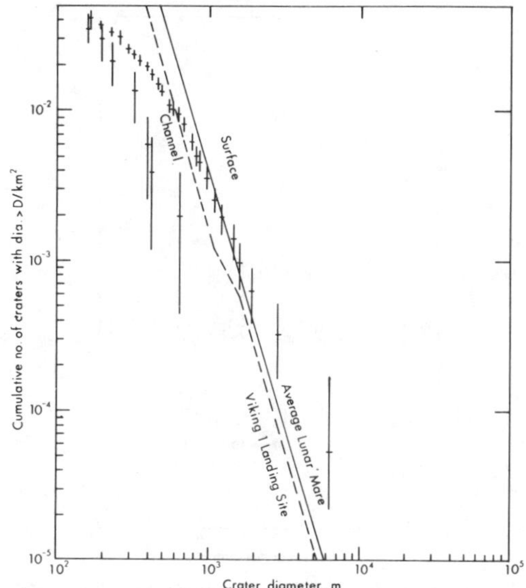

Fig. 4*d*. Crater size-frequency distribution curves based on photos 22A56, 22A76, 22A77, and 22A78 of the prechannel and channel surfaces showing the relatively younger age of the channel. The relatively steep slope of the crater curve for the old surface suggests that degradation is less than that at the other broad channels (Figures 5*d*, 6*d*, and 7*b*). The apparent age of the channel is younger than that of the Viking 1 landing site, but the area counted is not statistically significant.

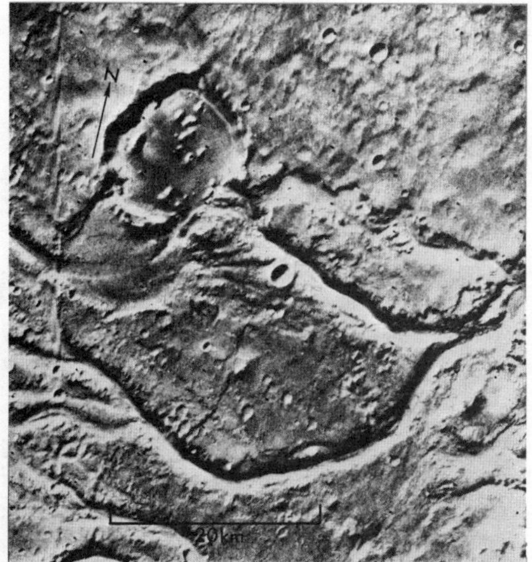

Fig. 4*c*. Viking photograph (22A76) showing part of the branched headwaters of Vedra Valles. The channel system drains the eastern flank of Lunae Planum. Note that the channel has cut the degraded rim of a crater 14 km in diameter.

total time needed to form each channel must have been relatively short; limited and sporadic rainfall is thus inferred. Only a few of the dendritic channel networks have been photographed by Viking to date, but crater density determinations have been obtained for several of these areas; one (Figure 9*d*) is shown here. The crater densities obtained for the surfaces these channels cut indicate that the surfaces are very old. The channels are clearly younger than the surface, but a crater count for the channels themselves cannot be obtained because of the restricted channel width and area. The crater size-frequency curve for the surface cut by the fluvial channel network on the flank of Alba Patera (Figures 10*c* and 10*d*) yields a much younger age than do the curves obtained for areas in the Martian highlands. At least two ages for dendritic channel formation are proposed, based on these counts.

Many channel networks in the southern highlands that were recognized in Mariner 9 pictures were assigned informal relative ages, based on measurements of the degree of degradation of the channel walls ('degradation index'). Some channels have very irregular edges; others have smooth margins. One hypothesis for this variation in channel morphology is that the irregularity of the walls, caused by cratering and slumping, is caused by an increase in age, the youngest channels having the smoothest edges. However, a similar effect would be produced by variations in physical properties or atmospheric conditions due to differences in elevation of the terrain: less cohesive material would produce irregular channels and highly cohesive material would produce smooth margins, even though the

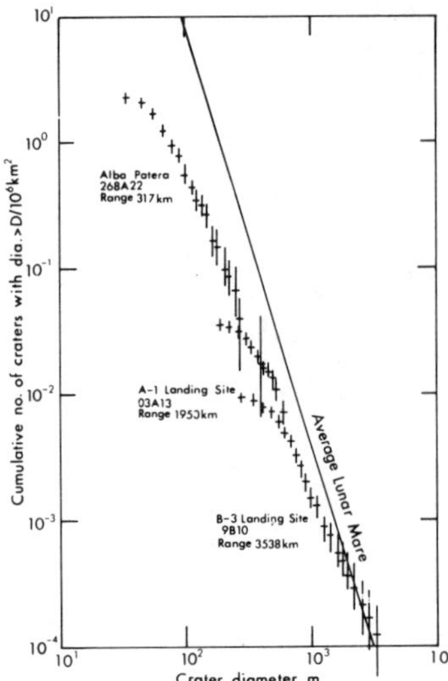

Fig. 5*a*. Crater size-frequency curves showing the effect of resolution. The slant range is indicated and varies from 3830 to 317 km. The rollover in the crater curves shows the true identification resolution, which is 5-6 times as great as the pixel resolution. The pixel resolution varies from 400 m at the B-3 site, to 40 m at A-1, to 8 m at Alba Patera.

channels might have been formed at the same time. Similarly, channels at a lower elevation, with the corresponding higher atmospheric pressures, would be subject to greater erosion. We are therefore concluding that variations in age of the channels and surfaces are best measured by variations in crater densities.

The small-scale dendritic channels may have formed during an interglacial period when warmer climatic conditions on the planet resulted in a denser atmosphere; under such conditions,

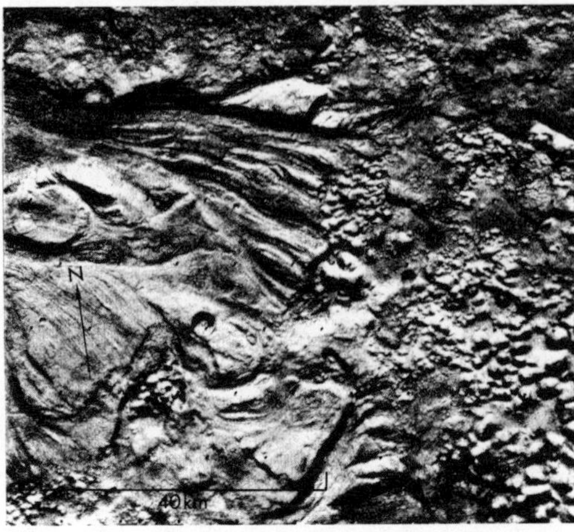

Fig. 5*c*. Photograph (83A37) of Hydaspis Chaos and Tiu Vallis extending north from it. Hydaspis Chaos is an elongate area of collapsed terrain, nearly 100 km wide, where permafrost ice may have melted and the water escaped, modifying the valley floor.

rain could form, fall, drain, and dissect the ancient terrain. Some workers [*McElroy et al.*, 1976; *Owen and Biemann*, 1976] have postulated that this could have happened only during the early history of Mars, when the atmosphere was denser. The younger ages determined for some channel networks suggest that the history of the Martian atmosphere and climate may be too complex to have been the result of a monotonic pressure decrease since its formation. Warm periods may have been caused by variations in solar output [*Hartmann*, 1974], variations in obliquity of the planet [*Ward*, 1973, 1974], or critical variations in the CO_2 and other gas abundances [*Sagan et al.*, 1973]. Additional photography and crater counts may allow

Fig. 5*b*. Photograph (268A22) of an area near Alba Patera taken at a slant range of 317 km with a pixel resolution of 8 m. Degraded microchannels are shown; small craters down to the resolution limit are common.

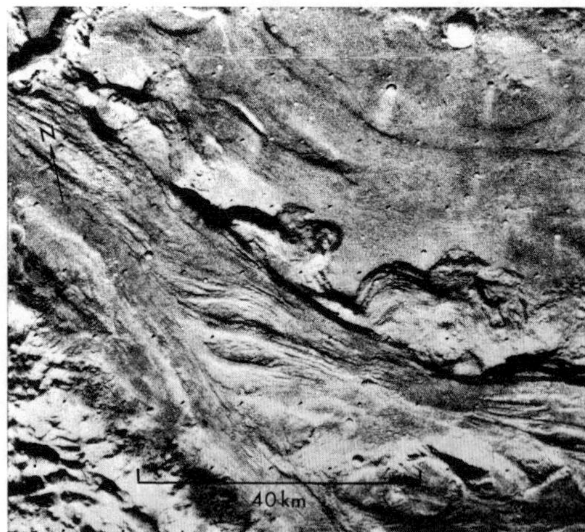

Fig. 5*d*. Photograph (83A38) of Tiu Vallis. In this area the channel is 30 km wide.

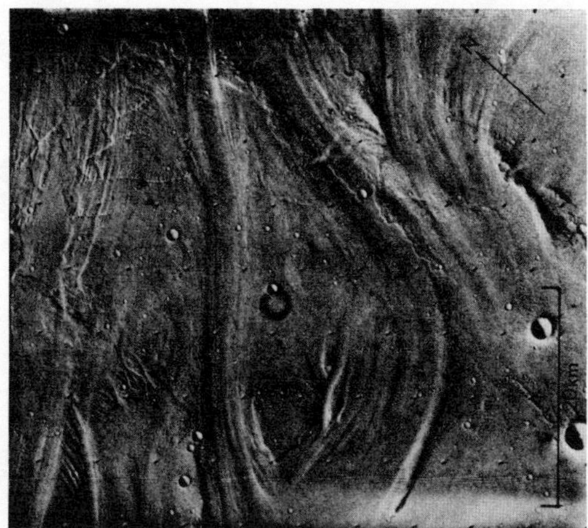

Fig. 6*a*. Photograph (20A56) showing braided channel pattern formed by water erosion at the mouth of Maja Vallis. This channel rises in Lunae Planum, spreads out in Chryse Planitia, and runs east, disappearing before it reaches the area of the Viking 1 landing site.

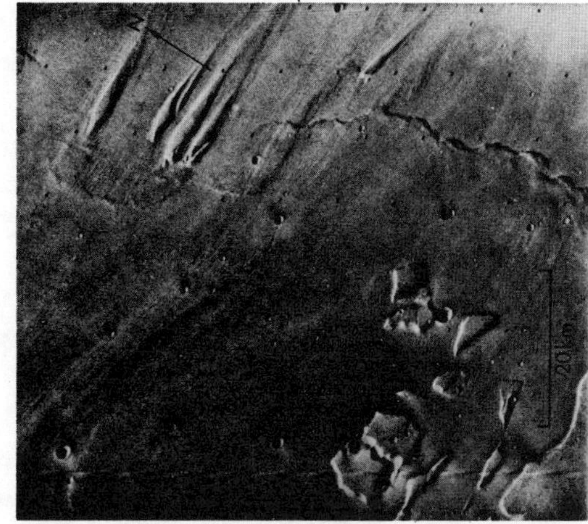

Fig. 6*c*. Photograph (20A26) of mouth of Kasei Vallis showing grooves on the floor of the channel and streamlined islands, which were probably shaped by flowing water. The southern shore of the channel is seen in the lower part of the picture. A mare ridge is crossed by the channel. Abundant small impact craters pepper the floor of the channel.

us to determine whether periods of intensive volcanic eruptions and channel formation coincide.

PERSISTENCE OF WATER ON THE MARTIAN SURFACE

In this section we will first consider a possible method by which running water might exist on the Martian surface under present atmospheric conditions, and then we will present the evidence given to date that more favorable conditions existed for surface running water during various episodes of Martian history.

Surface Water Under Present Martian Conditions

Wallace and Sagan [1977] have calculated the rate of evaporation for water and ice under present Martian atmospheric conditions. They showed that although liquid water would evaporate very rapidly, evaporative cooling would cause a layer of ice (1-m maximum thickness) to form on top of the water; further evaporation would thus be reduced to a level of about 10^{-6} g cm² s. It is difficult to envision subice flow as the mechanism of formation for any of the three types of Martian

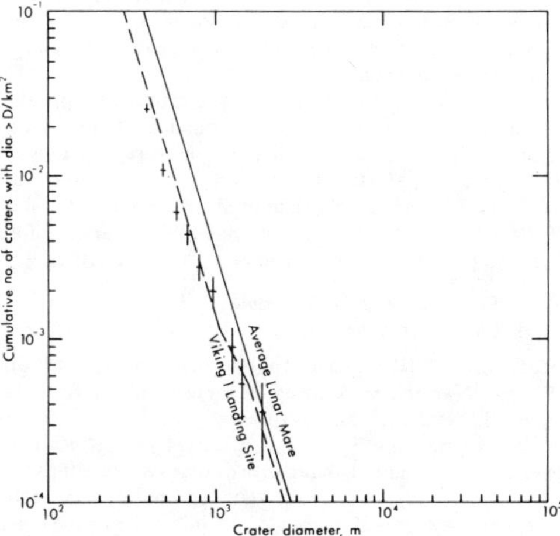

Fig. 6*b*. Composite crater size-frequency distribution curve for photographs 20A55, 20A56, 20A58, 44A43, and 44A44 of the Maja channel; the similarity of the curve to that for the Viking 1 landing site implies a similar age. It is also similar to the curve at Kasei Vallis.

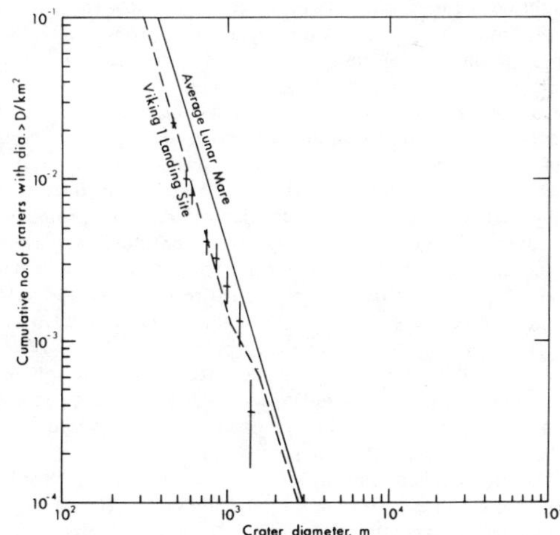

Fig. 6*d*. Crater size-frequency curve of the Kasei Vallis channel floor. The curve follows that of the surface at the Viking 1 landing site, southeast of this area.

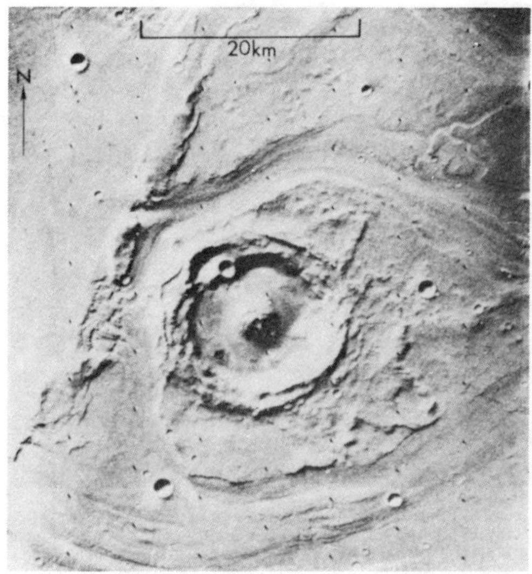

Fig. 7*a*. Photograph (20A62) of Maja Vallis where it enters Chryse Planitia, 150 km southwest of the Viking 1 landing site. Terrain slopes to the right (east). Water ponded west of the mare ridge, and then cut gaps as it overflowed the low points on the ridges; similar terrestrial features are seen in the channeled scabland area in Washington state. The impact crater in the center is 15 km in diameter; the gap is about 2 km wide. After crossing the ridge the water spread widely and cut many grooves in the channel floor. The high-water mark can be seen clearly where it cuts the ejecta blanket on both sides of the large crater. The channel erodes the ejecta blanket of a large crater and, in turn, is pockmarked by younger craters.

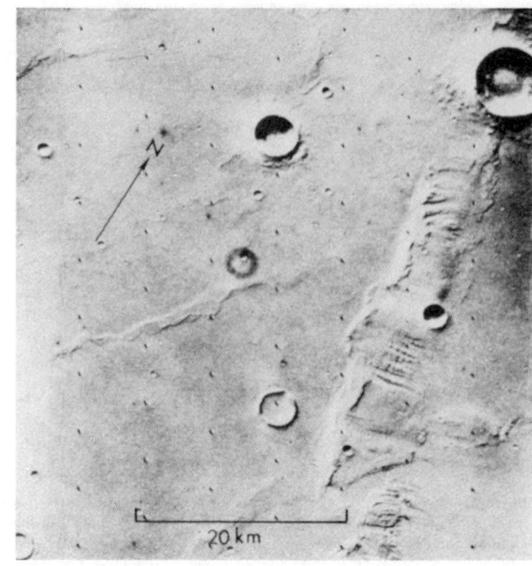

Fig. 7*b*. Photograph (20A40) of Maja Vallis in Chryse Planitia, 130 km west southwest of the Viking 1 landing site. The mare ridges are part of the Xanthe Dorsa system. The largest impact crater is 7 km in diameter and displays a hummocky ejecta blanket. Two smaller craters 2.5–3.5 km in diameter have been flooded by younger lavas. The mare ridges have been incised on the downhill side; grooves 200–400 m across apparently were cut where the water accelerated in crossing the ridge, like a dam spillway; cavitation may have helped cut the grooves. Obscure signs of water erosion are present 80 km east of this area. Perhaps higher-resolution pictures at lower sun elevation angles could bring out more subtle erosion features on the channel floor.

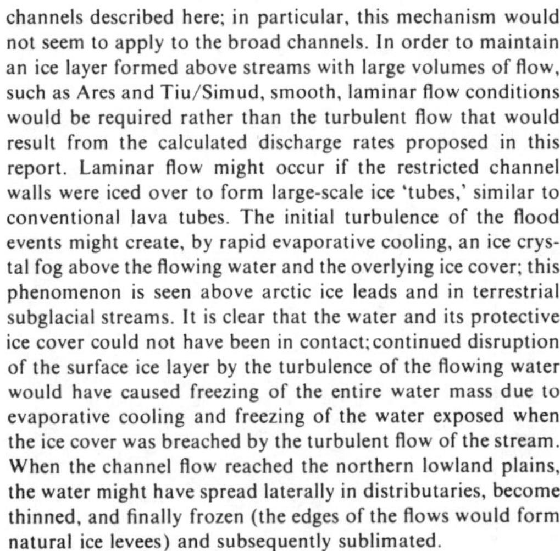

channels described here; in particular, this mechanism would not seem to apply to the broad channels. In order to maintain an ice layer formed above streams with large volumes of flow, such as Ares and Tiu/Simud, smooth, laminar flow conditions would be required rather than the turbulent flow that would result from the calculated discharge rates proposed in this report. Laminar flow might occur if the restricted channel walls were iced over to form large-scale ice 'tubes,' similar to conventional lava tubes. The initial turbulence of the flood events might create, by rapid evaporative cooling, an ice crystal fog above the flowing water and the overlying ice cover; this phenomenon is seen above arctic ice leads and in terrestrial subglacial streams. It is clear that the water and its protective ice cover could not have been in contact; continued disruption of the surface ice layer by the turbulence of the flowing water would have caused freezing of the entire water mass due to evaporative cooling and freezing of the water exposed when the ice cover was breached by the turbulent flow of the stream. When the channel flow reached the northern lowland plains, the water might have spread laterally in distributaries, become thinned, and finally frozen (the edges of the flows would form natural ice levees) and subsequently sublimated.

Moisture could also be added to the atmosphere by creating fog banks of ice crystals above the floods (S. Hess and C. Leovy, personal communication, 1977), which if blown southward against the highlands might precipitate as snow; it is difficult to envision transport of the moisture in an unfrozen state so that rain would fall. The fallen snow might later melt and erode the channel network if it were first covered by a layer of volcanic or eolian dust that would darken it and cause melting. Only a small part of the water involved in one of the great floods would have to be evaporated and precipitated during the warmest part of each day to engender enough flow to carve the small channels.

A more logical, less complicated, and thus more appealing mechanism for forming the Martian channels of intermediate and small size is to assume that temperature and pressure conditions of the Martian atmosphere in the past were such that liquid water could flow uninhibited for at least short periods of time. What, then, is the evidence that varying atmospheric and thermal conditions have existed in the past?

Surface Water Under More Favorable Martian Conditions in the Past

McElroy et al. [1976] have reported that the observed ratio of ^{15}N to ^{14}N in the Martian upper atmosphere suggests that the partial pressure of this gas (presently 2.5% of the Martian atmosphere) may have reached 2–30 mbars. *Nier et al.* [1976] suggested that the initial abundance of oxygen, present as CO_2 or H_2O, must have been equivalent to an interchangeable atmospheric pressure of at least 2 bars in order to have inhibited escape-related enrichment of ^{18}O. *Pollack* [1977], using equations for greenhouse heating, has calculated that a Martian atmospheric pressure in the range of 1–2 bars (CO_2 plus a trace of water vapor) is required to reach atmospheric temper-

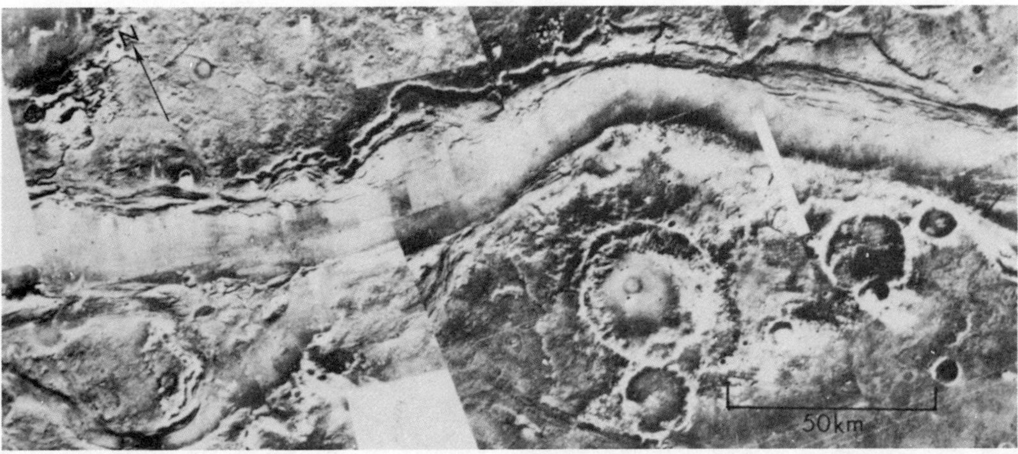

Fig. 7*c*. Photomosaic of part of Ares Vallis, where it is incised in the heavily cratered upland. The channel is 25 km wide and about a kilometer deep; stereoscopic photographs that would provide more accurate measurements are not yet available. Four to eight layers, probably lava flows, are exposed in the walls of the canyon where they have been eroded by flowing water and mass wasting processes. Abundant impact craters up to 50 km in diameter mark the upland; the younger channel floor is sparsely cratered.

atures above 273°K and permit liquid water to exist on the surface for at least short periods of time.

The large amounts of CO_2 reported from initial Viking results must represent the total amount of carbon gases outgassed over the lifetime of Mars [*Pollack*, 1977]. *Fanale* [1977] has suggested that if the total amount of CO_2 in the atmosphere-regolith system has remained constant and the temperature of any regolith column (or of Mars as a whole) has increased by 40°C over a period of $\sim 10^6$ years or more, equilibrium pressures of only 25–35 mbars are likely to result. It is clear that additional feedback mechanisms, such as global heat transport, greenhouse effects, additional atmospheric water vapor, and presence of reducing compounds in the atmosphere, are required. *Pollack* [1977] has shown that liquid water would flow on Mars under substantially lower atmospheric pressure conditions (~ 150 mbars) if the Martian atmosphere were enriched in reducing compounds such as ammonia and methane. Pollack points out that this type of atmosphere, formed during the early history of earth, provided the chemical basis from which life originated. The atmospheric reducing agents probably resulted from extensive volcanism on both planets.

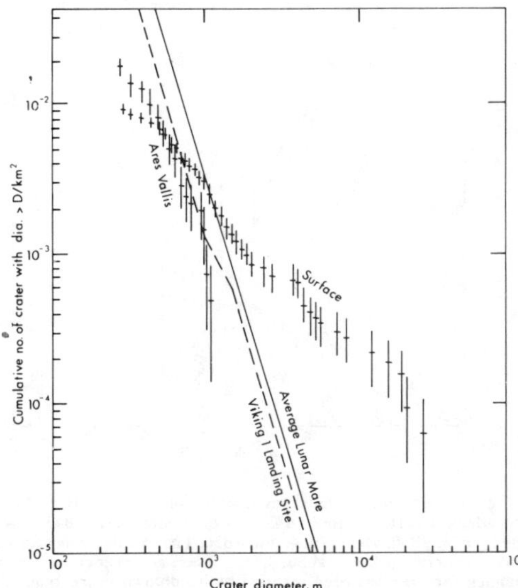

Fig. 7*d*. Crater size-frequency distribution curves for the old cratered upland surface adjacent to Ares Vallis and its channel floor. The surface crater curve has a low slope, indicating a steady degradation of essentially the entire crater population. The channel curve closely follows the Viking 1 landing site curve and probably has a similar age.

We have concluded from photogeologic observations that both the dendritic and the large channel system may have formed during relatively short periods of time at several different times; the larger channel systems have few tributary streams, and the smaller, older dendritic channels that are found primarily in heavily cratered or sloping terrain are not deeply incised. Flowing water on the Martian surface caused by the above or other conditions may have eventually frozen and then sublimated after a short period of time, may have percolated into the ground, or may have evaporated rapidly under conditions of elevated wind velocities and low water vapor pressures in the atmosphere. Depending on the initial temperature of the released water the flowing streams may also have melted the near-surface permafrost layers, so that additional volumes of meltwater were dispersed on the channel floors.

In the following discussion of Martian channel formation by flowing water, we will assume that one or more of the above criteria for increased atmospheric pressure and temperature were achieved during short periods of Martian history. The rate of evaporation of water under such conditions can be at

Fig. 8*a*. Photomosaic of Tiu Vallis near its source area, Hydaspis Chaos. Relative youthfulness of this channel is inferred from the detailed structure seen within the channel.

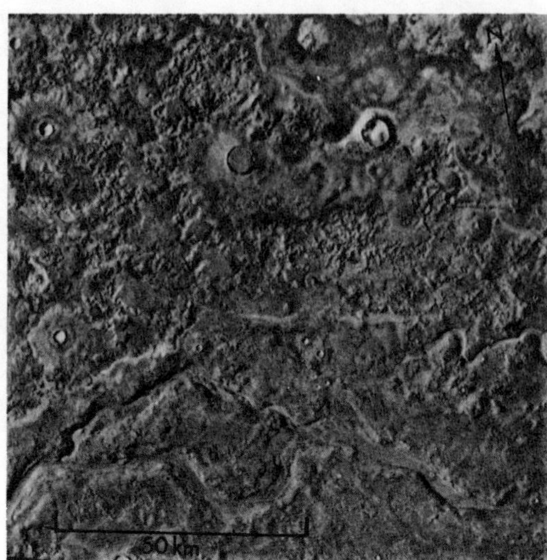

Fig. 8*c*. Photograph (9B54) of Hrad Vallis area (~350 km) south of the Viking 2 landing site. The tributaries head in box canyons that probably formed by spring sapping of nonresistant rock materials by meltwater derived from subsurface ice. Melting may have been caused by geothermal heating or climatic warming.

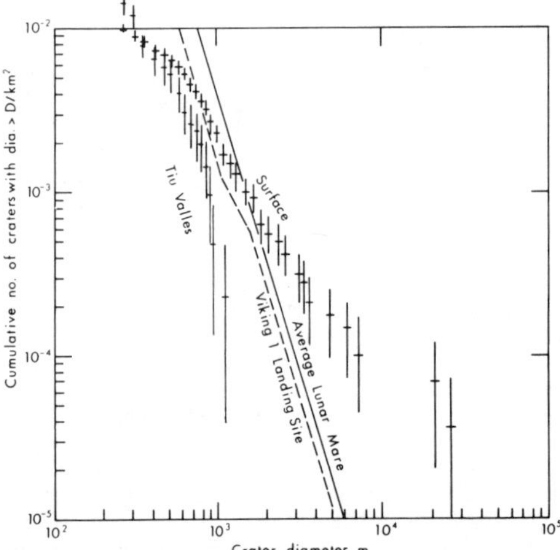

Fig. 8*b*. Crater size-frequency distribution curves for the Tiu channel and the cratered upland surface it cuts. Crater curve of the cratered upland surface is similar to that at Ares Vallis. The position of the channel curve, well below that of the Viking 1 landing site curve, may indicate that the feature is younger than any of the other large channels on Mars.

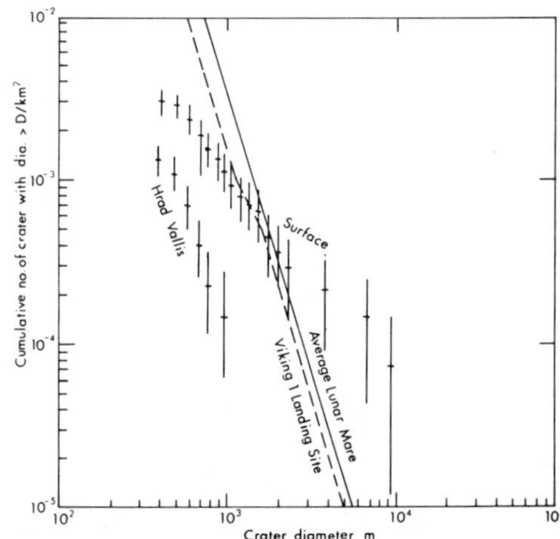

Fig. 8*d*. Crater size-frequency distribution curves of Hrad Vallis and adjacent surface based on counts of frames 9B22, 9B43, 9B45, 9B47, 9B48, 9B50, 9B52, 9B54, and 9B56. The calculated age for the heavily cratered surface is about 3.5 Gy. There are fewer craters on the channel; the area has been mantled by windblown debris that subsequently has been partially stripped. Position of the crater curve may be biased by this blanket of younger sediments.

Fig. 9*a*. Photomosaic of an unnamed channel in Hellas. The sinuous channel coming in from the north cuts through only one layer. The deeper east-west part of the channel heads in a box canyon where at least three layers are exposed. Many small tributaries occur along the channel; most head in box canyons, indicating spring sapping. However, small dendritic networks at the headwaters of two tributaries may indicate rainfall. The channel flows westward into Hellas Planitia. A rock layer etched by the wind crops out on the plain shown; all of these layers may be lava flows. The debris apron around the hill and ridge in the center of the mosaic was probably caused by mass wasting. (Photographs 97A48–97A54 and 97A60–97A68; 39°–48°S, 260°–270°W.)

least estimated using a basic equation [*Dalton*, 1798–1802] of the form

$$E = C(P_w - P_a) \tag{1}$$

where

E rate of evaporation;
P_w vapor pressure just above the water surface;
P_a vapor pressure in the lower atmosphere;
C coefficient dependent on barometric pressure, wind velocity, and other variables.

Dalton's original equation was modified by many workers, including *Fitzgerald* [1886], *Carpenter* [1891], and *Magin and Randall* [1960], by introducing an empirical calculation of the constant and the actual wind velocity:

$$E = (0.6)(W)(P_{H_2O} - P_{atm}) \tag{2}$$

where

E evaporation rate, mm/day;
0.6 empirically derived constant;
W wind velocity, m/s;
P_{H_2O} vapor pressure immediately in contact with water, mm Hg;
P_{atm} vapor pressure of the atmosphere near the surface.

Figure 11*a* shows the relation between wind velocity and water temperature (the latter being the controlling factor for P_{H_2O}) for an arbitrarily high atmospheric vapor pressure of 5 mm Hg and on the assumption of a near-surface atmospheric temperature of slightly over 0°C (273°K). Note that for water temperature ranging from 5°C to 80°C (due to volcanic heating at depth) and wind velocities between 2 and 20 m/s, evaporation rates from 2 mm/day to over 4 m/day can be achieved. The magnitude of the atmospheric vapor pressure P_{atm} is insignificant as long as it does not approach the P_{H_2O} value, indicating a saturated condition. These preliminary calculations do not address the rather important effects of evaporative cooling on the water and the presence of solute or sediment in the water that might alter its freezing temperature or evaporation rate.

Another problem that must be addressed is concerned with the dispersal of part of the floodwaters so that the channel

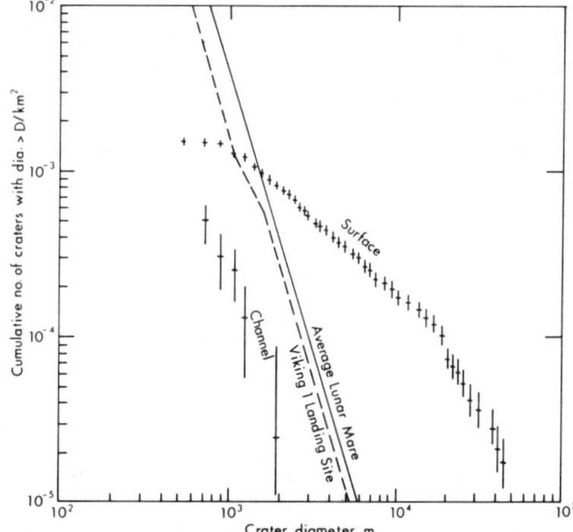

Fig. 9*b*. Crater size-frequency distribution curves for the floor of a channel east of Hellas Planitia and the older surface which it cuts. The curve for the surface has a low slope—possibly due to degradation of the smaller craters by wind erosion. The curve for the channel floor lies below the Viking 1 site curve, indicating that it may be much younger than the Bahram and Vedra valles. However, the channel floor may have been covered by subsequent windblown or other volcanic basalt deposits; if so, the curve shows the age of these deposits rather than the time of channel formation. (Photographs 97A48–97A54 and 97A60–97A68; 39°–48°S, 260°– 270°W.)

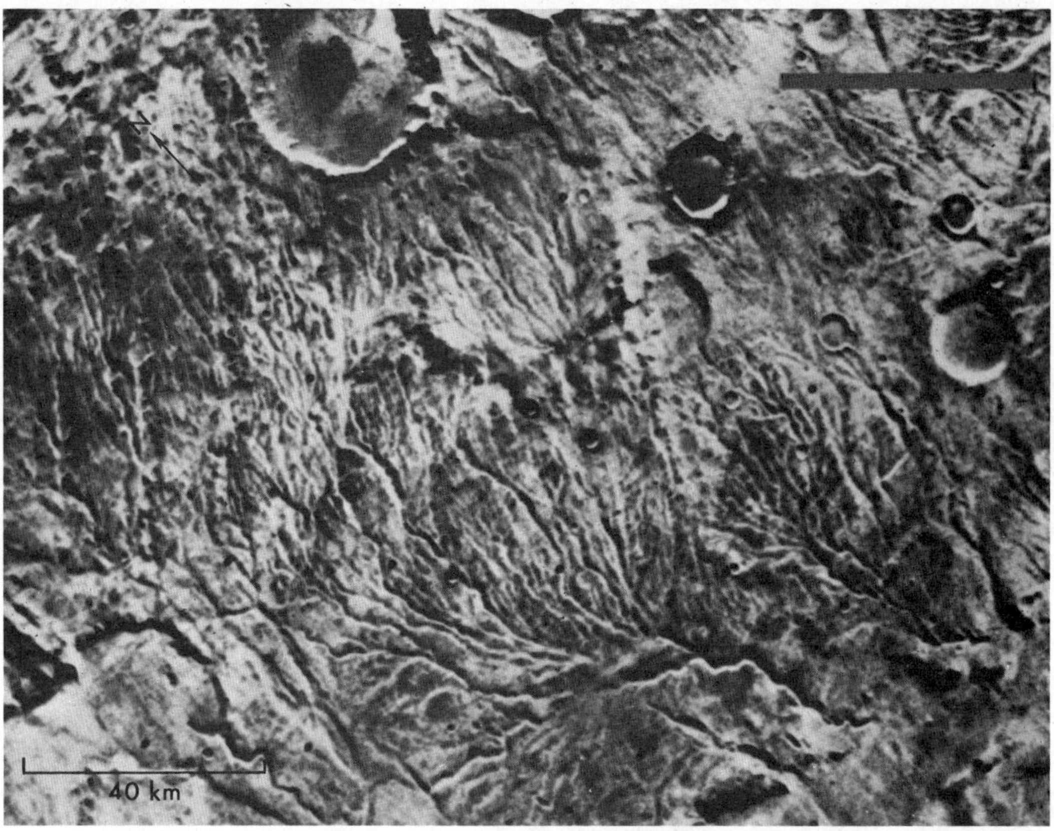

Fig. 9c. Part of Viking photograph (63A08) showing small dendritic ('filamentous') channels in the cratered terrain of the southern Martian highlands. Channels intersect and erode some large craters. The trunk stream of the channel system is deeply incised.

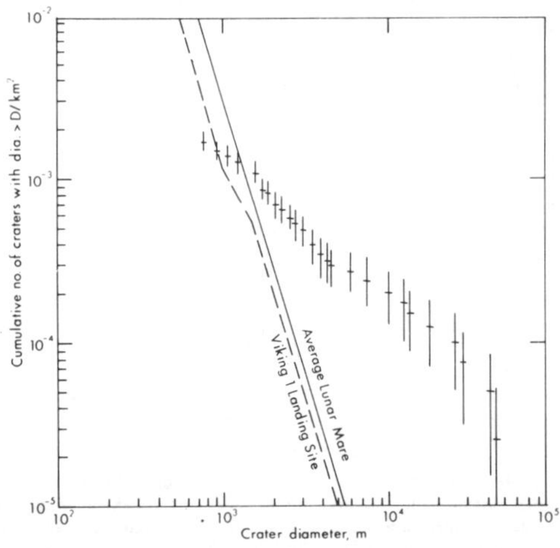

Fig. 9d. Crater size-frequency distribution curve of an old surface on which the dendritic channel was superposed. No crater counts were attempted for the channel because of its small size and the restricted number of postchannel craters. The low slope of the curve for the heavily cratered surface suggests degradation and aggradation of the surface; photogeologic interpretation confirms this interpretation. (Part of photograph 63A09; 43°S, 94°W.)

erodes the plains surface for only about 350 km beyond the end of the channel, as is the case with the larger Martian features. The peak discharge in the flood would last from 2 to 4 days, and the flood could continue for 2 weeks if it were similar to the Missoula flood [*Bretz*, 1923; *Malde*, 1968; *Baker*, 1973*a*]. If the atmosphere were only slightly above 0°C and the water in the channel were warm owing to the volcanic heat, 10–70% could be lost to the atmosphere with sufficient winds and the rest could be lost by percolation into the regolith or plains surface. Much of the water that spread thinly over the plains might have frozen temporarily, owing to increased area available for evaporative cooling, but could then be lost to the atmosphere by slow ice sublimation.

Channel Parameter Estimates

Preliminary estimates of the open channel flow in the Ares channel where it spreads out in Chryse Planitia (21°N, 31°–34°W) have been computed by Lawrence Mann and C. H. Bell (personal communication, 1977) of the U.S. Geological Survey, using the Manning formula [*Chow*, 1959] adjusted for Mars gravity. Ares channel measurements obtained photographically from Viking pictures by S. S. C. Wu, R. Jordan, and F. Schafer, also of the U.S. Geological Survey, indicate that the channel is 300 km wide and 100 m deep. The calculated slope of the channel floor is 10 m/km to the north, and the slope of the channel banks is ~20°. Boulders as large

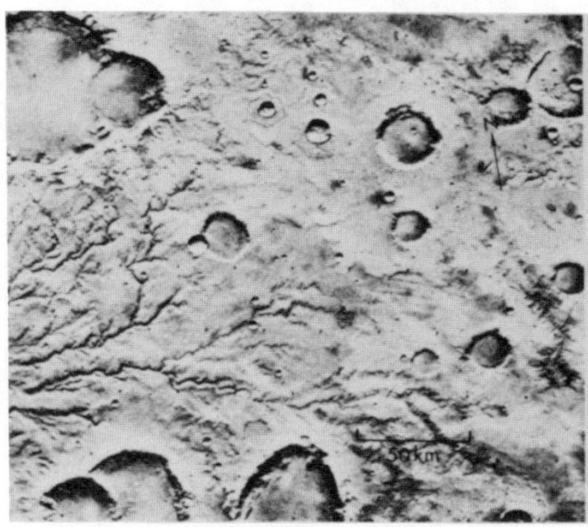

Fig. 10*a*. Photograph (84A47) showing dendritic channel network in southern highlands 2000 km northeast of Argyre Planitia. These channels are part of a network of streams that extend for more than 200 km; the channel network is almost as long as channels in the intermediate size range. The trunk channels are about 2 km wide.

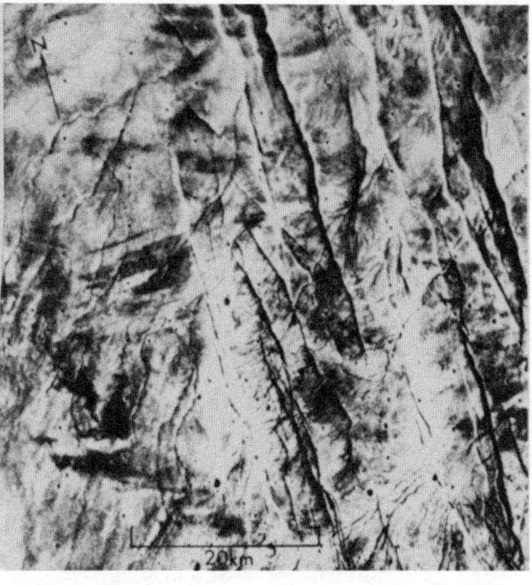

Fig. 10*c*. Photograph (4B57) of fluviatile channels southeast of Alba Patera. These form a dendritic channel network; many filamentous channels join the trunk stream. The headwater area of the channels does not begin as a box canyon or in collapsed terrain. The channels are offset by en echelon faults that are the most sharply defined, least modified ones on Mars; they are therefore thought to be quite young in age. These dendritic channels, at 44° N, 104° W, are the most northerly ones observed.

Fig. 10*b*. Photograph (84A43) of several channel networks; some are tributaries to an intermediate size sinuous channel that is nearly 5 km wide. This part of the channel is located about 1400 km northeast of Argyre Planitia in the southern highlands. Old craters ranging up to 50 km in diameter pockmark the surface. There may have been two channeling episodes in this area.

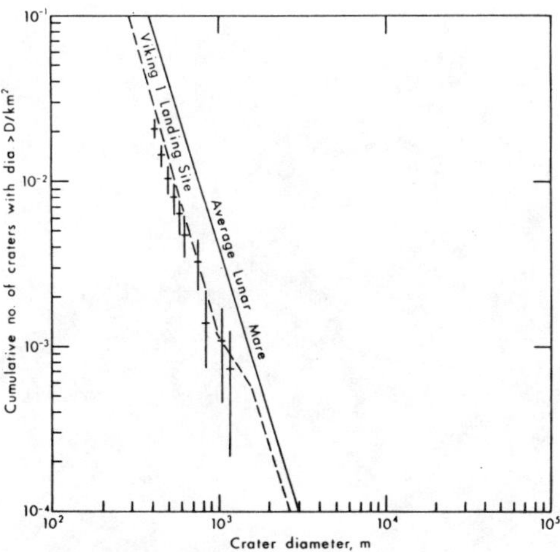

Fig. 10*d*. Crater size-frequency distribution curve for the northeast side of Alba Patera. The crater count curve for this volcanic surface coincides with that for the Viking 1 landing site, and this surface is probably the same age or slightly younger. These dendritic channels are superimposed on a younger surface than those that occur in the southern highlands where the dendritic channels cut ancient cratered terrain. (Photograph 4B57; 44° N, 104° W.)

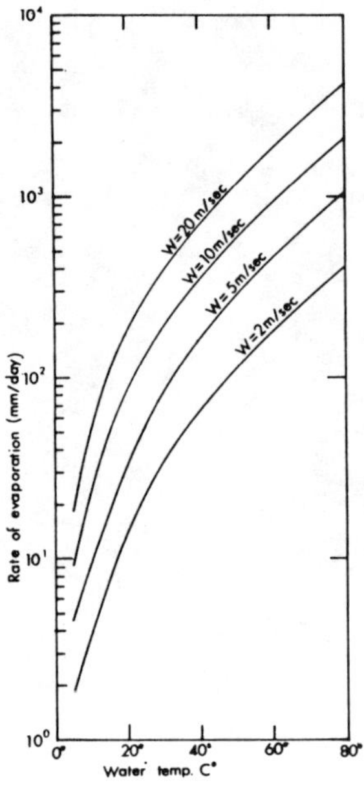

Fig. 11*a*. Graph relating water temperature and wind velocity to evaporation rate given an atmospheric vapor pressure of 5 mm Hg (see equation (2) in text).

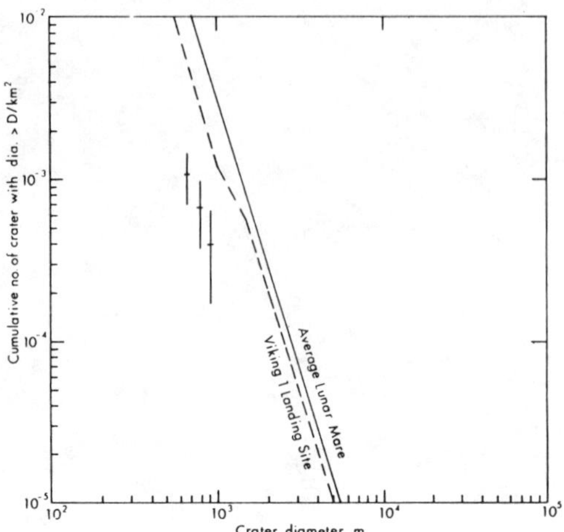

Fig. 11*c*. Crater size-frequency distribution curve for photograph (39A09) near the summit caldera of Arsia Mons. Few impact craters are visible; the lava channels and flows therefore are younger than those on Alba Patera.

Fig. 11*b*. Part of photograph (52A04) of lava channels near the summit caldera of Arsia Mons. Two types are shown: those that originate in source craters and become narrower and shallower down the shield slope and those with marginal natural levees formed when the lava freezes along the borders of the long, thin, narrow flow.

Fig. 11*d*. Photograph of the west flank of Alba Patera showing lava flows and many lava channels. Several channels, particularly those seen in the lower left corner of the picture, lie along the crests of the ridges. The ridges apparently were formed by the cooling flows in which the center top remained mobile longest and drained the crest of the flow. (Photograph 39A09; 42°N, 118°W.)

Fig. 12*a*. Viking photographs (7B60 and 7B62) of lava flows west of the summit of Alba Patera. The lava flows exhibit many lava channels with marginal natural levees. Similarities between these flows and terrestrial lava flows in Hawaii and the Galapagos Islands suggest that they are basaltic in composition. Many small impact craters are visible; the surface is little modified by wind or water.

as 2 m in diameter, like the boulder observed in the Viking 1 landing site [*Mutch et al.*, 1976], are inferred to lie on the channel floor. From these parameters the velocity is estimated to be about 3 m/s, and peak discharge is estimated to be 8.6 $\times$ 10^7 m^3/s. The velocity is comparable to derived velocities of 11–30 m/s and peak discharge of 2.1 $\times$ 10^7 m^3/s for the Pleistocene Missoula flood [*Baker*, 1971, 1973*a*, *b*] and the Bonneville flood [*Malde*, 1968]. Total flood volume for the Missoula flood is 7 $\times$ 10^{13} m^3/s, and for the Bonneville flood, 3 $\times$ 10^{12} m^3. Volume of release for recent Icelandic glacial bursts is 7 $\times$ 10^9 m^3/s; *Thorarinsson* [1957] quotes a peak discharge of 3–4 $\times$ 10^5 m/s for the Katla Jökulhlaup. The volume for Ares is 1.5 $\times$ 10^{13} m^3, a flood duration on Mars of 4 days with lower stages for about 2 weeks being assumed. These values appear to be comparable to those inferred for the terrestrial floods, but these estimates are generalized. There is considerable variation in the infrared durations of Pleistocene and recent floods.

We calculate the volume of water that could have been derived from chaotic terrain in the headwaters of Ares Valles to be 2.6 $\times$ 10^{14} m^3 and that for the Simud/Tiu valles to be 4.4 $\times$ 10^{14} m^3. Because stereoscopic measurements from Viking photographs are not yet available for the collapsed terrain, our calculated depth is based on Mariner 9 ultraviolet spectrometer and Goldstone radar measurements. Three ultraviolet spec-

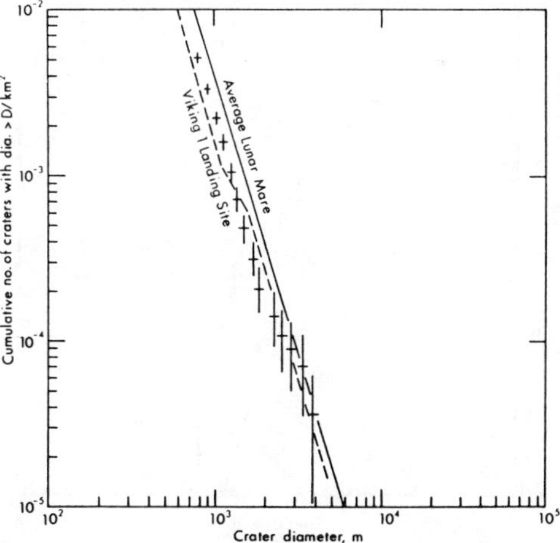

Fig. 12*b*. Crater size-frequency distribution curve of the volcanic surface west of the summit of Alba Patera. The curve is coincident with the one for the Viking 1 landing site; the two surfaces are probably similar in age. (Photographs 7B60 and 7B62; 43°–44°N, 108°–111°W.)

Fig. 12*c*. Photomosaic of Hadriaca Patera on the northeast rim of Hellas Planitia. This volcanic center has a central caldera 60 km in diameter and radiating lava flows and channels that are degraded and impact cratered. The western of two unnamed channels originates in the collapsed area in the upper right part of the picture. These are intermediate between the spring sapped box canyons, where most intermediate size sinuous channels start, and the chaotic terrain, where the large channels originate. The streams may be fed by water coming from melting permafrost. Because these channels originate in a volcanic area they may be activated by geothermal heating. Alternatively, they may be due to climatic warming. (Photographs 97A40 and 97A42; 28°–34°S, 265°–274°W.)

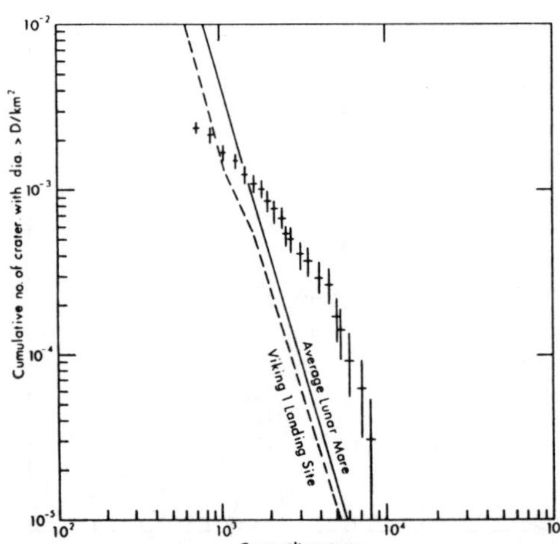

Fig. 12*d*. Crater size-frequency distribution curve of Hadriaca Patera (photograph 97A42). The crater densities on this volcanic center are much higher than are those on Alba Patera. This curve indicates an age of 3.5 Gy. A less-cratered fluviatile channel crosses the flows that emanate from Hadriaca Patera; it is therefore thought to be younger in age.

trometer transects of the collapsed terrain indicate a depth of about 1 km [*Barth and Hord*, 1971; *Hord et al.*, 1974]. Two Goldstone radar (R. Goldstein, personal communication, 1973) traverses indicate depths of 1 km and 0.5 km, respectively. We therefore have used a 1.0-km nominal depth and assumed that the total maximum volume represents ice that was melted by the volcanic heat to form the water released in the floods. There are three separate areas of collapsed terrain in the headwaters of Ares Valles; collapse in these areas probably did not occur at the same time, and there may have been three separate episodes of flooding. It is apparent that the volume of water contained as permafrost was large enough to form floods on Mars comparable to terrestrial late Pleistocene and recent floods on the earth.

Volcanic Channels

Three types of volcanic channels have been observed. The first type of channel originates in volcanic craters and becomes narrower and shallower as it flows downslope; it resembles the lava channels and collapsed lava tubes seen on the earth and the moon. Channels of this type are seen near the summit of Arsia Mons, a great shield volcano (Figure 11*b*). Both the gross shape of the volcanic shield and the digitate shape of the flows suggest that the flows may be basaltic in composition. The number of impact craters in Figure 11*b* is not sufficient to obtain a good size-frequency curve (Figure 11*c*) using the crater counting technique; however, the small number of impact craters clearly indicates a very young surface—perhaps

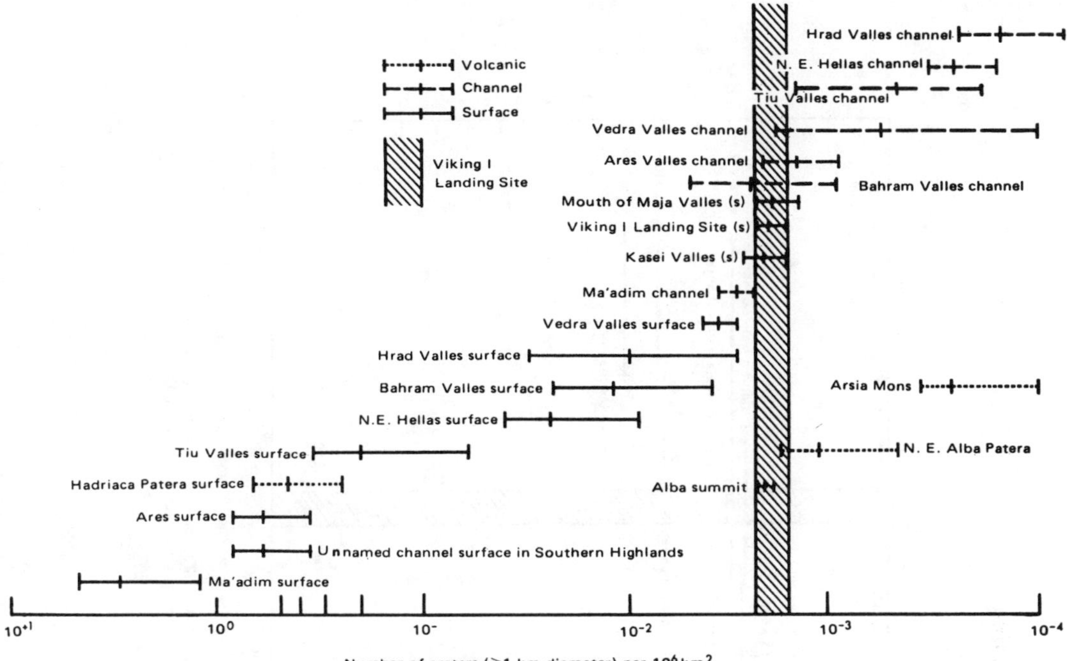

Fig. 13*a*. Diagram showing variations in crater densities among fluviatile channels and volcanic channels and the surfaces they cut. The wide variation in crater densities implies widely varying ages. The fluviatile channels have substantially lower crater densities than the surfaces into which they are cut. Both fluvial and volcanic channels vary greatly in age, but those plotted here are only a small subset of the Mars occurrences. The position of the Viking 1 landing site is plotted as a reference.

about 200 m.y. Examples of this channel type were also identified in Mariner 9 photographs of Elysium Mons [*Masursky*, 1973]. The second type of volcanic channel is characterized by natural levees that occur along the edges of the channels. This type was observed in Mariner 9 pictures of Olympus Mons [*Carr*, 1973] and is also seen in Figure 11*b*. The third type of lava channel occurs on the crests of ridges that were built up by the cooling lava. This type also was recognized in Mariner photographs of Olympus Mons and in Viking photographs of the western flank of Alba Patera (Figure 11*d*).

Photographs of Hadriaca Patera (Figure 12*a*) show a volcanic center with radiating flows and channels that are heavily impact cratered (Figure 12*b*). Relative age determinations indicate that this feature may be very old. The crater densities for the channels on Alba Patera are lower, and those for Arsia Mons are even lower, implying very different ages for these volcanic units.

The channels that have natural levees and those that lie along the crests of ridges are clearly volcanic in origin; only in the complex channels with distributaries seen near Alba (Figure 10*d*) is there a possibility of confusion with the fluviatile channels.

FUTURE WORK

Viking orbital photography during the nominal mission covers only a small part of the area where channels have been incised. Stereoscopic photographs that allow measurements to be made of channel cross sections and slopes cover only a few of these channel areas. In a future report we will discuss additional channels and channel measurements elucidated by extended mission photography. In addition, the increased photographic resolution resulting from the lowered periapsis (from 1500 km to 300 km) should clarify the details of the tributary network source areas and the distributaries where the channels spread out in the plains. Finally, we hope to construct a new Martian crater flux curve so that we can estimate more accurately the relative ages of the fluviatile and lava channels, using the crater counting technique described above. These counts should demonstrate whether the water and lava channels are related in time and genesis or whether water episodes and volcanism have independent histories, one controlled by internal dynamics, the other possibly by varying solar output.

SUMMARY

Crater densities, both for fluvial channel surfaces and lava channels and for the surfaces incised by the channels, are summarized in Figure 13*a*. The wide variation in crater densities shown by the Martian channels discussed in this paper strongly implies widely differing ages for both fluviatile and lava channels. Plots of fluvial channel occurrences based on Mariner 9 photographs are latitude dependent and peak south of the equator [*Sagan et al.*, 1973; G. Granata and H. Masursky, personal communication, 1974]. It is unclear

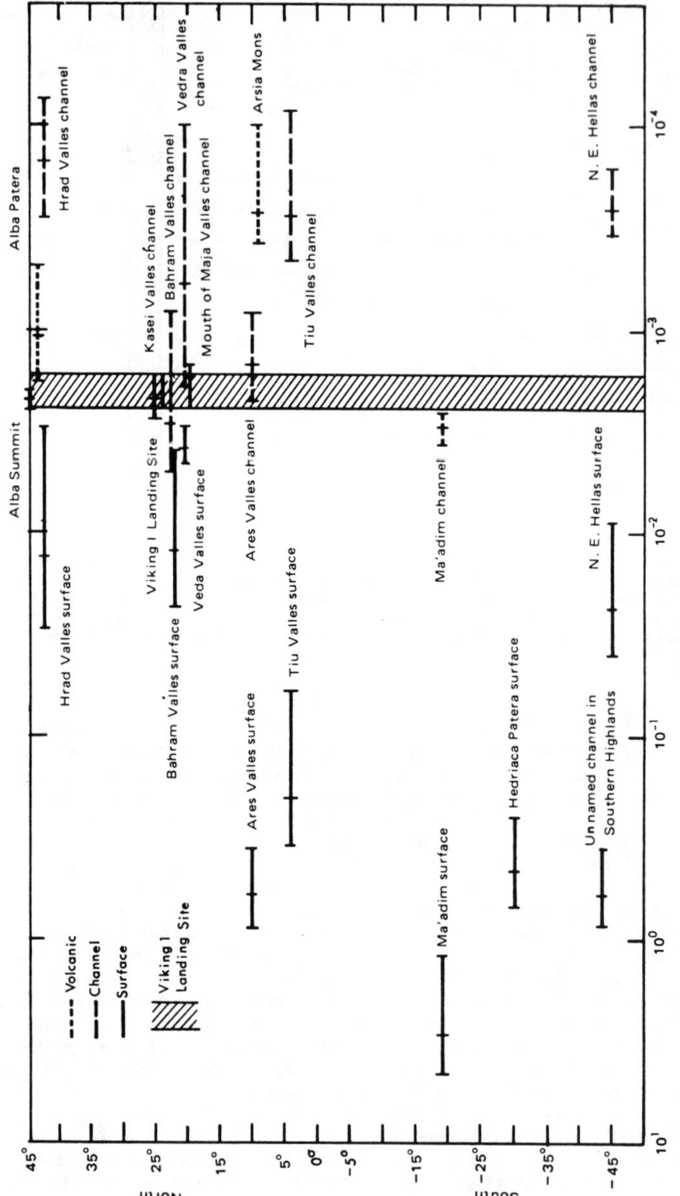

Fig. 13*b*. Crater densities plotted against latitude of occurrence. The data set is too small to show whether channels were never formed at higher latitudes or whether they are buried in the higher latitudes by younger eolian deposits.

whether this distribution is a true reflection of original channel locations or whether channels north and south in the equatorial region have been eroded and/or covered by eolian material [*Soderblom et al.*, 1974]. In Figure 13*b* the channels discussed in this paper are plotted by latitude. No distributional pattern is recognized, but our data set is too small to confirm or refute the bias toward channel formation or location in the equatorial latitudes that was shown in the Mariner photographs. Additional Viking photography should help clarify this issue.

Even the small number of channels imaged to date show that a considerable variety of crater densities are observed, implying a variety of ages.

Using the *Soderblom et al.* [1974] curve and age dating technique, it appears that the summit of Arsia Mons and the Hrad Vallis channel surface may be about 0.5 Gy old. Ares Vallis and the Viking 1 landing site may be 1–2.5 Gy old. The heavily cratered upland surfaces, like Hadriaca Patera and the surface cut by the southern highland channels, may be 3.5–4.5 Gy old. If these latter counts, where channel networks have modified the surface considerably, are accurate representations of channel ages, as well as of surface ages, then we have documented channel ages (by crater counts) that span the entire time interval that can be analyzed. The important results of this study are that the greatly varying crater densities imply variations in time for channel formation and that water and volcanic episodes must have spanned much of the decipherable Mars history.

Acknowledgments. Thanks are due to Lawrence Soderblom, David Scott, and Victor Baker for helpful review of the manuscript. The work was carried out under NASA contract WO-8259.

REFERENCES

Baker, V. R., Paleohydrology of catastrophic Pleistocene flooding in eastern Washington, *Geol. Soc. Amer. Abstr. Programs, 3,* 497, 1971.

Baker, V. R., Paleohydrology and sedimentology of Lake Missoula flooding in eastern Washington, *Geol. Soc. Amer. Spec. Pap., 144,* 79, 1973*a*.

Baker, V. R., Erosional forms and processes for the catastrophic Pleistocene Missoula floods in eastern Washington, in *Fluvial Geomorphology, A Proceedings Volume of the 4th Annual Geomorphology Symposium Series,* edited by M. Morisawa, State Univ. of N. Y., Binghamton, 1973*b*.

Baker, V. R., Viking—Slashing at the Martian scabland problem (abstract), Reports of the Planetary Geology Program, 1976–1977, *NASA Tech. Memo., X-3511,* 169, 1977.

Baker, V. R., and D. J. Milton, Erosion by catastrophic floods on Mars and earth, *Icarus, 23,* 27–41, 1974.

Baldwin, R. B., Lunar crater counts, *Astron. J., 69,* 377–392, 1964.

Barth, C. A., and C. W. Hord, Mariner ultraviolet spectrometer—Topography and polar cap, *Science, 173,* 197–201, 1971.

Bretz, J. H., The channeled scablands of the Columbia Plateau, *J. Geol., 31,* 617–649, 1923.

Carpenter, L. D., Section of meteorology and irrigation engineering, *Colo. Agr. Exp. Sta. Annu. Rep., 4,* 1891.

Carr, M. H., Volcanism on Mars, *J. Geophys. Res., 78,* 4049–4062, 1973.

Carr, M. H., and G. G. Schaber, Mars permafrost features, *J. Geophys. Res., 82,* this issue, 1977.

Carr, M. H., H. Masursky, W. A. Baum, K. R. Blasius, G. A. Briggs, J. A. Cutts, T. Duxbury, R. Greeley, J. E. Guest, B. A. Smith, L. A. Soderblom, J. Veverka, and J. B. Wellman, Preliminary results from Viking orbiter imaging experiment, *Science, 193,* 766–776, 1976.

Chapman, C. R., Asteroids as meteorite parent bodies: An astronomical perspective, *Geochim. Cosmochim. Acta, 40,* 701–719, 1976.

Chow, V. T., *Open-Channel Hydraulics,* p. 680, McGraw-Hill, New York, 1959.

Cutts, J. A., The origin of the channels of the Chryse basin region, paper presented at the 7th Lunar Science Conference, Lunar Sci. Inst., Houston, Tex., March 15, 1977.

Cutts, J. A., K. Blasius, and K. W. Farrell, Mars: New data on Chryse basin land forms (abstract), *Bull. Amer. Astron. Soc., 8,* 480, 1976.

Dalton, J., Experimental essay on the constitution of mixed gases; on the force of steam or vapor from water and other liquids in different temperatures, both in a Torricellian vacuum and in air; on evaporation; and on the expansion of gases by heat, *Mem. Proc. Manchester Lit. Phil. Soc., 5,* 535–602, 1798–1802.

Driscoll, E., Mariner views a dynamic, volcanic Mars, *Sci. News, 101,* 106–107, 1972.

Fanale, F. P., Volatile evolution, Reports of the Planetary Geology Program, 1976–1977, *NASA Tech. Memo., X-3511,* 183–186, 1977.

Farmer, C. B., D. W. Davies, and D. D. LaPorte, Mars: Northern summer ice cap-water vapor observations from Viking 2, *Science, 194,* 1339–1341, 1976.

Fitzgerald, D., Evaporation, *Trans. Amer. Soc. Civil Eng., 15,* 581–646, 1886.

Hack, J. T., Studies of longitudinal stream profiles in Virginia and Maryland, shorter contributions to general geology, *U.S. Geol. Surv. Prof. Pap., 294B,* 45–94, 1956.

Hartmann, W. K., Martian cratering, 4, Mariner 9 initial analysis of cratering chronology, *J. Geophys. Res., 78,* 4096–4116, 1973.

Hartmann, W. K., Geological observations of Martian arroyos, *J. Geophys. Res., 79,* 3951–3957, 1974.

Hartmann, W. K., Cratering in the solar system, *Sci. Amer., 236,* 84–99, 1977.

Hord, C. W., K. E. Simmons, and L. K. McLaughlin, Mariner ultraviolet-spectrometer experiment pressure-altitude measurements on Mars, *Icarus, 21,* 293–302, 1974.

Kieffer, H. H., Soil and surface temperatures at the Viking landing sites, *Science, 194,* 1344–1346, 1976.

Kieffer, H. H., S. C. Chase, T. Z. Martin, E. D. Miner, and F. D. Palluconi, Martian north polar summer temperature: Dirty water ice, *Science, 194,* 1341–1343, 1976*a*.

Kieffer, H. H., S. C. Chase, Jr., E. D. Miner, F. D. Palluconi, G. Munch, G. Neugebauer, and T. Z. Martin, Infrared thermal mapping of the Martian surface and atmosphere: First results, *Science, 193,* 780–785, 1976*b*.

Kieffer, H. H., P. R. Christensen, T. Z. Martin, E. D. Miner, and F. D. Palluconi, Temperatures of the Martian surface and atmosphere: Viking observations of diurnal and geometric variations, *Science, 194,* 1346–1351, 1976*c*.

Leopold, L. B., and T. Maddock, The hydraulic geometry of stream channels and some physiographic implications, *U.S. Geol. Surv. Prof. Pap., 252,* 57 pp., 1953.

Magin, G. B., Jr., and L. E. Randall, Review of literature on evaporation suppression, Studies of Evaporation, *U.S. Geol. Surv. Prof. Pap., 272C,* 69 pp., 1960.

Malde, H. E., The catastrophic late Pleistocene Bonneville flood in the Snake River plain, Idaho, *U.S. Geol. Surv. Prof. Pap., 596,* 52 pp., 1968.

Malin, M. C., Age of Martian channels, *J. Geophys. Res., 81,* 4825–4845, 1976.

Masursky, H., An overview of geological results from Mariner 9, *J. Geophys. Res., 78,* 4009–4030, 1973.

Masursky, H., Martian channels, *NASA Tech. Memo., X-3364,* 169–171, 1976.

Masursky, H., and N. L. Crabill, Search for the Viking 1 landing site, *Science, 194,* 62–68, 1976*a*.

Masursky, H., and N. L. Crabill, Search for the Viking 2 landing site, *Science, 194,* 809–812, 1976*b*.

McCauley, J. F., Mariner 9 evidence for wind erosion in the equatorial and mid-latitude regions of Mars, *J. Geophys. Res., 78,* 4123–4137, 1973.

McElroy, M. B., Y. L. Yung, and A. O. Nier, Isotopic composition of nitrogen: Implications for the past history of Mars atmosphere, *Science, 194,* 70–72, 1976.

Milton, D. J., Water and processes of degradation in the Martian landscape, *J. Geophys. Res., 78,* 4037–4047, 1973.

Mutch, T. A., R. E. Arvidson, A. B. Binder, F. O. Huck, E. C. Levinthal, S. Liebes, Jr., E. C. Morris, D. Nummedal, J. B. Pollack, and C. Sagan, Fine particles on Mars: Observations with the Viking 1 lander cameras, *Science, 194,* 87–97, 1976.

Neukum, G., and P. Horn, Effects of lava flows on lunar crater populations, *Moon, 15,* 205–222, 1976.

Neukum, G., and D. U. Wise, A standard crater curve and possible new time scale, *Science, 194,* 1381–1387, 1976.

Neukum, G., B. Konig, H. Fechtig, and D. Storzer, Cratering in the earth-moon system: Consequences for age determination by crater counting, *Proc. Lunar Sci. Conf. 6th*, 239–263, 1974.
Nier, A. O., M. B. McElroy, and Y. L. Yung, Isotopic composition of the Martian atmosphere, *Science*, *194*, 68–70, 1976.
Nummedal, D., Fluvial erosion on Mars: A review, paper presented at Colloquium on Water in Planetary Regoliths, NASA, Hanover, N. H., Oct. 5–7, 1976.
Owen, T., and K. Biemann, Composition of the atmosphere at the surface of Mars: Detection of argon-36 and preliminary analysis, *Science*, *193*, 801–803, 1976.
Pieri, D., Martian channels: Distribution of small channels on the Martian surface, *Icarus*, *27*, 25–50, 1976.
Playfair, J., Illustrations of the Huttonian theory of the earth, pp. 102, 116, 124, and 354, Edinburgh, 1802. (Reproduced in *Geology From Original Sources*, edited by W. Agar, R. P. Flint, and C. R. Longwell, Henry Holt, New York, 1925.)
Pollack, J. B., Climatic changes on Mars: Inferences based on Viking and Mariner data, Reports of the Planetary Geology Program, 1976–1977, *NASA Tech. Memo.*, *X-3511*, 187–188, 1977.
Sagan, C., O. B. Toon, and P. J. Gierasch, Climatic change on Mars, *Science*, *181*, 1048–1049, 1973.
Schumm, S. A., Structural origin of large Martian channels, *Icarus*, *22*, 371–384, 1974.
Sharp, R. P., and M. C. Malin, Channels on Mars, *Geol. Soc. Amer. Bull.*, *86*, 593–609, 1975.
Shoemaker, E. M., Interpretation of lunar craters, in *Physics and Astronomy of the Moon*, edited by Z. Kopal, pp. 283–360, Academic, New York, 1962.
Shoemaker, E. M., Present cratering rates on the terrestrial planets and the moon, Reports of the Planetary Geology Program, 1976–1977, *NASA Tech. Memo.*, *X-3511*, May 1977.
Shoemaker, E. M., and R. J. Hackman, Stratigraphic basis for lunar time scale, in *The Moon: Symposium of the International Astronomical Union*, edited by Z. Kopal and Z. K. Mikhailov, pp. 289–300, Academic, New York, 1962.
Soderblom, L. A., R. A. West, B. M. Herman, T. J. Kreidler, and C. D. Condit, Martian planet-wide crater distribution: Implications for geologic history and surface processes, *Icarus*, *22*, 239–263, 1974.
Thorarinsson, S., The jökulhlaup from the Katla area in 1955 compared with other jökulhlaups in Iceland, *Reykjavik Mus. Natur. Hist. Misc. Pap.*, *18*, 21–25, 1957.
Wallace, D., and C. Sagan, Evaporation of ice-choked rivers: Application to Martian channels, Reports of the Planetary Geology Program, 1976–1977, *NASA Tech. Memo.*, *X-3511*, 161, 1977.
Ward, W. R., Large scale obliquity variations on Mars, *Science*, *181*, 260–262, 1973.
Ward, W. R., Climatic variations on Mars: Astronomical theory of insolation, *J. Geophys. Res.*, *79*, 3375–3395, 1974.

(Received April 13, 1977;
revised June 3, 1977;
accepted June 3, 1977.)

Editor's Comments on Paper 22

22 **HUGUENIN**
The Formation of Goethite and Hydrated Clay Minerals on Mars

The reddish color of Mars is conventionally attributed to the presence of iron oxides in the surface materials. Their presence implies that an oxidizing environment existed during some past period, when the atmosphere was denser and the climate more Earth-like (Sagan 1977; Sagan et al. 1965). R. L. Huguenin, in a series of papers, has developed a weathering model in which UV radiation can oxidize iron under present environmental conditions, without needing to invoke past climatic changes. Surface oxidation can also provide a major sink for water (Huguenin 1976).

Paper 22 presents an outline of Huguenin's model. Traces of atmospheric O_2 (0.13 percent), UV radiation, and eolian abrasion (to expose fresh grain surfaces) are sufficient to convert Fe^{II} to Fe^{III}. The model predicts the decomposition of basaltic rocks into Ca and Mg carbonates, goethite, and clays. The clays are expected to be depleted in Fe^{II}, but should be more SiO_2-rich than the parent rock.

Comparison with Viking results indicates that while the model is basically correct, several details may have to be modified. X-ray fluorescence data shows a soil derived from rocks of mafic composition (Clark et al. 1976, 1977). However, the composition implies iron-rich, rather than iron-depleted clays. Although the soil SiO_2 content is low (<45 percent), airborne dust probably consists of clay particles with higher SiO_2 content (Hunt et al. 1973). Goethite is probably present only as thin films (Clark et al. 1976). Carbon dioxide is present in the soils (Biemann et al. 1977). Simulated tests have shown that carbonates could form in the martian environment (Booth and Kieffer 1977). However, derivation of clay minerals from feldspars by UV radiation has not yet been experimentally determined. Huguenin (1977, AGU abstr.) has shown that photochemical reactions with surface H_2O could produce the peroxides inferred from the results of the gas exchange experiment (Paper 35, Part XIII). It may be of interest to note that Hunten (1974) predicted the presence of peroxides and superoxides in the soil from rapid seasonal changes of CO and ozone.

22

Reprinted from *J. Geophys. Res.* **79**:3895–3905 (1974)

The Formation of Goethite and Hydrated Clay Minerals on Mars

ROBERT L. HUGUENIN

Planetary Astronomy Laboratory, Department of Earth and Planetary Sciences
Massachusetts Institute of Technology, Cambridge, Massachusetts 02139

Laboratory studies reported by Huguenin (1973*a*, *b*) on the kinetics and mechanism of the photostimulated oxidation of magnetite and preliminary laboratory data on the weathering of silicates, reported herein, are applied to Mars. Basalts in the Martian dark areas are predicted to alter to hydrated $Fe^{2\pm}$ depleted clay minerals, minor goethite, and minor to trace amounts of transition metal oxides such as TiO_2, MnO_2, and Cr_2O_3 at a rate of $10^{-1.5\pm1.5}$ μm yr^{-1}. Some Ca-Mg carbonates are also expected to be formed. The clay minerals are predicted to be more silica rich than the silicate source material, SiO_2 contents of 60% or higher being expected, and strongly depleted in Fe^{2+}. The oxygen, OH, and H_2O contents of the bulk weathering product are predicted to be significantly greater than those of the dark-area source materials, whereas the relative bulk metal abundances should be the same. The weathering products are expected to be in the form of aggregates bound by adsorbed H_2O with diameters of the order of tens of microns or less. If it is assumed that the Martian surface environment has been the same as it is today for the past 10^9 years, a surface layer of alteration products $10^{1.5\pm1.5}$ m in mean global thickness is predicted.

For years the red to ochre color of Mars has been attributed to the presence of ferric oxides and oxyhydroxides in the surface materials. Many proposed that hydrated ferric oxides (limonite) were the principal constituents [*Dollfus*, 1957, 1961; *Sinton and Strong*, 1960; *Sharonov*, 1961; *Moroz*, 1964; *Draper et al.*, 1964; *Rea*, 1964; *Sagan et al.*, 1965], whereas now most interpret the optical properties to indicate that the surface materials are composed primarily of silicates with only a minor amount of ferric oxides and oxyhydroxides [cf. *Binder and Cruikshank*, 1963, 1966; *Van Tassel and Salisbury*, 1964; *Salisbury*, 1966; *Adams*, 1968; *Adams and McCord*, 1969; *Salisbury and Hunt*, 1969; *McCord and Westphal*, 1971; *McCord et al.*, 1971; *Binder and Jones*, 1972].

The presence of ferric oxides in the surface materials indicates that they have been exposed to an oxidizing environment at some time in their history. *Sagan et al.* [1965] proposed that the ferric oxides have been metastably preserved from an earlier epoch when Mars had a surface environment similar to the oxidizing environment of the earth. *Huguenin* [1973*a*, *b*] proposed alternatively that the high oxidation state of the surface is due to a photostimulated oxidation process that has been operating on the surface materials for the past 10^9 years.

According to the photostimulated oxidation model, Fe^{2+} ions that become exposed to atmospheric O_2 and UV illumination ($\lambda \leq 0.350$ μm) undergo photostimulated oxidation to ferric oxide. The UV illumination photoejects electrons from the Fe^{2+} ions, and the electrons attach to adsorbed oxygen. The resultant chemisorbed O^{2-} ions combine with surface Fe^{3+} ions (photooxidized Fe^{2+} ions) to form a surface layer of Fe_2O_3. Photostimulated oxidation is a surface process, and its rate depends on (1) the atmospheric O_2 partial pressure, (2) the UV illumination intensity, and (3) the accessibility of surface Fe^{2+} ions to the UV and atmospheric O_2. The Fe^{2+} ions can be brought to the grain surface under the action of adsorbed H_2O (cation migration), or they can be exposed at the surface by removal of the oxidized surface layers (abrasion, particle disintegration, etc.). Even at the grain surface, however, the accessibility of the Fe^{2+} ions to atmospheric O_2 can be limited by the presence of adsorbed species (H_2O, CO_2, CO, etc.).

In this paper photostimulated oxidation as a natural weathering mechanism on Mars will be examined in more detail. Photostimulated oxidation weathering rates and products will be derived on the basis of (1) what we know about the surface environment (O_2 partial pressures, yearly average UV illumination intensities, and the various parameters affecting the accessibility of Fe^{2+} ions to atmospheric O_2 and UV radiation), (2) expected Fe^{2+} source materials, (3) kinetics and mechanism of the photostimulated oxidation of magnetite, and (4) preliminary results of subsequent experimental work with ferrosilicates. The derived rates and products will then be compared with published estimates of the volume and composition of the oxidized soils on Mars to determine whether photostimulated oxidation weathering alone can account for the high oxidation state of the surface materials.

SURFACE ENVIRONMENT

Atmospheric O_2 partial pressure. The atmosphere is composed primarily of CO_2 [*Belton et al.*, 1968; *Carleton et al.*, 1969; *Hunten*, 1971], surface pressures ranging between about 1 and 10 mbar [*Belton and Hunten*, 1971; *Hord et al.*, 1972]. Molecular O_2 is produced primarily in the upper atmosphere from photodissociation products of CO_2 and H_2O, and it diffuses downward [*Belton and Hunten*, 1968], an O_2/CO_2 mixing ratio at the surface of 1.3×10^{-3} thus being yielded [*Carleton and Traub*, 1972; *Barker*, 1972].

Adsorbed H_2O. The precipitable H_2O content of the atmosphere varies from about 10 to 45 μm, depending on the season, latitude, and time of day [*Spinrad et al.*, 1963; *Kaplan et al.*, 1964; *Schorn et al.*, 1966; *Barker et al.*, 1970; *Tull*, 1970; *Hanel et al.*, 1972; *Conrath et al.*, 1973; *Barker*, 1973]. At nearly all Martian latitudes the daily minimum surface temperatures are low enough for H_2O to condense on the surface [*Morrison et al.*, 1969; *Balsamo and Salisbury*, 1973]. With the onset of adsorption occurring at around 200°–225°K [*Leighton et al.*, 1969; *Pollack et al.*, 1970*a*, *b*; *Fanale and Cannon*, 1971, also unpublished data, 1974], bulk condensations will be in the form of a frost. *Farmer* [1973] suggests that the atmospheric H_2O may become concentrated in a layer near the surface during the coldest part of the day, and up to 45 μm of precipitable H_2O can probably condense out on the surface.

Farmer's model accounts for *Barker*'s [1973] observation that the atmospheric H_2O partial pressure drops to minimum levels at the evening terminator.

Between approximately 50°S and 50°N latitude the daily maximum temperatures exceed the H_2O condensation temperatures throughout the year [*Morrison et al.*, 1969; *Balsamo and Salisbury*, 1973], and all the physically adsorbed H_2O should desorb from the surface [*Fanale and Cannon*, 1971, also unpublished data, 1974]. Just below the surface the desorption becomes diffusion limited, however, and when the temperature exceeds 273°K, the residual H_2O may go into the liquid phase [*Farmer*, 1973]. Farmer argues that at a depth of about 1 cm the H_2O can survive the entire diurnal evaporation period, and at that depth the soil may become damp during the day and freeze at night. Photostimulated oxidation proceeds only when the relative humidity falls below 5% [*Huguenin*, 1973*a*, *b*]. Since the atmospheric H_2O/CO_2 abundance ratio is 10^{-4}, the humidity at the surface falls below 5% for temperatures greater than about 225°K at the 1-mbar elevation level and greater than 250°K at the 10-mbar level.

The fraction of the day during which the surface is free of adsorbed H_2O, i.e., the humidity is <5%, can be calculated by means of temperature expressions derived by *Morrison et al.* [1969] and *Sagan and Veverka* [1971]. They found that the temperature as a function of latitude λ (λ is positive in the north) and hour angle ϕ ($\phi = 0°$ at noon) is

$$T_s(\lambda, \phi) = [197 + 104 \cos(\phi - 7°)] \cos^{1/2} \lambda \quad (1)$$

Expression (1) applies only for the case in which the subsolar latitude γ is at the equator, however, and it must be generalized for all γ. For $\gamma = 0$ the daily illumination period P is π at all latitudes. For $\sin\lambda/\cos\gamma \le 1$, $P = 0$ for $\gamma > 0$ and $P = 2\pi$ for $\gamma < 0$, whereas for $\sin\lambda/\cos\gamma \ge 1$, $P = 2\pi$ for $\gamma > 0$ and $P = 0$ for $\gamma < 0$. For $-1 < \sin\lambda/\cos\gamma < 1$ the diurnal illumination period is $P = \pi + 2\sin^{-1}(\tan\lambda\tan\gamma)$. Sunrise and sunset occur only for $-1 < \sin\lambda/\cos\gamma < 1$ at $\phi = \pm P/2$. Clearly, for $-1 < \sin\lambda/\cos\gamma < 1$ and $\gamma \neq 0$, (1) becomes

$$T_s(\lambda, \gamma, \phi) = \{197° + 104 \cos[(\pi\phi/P) - 7°]\} \cos^{1/2}(\lambda - \gamma) \quad (2)$$

From (2), the onset of photostimulated oxidation occurs at

$$\phi_{\min} = 2\pi + (P/\pi)(7° - \cos^{-1}\{[T_s(P_s)_{\min}/104°\text{K}] \cos^{1/2}(\lambda - \gamma) - 1.9\}) \quad (3a)$$

and it will cease at

$$\phi_{\max} = P/\pi\,(7° + \cos^{-1}\{[T_s(P_s)_{\min}/104°\text{K}] \cos^{1/2}(\lambda - \gamma) - 1.9\}) \quad (3b)$$

where $T_s(P_s)_{\min}$ is the temperature at surface pressure P_s, for which the relative humidity is 5% (e.g., $T_s(1\text{ mbar})_{\min} = 225°$K and $T_s(10\text{ mbar})_{\min} = 250°$K). The fraction of the day during which photostimulated oxidation occurs is therefore

$$\frac{\phi_{\max} - \phi_{\min} + 2\pi}{2\pi} = \frac{P}{\pi^2}\left\{\cos^{-1}\left[\frac{T_s(P_s)_{\min}\cos^{-1/2}(\lambda - \gamma)}{104°\text{K}} - 1.9\right]\right\} \quad (4)$$

During the remaining fraction of the day, adsorbed H_2O limits the access of surface Fe^{2+} ions to atmospheric O_2, and photostimulated oxidation cannot occur.

The fraction of the year during which photostimulated oxidation can occur, obtained by integrating (4), ranges from about 0.4 at the equator for $P_s = 5$ mbar to about 0.1 at ±60° and 5 mbar for a Mars-to-sun distance of 1.5 AU.

UV illumination intensity. The relative incident spectral radiation intensity distribution at the subsolar point on the surface, scaled to 1.0 at $\lambda = 0.200\ \mu$m, is presented in Figure 1. This flux assumes a dust-free atmosphere. It is essentially the same as the flux outside the atmosphere, since the atmosphere is composed almost entirely of CO_2, which is relatively transparent down to about 0.195 μm [*Thompson et al.*, 1963]. The spectral flux is taken from *Allen* [1955, p. 172], with $\Phi_{0.200} = 10^{10}$ photons cm^{-2} s^{-1} Å^{-1} at 1.5 AU. The parameter $\Phi_{0.200}$ at hour angle ϕ and latitude λ is related to the intensity at the subsolar point by the expression

$$\Phi_{0.200}(\phi, \lambda, \gamma) = \Phi_{0.200}(0, 0, \gamma) \cos a \cos(\lambda - \gamma) \quad (5)$$

where $a = \phi$ for $0 \le \phi \le \pi$ and $a = \phi - 2\pi$ for $\pi \le \phi \le 2\pi$. The yearly average value of $\Phi_{0.200}$ during the daily oxidation intervals for a region at λ and P_s is

$$\Phi_{0.200}(\lambda, P_s)_{\text{ave}} = \int_{\gamma_{\min}}^{\gamma_{\max}} \int_{\phi_{\min}}^{\phi_{\max}} \frac{\Phi_{0.200}(0, 0, \gamma) \cos a \cos(\lambda - \gamma)}{(\phi_{\max} - \phi_{\min} + 2\pi)(\gamma_{\max} - \gamma_{\min})} \, d\phi \, d\gamma \quad (6)$$

The value of $\Phi_{0.200}(\lambda, P_s)_{\text{ave}}$ drops by a factor of about 3 at $\lambda = \pm 60°$ from the average daily flux at the equator.

Eolian abrasion. There are several mechanical weathering processes on Mars that insure access of Fe^{2+} ions to atmospheric O_2 and UV by removing weathering rinds from the grain surfaces. The most important of these is eolian abrasion. There is abundant evidence that wind erosion and deposition have strongly modified the Martian terrain and that eolian abrasion is probably many times more efficient than it is on the earth [*Rea*, 1964; *Sharp*, 1968; *Sagan and Pollack*, 1969; *Murray et al.*, 1971; *Cutts et al.*, 1971; *Sagan et al.*, 1972, 1973; *Cutts*, 1973; *Cutts and Smith*, 1973; *McCauley*, 1973; *Sagan*, 1973; *Soderblom et al.*, 1973]. *Sagan* [1973] has deduced abrasion rates of the order of $10^{-6\pm2}$ cm s^{-1} ($10^{1\pm2}$ cm yr^{-1}).

Eolian abrasion should have a greater effect on surface clogging than any other physical weathering process on Mars,

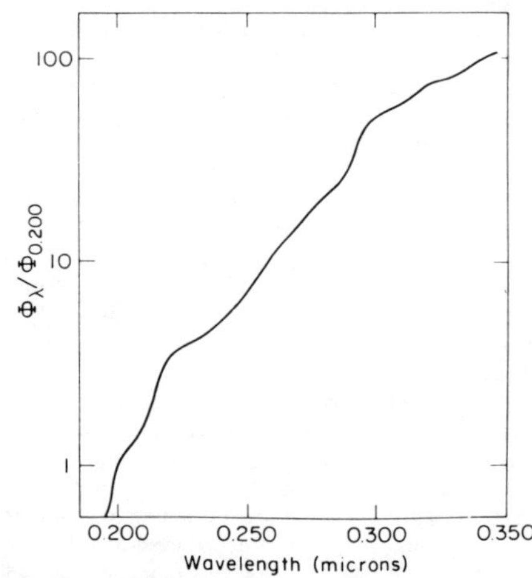

Fig. 1. Relative solar spectral radiation distribution at the Martian surface, scaled to 1.0 at $\lambda = 0.200\ \mu$m. Spectral flux is given by *Allen* [1955, p. 172].

not only because it is more aggressive than it is anywhere on the earth but also because it acts directly on the surface clogging layers *Glennie*, 1970, p. 19]. Most of the other processes, instead of abrading away the weathering layers, counteract the surface clogging effects indirectly by creating new surfaces through splitting grains along cleavage and fracture planes [*Reed*, 1930; *Ollier*, 1969, pp. 23, 62].

Hydrolytic disintegration. Among these weathering processes, hydrolytic disintegration is probably the most important, as it is in the terrestrial arctic tundras and deserts [*Kelley and Zumberge*, 1961; *Rudolph*, 1966; *Tedrow and Ugolini*, 1966]. The H_2O adsorbs to the grain surfaces every night on Mars, and as the temperature decreases, the interaction potential increases, and thus an increase is caused in the density and thickness of the H_2O film. The H_2O that adsorbs along fracture planes and in crevices reaches a maximum density and thickness during the early morning, and then it begins to decrease. If the fissures are not straight walled or V shaped, desorption can become sufficiently diffusion limited that the decrease in interaction potential is accompanied by the formation of hydrogen bonds and a conversion to ice, which could exert pore pressures as high as 10^3 kbar [*Harkins and Ewing*, 1921; *Hennicker*, 1949; *Kitchener*, 1963; *Tyutyunov*, 1964, chapter 1; *Adamson*, 1967, 1968; *Somorjai and Szalkowski*, 1971]. Such pressures are generally never attained, however, since the H_2O partly extrudes out of the open fissures during its conversion to ice. In natural systems pore pressures generally do not exceed about 10^{-1} kbar [*Blatt et al.*, 1972, p. 224], but this is still greater than the average tensile strengths of even the most massive and homogeneous rocks.

If *Farmer*'s [1973] model is correct (i.e., if H_2O films immediately under the surface pass into the liquid phase during the day and freeze during the night), the conversion of pore liquid to ice may also bring about the hydrolytic disintegration of soil particles. The conversion of water to ice involves a 9.2% volume increase, and as the temperature decreases, the ice continues to expand until about 253°K. Pore pressures of up to 2 kbar are theoretically possible, but again, the actual pressures would probably never exceed about 10^{-1} kbar.

Salt weathering. A third mechanism known to disintegrate rocks and soil particles in terrestrial deserts and arctic regions is the crystallization of salts within cracks and pores [*Tricart*, 1960; *Wellman and Wilson*, 1965; *Coleman et al.*, 1966]. *Malin* [1973] suggests that volcanic salts may combine with water and be available for weathering on Mars, and *Soderblom* [1973] and *Murray* [1973] propose that additional salt solutions may form at about 1 km below the Martian surface. At this depth they propose that permafrost will melt and combine with gases evolving from the interior to form acids, which can react with soils and form saline solutions. Although the extent to which salts may have formed on Mars is unknown, it is expected that salt weathering plays a very secondary role to eolian abrasion in counteracting the effects of surface clogging.

Other disintegration processes. Insolation weathering is still another mechanism that may contribute to particle disintegration and counteract the effects of surface clogging. In this process the diurnal temperature variations cause minerals to expand and contract, and since different minerals have different albedos, different coefficients of expansion, and different thermal conductivities, thermal stresses are produced that may lead to fracture and spallation. The extent to which insolation variations can cause such fracturing is controversial, and the various arguments for and against its importance are reviewed by *Ryan* [1962] and *Ollier* [1969, p. 14].

There are numerous other possible physical weathering processes. They are reviewed by *Ollier* [1969, chapter 2] and include such processes as unloading, expansion during alteration, and wedging by dirt in fractures. Again, these disintegration processes will play only a minor role relative to eolian abrasion in counteracting the surface clogging effects of weathering rinds.

Eolian deflation. There is abundant observational evidence for an efficient debris segregation process that prevents fine-grained weathering products from accumulating in some regions on Mars. Strong winds apparently scour fine-grained high-albedo soil from the low-albedo regions and deposit it in the high-albedo regions. The debris segregation process was observed directly during the Mariner 9 mission when a planetwide dust storm blanketed most of the surface with high-albedo silt and clay-sized particles. At the end of the storm the equatorial and mid-latitude dark regions were totally obscured, but as the mission progressed, they gradually darkened. Examination of the Mariner 9 photographs confirmed earlier predictions [*Rea*, 1964; *Sagan and Pollack*, 1969; *Cutts et al.*, 1971] that the darkening was due to eolian deflation of the bright dust from the darker, more coarse grained underlying soil [*Sagan et al.*, 1972, 1973; *Cutts and Smith*, 1973]. The dust that deflates from these regions is deposited in large basins and other debris traps [*Murray et al.*, 1971; *Cutts*, 1973; *McCauley*, 1973; *Soderblom et al.*, 1973], and it apparently remains partitioned from the coarser-grained material until the next dust storm occurs.

Summary. The Martian surface environment should definitely support photostimulated oxidation of Fe^{2+} surface materials. The atmosphere contains about 0.13% O_2, and UV radiation penetrates to the surface nearly unattenuated at wavelengths as short as 0.195 μm. Relative humidities fall below the critical 5% level during a part of every day between latitudes of $\pm 50°$. Eolian abrasion and deflation insure access of surface Fe^{2+} ions to atmospheric O_2 and UV through removal of weathering material from the unweathered soils.

Fe^{2+} SOURCE MATERIALS

The spectral reflectance properties of the high- and low-albedo regions [*McCord and Westphal*, 1971; *McCord et al.*, 1971; also T. B. McCord, R. L. Huguenin, and J. B. Adams, manuscripts in preparation, 1974] indicate that the soils in bright areas have smaller mean particle diameters than those in the dark areas and that they contain significantly less Fe^{2+} and more Fe^{3+} than the dark-area soils [*Adams and McCord*, 1969; R. L. Huguenin, J. B. Adams, and T. B. McCord, manuscripts in preparation, 1974]. Laboratory modeling studies by *Adams and McCord* [1969] indicate that the dark-area soils are basaltic in composition, the principal Fe^{2+} phases being pyroxene and olivine along with some magnetite. The Little Lake basalt used in their study contained about 10–15% magnetite. *Binder and Jones* [1972] similarly conclude that the dark areas are richer in pyroxene or olivine or both than the bright areas. Subsequent analysis by *Huguenin et al.* [1974] indicates that the bright areas may be composed primarily of Fe^{2+}-depleted clay minerals plus minor goethite and that the dark areas contain essentially the only Fe^{2+} minerals.

MAGNETITE OXIDATION RATE

The magnetite oxidation kinetics presented by *Huguenin* [1973*a*] can be used to estimate the magnetite oxidation rate on Mars. Neglecting the effects of adsorbed H_2O, eolian abrasion, and particle size variations, substituting the average surface O_2 partial pressure, total atmospheric pressure, and yearly average daily illumination intensity into the kinetic rate equation of *Huguenin* [1973*a*], and multiplying the result by the fraction of the Martian year during which photostimulated oxidation can occur, one obtains a ferric oxide production rate of 10^{-4} μm yr^{-1} for a 5-mbar region at the equator. For a 2-mbar region at the equator the rate rises to 10^{-3} μm yr^{-1}. The rate at $\lambda = \pm 30°$ is essentially the same as it is at the equator, whereas at $\lambda = \pm 60°$ the rate drops to 10^{-5} μm yr^{-1}. With most of the low-albedo (Fe^{2+} source) regions located in the latitude range $-30° < \lambda < 30°$ and with their surface pressures ranging between 3 and 5 mbar [*Belton and Hunten*, 1971; *Hord et al.*, 1972], the yearly average magnetite oxidation rate would be 10^{-3}–10^{-4} μm yr^{-1}.

Effect of adsorbed H_2O on rate. At 300°K adsorbed H_2O has two principal effects on the oxidation rate: (1) during illumination it poisons the reaction by limiting access of the Fe^{2+} ions to atmospheric O_2, and (2) the interruption of the illumination interval with periods of H_2O adsorption increases the rate markedly [*Huguenin*, 1973*a*, *b*]. Thirty-second illumination intervals alternated with 1-min H_2O adsorption intervals reduced the total illumination time required to form a 1-μ thick Fe_2O_3 layer from $\Delta t_0 = 8.0$ ($\pm$0.4) min to 6.5 ($\pm$3) min, a 19% ($\pm$7) increase in oxidation rate.

In a subsequent experiment, not reported by *Huguenin* [1973*a*], the effect of adsorbed H_2O was further tested. The same illumination intensity, O_2 partial pressure, total atmospheric pressure, and temperature employed in the above experiment were used, and 10-s illumination intervals were alternated with 1-min H_2O adsorption intervals. The value of Δt_0 was reduced from 8.0 ($\pm$0.4) min to 5.2 ($\pm$0.3) min, a 35% ($\pm$4) increase in the photostimulated oxidation rate.

These new experimental results support the prediction of *Huguenin* [1973*b*] that the H_2O promotes migration of Fe^{2+} ions to the grain surface. It was predicted that the oxidation rate should increase by an amount proportional to the number of equal length adsorption intervals in Δt_0 since the rate is proportional to the surface density of Fe^{2+} ions. Together, the earlier and the newly reported data indicate that for T = 300°K and for adsorbed H_2O layer thicknesses in equilibrium with relative humidities of 5% the rate increases by 1% per adsorption interval in Δt_0.

These results cannot be directly applied to Mars, however, since (1) the H_2O reaches temperatures as low as 150°–190°K, (2) the adsorbed H_2O layers are much thicker [*Fanale and Cannon*, 1971, also unpublished data, 1974; *Balsamo and Salisbury*, 1973], and (3) the adsorption periods are of the order of hours to tens of hours in length [*Balsamo and Salisbury*, 1973]. The effects of the subzero temperatures and longer adsorption intervals are now discussed.

Cryogenic cation migration. It is well known that many chemical processes follow the Van't Hoff–Ostwald law (i.e., a drop in temperature by 10°K reduces the rate of the reaction by a factor of 2–3), and it is assumed by most authors [*Loughnan*, 1969, pp. 121, 167] that reactions between soil particles and H_2O would be either very retarded or arrested at temperatures below 273°K. This assumption is not correct for cation migration, however. In fact, there is abundant experimental and field evidence that cation migration under the action of adsorbed H_2O intensifies at low temperatures [*Tyutyunov*, 1964, chapter 1 and references therein]. Tyutyunov reports several studies in which it was demonstrated that mineralogical transformations of silicate dust suspended in laboratory and permafrost ice proceed more rapidly than those in water at 300°K and that the cations involved in these transformations migrate into the ice and form abundant dispersed colloids. The migration of these cations into the ice has been observed to set up spontaneous electrical potentials in excess of 1 V, the potentials increasing as the temperature decreases.

The migration of cations to the grain surface occurs under the action of an interaction potential Ψ_i exerted by hydroxyl ions in the H_2O film. Migration occurs for those cations that possess sufficient vibrational kinetic energies to overcome the migrational barrier potential β, i.e., for those with vibrational energies in excess of the activation energy

$$k = \beta - \Psi_i \tag{7}$$

Clearly, if β and Ψ_i remained constant during a drop in temperature, as they generally do in solution, the cation migration rate would follow the Van't Hoff–Ostwald law. For the adsorbent-adsorbate system, however, the interaction potential does not remain constant. Rather, it is proportional to the specific free energy (or surface tension) σ of the grain surface, which increases with a decrease in temperature,

$$\sigma = \sigma_0 - bT \tag{8}$$

where σ_0 is the specific surface free energy at absolute zero and b is a constant [*Swalin*, 1962, p. 180].

As σ increases, the H_2O becomes more strongly bound to the surface, more hydrogen bonds are broken, and the H_2O-cation system becomes more elastic. This increase in interaction potential with decrease in temperature is responsible for the observed increases in heat of sorption and thickness of adsorbed H_2O layers with decreasing temperature at constant H_2O partial pressure [*Dushman and Lafferty*, 1966, chapter 6 and references therein].

If it is assumed that the decrease in temperature does not significantly alter the migrational barrier potential, the increase in the interaction potential $\Delta\Psi_i$ decreases the migrational activation energy

$$k' = \beta - (\Psi_i + \Delta\Psi_i) \tag{9}$$

The decrease in activation energy apparently more than compensates for the decrease in vibrational kinetic energy with decrease in temperature for $T < 273°K$ [*Tyutyunov*, 1964, chapter 1].

For a sample of magnetite in the Martian surface environment it is therefore expected that the cation migration rates are greater than the migration rates in the illumination interruption experiments described above. Further, with the thicker H_2O layers and longer adsorption periods the number of Fe^{2+} ions available for photostimulated oxidation per square centimeter of grain surface area will be greater at the end of the Martian adsorption periods than at the end of the laboratory adsorption periods.

A first-order estimate of how many more cations would be available for photostimulated oxidation at the end of the Martian interval can be made on the basis of data presented by *Tyutyunov* [1964, chapter 1]. In one experiment involving the exchange of Ca^{2+} ions for Na^+ in feldspar suspended in an NaCl solution, it was shown that between 293° and 265°K the exchange rate increased by between 0 and 2% per degree drop in temperature. In another experiment it was shown that the spontaneous electrical potentials set up by cations migrating from soil particles into H_2O films were proportional to the decrease in temperature for temperatures between 323° and 223°K. The potential increased by approximately 3% per degree drop in temperature. Since spontaneous electrical potentials are proportional to the concentration of cations in the H_2O film, it can be concluded that between 323° and 223°K the cation migration rate is approximately inversely proportional to the temperature. The stability against cation migration for magnetite is somewhat greater than that for the sodic feldspars used in the above experiments, but the changes in migration rates with change in temperature should be very similar [*Tyutyunov*, 1964, chapter 1].

An upper limit on the change of migration rate with temperature is provided by assuming that the cation migration increases on the order of 1% per degree drop in temperature and that the migration continues on the order of 10 hours per adsorption interval without saturating the H_2O film. At 300°K each adsorption interval increases the photostimulated oxidation rate by 1%. For the upper limit case the migration rate would be a factor of 10^2 higher, and it would occur 10^3 times longer on Mars than in the laboratory; i.e., each adsorption interval would increase the magnetite oxidation rate by a factor of 10^3 on Mars.

The lower limit is provided by assuming that the decrease in temperature, thicker films, and longer adsorption intervals on Mars simply have no effect on the cation migration rate. In this case each adsorption interval would increase the oxidation rate by only about 1%.

Effects of eolian abrasion on rate. *Sagan*'s [1973] estimate of the erosion rate on Mars (~ 1 cm yr^{-1}) is orders of magnitude higher than the estimated photostimulated oxidation rate; consequently, alteration rinds will be prevented from forming on the surfaces of the unweathered soil particles. It is thus insured that the magnetite oxidation kinetics remain surface controlled at the low rates predicted for Mars and that the oxidation product will be extremely fine grained (of the order of tens of microns or less).

Such high eolian abrasion rates similarly insure the production of an extremely fine grained fraction of unweathered soils. Surface areas per gram of this fraction should be similar to those used in the laboratory kinetic studies [*Huguenin*, 1973*a*], and the oxidation of this fraction should dominate the rate.

Summary. The yearly average photostimulated oxidation rate of magnetite, if the effects of adsorbed H_2O are neglected, is of the order of 10^{-4} $\mu m\ yr^{-1}$ in the equatorial and mid-latitude dark areas. Since there are 10^2 adsorption intervals per year, however, the oxidation rate may be as much as 10^3 times higher. The effects of adsorbed H_2O introduce the greatest uncertainty in the oxidation rate, and the predicted oxidation rate is $10^{-2.5\pm1.5}$ $\mu m\ yr^{-1}$.

SILICATE ALTERATION

Previous work. The experimental work with magnetite demonstrates that the diurnal H_2O adsorption intervals promote the migration of Fe^{2+} ions to the grain surface. These surface Fe^{2+} ions are subsequently photooxidized to Fe_2O_3 upon exposure to UV illumination and atmospheric O_2. Clearly, the diurnal H_2O adsorption intervals should bring about the breakdown of ferrosilicates as well. It is well known that surface hydroxylation, penetration of H^+ ions into the crystal, and migration of cations to the grain surface under the action of adsorbed H_2O are a primary mechanism for the breakdown of most oxide and silicate structures [*Tamm*, 1930; *Stevens and Carron*, 1948; *Jenny*, 1950; *Frederickson*, 1951; *Garrels and Howard*, 1959; *Loughnan*, 1969, p. 28; *Carroll*, 1970, chapter 8].

Most silicates are far less resistant to breakdown by adsorbed H_2O than magnetite. The relative stabilities of the silicates and oxides to breakdown by H_2O under normal pH conditions ($4 < pH < 10$) generally follow Bowen's reaction series [*Goldich*, 1938; *Pettijohn*, 1941; *Reiche*, 1943; *Jackson et al.*, 1948, 1952; *Gruner*, 1950; *Keller*, 1954; *Hay*, 1959; *Bates*, 1962; *Loughnan*, 1962]. Glasses containing Fe^{2+} and other mobile cations generally display the least stability. Olivine and Ca feldspars are slightly more stable. Augite, diopside, hornblende, and the sodic and potash feldspars display moderate stability. Biotite and ilmenite generally display greater resistance to breakdown by adsorbed H_2O and alter at rates similar to those of magnetite for $4 < pH < 10$. Minerals such as muscovite, SiO_2, TiO_2, and Fe_2O_3 are very stable.

The cations that readily migrate to the grain surfaces of these minerals include Ca^{2+}, Mg^{2+}, Fe^{2+}, Mn^{2+}, K^+, Na^+, and other alkali and alkaline earths under normal pH conditions ($4 < pH < 10$) [*Stevens and Carron*, 1948; *Krauskopf*, 1956; *Anderson and Hawkes*, 1958; *Miller*, 1961; *Keller et al.*, 1963]. Those displaying only limited mobility under normal pH conditions include Ti^{4+} [*Sherman*, 1952*a*; *Bardossy*, 1959; *Craig and Loughnan*, 1964], Fe^{3+} [*Beadle and Burgess*, 1953; *Anderson and Hawkes*, 1958; *Miller*, 1961], Si^{4+} [*Krauskopf*, 1956, 1959; *Siever*, 1962, 1971], and Al^{3+} [*Sherman*, 1949, 1952*b*; *Hem*, 1959; *Hem and Roberson*, 1967].

Those Fe^{2+} ions that migrate to the grain surface and become oxidized generally form a separate ferric oxide phase rather than occupy sites in the parent mineral structures [*Beadle and Burgess*, 1953; *Keller*, 1954; *Loughnan*, 1969, p. 52; *Carroll*, 1970, p. 112]. When the Fe^{2+} ions are removed from substances such as volcanic glass [*Hay*, 1959], olivine [*Sherman et al.*, 1962; *Von Schellmann*, 1964; *Craig and Loughnan*, 1964], pyroxenes [*Craig and Loughnan*, 1964], hornblende [*Alexander et al.*, 1941], and biotite [*Walker*, 1949; *Stephen*, 1952; *MacEwan*, 1954; *Barshad*, 1954; *Bassett*, 1960; *M. J. Wilson*, 1966], their structures are generally rendered unstable, and they alter to the more stable secondary clays plus ferric oxide, whereas the removal of Fe^{2+} from titanomagnetite, ilmenite, and some titaniferous silicates generally yields TiO_2 as a residual weathering product [*Goldich*, 1938; *Sherman*, 1952*a*; *Loughnan and Golding*, 1957; *Craig and Loughnan*, 1964; *Loughnan*, 1969, pp. 80, 97].

The degree to which feldspars participate in the formation of alteration clays is uncertain. The highly mobile cations Ca^{2+}, Na^+, and K^+ can migrate over limited distances in the H_2O film, especially at low temperatures (see discussion on cryogenic cation migration above), and occupy sites left vacant by the cations that have undergone photostimulated oxidation. Such an exchange would render the surface layers of the feldspars unstable and alter them to the more stable secondary clays.

The types of clays that would be formed on Mars can be predicted from studies of clays formed in warm terrestrial deserts. Like the H_2O adsorption intervals on Mars, the evening dew and periodic rainstorms in the warm deserts promote cation migration to the grain surfaces [*Opdyke*, 1961; *Glennie*, 1970, p. 19]. The high temperatures and ensuing dry spells cause any water that manages to penetrate into the soils to evaporate. Thus removal of the soluble constituents is prevented. Under such mild leaching conditions all of the cations, except those tied up as oxides and salts, are incorporated into the alteration clays [*Beadle and Burgess*, 1953; *Engel and Sharp*, 1958; *Opdyke*, 1961; *Degens*, 1965, chapter 3; *Carroll*, 1970, chapter 8; *Blatt et al.*, 1972, chapter 15], and those that form salts during the dry spells can participate in cation exchange reactions with the clays during the wet spells.

The alteration clays formed under such mild leaching conditions generally have the chlorite, montmorillonite, and illite structures [*Alexander et al.*, 1941; *McKenzie et al.*, 1949; *Hseung and Jackson*, 1952; *Buol*, 1965; *Loughnan*, 1969, chapter 5, p. 121], although some of the feldspars [*DeVore*, 1959] and glasses [*Hay*, 1959] can produce kaolins as well. The composition of the clays depends on the composition of the parent materials and on the extent to which cations are tied up as oxides and salts. On Mars the alteration between intervals of H_2O adsorption and UV illumination occurs at such a high rate that the alteration clays should be strongly depleted in Fe^{2+}.

Preliminary experimental work on ferrosilicates. The effects of adsorbed H_2O on the magnetite oxidation rate (investigated by the author) and previous work by other investigators on the breakdown of silicates by adsorbed H_2O and oxidation (reviewed above) clearly indicate that if magnetite undergoes photostimulated oxidation on Mars, so do ferrosilicates. The studies indicate that ferrosilicate alteration rates will be greater than magnetite alteration rates.

These predictions are currently being investigated experimentally by the author. The kinetics and mechanisms of the production of Fe_2O_3 from ferrosilicates are being determined by following procedures similar to those used by *Huguenin* [1973*a*, *b*]. Several preliminary results are particularly relevant to the present paper.

1. The production of Fe_2O_3 from samples of fresh olivine and basaltic glass from Hawaii by photostimulated oxidation has been observed but only when the illumination interval is interrupted by periods of H_2O adsorption. The extent of alteration depended directly on the number and length of the adsorption intervals at 300°K. In particular, exposure of pulverized basaltic glass and olivine samples to uninterrupted high-intensity UV illumination in 100-torr atmospheres composed of up to 60% O_2 had no effect on the samples for exposure intervals of up to 120 hours. Interruption of the illumination interval with 10 12-hour H_2O adsorption intervals (H_2O-saturated N_2 atmosphere at 300°K) resulted in a reddening of the powdered (40 μm) basaltic glass after 12 hours of total illumination time in an environment for which $\Delta t_0 = 0.8$ min for magnetite [*Huguenin*, 1973*a*]. Similar results were found for olivine.

2. The extent of alteration was not directly proportional to the number and length of the H_2O adsorption intervals, however. Increasing the number of 12-hour adsorption intervals from 10 to 15 decreased the reddening time (Δt_0 from *Huguenin* [1973*a*]) to 10 hours, 20 min, whereas increasing to 20 intervals reduced Δt_0 to 10 hours, 5 min. Decreasing the length of the H_2O adsorption intervals to 6 hours had no appreciable effect on Δt_0.

3. The rate decrease was partially counteracted by grinding the samples after each adsorption interval. After each UV illumination interval the pulverized basaltic glass sample was ground in a mortar for 5 min. It was then placed in the H_2O adsorption environment. With the addition of the grinding intervals, Δt_0 decreased from 12 hours to 11 hours, 15 min, for 10 12-hour adsorption intervals. The value of Δt_0 decreased from 10 hours, 20 min, to 9 hours, 30 min, for 15 adsorption intervals, and it dropped from 10 hours, 5 min, to 8 hours, 50 min, for 20 adsorption intervals. By interrupting the illumination interval with two 5-min grinding intervals, Δt_0 was noticeably decreased, whereas doubling the length of the grinding intervals had no appreciable effect on Δt_0.

Discussion of laboratory results. These laboratory results, although still preliminary, indicate that ferrosilicates can be decomposed upon exposure to UV illumination, atmospheric O_2, intervals of H_2O adsorption, and abrasion. Each of these constituents is necessary for the decomposition of the silicates to proceed, whereas only the UV and O_2 are necessary for the oxidation of magnetite.

Several investigators have noted that in closed systems, where decomposition products are allowed to accumulate, the alteration of silicates can proceed only if they are continually subjected to grinding [*Tamm*, 1930; *Kelley and Jenny*, 1936; *Laws and Page*, 1946; *Perkins*, 1948; *McClelland*, 1950; *Takahashi*, 1959]. They attributed the rate decrease to surface clogging effects, ferric oxide apparently being far more effective as a surface clogging agent than most of the other decomposition products [*Dion*, 1944; *Fripiat and Gastuche*, 1952]. The fact that the rate decrease in my laboratory work was more sensitive to the frequency of grinding than to the total amount of grinding suggests that it was probably also due primarily to a surface clogging effect.

Summary. All the necessary constituents for the breakdown of ferrosilicates to ferric oxides and clay minerals are present on Mars (i.e., UV illumination, atmospheric O_2, intense eolian abrasion, and a high frequency of H_2O adsorption intervals ($\sim 10^2$ year^{-1})). The breakdown of ferrosilicates should occur more rapidly than the alteration of magnetite on Mars. The laboratory modeling studies of *Adams and McCord* [1969], discussed above, have shown that magnetite comprises of the order of only 10% of the dark-area soils. From the depth of the Fe^{2+} absorption features in the dark-area reflectance spectra [*McCord and Westphal*, 1971; *McCord et al.*, 1971] an FeO content of the order of 10% in the ferrosilicates (olivine and pyroxene) is inferred (R. L. Huguenin, T. B. McCord, and J. B. Adams, manuscript in preparation, 1974). Consequently, the production rates of Fe_2O_3 from the ferrosilicates are predicted to be similar to or greater than the production rate of Fe_2O_3 from magnetite ($\sim 10^{-2.5 \pm 1.5} \mu$m yr^{-1}). The production rate of clay minerals would be at least an order of magnitude higher than the Fe_2O_3 production rate from magnetite (i.e., $\sim 10^{-1.5 \pm 1.5}$ μm yr^{-1}), if a ferrosilicate FeO content of 10 wt % is assumed.

PREDICTED COMPOSITION, PARTICLE SIZE, AND VOLUME OF ALTERATION PRODUCTS

Composition. The photostimulated oxidation weathering products are predicted to be composed primarily of Fe^{2+}-depleted clay minerals with a minor amount of ferric oxide. The rapid alteration between H_2O adsorption and photostimulated oxidation ($\sim 10^2$ cycles yr^{-1}) insures that the Fe^{2+} depletion is extensive.

The depletion of Fe^{2+} may be accompanied by the depletion of other oxidizable cations, including Ti^{3+}, Mn^{2+}, V^{2+}, Cr^{2+}, Co^{2+}, Ni^{3+}, and Ni^{2+}. These are found primarily in oxides and in those silicates in which Al^{3+} has substituted for Si^{4+} [*Verhoogen*, 1962]. Each displays relatively high mobility, and in most terrestrial weathering environments they readily oxidize upon exposure to dissolved oxygen. The photoelectric work functions of these cations should all be low enough to undergo photostimulated oxidation on Mars. Their work functions in the common minerals have not been measured, but they can be estimated from the metallic work functions. In general, the work functions of transition metal cations in oxygen sites are lower than the metallic work functions by about 0.2–0.6 eV [*Burshtein and Shurmovskaya*, 1964]. For example, the measured work function of metallic Ti is 4.33 ± 0.1 eV [*Eastman*, 1970], and therefore the work function of Ti^{3+} in an oxygen environment should be of the order of 3.9 ± 0.3 eV. This is consistent with the work function measurements of oxidized Ti films by *Schulze* [1934] and *R. G. Wilson* [1966], who obtained values of 3.95 and 3.6 eV, respectively. Similarly, the work functions of the other cations, based on the metallic work functions [*Eastman*, 1970] and confirmed by the early work function measurements using oxidized films [*Sommer*, 1968, p. 20], are $\phi(Mn^{2+}) = 3.7 \pm 0.4$ eV, $\phi(V^{2+}) = 3.8 \pm 0.3$ eV, $\phi(Cr^{2+}) = 3.9 \pm 0.3$ eV, $\phi(Co^{2+}) = 4.3 \pm 0.3$ eV, and $\phi(Ni^{2+}, Ni^{3+}) = 4.6 \pm 0.4$ eV. Like Fe^{2+}, each of these cations should undergo photostimulated oxidation on Mars since the solar radiation penetrates to the surface nearly unattenuated at energies as high as 6.3 eV. It is expected that these oxides would be only minor or trace constituents in the weathered surface materials, as they are in most weathered terrestrial soils.

It should be noted that Ca and Mg should also be depleted from the source silicates through the formation of carbonates. The diurnal desorption of H_2O exposes Ca^{2+} and Mg^{2+} to atmospheric CO_2; therefore Ca-Mg carbonates should be formed in addition to the photostimulated oxidation weathering products [*O'Connor*, 1968]. Other salts may be produced by circulating groundwater [*Malin*, 1973; *Murray*, 1973; *Soderblom*, 1973]. Further laboratory work is required to predict the extent to which carbonates and other salts have formed on Mars.

The depletion of Fe^{2+} and, to a lesser extent, of these other cations from the source silicates increases the relative SiO_2 content of the residual silicates. With olivine, pyroxene, and probably some basaltic glass assumed to be the source silicates, alteration clays with SiO_2 contents of 60% or even higher would be expected.

The relative metal abundances of the bulk weathering product should be the same as those of the bulk source material. The oxygen content of the weathering products will be greater, however, as a result of photostimulated oxidation. The OH and H_2O contents of the weathering products will be greater as well. Upon exposure to H_2O, the surface of Fe_2O_3 becomes fully and rapidly hydroxylated [*Zettlemoyer et al.*, 1966; *McCafferty et al.*, 1970], this hydroxylation being followed by the formation of a molecular H_2O layer that is rigidly bonded to the hydroxyl layer. The H_2O is ordered and retained with a force of the order of 10^6 atm by the Fe_2O_3 surface potential [*Harkins and Ewing*, 1921; *Hennicker*, 1949; *Kitchener*, 1963; *Adamson*, 1967, 1968; *Somorjai and Szalkowski*, 1971], which is so strong that the activation energy for dipole rotation at 300°K is about 23 kcal mol^{-1} [*McCafferty et al.*, 1970]. The induced packing and ordering destroy bonds between and within the H_2O molecules, some OH^- ions bonding to surface cations and some H^+ ions penetrating into the crystal lattice [*Zettlemoyer and McCafferty*, 1969]. The rapid alteration between intervals of H_2O adsorption and photostimulated oxidation should result in the formation of goethite rather than hematite on Mars. The clay minerals will similarly be hydroxylated and hydrated.

Particle size. Although the FeOOH formation rate is predicted to be of the order of $10^{-2.5\pm1.5}$ μm yr^{-1} and the clay mineral formation rate is $\sim10^{-1.5\pm1.5}$ μm yr^{-1}, the eolian abrasion rate is estimated to be orders of magnitude higher ($\sim$1 cm yr^{-1}, according to *Sagan* [1973]). Under these conditions the relatively soft FeOOH and clay mineral alteration surface layers would be removed as fast as they are formed. The weathering products would consequently be colloidal in particle size. Owing to the presence of adsorbed H_2O, the individual particles would be bonded and ordered into larger aggregates, but eolian abrasion would probably limit the size of the aggregates on the surface to tens of microns or less.

Volume. The volume of weathering products can be estimated from the derived magnetite and silicate alteration rates, if it is assumed (1) that the dark-area soils are the principal source material for photostimulated oxidation weathering and that they are basaltic in composition, as is indicated by reflectance spectra; (2) that the dark areas have occupied one-fourth of the planetary surface area or greater for the past 10^9 years; and (3) that the surface environment has been the same as it is today for the past 10^9 years. Under these conditions the weathering products would form a surface layer of the order of $10^{1.5\pm1.5}$ m in mean global thickness.

Summary. The alteration products are predicted to be composed primarily of hydrated and hydroxylated clay minerals along with minor amounts of goethite (FeOOH). Carbonates, minor or trace amounts of TiO_2, MnO_2, Cr_2O_3, and other transition metal oxides may also be formed. The clay minerals are predicted to be more silica rich than the silicate source material, SiO_2 contents of 60% or higher being expected. Although the relative metal abundances of the bulk weathering product should be the same as the metal abundances of the bulk source material, the oxygen, OH, and H_2O contents will be greater. Aggregates should have particle sizes of tens of microns or less, and the mean global thickness of the weathering products is estimated to be $10^{1.5\pm1.5}$ m.

COMPARISON OF PREDICTIONS WITH OBSERVATIONS

Composition. Evidence that the soils in the bright areas contain abundant alteration phases comes from reflectance spectra and the Mariner 9 Iris data. *Adams and McCord* [1969] deduce Fe_2O_3 contents of the order of 5 wt % and FeO contents much lower than those of the soils in the dark areas. *Huguenin et al.* [1974] indicate that the bright-area spectra are consistent with the soils' being composed primarily of strongly Fe^{2+} depleted or Fe^{2+} free clay minerals plus minor goethite. *Binder and Jones* [1972] similarly deduce much lower relative FeO contents in the high-albedo soils. *Houck et al.* [1973] and *Pimentel et al.* [1974] deduce the presence of widespread hydrated minerals on the surface from 2- to 4-μm reflectance spectra. *Hunt et al.* [1973] deduce from the position and shape of the restrahlen band in the Mariner 9 Iris spectra that the dust raised from the surface into the planetwide dust storm of 1971 was composed primarily of the clay mineral montmorillonite. *Hanel et al.* [1972] deduce an SiO_2 content of 60 ± 10 wt % from the Iris spectra.

Particle size. Independent observational evidence from a wide variety of sources indicates that the particle diameter of the proposed weathered bright-area soils is much less than 100 μm [*Adams and McCord*, 1969; *Egan*, 1969; *Pollack and Sagan*, 1969; *Leovy et al.*, 1972; *Sagan et al.*, 1972, 1973; *Conrath et al.*, 1973; *Hunt et al.*, 1973; *Kieffer et al.*, 1973]. *Conrath et al.* [1973] deduce particle sizes as small as 1–10 μm. In contrast, the low-albedo soils apparently have mean particle diameters in excess of 200 μm [*Adams and McCord*, 1969; *Pollack and Sagan*, 1969; *Sagan and Pollack*, 1969; *Sagan et al.*, 1972, 1973; *Cutts and Smith*, 1973; *Kieffer et al.*, 1973].

Volume. Estimates of the thicknesses of the eolian deposits in the high-albedo regions have been made by *Masursky* [1973] and *Soderblom et al.* [1973]. The eolian deposits in the polar regions average on the order of tens to hundreds of meters in thickness, although in some debris traps the thickness may be as great as 1 km. In the equatorial and mid-latitude high-albedo regions the debris probably does not exceed 1–10 m in thickness, except in some local debris traps. These estimates are in good agreement with the volume predicted by the photostimulated oxidation model.

SUMMARY AND CONCLUSIONS

Origin of high-albedo soils. It is proposed that the high-albedo soils on Mars are photostimulated oxidation weathering products of the dark-area soils. Primarily on the basis of the kinetics and mechanism of the photostimulated oxidation of magnetite, presented by *Huguenin* [1973*a*, *b*], along with some preliminary experimental work on silicates described herein, the photostimulated oxidation rates of magnetite and ferrosilicates on Mars have been derived. Upon exposure to UV illumination, atmospheric O_2, adsorbed H_2O, and eolian abrasion, ferrosilicates such as olivine, pyroxene, and basaltic glass alter at rates of the order of $10^{-1.5\pm1.5}$ μm yr^{-1}, whereas magnetite alters at a rate of $10^{-2.5\pm1.5}$ μm yr^{-1}. If a basaltic composition for the dark-area soils is assumed and if it is also assumed that the environment has been the same as it is today for the past 10^9 years, a surface layer of alteration products $10^{1.5\pm1.5}$ m in mean global thickness is predicted. This value is in good agreement with volume estimates by *Masursky* [1973] and *Soderblom et al.* [1973] from Mariner 9 photographs. It can be concluded that models accounting for the presence of the high-albedo soils as artifacts from an earlier, more oxidizing epoch are unnecessary, although they are not excluded.

Composition and particle size. The weathering products of the dark-area soils are predicted to be composed primarily of hydroxylated and hydrated clay minerals along with minor amounts of goethite and traces of TiO_2, MnO_2, Cr_2O_3, and other transition metal oxides. Some carbonates are also expected to be formed. The clay minerals are predicted to be more silica rich than the silicate source material, SiO_2 contents of 60% or higher being expected. The oxygen, OH, and H_2O contents of the bulk weathering product are predicted to be significantly greater than those in the dark-area soils, whereas the metal abundances should be the same. The clays, oxides, and carbonates should be in the form of aggregates bound by adsorbed H_2O with diameters of the order of tens of microns or less. Within the aggregates, constituent particles should be colloidal in size. A variety of independent observational evidence supports the composition and particle size estimates.

It is not expected that all of the soils in the high-albedo regions are photostimulated oxidation weathering products. Owing to the intense eolian abrasion, it is expected that some mobile fine-grained unweathered soils deflate from the dark areas along with the weathering products. Of course, they would not have the extremely small particle diameters of the weathering products and may be sorted from them. Evidence for such mobile low-albedo soils has been discussed by *Sagan et al.* [1973].

Acknowledgments. This paper and two earlier papers [*Huguenin*, 1973*a*, *b*] are a three-part publication of my doctoral dissertation, completed in 1972 at the Massachusetts Institute of Technology (MIT); consequently, the acknowledgments presented herein apply to all three papers. The dissertation was an outgrowth of a pilot study carried out by myself and Bruce Hapke in 1969 while I was an undergraduate student at the University of Pittsburgh. The pilot study, formulated by Hapke, was designed to test a speculation that iron-bearing minerals on the surface of Mars may be oxidized by ozone as a photodissociation product of CO_2. We placed samples of magnetite in CO_2 atmospheres and subjected them to intense UV visible IR radiation from a 200-W Oriel xenon-mercury lamp, and we discovered that the magnetite turned red. We assumed that the reddening phenomenon was due to the oxidation of the magnetite, but the CO_2 was contaminated by unknown amounts of air, and the nature of the phenomenon could not be determined. In 1969 I went to MIT to become a graduate student, and Hapke and his associates studied the phenomenon further, using Apollo 11 samples. They proposed a mechanism, which they published in 1970 and which was discussed by *Huguenin* [1973*a*, *b*]. In 1971 I decided to redesign the experiment to study the kinetics of the phenomenon and determine its mechanism. These results, along with their application to Mars, were the subject of my dissertation. I wish to acknowledge Hapke's primary rôle in the discovery of the reddening phenomenon. I wish to thank Thomas McCord, my thesis advisor, for his guidance, encouragement, many useful discussions, and his sponsorship of the experiment. I would also like to acknowledge useful discussions with Jim Westphal (California Institute of Technology) and Grant Snellen and Paul Gabriel (MIT) concerning the design of the apparatus. I would further like to thank John Lewis and Ronald Prinn (MIT) for their valuable discussions concerning the derivation of the mechanism. Finally, I would like to acknowledge many useful comments by John Adams and Roger Burns (MIT), Walter Egan (Grumann), James Cutts, Torrence Johnson and Fraser Fanale (Jet Propulsion Laboratory), Bill Hartmann (Planetary Science Institute), Thomas McGetchin (MIT), Carl Sagan (Cornell), John Salisbury (Air Force Cambridge Research Laboratories), Ray Siever (Harvard), and John Wood (Smithsonian Astrophysical Observatory). This work and the work by *Huguenin* [1973*a*, *b*] were supported by NASA grants #NGR 22-009-473 and NGL 22-009-350. Contribution 97, Planetary Astronomy Laboratory, Massachusetts Institute of Technology.

REFERENCES

Adams, J. B., Lunar and Martian surfaces: Petrological significance of absorption bands in the near-infrared, *Science, 159,* 1453, 1968.

Adams, J. B., and T. B. McCord, Mars: Interpretation of the spectral reflectivity of light and dark regions, *J. Geophys. Res., 74,* 4851, 1969.

Adamson, A. W., Physical adsorption—A tool in the study of the frontiers of matter, *J. Chem. Educ., 44,* 710, 1967.

Adamson, A. W., Physical adsorption of vapors on ice, 1, Nitrogen, *J. Colloid Interface Sci., 27,* 180, 1968.

Alexander, L. T., S. B. Hendricks, and G. T. Faust, Occurrence of gibbsite in some soil forming minerals, *Soil Sci. Soc. Amer. Proc., 6,* 52, 1941.

Allen, C. W., *Astrophysical Quantities,* p. 172, Athlone, London, 1955.

Anderson, D. H., and H. E. Hawkes, Relative solubility of the common elements in weathering of some schist and granite areas, *Geochim. Cosmochim. Acta, 14,* 204, 1958.

Balsamo, S. R., and J. W. Salisbury, Slope angle and frost formation on Mars, *Icarus, 18,* 156, 1973.

Bardossy, G., The geochemistry of Hungarian bauxites, *Acta Geol. Acad. Sci. Hung., 6,* 1, 1959.

Barker, E. S., Detection of O_2 in the Martian atmosphere with the eschelle-coude scanner of the 107-inch telescope, *Bull. Amer. Astron. Soc., 4,* 371, 1972.

Barker, E. S., Atmospheric water: Ground based observations, paper presented at Symposium on Water on Mars, NASA, Harvard Univ., Cambridge, Mass., November 2, 1973.

Barker, E. S., R. A. Schorn, A. Woszcyk, R. G. Tull, and S. J. Little, Mars: Detection of water vapor during southern hemisphere spring and summer seasons, *Science, 170,* 1308, 1970.

Barshad, I., Cation exchange in micaceous minerals, 2, Replaceability of ammonium and potassium from vermiculite, biotite, and montmorillonite, *Soil Sci., 78,* 57, 1954.

Bassett, W. A., Role of hydroxyl orientation in mica alteration, *Geol. Soc. Amer. Bull., 71,* 449, 1960.

Bates, T. F., Halloysite and gibbsite formation in Hawaii, *Proc. Nat. Conf. Clays Clay Miner., 9,* 307, 1962.

Beadle, N. C. W., and A. Burgess, Further note on laterites, *Aust. J. Sci., 15,* 170, 1953.

Belton, M. J. S., and D. M. Hunten, A search for O_2 on Mars and Venus: A possible detection of oxygen in the atmosphere of Mars, *Astrophys. J., 153,* 963, 1968.

Belton, M. J. S., and D. M. Hunten, The detection of CO_2 on Mars: A spectroscopic determination of surface topography, *Icarus, 15,* 204, 1971.

Belton, M. J. S., A. L. Broadfoot, and D. M. Hunten, Abundance and

temperature of CO_2 on Mars during the 1967 opposition, *J. Geophys. Res.*, *73*, 4795, 1968.
Binder, A. B., and D. P. Cruikshank, Comparison of the infrared spectrum of Mars with the spectra of selected terrestrial rocks and minerals, *Commun. Lunar Planet. Lab.*, *2*, 193, 1963.
Binder, A. B., and D. P. Cruikshank, Lithological and morphological investigation of the surface of Mars, *Icarus*, *5*, 521, 1966.
Binder, A. B., and J. C. Jones, Spectrophotometric studies of the photometric function, composition, and distribution of the surface materials of Mars, *J. Geophys. Res.*, *77*, 3003, 1972.
Blatt, H., G. Middleton, and R. Murray, *Origin of Sedimentary Rocks*, 634 pp., Prentice-Hall, Englewood Cliffs, N. J., 1972.
Buol, S. W., Present soil forming factors and processes in arid and semiarid regions, *Soil Sci.*, *99*, 45, 1965.
Burshtein, R. Kh., and N. A. Shurmovskaya, The effect of electronegative gases on the work function of a metal, *Surface Sci.*, *2*, 210, 1964.
Carleton, N. P., and W. A. Traub, Detection of molecular oxygen on Mars, *Science*, *177*, 988, 1972.
Carleton, N. P., A. Sharma, R. M. Goody, W. L. Liller, and F. L. Roesler, Measurement of the abundance of CO_2 in the Martian atmosphere, *Astrophys. J.*, *155*, 323, 1969.
Carroll, D., *Rock Weathering*, 203 pp., Plenum, New York, 1970.
Coleman, J. M., S. M. Gagliano, and W. G. Smith, Chemical and physical weathering on saline high tidal flats, northern Queensland, Australia, *Geol. Soc. Amer. Bull.*, *77*, 205, 1966.
Conrath, B., R. Curran, R. Hanel, V. Kunde, W. Maguire, J. Pearl, J. Pirraglia, and J. Welker, Atmospheric and surface properties of Mars obtained by infrared spectroscopy on Mariner 9, *J. Geophys. Res.*, *78*, 4267, 1973.
Craig, D. C., and F. C. Loughnan, Chemical and mineralogical transformations accompanying the weathering of basic volcanic rocks from New South Wales, *Aust. J. Soil Res.*, *2*, 218, 1964.
Cutts, J. A., Wind erosion in the Martian polar regions, *J. Geophys. Res.*, *78*, 4211, 1973.
Cutts, J. A., and R. S. U. Smith, Eolian deposits and dunes on Mars, *J. Geophys. Res.*, *78*, 4139, 1973.
Cutts, J. A., L. A. Soderblom, R. P. Sharp, B. A. Smith, and B. C. Murray, The surface of Mars, 3, Light and dark markings, *J. Geophys. Res.*, *76*, 343, 1971.
Degens, E. T., *Geochemistry of Sediments*, 342 pp., Prentice-Hall, Englewood Cliffs, N. J., 1965.
DeVore, G. W., The surface chemistry of feldspars as an influence on their decomposition products, *Proc. Nat. Conf. Clays Clay Miner.*, *6*, 26, 1959.
Dion, H. G., Iron oxide removal from clays and its influence on base exchange properties and X-ray diffraction patterns of the clays, *Soil Sci.*, *58*, 411, 1944.
Dollfus, A., Propriétés photometrique des contrées desertique sur la planète Mars, *C. R. Acad. Sci.*, *244*, 162, 1957.
Dollfus, A., Polarization studies of planets, in *Planets and Satellites*, vol. 3, *The Solar System*, edited by G. P. Kuiper and B. M. Middlehurst, p. 379, University of Chicago Press, Chicago, Ill., 1961.
Draper, A. L., J. A. Adamcik, and E. K. Gibson, Comparison of the spectra of Mars and a goethite-hematite mixture in the 1 to 2 micron region, *Icarus*, *3*, 63, 1964.
Dushman, S., and J. M. Lafferty, *Scientific Foundations of Vacuum Technique*, 806 pp., John Wiley, New York, 1966.
Eastman, D. E., Photoelectric work function of transition, rare-earth, and noble metals, *Phys. Rev., Sect. B*, *2*, 1, 1970.
Egan, W. G., Polarimetric and photometric simulation of the Martian surface, *Icarus*, *10*, 223, 1969.
Engel, C. G., and R. P. Sharp, Chemical data on desert varnish, *Geol. Soc. Amer. Bull.*, *69*, 487, 1958.
Fanale, F. P., and W. A. Cannon, Adsorption on the Martian regolith, *Nature*, *230*, 502, 1971.
Farmer, C. B., Possibilities for liquid water and expectations for Viking '75, paper presented at Symposium on Water on Mars, NASA, Harvard Univ., Cambridge, Mass., Nov. 2, 1973.
Frederickson, A. F., Mechanism of weathering, *Geol. Soc. Amer. Bull.*, *62*, 221, 1951.
Fripiat, J. J., and M. Gastuche, Etude physico-chimique des surfaces des argiles, *Inst. Nat. Agron. Congo Belge, Ser. Sci.*, *54*, 1, 1952.
Garrels, R. M., and P. Howard, Reactions of feldspars and mica with water at low temperature and pressures, *Proc. Nat. Conf. Clays Clay Miner.*, *6*, 68, 1959.
Glennie, K. W., *Desert Sedimentary Environments*, p. 19, Elsevier, New York, 1970.
Goldich, S. S., A study of rock weathering, *J. Geol.*, *46*, 17, 1938.
Gruner, J. W., An attempt to arrange silicates in the order of reaction energies at relatively low temperatures, *Amer. Mineral.*, *35*, 137, 1950.
Hanel, R., B. Conrath, W. Hovis, W. Kunde, P. Lowman, W. Maguire, J. Pearl, J. Pirraglia, C. Prabhakara, B. Schlachman, G. Levin, P. Straat, and T. Burke, Investigation of the Martian environment by infrared spectroscopy on Mariner 9, *Icarus*, *17*, 423, 1972.
Harkins, W. D., and D. T. Ewing, A high-pressure due to adsorption and the density and volume relations of charcoal, *J. Amer. Chem. Soc.*, *23*, 1787, 1921.
Hay, R. L., Origin and weathering of late Pleistocene ash deposits on St. Vincent, B. W. I., *J. Geol.*, *67*, 65, 1959.
Hem, J. D., Study and interpretation of the chemical characteristics of natural waters, *U.S. Geol. Surv. Water Supply Pap.*, *1473*, 1, 1959.
Hem, J. D., and C. E. Roberson, Form and stability of aluminum hydroxide complexes in dilute solution, *U.S. Geol. Surv. Water Supply Pap.*, *1827-A*, 1, 1967.
Hennicker, J. C., Depths of the surface zone of a liquid, *Rev. Mod. Phys.*, *21*, 322, 1949.
Hord, C. W., C. A. Barth, A. I. Stewart, and A. L. Lane, Mariner 9 ultraviolet spectrometer experiment: Photometry and topography of Mars, *Icarus*, *17*, 443, 1972.
Houck, J. R., J. B. Pollack, C. Sagan, D. Schaack, and J. A. Decker, Jr., High altitude infrared spectroscopic evidence for bound water on Mars, *Icarus*, *18*, 470, 1973.
Hseung, Y., and M. L. Jackson, Mineral composition of the clay fraction of some main soil groups of China, *Soil Sci. Soc. Amer. Proc.*, *16*, 294, 1952.
Huguenin, R. L., Photostimulated oxidation of magnetite, 1, Kinetics and alteration phase identification, *J. Geophys. Res.*, *78*, 8481, 1973*a*.
Huguenin, R. L., Photostimulated oxidation of magnetite, 2, Mechanism, *J. Geophys. Res.*, *78*, 8495, 1973*b*.
Huguenin, R. L., T. B. McCord, and J. B. Adams, Mars: Origin of the high-albedo materials, paper presented at 55th Annual Meeting, AGU, Washington, D. C., April 8–12, 1974.
Hunt, G. R., L. M. Logan, and J. W. Salisbury, Mars: Components of infrared spectra and the composition of the dust cloud, *Icarus*, *18*, 459, 1973.
Hunten, D. M., Composition and structure of planetary atmospheres, *Space Sci. Rev.*, *12*, 539, 1971.
Jackson, M. L., S. A. Tyler, A. L. Willis, G. A. Bourbeau, and R. P. Pennington, Weathering sequence of clay-size minerals in soils and sediments, 1, Fundamental generalizations, *J. Phys. Colloid Chem.*, *52*, 1237, 1948.
Jackson, M. L., Y. Hseung, R. B. Corey, E. J. Evans, and R. C. Heuval, Weathering sequence of clay-size minerals in soils and sediments, 2, Chemical weathering of layer silicates, *Soil Sci. Soc. Amer. Proc.*, *16*, 3, 1952.
Jenny, H., Origin of soils, in *Applied Sedimentation*, edited by P. D. Trask, p. 41, John Wiley, New York, 1950.
Kaplan, L. D., G. Münch, and H. Spinrad, An analysis of the spectrum of Mars, *Astrophys. J.*, *139*, 1, 1964.
Keller, W. D., Bonding energies of some silicate minerals, *Amer. Mineral.*, *39*, 783, 1954.
Keller, W. D., W. D. Balgord, and A. L. Reesman, Dissolved products of artificially pulverized silicate minerals and rocks, 1, *J. Sediment. Petrology*, *33*, 191, 1963.
Kelley, W. C., and J. H. Zumberge, Weathering of a quartz diorite at Marble Point, McMurdo Sound, Antarctica, *J. Geol.*; *69*, 433, 1961.
Kelley, W. P., and H. Jenny, The relation of crystal structure to base-exchange and its bearing on base-exchange in soils, *Soil Sci.*, *41*, 367, 1936.
Kieffer, H. H., S. C. Chase, Jr., E. Miner, G. Münch, and G. Neugebauer, Preliminary report on infrared radiometric measurements from the Mariner 9 spacecraft, *J. Geophys. Res.*, *78*, 4291, 1973.
Kitchener, J. A., Surface forces in thin liquid films, *Endeavour*, *22*, 118, 1963.
Krauskopf, K. B., Dissolution and precipitation of silica at low temperatures, *Geochim. Cosmochim. Acta*, *10*, 1, 1956.
Krauskopf, K. B., The geochemistry of silica in sedimentary environments: Silica in sediments, *Soc. Econ. Paleontol. Mineral Spec. Publ.*, *7*, 4, 1959.
Laws, W. D., and J. B. Page, Changes produced in kaolinite by dry grinding, *Soil Sci.*, *62*, 319, 1946.

Leighton, R. B., N. H. Horowitz, B. C. Murray, R. P. Sharp, A. H. Herriman, A. T. Young, B. A. Smith, M. E. Davies, and C. B. Leovy, Television observations from Mariners 6 and 7, *Science, 166,* 49, 1969.

Leovy, C. B., G. A. Briggs, A. T. Young, B. A. Smith, J. B. Pollack, E. N. Shipley, and R. L. Wildey, Mariner 9 TV experiment: Progress report on studies of Mars atmosphere, *Icarus, 17,* 373, 1972.

Loughnan, F. C., Some considerations in the weathering of the silicate minerals, *J. Sediment. Petrology, 32,* 289, 1962.

Loughnan, F. C., *Chemical Weathering of the Silicate Minerals,* 154 pp., Elsevier, New York, 1969.

Loughnan, F. C., and H. G. Golding, The mineralogy of the commercial dike clays in the Sydney district, *J. Proc. Roy. Soc. N. S. W., 91,* 85, 1957.

MacEwan, D. M. C., 'Cardenite,' a trioctahedral montmorillonite derived from biotite, *Clay Miner. Bull., 2,* 120, 1954.

Malin, M. C., Salt weathering on Mars, paper presented at 3rd Annual Meeting, Div. of Planet. Sci., Amer. Astron. Soc., Tucson, Ariz., March 20–23, 1973.

Masursky, H., An overview of geological results from Mariner 9, *J. Geophys. Res., 78,* 4009, 1973.

McCafferty, E., V. Pravdic, and A. C. Zettlemoyer, Dielectric behavior of adsorbed water films on the α-Fe_2O_3 surface, *Trans. Faraday Soc., 66,* 1720, 1970.

McCauley, J. F., Mariner 9 evidence for wind erosion in the equatorial and mid-latitude regions of Mars, *J. Geophys. Res., 78,* 4123, 1973.

McClelland, J. E., The effect of time, temperature, and particle size on the release of bases from some common soil forming minerals of different crystal structures, *Soil Sci. Soc. Amer. Proc., 15,* 302, 1950.

McCord, T. B., and J. A. Westphal, Mars: Narrow-band photometry, from 0.3 to 2.5 microns, of surface regions during the 1969 apparition, *Astrophys. J., 168,* 141, 1971.

McCord, T. B., J. H. Elias, and J. A. Westphal, Mars: The spectral albedo (0.3–2.5μ) of small bright and dark regions, *Icarus, 14,* 245, 1971.

McKenzie, R. C., G. F. Walker, and R. Hart, Illite occurring in decomposed granite at Ballater, Aberdeenshire, *Mineral. Mag., 28,* 704, 1949.

Miller, J. P., Solutes in small streams draining single rock types, Sangre de Cristo range, New Mexico, *U.S. Geol. Surv. Water Supply Pap., 1535-F,* 1, 1961.

Moroz, V. I., The infrared spectrum of Mars (1.1–4.1 μm), *Astron. Zh., 41,* 350, 1964.

Morrison, D., C. Sagan, and J. B. Pollack, Martian temperatures and thermal properties, *Icarus, 11,* 36, 1969.

Murray, B. C., Martian uniqueness, paper presented at the International Colloquium on Mars, Jet Propul. Lab., Pasadena, Calif., Nov. 28–Dec. 1, 1973.

Murray, B. C., L. A. Soderblom, R. P. Sharp, and J. A. Cutts, The surface of Mars, 1, Cratered terrains, *J. Geophys. Res., 76,* 313, 1971.

O'Connor, J. T., Mineral stability at the Martian surface, *J. Geophys. Res., 73,* 5301, 1968.

Ollier, C., *Weathering,* 304 pp., Elsevier, New York, 1969.

Opdyke, N. D., The paleoclimatological significance of desert sandstone, in *Descriptive Paleoclimatology,* edited by A. E. M. Nairn, p. 45, Interscience, New York, 1961.

Perkins, A. T., Kaolin and treated kaolins and their reactions, *Soil Sci., 65,* 185, 1948.

Pettijohn, F. J., Persistence of heavy minerals and geologic age, *J. Geol., 49,* 610, 1941.

Pimentel, G. C., P. B. Forney, and K. C. Herr, Evidence about hydrate and solid water in the Martian surface from the 1969 Mariner infrared spectrometer, *J. Geophys. Res., 79,* 1623, 1974.

Pollack, J. B., and C. Sagan, An analysis of Martian photometry and polarimetry, *Space Sci. Rev., 9,* 243, 1969.

Pollack, J. B., D. Pitman, B. N. Khare, and C. Sagan, Goethite on Mars: A laboratory study of physically and chemically bound water in ferric oxides, *J. Geophys. Res., 75,* 7480, 1970*a*.

Pollack, J. B., R. N. Wilson, and G. G. Goles, A reexamination of the stability of goethite on Mars, *J. Geophys. Res., 75,* 7491, 1970*b*.

Rea, D. G., The darkening wave on Mars, *Nature, 201,* 1014, 1964.

Reed, R. D., Recent sands of California, *J. Geol., 38,* 223, 1930.

Reiche, P., Graphic representation of chemical weathering, *J. Sediment. Petrology, 13,* 58, 1943.

Rudolph, E. D., Terrestrial vegetation of Antarctica: Past and present studies, in *Antarctic Soils and Soil Forming Processes, Antarctic Res. Ser.,* vol. 8, edited by J. C. F. Tedrow, p. 109, AGU, Washington, D. C., 1966.

Ryan, J. A., The case against thermal fracturing on the lunar surface, *J. Geophys. Res., 67,* 2544, 1962.

Sagan, C., Sandstorms and eolian erosion on Mars, *J. Geophys. Res., 78,* 4155, 1973.

Sagan, C., and J. B. Pollack, Windblown dust on Mars, *Nature, 223,* 791, 1969.

Sagan, C., and J. Veverka, The microwave spectrum of Mars: An analysis, *Icarus, 14,* 222, 1971.

Sagan, C., J. P. Phaneuf, and M. Ihnat, Total reflection spectrophotometry and thermogravimetric analyses of simulated Martian surface materials, *Icarus, 4,* 43, 1965.

Sagan, C., J. Veverka, P. Fox, R. Dubisch, J. Lederberg, E. Levinthal, L. Quam, R. Tucker, J. B. Pollack, and B. A. Smith, Variable features on Mars: Preliminary Mariner 9 television results, *Icarus, 17,* 346, 1972.

Sagan, C., J. Veverka, P. Fox, R. Dubisch, R. French, P. Gierasch, L. Quam, J. Lederberg, E. Levinthal, R. Tucker, B. Eross, and J. B. Pollack, Variable features on Mars, 2, Mariner 9 global results, *J. Geophys. Res., 78,* 4163, 1973.

Salisbury, J. W., The light and dark areas of Mars, *Icarus, 5,* 291, 1966.

Salisbury, J. W., and G. R. Hunt, Compositional implications of the spectral behavior of the Martian surface, *Nature, 222,* 132, 1969.

Schorn, R. A., H. Spinrad, R. C. Moore, H. J. Smith, and L. P. Giver, High dispersion spectroscopic observations of Mars, 2, Water vapor variations, *Astrophys. J., 147,* 743, 1966.

Schulze, R., Photoelectric effect, *Z. Phys., 92,* 212, 1934.

Sharonov, V. V., A lithological interpretation of the photometric and colorimetric studies of Mars, *Astron. Zh., 38,* 267, 1961.

Sharp, R. P., Surface processes modifying Martian craters, *Icarus, 8,* 472, 1968.

Sherman, G. D., Factors influencing the development of lateritic and laterite soils in the Hawaiian Islands, *Pac. Sci., 3,* 307, 1949.

Sherman, G. D., The titanium content of Hawaiian soils and its significance, *Soil Sci. Soc. Amer. Proc., 16,* 15, 1952*a*.

Sherman, G. D., The genesis and morphology of the alumina-rich laterite clays, in *Problems in Clay and Laterite Genesis,* p. 154, American Institute of Mining and Metallurgical Engineers, Washington, D. C., 1952*b*.

Sherman, G. D., H. Ikawa, G. Uehara, and E. Okasaki, Types and occurrence of nontronite and nontronite-like minerals in soils, *Pac. Sci., 16,* 57, 1962.

Siever, R., Silica solubility, 0°–200°C, and the diagenesis of siliceous sediments, *J. Geol., 70,* 127, 1962.

Siever, R., Low temperature geochemistry of silicon, in *Handbook of Geochemistry,* vol. 2, edited by K. H. Wedepohl, sect. 14, Springer, New York, 1971.

Sinton, W. M., and J. Strong, Radiometric observations of Mars, *Astrophys. J., 131,* 459, 1960.

Soderblom, L. A., Mariner 9 imagery experiment: Geological evidence for water, paper presented at Symposium on Water on Mars, NASA, Harvard Univ., Cambridge, Mass., Nov. 2, 1973.

Soderblom, L. A., T. J. Kreidler, and H. Masursky, Latitudinal distribution of a debris mantle on the Martian surface, *J. Geophys. Res., 78,* 4117, 1973.

Sommer, A. H., *Photoemissive Materials,* 256 pp., John Wiley, New York, 1968.

Somorjai, G. A., and F. J. Szalkowski, Simple rules to predict the structure of adsorbed gases on crystal faces, *J. Chem. Phys., 54,* 389, 1971.

Spinrad, H. C., G. Münch, and L. D. Kaplan, The detection of water vapor on Mars, *Astrophys. J., 137,* 1319, 1963.

Stephen, I., A study of rock weathering with reference to the rocks of the Malvern Hills, *J. Soil Sci., 3,* 20, 1952.

Stevens, R. E., and M. K. Carron, Simple field test for distinguishing minerals by abrasion pH, *Amer. Mineral., 33,* 31, 1948.

Swalin, R. A., *Thermodynamics of Solids,* p. 180, John Wiley, New York, 1962.

Takahashi, H., Effects of dry grinding on kaolin minerals, *Proc. Nat. Conf. Clays Clay Miner., 6,* 279, 1959.

Tamm, O., Experimentelle Studien über die Vertwitterung und Tonbildung von Feldspaten, *Chem. Erde, 4,* 420, 1930.

Tedrow, J. C. F., and F. C. Ugolini, Antarctic soils, in *Antarctic Soils and Soil Forming Processes, Antarctic Res. Ser.,* vol. 8, edited by J. C. F. Tedrow, p. 161, AGU, Washington, D. C., 1966.

Thompson, B. A., P. Harteck, and R. R. Reeves, Jr., Ultraviolet absorption coefficients of CO_2, CO, O_2, H_2O, N_2O, NH_3, NO, SO_2, and CH_4 between 1850 and 4000 Å, *J. Geophys. Res.*, *68*, 6431, 1963.
Tricart, J., Experience de désagregation de roches granitiques par la cristallisation du sel marin, *Z. Geomorph., Suppl. 1*, 239, 1960.
Tull, R. G., High dispersion spectroscopic observations of Mars, 4, The latitude distribution of atmospheric water, *Icarus*, *13*, 43, 1970.
Tyutyunov, I. A., *An Introduction to the Theory of the Formation of Frozen Rocks*, 94 pp., Macmillan, New York, 1964.
Van Tassel, R. A., and J. W. Salisbury, The composition of the Martian surface, *Icarus*, *3*, 264, 1964.
Verhoogen, J., Distribution of silicates between silicates and oxides in igneous rocks, *Amer. J. Sci.*, *260*, 211, 1962.
Von Schellmann, W., Zur lateritischen Verwitterung von Serpentinit, *Geol. Jahrb.*, *81*, 645, 1964.
Walker, G. F., Decomposition of biotite in the soil, *Mineral. Mag.*, *28*, 693, 1949.
Wellman, H. W., and A. T. Wilson, Salt weathering, a neglected erosive agent in coastal and arid environments, *Nature*, *205*, 1097, 1965.
Wilson, M. J., The weathering of biotite in some Aberdeenshire soils, *Mineral. Mag.*, *35*, 1080, 1966.
Wilson, R. G., Thermionic work functions, *J. Appl. Phys.*, *37*, 2261, 1966.
Zettlemoyer, A. C., and E. McCafferty, Heat of immersion of α-ferric oxide in water, *Z. Phys. Chem.*, *64*, 41, 1969.
Zettlemoyer, A. C., R. D. Ivengar, and P. Scheidt, Heats of immersion and water sorption studies on bare and silica-coated rutile surfaces, *J. Colloid Interface Sci.*, *22*, 172, 1966.

Part VII

VOLCANISM AND TECTONICS

Editor's Comments on Papers 23 and 24

23 CARR
Volcanism on Mars

24 MUTCH et al.
A Summary of Martian Geologic History

Both Mars and Earth exhibit the types of volcanism that develop over hot spots and that produce extensive flood eruptions. However, Mars lacks stratovolcanoes that typically form along subduction zones on Earth. The absence of plate tectonics on Mars has, in fact, been inferred by M. M. Carr of the U. S. Geological Survey, from the large sizes that martian shield volcanoes, such as Olympus Mons (formerly Nix Olympica), have attained (Paper 23). The martian lithosphere has not moved relative to a fixed mantle hot spot and thus vast piles of lava can accumulate—in contrast to the Hawaiian Islands. Assuming a similar rate of eruption as the Hawaiian Emperor chain, Carr estimates it would have taken at least 130 million years for Olympus Mons to grow to its present size. Crater counts, however, suggest that these volcanoes may be around a billion years old (Blasius 1976; Neukum and Wise 1976).

Volcanic and tectonic features appear to be closely related. In another paper (not reprinted here), Carr (1974) shows how most of the prominent martian fractures, including Valles Marineris, are radial to the Tharsis uplift, in a manner analogous to terrestrial doming (see also Hartmann 1973). The uplift may have been controlled by a pre-existing NE–SW trend, and calculation of stress trajectories predicts NE–SW trending tension fractures. The NE–SW and to a lesser extent NW–SE distribution of vents on the flanks of the major Tharsis constructs are consistent with the proposed stress field (Carr et al. 1977; Nakamura 1976, unpublished report).

Structural deformation on Mars occurred over a long period of time. The giant shield volcanoes are merely the latest signs of activity in the Tharsis region. Multiple episodes of deformation have been traced by changes in orientation of faults (Wise 1974, 1977; Masson 1977). A pervasive process of crustal disruption produced extensive fractures, now being eroded by wind and mass wasting along the smooth plains-cratered highlands boundary (Mutch et al. 1976). Preliminary data suggest the presence of a "martian grid" analogous to the "lunar grid" (a planet-wide system of NW–SE, NE–SW trending lineaments; see Binder and McCarthy 1972; Roland 1976). Paper 24, from Chapter 9 in Mutch et al.'s excellent book "The Geology of Mars," summarizes the complex tectonic events that have shaped the martian crust.

23

Reprinted from *J. Geophys. Res.* **78**:4049–4062 (1973)

Volcanism on Mars[1]

Michael H. Carr

U.S. Geological Survey, Menlo Park, California 94025

Two classes of volcanic features occur on Mars: sparsely cratered plains resembling the lunar maria and circular constructs such as shields, domes, and craters. They show a markedly nonuniform distribution, being largely but not exclusively confined to one hemisphere. Only analogs of terrestrial intraplate volcanic features are found; analogs of terrestrial subduction zone features are absent. The Mars shield volcanoes are larger than their terrestrial counterparts, because the Martian crust is fixed in relation to the mantle, allowing more time for shields to grow. Volcanic activity has occurred throughout all the decipherable history of the planet.

One of the earliest and most significant of the Mariner 9 results was the recognition of prominent volcanic features on Mars. At the height of the dust storm, which occurred at the beginning of the orbital phase of the mission, the only surface features clearly visible outside the polar areas were four dark spots in the Amazonis-Tharsis region. Each was soon seen to include, at its center, a complex summit pit. The morphology of the craters, and their position atop mountains high enough to be above much of the obscuring dust in the atmosphere, strongly suggested a volcanic origin. As the atmosphere cleared, the four spots were revealed as being centered on four enormous shield volcanoes with summit calderas [*Masursky et al.*, 1972; *McCauley et al.*, 1972]. Subsequent photography of the remainder of the planet has revealed a wide variety of volcanic features, indicating that volcanism has played a major role in the evolution of the planet.

The volcanic features have a markedly asymmetric distribution (Figure 1). The planet can be roughly divided into two hemispheres. One includes nearly all the central volcanoes and associated features and the sparsely cratered plains; the other, for the most part, is densely cratered terrain, superficially resembling the lunar highlands.

The volcanic areas are generally high. The three shield volcanoes in Tharsis appear to lie on a broad NE-SW trending ridge, approximately 3–5 km above the mean Mars level [*Kliore et al.*, 1973]; the largest of all the shields, Nix Olympica, lies on the western flank of this ridge. The volcanic features of Elysium are similarly in a region of high elevation.

Volcanic and tectonic features appear to be closely related. Fractures radiate in nearly all directions from the Tharsis region [*Carr et al.*, 1973], especially to the northeast, where one of the most intensely fractured regions of Mars occurs. Here numerous fractures trend NE-SW, the same direction along which the three Tharsis shield volcanoes are aligned. Additional highly fractured terrain occurs just south of Tharsis in the Phoenicus Lacus area, and the Coprates canyon ('Grand Canyon of Mars') begins on the eastern flank of the Tharsis ridge.

Because the Mariner 4, 6, and 7 photography was largely confined to the densely cratered terrain, the spectacular volcanic constructs of the Tharsis and Elysium regions were unsuspected. On the basis of Mariner 6 and 7 pictures, several authors [*McGill and Wise*, 1972; *Elston*, 1971], employing arguments similar to those applied to the lunar highlands, suggested that volcanic processes had been significant on Mars and were responsible for the formation of many of the craters in the densely cratered terrain. The problem of the origin of the densely cratered terrain is not, however, addressed in this paper. The concern here is with the obvious volcanic constructs that lie largely but not exclusively outside the cratered

[1] Publication authorized by the Director, U.S. Geological Survey, Menlo Park, California.

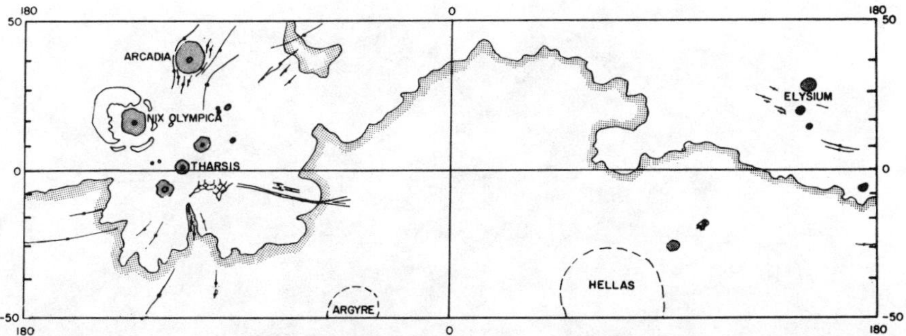

Fig. 1. Index map showing the location of the main volcanic features (patterned). Most lie in the Tharsis-Arcadia and Elysium regions, but isolated features occur elsewhere. The shaded line marks the boundary between the densely cratered province and the sparsely cratered plains. Major structures are also shown.

terrain. It is assumed that the densely cratered terrain is a primitive part of the Martian crust. If volcanic features are widely present there, they bear very little genetic relation to the younger less ambiguous volcanic features that abound elsewhere.

VOLCANIC FEATURES

Shield volcanoes. The largest volcanic features on Mars are shield volcanoes. The term shield is applied to terrestrial volcanoes that are built by successive outpourings of mainly basaltic lavas into broad gently sloping structures. Mauna Loa in Hawaii is the classic example and the largest active volcano on the earth. It is approximately 200 km across at its base and stands approximately 9 km above the sea floor. The main mode of eruption is relatively quiet effusion of lavas either along fissures on the flanks of the structure or from the summit caldera [*Macdonald and Abbott*, 1970].

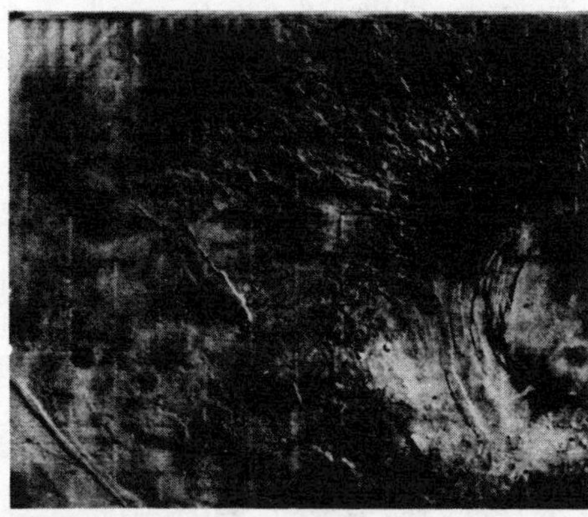

Fig. 2. West flank of the shield volcano at Nodus Gordii (9°S, 120°W), informally called South Spot. The smooth floored central caldera is approximately 140 km across and surrounded by a smooth zone with concentric graben. Beyond the smooth zone the surface has a marked radial pattern made up largely of numerous mutually intersecting radially elongate lobes. The surrounding plains appear to cover part of the shield to the southwest. (MTVS 4182-45, DAS 07038748.)

By analogy with terrestrial examples the term shield has been applied to several features on Mars that appear to be broad volcanic constructs with central calderas. The most prominent are Nix Olympica (18°N, 134°W), North Spot (11°N, 104°W), South Spot (9°S, 120°W), and Middle Spot (1°N, 113°W), all in the Amazonis-Tharsis region, and an unnamed feature at 24°N, 212°W in Elysium. The Martian shields are substantially larger than their terrestrial counterparts. The Nix Olympica shield is 600 km across, and there is some evidence that its size has been reduced significantly by erosion. North, South, and Middle Spot, the three Tharsis shields, are each approximately 400 km across. The heights are still uncertain, but present indications are that the summit of Nix Olympica stands 23 km above the surrounding terrain [*Wu et al.*, 1973] and that the other shields have similar but slightly smaller vertical dimensions.

Although they differ in detail, the shields all have the same general characteristics. Each is roughly circular in outline and has a central summit depression. The central pit may be simple or complex. South Spot (Figure 2), Middle Spot, and the Elysium shield all have

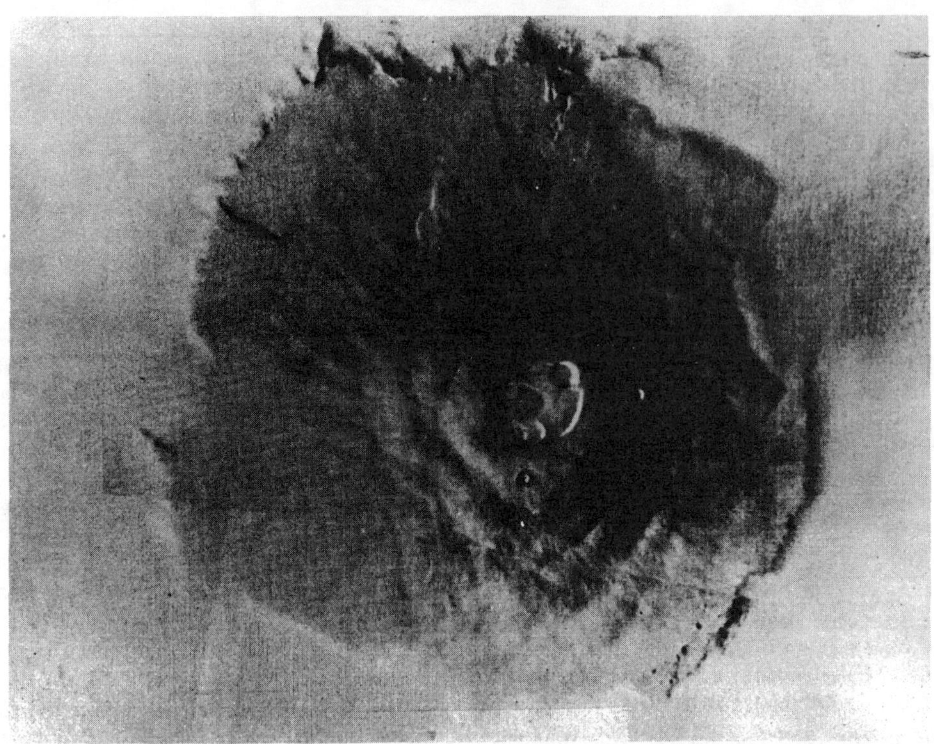

Fig. 3. Computer-processed mosaic of Nix Olympica (18°N, 134°W), the largest of the Martian shield volcanoes. The structure is approximately 600 km across. The summit caldera is 65 km in diameter and stands 23 km above the surrounding plains. The edge of the structure is marked by an escarpment approximately 2 km high. (IPL Roll 1288, 160226.)

simple summit craters. Nix Olympica (Figure 3) and North Spot have complex summit calderas in which successive collapse pits have formed about different centers. At the scale of the wide angle camera the flanks of the shields have a fine radial pattern that either feathers out into the surrounding terrain or terminates abruptly at a scarp marking the edge of the shield. The radial pattern at the outer edge of the South Spot shield is unusually coarse and results from numerous radially elongate lobes, superimposed on one another, and more rarely from short sinuous rillelike radial depressions (Figure 2). Narrow angle detail of the flanks of Nix Olympica (Figure 4) also reveals long radially elongate lobes and a narrow channellike feature, interpreted as a lava channel or collapsed lava tube, running along the crest of a low radial ridge. On all the shields a much coarser, crudely concentric pattern is superimposed on the radial pattern. The concentric pattern appears to result from slight breaks in slope that divide the shield flanks into a series of indistinct rounded sometimes mutually intersecting terraces.

The Tharsis shields have, in addition to the subtle slope changes, a well-developed concentric fracture pattern. South Spot has a prominent set of grabens immediately outside the central crater and several less distinct circular fractures as far out as 300 km from the center of the shield. Along parts of some of the fractures are lines of craters.

North Spot and the Elysium shield (Figure 5) both have concentric fractures as far out as 300 km from the central crater. Middle Spot has numerous concentric grabens on its flanks (Figure 6) in addition to concentric lines of craters (MTVS 4267-52, DAS 09917609). (The numbers in parentheses that appear in the text refer to Mariner 9 television pictures that are not included in this paper but that provide additional pertinent explanatory material. All Mariner 9 pictures may be ordered from the

Fig. 4. Narrow angle picture of the northwest flanks of Nix Olympica showing the fine surface detail. The arrows indicate radially elongate lobes. In the west half of the picture, on top of a low ridge, is a 250-meter-wide channel, interpreted as a lava channel or collapsed lava tube. (MTVS 4133-96, DAS 05492413.)

National Space Science Data Center, Greenbelt, Maryland.) Some concentric structures occur outside the shield itself, and very prominent concentric cracks occur just to the north. Near the central crater is a different type of circular structure, part scarp and part ridge. Its center is farther north than the center of the central crater (Figure 7). The origin of the structure is uncertain. It may have formed by movement along a circular fault or by ring-dike intrusion or by both.

The three Tharsis shields appear to have undergone modification that is in some way related to the major NE-SW structure along which the shields are aligned. The NE and SW edges of the Tharsis shields are embayed, and the shields are cut by numerous closely spaced linear almost crevasselike depressions aligned along the NE-SW direction. The embayment in the edge of South Spot almost reaches the central crater. Sparsely cratered terrain surrounds each of the shields, and the contacts take various forms. The edge of the Nix Olympica shield is almost everywhere a steep escarpment against which the surrounding plains abut. In contrast, plains materials appear to cover the edges of the three Tharsis shields, so that the radial and concentric structures of the shields are truncated as they dip beneath the plains (Figure 6). Scarps occur only to the NE and SW of the shields, around the embayments. Here plains materials appear to have encroached at the expense of the shield to form steep escarpments. The most prominent of the Elysium shields grades imperceptibly into the surrounding plains in such a way that its edge cannot be precisely determined.

Nix Olympica is unique among the Martian shield volcanoes in being surrounded by an aureole of what appears to be highly fractured terrain [*Carr et al.*, 1973]. This distinctive unit extends from the edge of the shield to as far as 1400–1500 km from the shield center. It is not continuous but complexly embayed by the plains. The terrain is made up of closely spaced

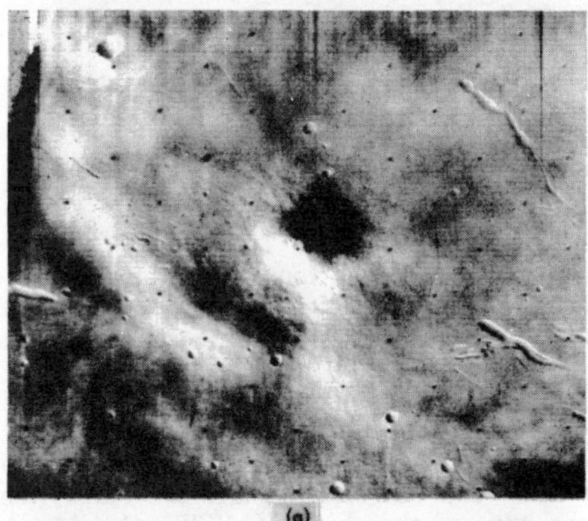

(a)

(b)

Fig. 5. (*a*) The Elysium shield at 24°N, 212°W. The structure is approximately 200–250 km across and markedly symmetrical. Concentric fractures occur as far as 300 km from the central crater. The surrounding plains are more heavily cratered than in the Tharsis region. (MTVS 4298-43, DAS 13496123.) (*b*) Narrow angle detail showing the central crater and the fine radial structure on the flanks. (MTVS 4298-39, DAS 13496088.)

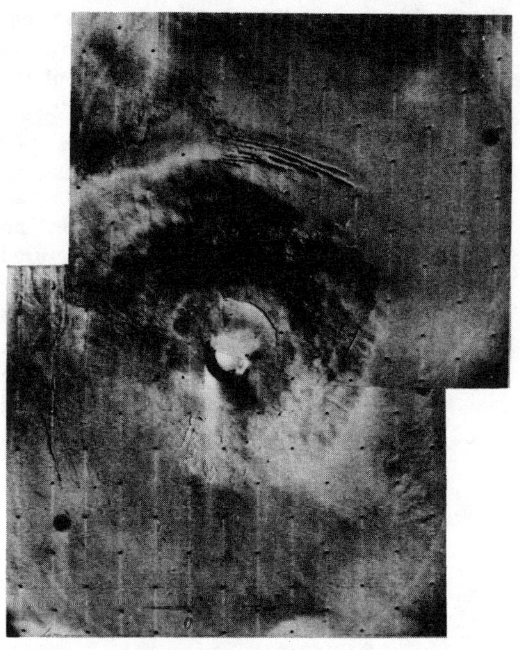

Fig. 6. Two-picture mosaic of the shield volcano at Pavonis Lacus (1°N, 113°W) informally named Middle Spot. The entire structure is approximately 400 km across. The central crater is 40 km in diameter and intersects a ring structure to the north. Concentric grabens occur both on the flanks of the shield and in the surrounding plains. To the NE and the SW are embayments in the edge of the shield. (MTVS 4184-54, DAS 07111128 and MTVS 4184-60, DAS 07111198.)

elongate subparallel ridges and troughs; the tops of the ridges form accordant summits at a somewhat higher elevation than the surrounding plains. The individual troughs are for the most part embayed by the plains materials. Movement appears to have occurred along arcuate faults, which, west of Nix Olympica, divide the fractured terrain into a series of blocks that dip gently northeast.

The origin of the aureole is unclear. One possibility is that it is a pediment that formed as the escarpment around Nix Olympica retreated toward the center of the shield. Another possibility is that the aureole is all that remains of one or more ancient shield volcanoes. The pediment hypothesis implies that the Nix Olympica shield was once 2–3 times its present size and that it has been reduced to its present size by erosion. There are several difficulties with this interpretation. The surface of the Nix Olympica shield appears to be relatively young [*Hartmann,* 1973]; a very fine surface texture is preserved on the flanks, and very few impact craters are present. Thus, if the aureole is a pediment, either it had to form in a relatively short time during which there was little modification to the surface of the shield itself or renewed eruptions created a relatively fresh surface on the shield. There is a strong contrast between the structure of the aureole and the structure of the shield. No traces of the large arcuate faults that cut the aureole can be found on the shield. The finer-scale fracturing of the aureole is also absent on the shield. The major structures of the aureole are so large that a considerable thickness of younger deposits would be required to achieve their complete burial. The implication is that the aureole is older than much of the uppermost portions of the shield, supporting the hypothesis that it represents the remnants of one or more old shield volcanoes similar to Nix Olympica but long since destroyed.

Domes. In the Tharsis region there are several volcanic constructs that differ from the shields by being smaller and having steeply sloping sides. Their genetic significance is uncertain. They may merely be the form that a shield takes with only a small volume of lava. On the other hand, they may indicate a different lava chemistry. They are sufficiently dis-

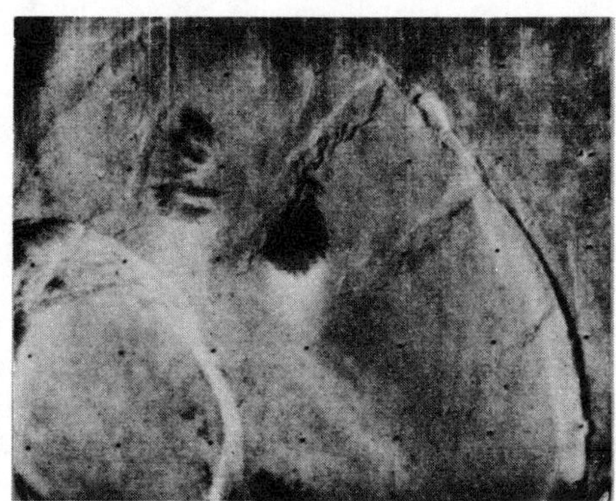

Fig. 7. Oblique narrow angle view of the central pit and ring structure of Middle Spot. The smooth crater-free floor and talus on the walls of the summit pit are clearly visible. Radial ridges similar to lunar mare ridges connect the central pit to the ring. The dark patches formed during the mission and are almost certainly eolian in origin. (MTVS 4267-44, DAS 09917329.)

tinctive morphologically to be separated from shield volcanoes and are here termed domes. At 13°N, 91°W there is a dome approximately 180 km across with a central summit pit 60 km across (MTVS 4189-72, DAS 07255398). The flanks are smooth, convex upward, and form a sharp contact with the surrounding plains [*McCauley et al.*, 1972]. The central pit is complex, consisting of several nested craters, each with steep walls and level sparsely cratered floors (MTVS 4231-39, DAS 08657889). Some larger, more subdued circular depressions occur on the flanks on either side of the central pit. Another 150-km-diameter dome at 24°N, 97°W differs markedly from the one just described (Figure 8). Numerous closely spaced fine channels run from the central pit down the flanks of the dome and disappear under the surrounding plains (MTVS 4271-52, DAS 10061459). A larger channel, 2 km across, begins in the central crater and ends in an irregular depression in the plains just beyond the edge of the dome. Several other more subdued domes also occur in this same north Tharsis region.

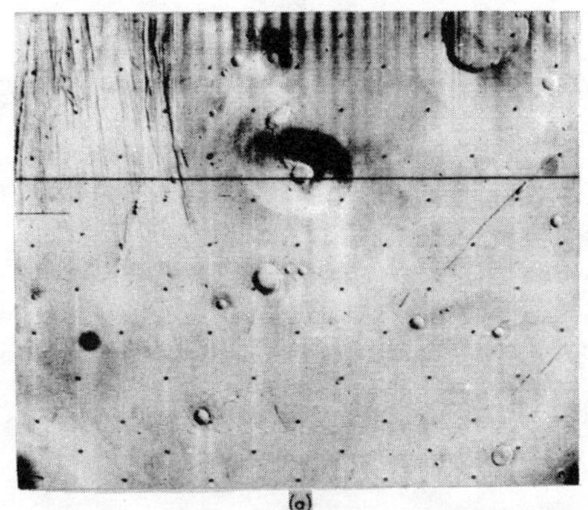

(a)

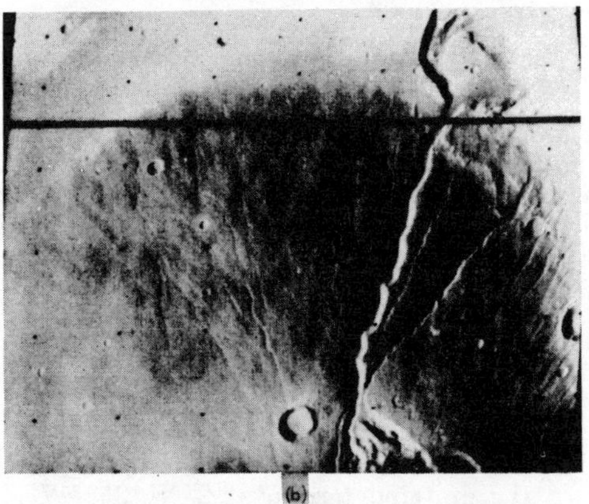

(b)

Fig. 8. (*a*) Dome, 150 km in diameter, at 24°N, 97°W. To the northeast corner of the photo is a large volcanic crater with a scalloped outline and smooth well-defined rim. (MTVS 4187-90, DAS 07183788.) (*b*) Narrow angle picture showing the channels on the flanks of the dome. A 2-km-wide channel begins in the central pit and ends in a shallow depression in the surrounding plains. Numerous other narrower channels extend down the flanks of the dome and are cut off at the base by the surrounding plains. (MTVS 4271-52, DAS 10061459.)

A different type of dome occurs in Elysium at 32°N, 209°W. It is approximately 200 km across and has a flat top but steep sides that terminate abruptly against the surrounding plains (Figure 9). At the resolution of the wide angle camera the top of the dome appears rough in contrast to the smooth appearance of the two domes previously described. The feature lacks a prominent central pit, but there are numerous small craters (<15 km) on the surface; the largest appears to be a caldera that has undergone several episodes of collapse (Figure 10*a*). Radiating from the crater are several narrow channels, chains of craters, and numerous fine striae. A narrow sinuous channel occurs at the eastern edge of the dome at its junction with the surrounding plains. The surface features of the dome appear to dip under the surrounding plains, except to the west, where an escarpment cuts into the dome (Figure 10*b*).

Craters. Several craters morphologically distinct from typical impact craters occur in the Tharsis and Elysium regions. These craters are presumed to be volcanic. Particularly distinctive because of their scalloped outlines are multiple craters at 40°N, 110°W (Figure 11), 36°N, 93°W (Figure 8*a*), and 11°N, 185°W. Crater rims, where present, have a smooth surface that terminates against the surrounding terrain with a sharp break in slope, giving the features very distinct circular outlines. These circular outlines contrast markedly with the irregular stellate outlines that characterize the rims of large fresh impact craters. Other probable volcanic craters to the west of Middle Spot have a simple outline like impact craters, but they also have smooth regularly shaped well-demarcated rims.

The craters at 40°N, 110°W are at the center

of a much larger circular structure. The region NE-SW fractures are deflected around the structure, so that they form an almost complete ring approximately 600 km in diameter (Figure 11). At the center of the ring is a complex of several intersecting craters. Faint radial ridges connect the craters to the fracture ring; otherwise the terrain within the ring is relatively featureless. Outside the ring is a faint radial pattern similar to that at the edges of the large Tharsis shield volcanoes. Narrow angle television pictures show the radial pattern to result partly from numerous fine channels with well-developed tributaries (MTVS 4182-096, DAS 07039903). Several impact craters occur both on and within the fracture ring. In the discussion below a case will be made for this large circular structure being an old partly collapsed shield volcano.

Other central volcanic structures. Throughout the cratered province there are many craters with morphologies difficult to reconcile with a simple impact origin. Many may have been so thoroughly changed by eolian processes that the characteristics of impact have been almost completely obliterated. Others, however, are so unlike typical impact craters that some form of volcanism must be invoked to explain their origin. There are widespread areas formed by volcanic plains in the cratered province (see below), so that the presence of volcanic craters in the region is not surprising. Two particularly striking examples are described here.

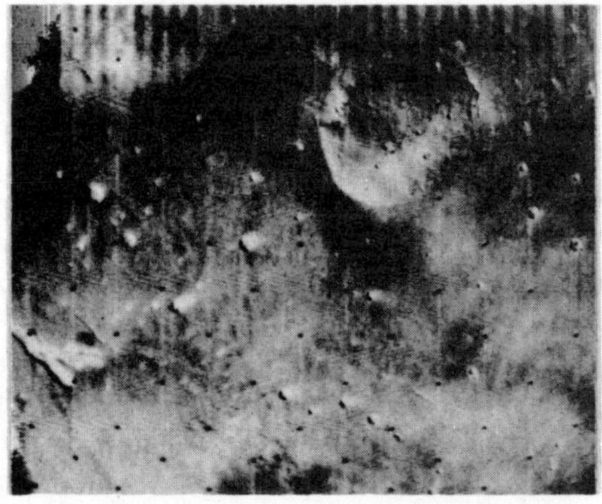

Fig. 9. Flat topped dome at 32°N, 209°W in Elysium. The structure is approximately 200 km across and has a very sharp outline and no large central summit pit. A channel runs along the northeast edge of the structure. To the west is an embayment around which an escarpment has developed. (MTVS 4242-56, DAS 09054544.)

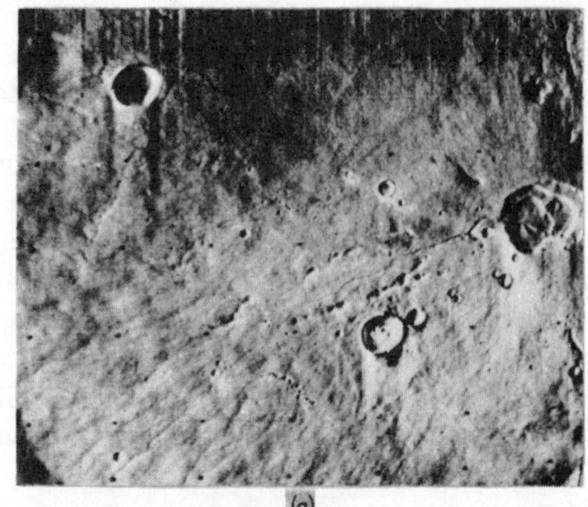

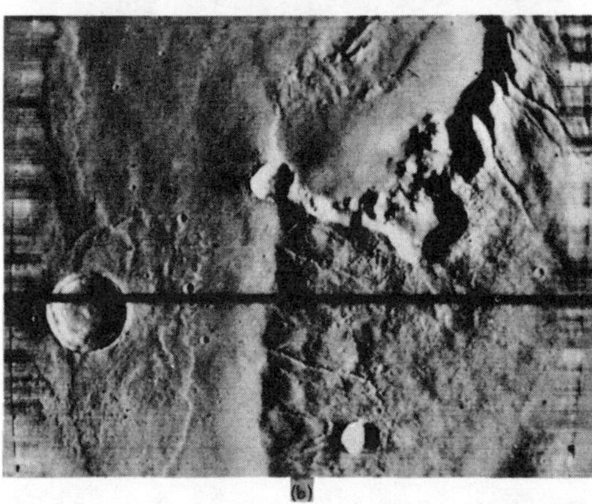

Fig. 10. (*a*) Narrow angle detail picture of the Elysium dome shown in Figure 9. Lines of craters and fine channels radiate from a 9-km-wide pit that has undergone multiple collapse. The fine channels disappear under the surrounding plains in the extreme southwest corner of the picture. (MTVS 4242-59, DAS 09054579.) (*b*) More narrow angle detail of the Elysium dome showing the escarpment at the west edge and the radial channels buried by the surrounding plains. Lobate structures on the plains are strongly suggestive of lava flows. (MTVS 4298-51, DAS 13496368.)

In the cratered terrain at 23°S, 253°W is a probable volcanic feature unusual both in its location outside the main volcanic regions and in its peculiar morphology (Figure 12). The feature lies well within the cratered terrain but in an area where there appears to be a relatively young sparsely cratered unit, which resembles the lunar maria. At the center of the structure is a circular depression, approximately 15 km across, connected by a channel to an elliptical depression of similar dimensions,

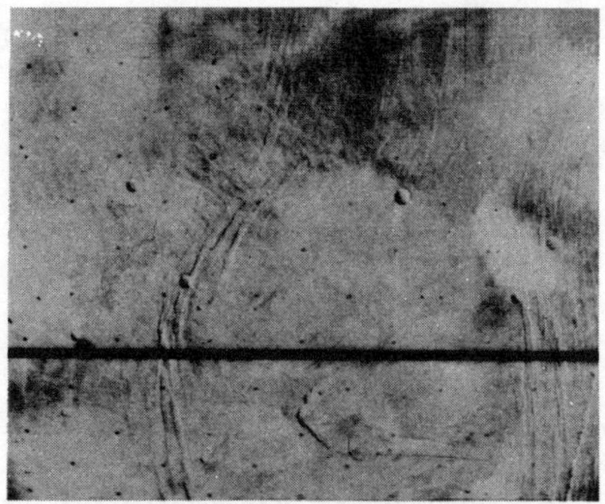

Fig. 11. Arcadia ring at 40°N, 110°W. A NE-SW trending set of fractures is deflected around a circular structure to form a fracture ring 600 km across. At the center of the ring just below the dark band are some shallow depressions with scalloped walls. Radial ridges connect the depression and the fracture ring. Outside the fracture ring the terrain has a radial pattern similar to that on the flanks of the Tharsis shields. The white patch to the northeast is a cloud. (MTVS 4222-69, DAS 08371134.)

35 km to the southwest. Around the central depression is a fracture ring approximately 45 km in diameter. Numerous ridges and depressions extend outward from the ring as far as 200 km. A 4-km-wide flat floored channel extends southwest from the ring, a narrower, more sharply defined channel occurs to the northwest, and several other channels occur to the west. The eastern part of the structure appears to be buried by material resembling the lunar maria. The structure has been given the informal name of dandelion because of its flowerlike appearance.

About 350 km southwest of the dandelion is another probable volcanic construct (Figure 13). An indistinct depression, 70 km in diameter, occurs in the center of a structure approximately 350 km across. The flanks have a pronounced radial pattern that terminates abruptly against the surrounding plains. Several irregular depressions suggestive of collapse occur in the plains beyond the edge of the central structure. The number of impact craters on the feature suggest that it is much older than the volcanic features of the Tharsis and Elysium regions.

Volcanic plains. Although the radially symmetric volcanic structures such as the shields and domes present the most spectacular evidence of volcanism on Mars, the materials that form the sparsely cratered plains may be volumetrically more significant. The plains, which surround the central structures, are relatively featureless at the resolution of the wide angle camera. Narrow angle pictures, however, commonly show long low lobate scarps (Figure 10*b*) that strongly resemble features in the Mare Imbrium on the moon. The lunar scarps, which have been widely interpreted as lava flow fronts, are difficult to discern, and relatively few

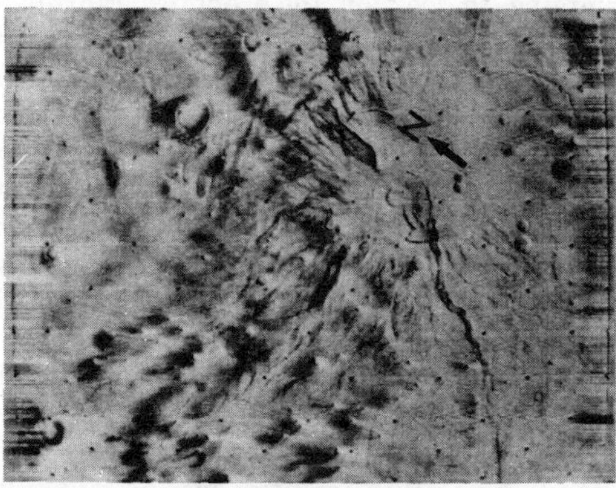

Fig. 12. Volcanic complex in Mare Tyrrhenum (23°S, 253°W). Numerous channels and low ridges radiate from a 15-km-diameter central depression surrounded by a fracture ring 45 km across. The eastern part of the structure is partially buried by the adjacent plains materials, which resemble lunar maria. (MTVS 4238-56, DAS 08909224.)

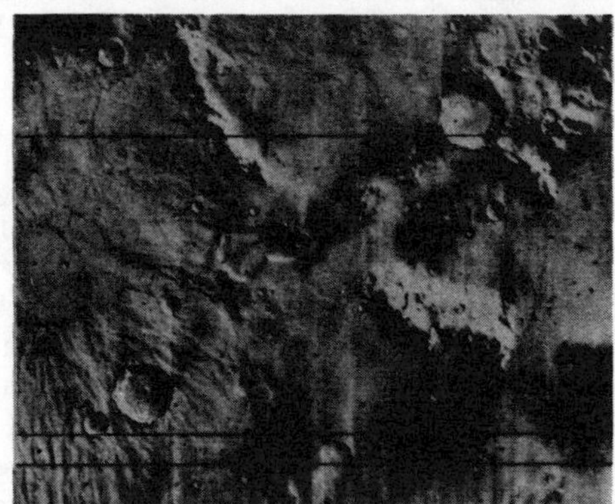

Fig. 13. Volcanic dome at 29°S, 264°W at the northeast edge of the Hellas basin. The central depression, which is at the western edge of the picture, is indistinct, but a strong radial pattern on the flanks is clearly visible. Superimposed craters suggest a relatively old age. (MTVS 4290-04, DAS 11974069.)

have been documented. In contrast, lobate scarps (possibly flow fronts) are visible on many, if not most, of the narrow angle pictures of the plains in the volcanic regions of Mars. Ridges similar to lunar mare ridges are also common, particularly on some of the more heavily cratered plains. By analogy with the lunar maria, then, the plains are probably largely volcanic in origin. Eolian deposits may be present locally and are particularly likely in those areas where little surface detail is apparent. It is clear from the different extent to which the plains have been cratered that they are not everywhere the same age.

In many areas of the densely cratered province the cratered surface appears to be partly or wholly covered by younger plains-forming materials. In some places, as around the volcanic feature at 23°S, 253°W, the cratered terrain appears to be completely buried. In other areas only a subdued impression of the largest craters is apparent. In yet other areas only the smaller craters are buried. Such effects could result from eolian deposition, and this almost certainly has occurred. However, volcanic activity also appears to have been widespread, as is indicated by lobate scarps (as at 5°S, 343°W) and lunar-mare-like ridges, and irregular depressions and products of this activity may cover much of the cratered surface. Thus, although the most spectacular volcanic features occur in the sparsely cratered regions, both volcanic plains and circular constructs are found within the densely cratered province, this fact indicating that volcanism has been in effect all over the planet.

Discussion

In this discussion it will be argued that the crust of Mars, in contrast to that of the earth, is stationary with respect to the rest of the planet. There are no apparent Martian equivalents of terrestrial subduction zones and mid-oceanic ridges that accompany crustal plate motions. The volcanic features typically associated with these zones on the earth are similarly absent on Mars. On the other hand, Martian equivalents of intraplate volcanic features, such as shield volcanoes and flood basalts, do occur, albeit in a somewhat different form because of the lack of plate movement. It will be further argued that volcanism has been an active process since very early in the history of Mars.

One of the most striking characteristics of the volcanic features on Mars is their distribution. Mars can be divided into two hemispheres by a plane dipping 50° to the equator such that on one hemisphere there are nearly all the volcanic features and sparsely cratered plains and on the other hemisphere there is nearly all the densely cratered terrain. The division is not exclusive, since some cratered terrain occurs in the volcanic hemisphere and vice versa; nevertheless, the difference between the two hemispheres is striking. By analogy with the earth one can invoke a simple convection cell within the mantle of Mars such that upwelling occurs in the volcanic province and downwelling occurs in the cratered province. Some support for this model is derived from the fracture pattern around the Tharsis region. Fractures extend several thousand kilometers out in different directions from Tharsis to form a pattern similar to that which occurs, on a much smaller scale, around terrestrial diapirs such as salt domes. The fractures are what one would expect from a broad domical uplift around Tharsis, as might be caused by upwelling of the underlying mantle.

Even if such convection has occurred in the Martian mantle, it does not appear to have caused different parts of the crust to move laterally with respect to one another. It is tempting to look on the Coprates canyon as a rift zone, away from which the crust has moved, and to interpret the Phoenicus Lacus rift zone as a transform fault. This interpretation does not, however, survive close scrutiny, since the complementary transform fault at the east end of the canyon is missing and there is no evidence of linear subduction zones anywhere on Mars. Terrestrial subduction zones, with their intensely folded arcuate mountain belts, are evident even on very poor satellite pictures, and on Mars any subduction zone should be similarly visible. The almost complete lack of other compressional or strike-slip structures also argues forcibly for a crust free of lateral movement. This conclusion is strengthened by the types of volcanic features on Mars, their distribution, and their size.

Factors contributing to incidence of volcanism. The incidence and the type of most terrestrial volcanoes are closely controlled by movement of the crustal plates with respect to one another [e.g., *Martin and Piwinskii*, 1972; *Dickinson*, 1970; *Gilluly*, 1971]. Most, but not all, volcanism is concentrated along the plate junctions, the type of volcanism depending on the type of junction. Along mid-oceanic ridges, where new crust is being formed, volcanic activity is mainly tholeiitic, accompanied by minor amounts of alkaline and olivine basalt [*Engel and Engel*, 1964]. At subduction zones, where the plates are consumed, volcanism is more varied. In the case of a continental plate overriding a downgoing oceanic plate, the volcanic belt may be zoned, either with tholeiites near the junction and high-alumina basalts and alkali basalts successively inland or with andesites near the junction and more potassic rocks inland [*Dickinson and Hatherton*, 1967]. At the junction of two oceanic plates, magmas are mainly andesitic; at continent to continent junctions very little volcanism occurs [*Gilluly*, 1971]. Several types of terrestrial volcanic activity are difficult to relate to plate margins and subduction zones, although they may have been affected by plate movement. Conspicuous among these are the volcanics of mid-oceanic islands such as Hawaii, the subaerial flood basalts such as those of the Columbia River plateau [*Waters*, 1962] and Greenland, and the confocal volcano-tectonic complexes such as those of the Mogollon plateau [*Elston*, 1971].

It is only these last three types of features that appear to have equivalents on Mars. There are certainly many resemblances between Hawaiian and Martian shield volcanoes and between the Columbia River basalts and the Martian plains. A less convincing case can be made for resemblance between some of the large ring structures on Mars and terrestrial ring complexes. Features resembling stratovolcanoes and breached structures that might be indicative of explosive activity are notable for their absence on Mars. Terrestrially, these types of volcanoes (pyroclastic and explosive) are most common inland of subduction zones in areas of andesitic or alkalic volcanism.

Comparison of Nix Olympica with Hawaiian shield volcanoes. Comparison of the terrestrial and Martian shield volcanoes will be explored in some detail, since each feature may contribute to the understanding of the other. The Martian shield volcanoes are much larger than their terrestrial counterparts. The large size of the Mars shields may result from being stationary over a magma source for a long period of time; in contrast, terrestrial shield volcanoes are almost certainly moving with respect to their magma source in the mantle and become extinct before achieving a comparable size.

The largest and best-known terrestrial shield volcanoes occur in Hawaii. Active shields lie at the southeast end of the Hawaiian Emperor chain, a line of extinct volcanoes that stretches several thousand kilometers across the Pacific Ocean [*Jackson et al.*, 1972]. The presently active Hawaiian volcanoes, Kilauea, Mauna Loa, and Hualalai, are at the southeastern end of the Hawaiian Archipelago; the other Hawaiian volcanoes become progressively older to the northwest [*McDougall*, 1964]. Similarly, the submarine volcanoes that make up the Hawaiian Emperor chain become progressively older to the northwest such that Koko seamount, about 5000 km from Hawaii, is 46 m.y. old [*Claque and Dalrymple*, 1973]. Each volcano of the chain appears to go through a similar evolutionary cycle: an initially rapid stage of building by the quiet effusion of tholeiitic lavas is followed by a less copious, more pyroclastic stage with eruption of more alkali basalts, hawaiites, and mugearites. In the final stages, nephelinitic and related rocks may be erupted, after which the volcano becomes inactive [*Powers*, 1955].

According to *Swanson* [1972], the rate of magma supply to Kilauea and Moana Loa since 1952 averages 0.01 km^3/yr, which could build the present Mauna Loa structure within 0.3 m.y. [*Macdonald and Abbott*, 1970]. This figure is consistent with K-Ar dates [*McDougall*, 1964] and paleomagnetic data [*Doell and Cox*, 1965]. There is, however, some evidence that eruption is episodic and that the very recent rates may be a maximum for the Hawaiian Emperor chain as a whole [*Shaw*, 1973]. The source of the magma is problematical, but earthquakes as deep as 50–60 km directly beneath Kilauea [*Eaton*, 1967] and exotic xenoliths in some of the lavas [*Jackson and Wright*, 1970] have been interpreted as indicating origin from depths of at least 60 km.

Wilson [1963] suggested that the Hawaiian Islands formed as the crust and rigid part of the upper mantle moved northwest over a fixed 'hot spot' in the mantle. *Morgan* [1971] extended the concept to include the entire Hawaiian Emperor chain as well as several other chains within the Pacific Basin. He also introduced the concept of 'convection plumes' in the lower mantle. According to Morgan, thermal instabilities near the core-mantle boundary cause upward convection of a plume of hot mantle rocks. Where the plume comes in contact with the rigid lithosphere, its constituent materials spread out over a large area to migrate slowly back down to the lower mantle. Shield volcanoes form over the plumes and become extinct as the plate on which the shield sits moves away from the site of the plume. Protracted movement of the Pacific plate over the stationary plume presently under Hawaii has resulted in the Hawaiian Emperor chain.

If we assume that in the Martian mantle, beneath Nix Olympica, a plume exists that is similar to the one proposed beneath the Hawaiian Emperor chain, we can calculate how long it would take to build Nix Olympica. Recent photogrammetric data (S. S. Wu, personal communication, 1972) indicate that the summit of Nix Olympica is 23 km above the surrounding plains, an elevation in good agreement with the ultraviolet spectrometer value of 25 km (C. W. Hord, personal communication, 1972). If 90 km is taken as the diameter of the summit pit and 600 km is taken as the diameter of the entire structure, the volume of Nix Olympica is 2.6×10^6 km^3. *Shaw* [1973] estimates that, in about the last 45 m.y., 8.5×10^5 km^3 of lava have erupted to form the Hawaiian segment of the Hawaiian Emperor chain. This gives a rate of accumulation of about 2×10^{-2} km^3/yr over the last 45 m.y. If we assume the same rate for Nix Olympica, it would take 130 m.y. to construct the entire shield. Clearly this number has value only for comparative purposes, since we have no idea what the accumulation rates actually are on Mars. One hundred and thirty m.y. may be a minimum accumulation age, because for comparison, we have chosen the Hawaiian chain, which is the most active volcanic feature on the earth. The choice was based partly on analogy and partly on the fact that more is known about Hawaii than any other active volcanic feature. A more valid comparison might be made with some of the less active mid-oceanic islands such as Reunion and the Galapagos, but accurate data on eruption rates are not available. It appears likely that Nix Olympica would take a few hundred million years to build at the lower eruption rates that are more typical of mid-oceanic islands.

Almost certainly we are not seeing all of Nix Olympica that once existed. Part has been destroyed by whatever mechanism is called on to form the escarpment around the structure (Figure 3); part must be hidden because of subsidence of the crust under the weight of the huge volcanic pile. Considering these probable hidden and destroyed parts of the shield, it is not inconceivable that Nix Olympica could have been accumulating for as long as a billion years. Furthermore, if the interpretation of the aureole of grooved terrain around Nix Olympica as remnants of ancient shields or extensions of the main structure is correct, volcanic activity may have been taking place at the site for considerably longer than this. The alternative to this reasoning is a high rate of eruption as compared with the most typical terrestrial analogs.

The probable long accumulation time implies that the Martian crust was stationary with respect to the mantle for long periods of time. Astonishing rates of volcanism would be required to build Nix Olympica if Martian crustal plates were moving with respect to the magma source. Other features of Martian volcanoes reflect this lateral stability. Their radial symmetry and lack of contiguous chains contrast with the Hawaiian shields, which are widely believed to form in chains as a result of relative shifts between terrestrial lithosphere and mantle. The apparent lack of Martian volcanic constructs indicative of pyroclastic activity is also worth noting. In the case of the Hawaiian shields the late pyroclastic alkalic stage has been attributed to differentiation in a lithospheric magma chamber that has been isolated from the magma source deep in the mantle as a result of plate movement [*Wilson*, 1963]. If a stable Mars crust is assumed, a magma chamber in the lithosphere beneath the volcano could not be cut off from the magma source to fractionate in isolation and produce alkalic differentiates. The late pyroclastic stage, therefore, would not be expected to occur. (It should, however, be noted that *Jackson and Wright*

[1970] do not agree with Wilson that the alkali lavas are differentiates that form as a result of plate movement. They believe that copious eruption of tholeiitic basalts leads to chemical heterogeneities in the mantle. Fractional melting of different variants leads to eruption of alkali lavas.)

The conclusion that plate movement does not occur on Mars and that Nix Olympica was stationary over its magma source has important implications for the origin of terrestrial shield volcanoes. Some hypotheses of origin invoke localized convection in the mantle to create the melting zone over which the volcano develops. In these hypotheses, plate motion is incidental and not necessary for the volcano to originate. In other theories, melting is caused by interaction between the moving lithospheric plate and the mantle beneath. Here plate motion is a prerequisite. Shield volcanoes appear to have developed on Mars without the help of plate tectonics, and thus credence is given to the hypotheses in which plate motion is not required.

The height of Nix Olympica has implications regarding the depth of origin of the magma. *Eaton and Murata* [1960] suggested that differences in density between the lava in its conduit and the surrounding rocks creates a hydrostatic head. They calculate that, for delivery of magma to the summit of Mauna Loa, a depth of origin of 57 km is required, which is consistent with the earthquake data. A similar calculation cannot be made for Mars because the densities are so uncertain. If the density differences were, however, the same as for Mauna Loa, a 130-km depth of origin would be required. Whatever the actual value, a rigid crust substantially thicker than that of earth is suggested. This conclusion is consistent with the gravity anomalies that occur in the Tharsis volcanic region [*Lorell et al.*, 1973].

Comparison of Nix Olympica with other Martian shield volcanoes. Nix Olympica has the freshest appearance of all the Martian shield volcanoes. Several lines of evidence suggest that the Tharsis volcanoes are older and in various stages of collapse. The surfaces lack the fine texture that occurs on the flanks of Nix Olympica, and thus an older age for their surfaces is suggested. All the Tharsis shields each have arcuate faults on their flanks and in the plains just beyond the edge of the shields. These are lacking at Nix Olympica. The structures are interpreted as faults along which the central part of the structure has subsided. Of the three Tharsis shields, South Spot appears to be in the most advanced state of collapse, and North Spot appears to be the freshest. The large size of the summit pit of South Spot as compared with the other shields suggests that it has been enlarged by collapse along arcuate faults similar to the ones that presently bound the central crater. An older age for South Spot is also indicated by the lack of definition of the central crater in the northwest. Of the three Tharsis shields, South Spot also appears to be in the most advanced state of destruction by erosion. The lobate structures at its outer margin are very coarse, embayment by the plains is extensive, and the embayments to the NE and the SW are the most well developed. All these factors, although not proof, are suggestive of an inactive structure certainly older than Nix Olympica and probably older than North and Middle Spots.

The large circular structure in Arcadia (Figure 11) may represent a yet more advanced state of collapse of a shield volcano. The structure consists of a fracture ring, 600 km in diameter, in the center of which are the remnants of some calderalike pits. Outside the fracture ring the terrain has a radial texture very similar to the Tharsis shields. It is suggested that this is a very old extinct shield, the center of which has collapsed along arcuate fractures. An alternative interpretation of the feature is that it is analagous to terrestrial confocal volcano-tectonic ring complexes [*Elston*, 1971].

The oldest central volcanic features occur within the cratered province. The features are too small to have a sufficient number of superimposed impact craters for statistically significant crater counts; nevertheless, the superimposed craters do provide some measure of relative age. For the shield volcano to the northeast of Hellas the number of superimposed craters up to 40 km in diameter is indistinguishable from the rest of the densely cratered province. At 40 km the curve for the densely cratered terrain crosses the 3-b.y. isochron

[*Hartmann,* 1973], this fact indicating a very old age for the feature. A similar shield at 9°S, 185°W has a crater density intermediate between the Tharsis shields and the Hellas shield.

Volcanic plains. Most of the plains have been divided into three broad units on the basis of the number of superimposed craters; a fourth plains unit, which occurs only in the high northern latitudes, appears to be mantled by younger material [*Carr et al.,* 1973]. The crater frequencies on the three nonmantled units are shown in Figure 14. Meaningful counts are not possible for the mantled unit because of the overlying materials. The most heavily cratered plains unit has a crater density that approaches that of the densely cratered province. The crater distribution suggests an age of 1–2 b.y. if the age estimates of *Hartmann* [1973] are correct. The less densely cratered units indicate ages of several hundred million years.

Caution should be exercised in interpreting these ages. Hartmann's isochrons are calculated on the assumption that most of the impacting bodies are asteroidal. He determines the flux history of the earth and the moon and then makes estimates of the cratering history of Mars by taking into account the distribution of asteroids between the earth and Mars and the different cratering effects on the earth, the moon, and Mars. He estimates his error as a factor of 3, but the error could be much larger than this if the large impact craters on the earth and the moon are primarily cometary in origin and not asteroidal. A high cometary component would result in older ages for Mars because of the smaller difference in the number of cometary impacts between the earth and Mars as compared with asteroidal impacts. Whatever the actual age, the point to be emphasized is that some of the volcanic plains appear to be old and may well be in the range of 1–4 b.y., and thus an early beginning to Martian volcanism is suggested. This is consistent with the long accumulation age suggested for Nix Olympica and the apparently old age of the volcanic feature at the edge of Hellas.

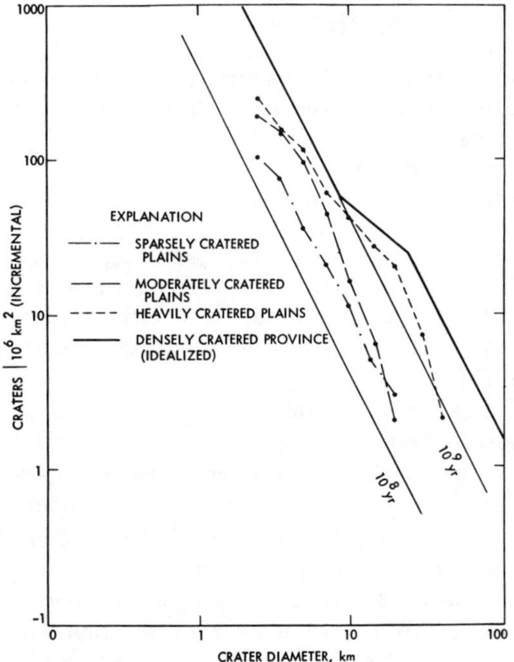

Fig. 14. Crater counts on different plains units are compared with the crater frequencies for the densely cratered province. Isochrons showing possible ages are from *Hartmann* [1973].

REFERENCES

Carr, M. H., H. Masursky, and R. S. Saunders, A generalized geologic map of Mars, *J. Geophys. Res., 78,* this issue, 1973.

Claque, D. A., and G. B. Dalrymple, Age of KoKo seamount, Emperor seamount chain, *Earth Planet. Sci. Lett., 17,* 411, 1973.

Dickinson, W. R., Relations of andesites granites and derivative sandstone to arc trench tectonics, *Rev. Geophys. Space Phys., 8,* 813, 1970.

Dickinson, W. R., and T. Hatherton, Andesitic volcanism and seismicity around the Pacific, *Science, 157,* 801, 1967.

Doell, R. R., and A. Cox, Paleomagnetism of Hawaiian lava flows, *J. Geophys. Res., 70,* 3377, 1965.

Eaton, J. P., Evidence of the source of magma in Hawaii from earthquakes, volcanic tremors and ground deformation (abstract), *Eos Trans. AGU, 48,* 254, 1967.

Eaton, J. P., and K. J. Murata, How volcanoes grow, *Science, 132,* 925, 1960.

Elston, W. E., Evidence for lunar volcano-tectonic features, *J. Geophys. Res., 76,* 5690, 1971.

Engel, A. E. J., and C. E. Engel, Composition of basalts from the mid-Atlantic ridge, *Science, 144,* 1330, 1964.

Gilluly, J., Plate tectonics and magmatic evolution, *Bull. Geol. Soc. Amer., 82,* 2382, 1971.

Hartmann, W. K., Martian cratering, 4, Mariner 9 initial analysis, *J. Geophys. Res., 78,* this issue, 1973.

Jackson, E. D., and T. L. Wright, Zenoliths in the Honolulu volcanic series, Hawaii, *J. Petrol., 11,* 405, 1970.

Jackson, E. D., E. A. Silver, and G. B. Dalrymple, Hawaiian-Emperor chain and its relation to Cenozoic circumpacific tectonics, *Bull. Geol. Soc. Amer., 83,* 601, 1972.

Kliore, A. J., G. Fjeldbo, B. L. Seidel, M. J. Sykes, and P. M. Woiceshyn, *S* band radio occultation measurements of the atmosphere and topography of Mars with Mariner 9: Extended mission coverage of polar and intermediate latitudes, *J. Geophys. Res., 78,* this issue, 1973.

Lorell, J., G. Born, E. Christensen, P. Esposito, J. Jordan, P. Laing, W. Sjogren, S. Wong, R. Reasenberg, and I. Shapiro, Gravity field of Mars from Mariner 9 tracking data, *Icarus, 18,* 304, 1973.

Martin, R. F., and A. J. Piwinskii, Magmatism and tectonic settings, *J. Geophys. Res., 77,* 4966, 1972.

Masursky, H., R. M. Batson, J. F. McCauley, L. A. Soderblom, R. L. Sildey, M. H. Carr, D. J. Milton, D. E. Wilhelms, B. A. Smith, T. B. Kirby, J. C. Robinson, C. B. Leovy, G. A. Briggs, A. T. Young, T. C. Duxbury, C. H. Acton, Jr., B. C. Murray, J. A. Cutts, R. P. Sharp, S. Smith, R. B. Leighton, C. Sagan, J. Veverka, M. Noland, J. Lederberg, E. Levinthal, J. B. Pollack, J. T. Moore, Jr., W. K. Hartmann, E. N. Shipley, G. de Vaucouleurs, and M. E. Davies, Mariner 9 television reconnaissance of Mars and its satellites: Preliminary results, *Science, 175,* 294, 1972.

Macdonald, G. A., and A. T. Abbott, *Volcanoes in the Sea,* University of Hawaii Press, Honolulu, 1970.

McCauley, J. F., M. H. Carr, J. A. Cutts, W. K. Hartmann, H. Masursky, D. J. Milton, R. P. Sharp, and D. E. Wilhelms, Preliminary Mariner report on the geology of Mars, *Icarus, 17,* 289, 1972.

McDougall, I., Potassium-argon ages from lavas of the Hawaiian Islands, *Bull. Geol. Soc. Amer., 75,* 107, 1964.

McGill, G. E., and D. U. Wise, Regional variations in degradation and density of Martian craters, *J. Geophys. Res., 77,* 2433, 1972.

Morgan, W. J., Convection plumes in the lower mantle, *Nature, 230,* 42, 1971.

Powers, H. A., Composition and origin of basaltic magma of the Hawaii Islands, *Geochim. Cosmochim. Acta, 7,* 77, 1955.

Shaw, H. R., Mantle convection and volcanic periodicity in the Pacific, Evidence from Hawaii, *Bull. Geol. Soc. Amer.,* in press, 1973.

Swanson, D. A., Magma supply rate at Kilauea volcano, 1952–1971, *Science, 177,* 169, 1972.

Waters, A. C., Basaltic magma types and their tectonic associations—Pacific Northwest of the United States, in *The Crust of the Pacific Basin, Geophys. Monogr. Ser.,* vol. 6, 158, AGU, Washington, D.C., 1962.

Wilson, J. T., A possible origin of the Hawaiian Islands, *Can. J. Phys., 41,* 863, 1963.

Wu, S. S. C., F. J. Shafer, G. M. Nakata, and R. Jordan, Photogrammetric evaluation of Mariner 9 photography, *J. Geophys. Res., 78,* this issue, 1973.

24

Reprinted from pp. 316–319 of *The Geology of Mars,* T. Mutch, R. E. Arvidson, J. W. Head, K. L. Jones, and R. S. Saunders, Princeton, N. J.: Princeton University Press (1976)

A SUMMARY OF MARTIAN GEOLOGIC HISTORY

T. Mutch, R. E. Arvidson, J. W. Head, K. L. Jones, and R. S. Saunders

DRAWING ON the data and interpretations of previous chapters it is possible to construct a sequence of events outlining the geologic history of Mars. The evolutionary model is illustrated by figure 9.1, which contains five paleogeologic maps of Mars, showing the progressive development of the surface.

The planet formed by accretion of many smaller objects over a relatively short time, probably a few hundred thousands of years. During the first billion years of Martian history, the meteoroid-asteroid flux decreased to a level that has remained generally constant for the past 3.5 b.y. The final stages of heavy bombardment that occurred during the first billion years produced a densely cratered crust (fig. 9.1a). Similar surfaces occur in the lunar and Mercurian highlands. On Earth the same type of surface probably was formerly present, but has been destroyed by remobilization of crustal materials.

Shortly after the accretionary phase, planet-wide differentiation occurred—driven largely by the kinetic energy of accretion—with formation of crust, mantle, and core. If the crust varied in thickness over the globe (Phillips et al., 1973) and if the more ancient regions were essentially in isostatic equilibrium (Phillips and Saunders, 1975) then it might be expected that elevation and crustal thickness would be closely correlated. Regions of thicker crust would stand at higher elevations. It is postulated that the topographic division of the planet into a northern topographically low hemisphere and a southern topographically high hemisphere reflects this type of crustal variation, thin in the north and thick in the south.

Figure 9.1b depicts the tectonic breakup of the crust by two processes: pervasive fracturing of thin crust and radial faulting around the Tharsis uplift. Evidence for disruption of the northern crust is present along the boundary between cratered terrain and plains where the cratered terrain has been broken up to form the fretted terrain. The same fretted pattern can be seen considerable distances north of the plains-cratered terrain contact, where remnants of ancient crust are incompletely covered by plains. This suggests that fracturing and modification of crust was not restricted to the zone where it is now best displayed.

A possible mechanism for break-up of the crust has been previously outlined. This involves expansion of the mantle as the olivine-to-spinel structure phase change migrates downward with increasing internal temperatures. Regions of thin crust would be weakest and hence most susceptible to rifting over the swelling mantle. The originally thin crust would be made even thinner by this extension, and subsequent isostatic adjustments would cause it to stand at progressively lower elevations.

The Tharsis uplift is apparently an ancient fea-

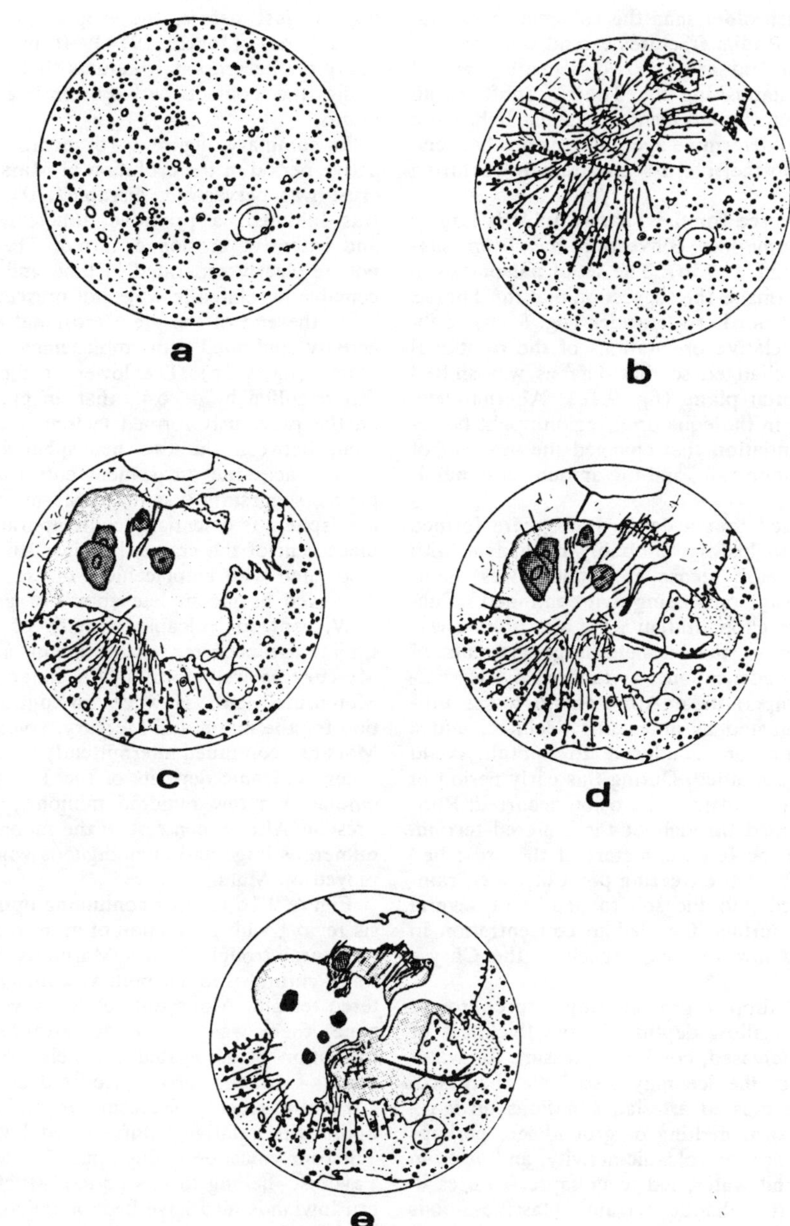

'gure 9.1. Schematic paleomaps showing geological evo-
tion of Mars. The eastern hemisphere, from 0°W to 180°W,
depicted. The sequence of events is further described in
e text.

(a). Last stage of early heavy bombardment.

(b). Uplift of proto-Tharsis region, with radial extension.
acturing of crust throughout northern hemisphere.

(c). Widespread emplacement of volcanic plains.

(d). Renewed uplift and radial extension in the Tharsis region. Partial mantling of fractured cratered terrain throughout the northern hemisphere.

(e). Emplacement of relatively recent volcanic plains, with formation of volcanoes. Virtually complete mantling of fractured cratered terrain in northern hemisphere with volcanic and/or eolian deposits.

ture, surely much older than the volcanic rocks that cap the region. Radial fractures extend well into the cratered terrain and are exposed beneath cratered plains (pc), establishing the Tharsis uplift as an event older than emplacement of plains. It has not been possible to determine age relationships between fracturing of northern crust and radial fracturing around Tharsis.

The driving force for the Tharsis uplift is largely speculative. Plume-like convection has been suggested (Hartmann, 1973b). The set of paleomaps in figure 9.1 conforms to the suggestion that the Tharsis uplift formed at northern latitudes (fig. 9.1b). Subsequently, the relative orientations of the rotational axis and crust changed so that Tharsis was shifted into the equatorial plane (fig. 9.1c). Alternatively, broad buckling in the equatorial region might be related to differentiation that changed the moment of inertia, the rotation rate, and the amount of dynamical flattening.

It is postulated that a dense atmosphere formed early in Martian history, probably coincident with accretion. The early dense atmosphere was maintained by extensive outgassing that continued immediately after accretion. Initially, the atmosphere was relatively warm so that it held large amounts of water vapor. Condensation was favored by decreasing surface temperatures. Atmospheric losses ultimately predominated as outgassing declined, and a greenhouse condition, conducive to rainfall, could no longer be maintained. During this early period of extensive rainfall, substantial erosion occurred. Runoff furrows formed throughout the cratered terrain. As long as the top few kilometers of the crust had temperatures above the freezing point of water, rainwater percolated into the soil to depths of several kilometers. Subsurface flow led to concentration in topographically low regions, especially the Chryse trough.

With a small drop in ground temperature, ground ice formed at shallow depths. As the thickness of this ice layer increased, confining pressure for water trapped beneath the ice may also have increased. This may have created artesian situations on a regional scale. Local melting of ground ice, perhaps related to the onset of volcanic activity, and artesian release of ground water, led to collapse features in what is now the chaotic terrain. Massive floods formed some of the large channels, particularly in the equatorial zone along the previously established slope between the high southern and low northern hemisphere. With continued cooling of the atmosphere most of the remaining H_2O and CO_2 migrated to the polar caps and to subsurface reservoirs where it remains to the present as ice and possibly as chemically bound $(OH)^-$ in goethite and in clays.

This great erosional event probably occurred during, or just before, the emplacement of cratered plains (pc) (Jones, 1974). It is responsible for many of the erosional features that are now visible along the contact between cratered terrain and plains.

In summary, figure 9.1b records a particularly active period in the evolution of Mars. The northern crust was pervasively disrupted; the Tharsis uplift was initiated; a primitive atmosphere condensed; and extensive erosion occurred. These events cannot be clearly separated in time and they might be genetically related in ways not presently understood.

At the end of the great erosional event, volcanic activity continued with emplacement of the cratered plains (fig. 9.1c). The lower northern hemisphere was engulfed by deposits that, in places, lapped up on the previously formed tectonic-erosional escarpment between the two hemispheres. However, in some places the escarpment cuts through cratered plains, suggesting that the break-up of the northern hemisphere is essentially contemporaneous with emplacement of the cratered plains. Ancient volcanoes associated with emplacement of cratered plains are preserved as incomplete annuli of grooved terrain.

Widespread volcanic activity of approximately similar age occurred on the moon, and possibly on Mercury as well, forming the lunar maria and the Mercurian plains. However, in contrast to the situation for the moon and Mercury, volcanic activity on Mars has continued intermittently to relatively recent times. Volcanic deposits of the Tharsis region were emplaced a few hundred million years before the present. Also in contrast to the moon and Mercury, numerous large and unambiguous volcanoes are displayed on Mars.

Figure 9.1d records continuing uplift of the Tharsis region, with generation of more radial faults. The structural trough, Valles Marineris, formed at this time, cutting through both volcanic plains and cratered terrain. Along this relatively youthful fracture zone, there was release of subsurface water with formation of associated channels. Further development of chaotic terrain occurred at this time.

Finally, recent volcanism has mantled the Tharsis uplift with relatively unfractured lavas (fig. 9.1e) The four volcanoes—Olympus, Ascraeus, Arsia, and Pavonis—belong to this period. Much of the northern lowlands also have been mantled, either by volcanic plains or eolian deposits. On Map 9.1e recent lavas are shown only in those regions where B frames contain diagnostic flow fronts.

The volcanic evolution of Mars places it between Earth and the moon. On the moon, volcanic activity was confined to an early stage of planetary evolution, with the younger magmas apparently generated at progressively greater depths, coincident with a general decrease in upper mantle temperature. On

Earth volcanic activity continues to the present with no sign of an end. In an active environment of plate tectonics, magma is generated at depths of a few tens of kilometers. On Mars, the crustal thickness and depth of melting is intermediate between Earth and the moon. The thermal environment is insufficient to generate crustal plate movement, but adequate to fuel widespread volcanism.

The structural features of Mars differ from those of the moon, Mercury, and Earth. The moon is largely devoid of compressional structural features. Linear valleys, apparently graben, are generally related to impact basins. The Mercurian landscape is dominated by huge ridge systems that may have formed during an early stage of crustal contraction. Only on Earth is there persuasive evidence for plate tectonics, with attendant zones of separation and subduction. Great rift systems are observed, along with mountainous fold belts.

During relatively recent times eolian activity has played a dominant role in shaping the Martian landscape. It is probable that a great amount of material has been eroded from equatorial regions and has been deposited at the poles. The polar layered deposits are of uncertain age, but undoubtedly record some climatic change—ancient or recent—that led to episodic precipitation of H_2O and CO_2 frost, intermixed with dust. More or less arbitrarily we show the polar deposits first appearing in figure 9.1c, tentatively related to early atmospheric changes. Subsequent to their formation, the polar deposits have been eroded by surface winds. Mantles of debris have spread equatorward to approximately $\pm$ 30° latitude (Soderblom et al., 1973b). Smaller craters have been covered and larger craters partly buried.

At lower latitudes some depressions, such as the Hellas basin, have trapped substantial amounts of wind-blown dust. Elsewhere a thin veneer of mobile sediments has been swept across volcanic bedrock. The shifting pattern of dark basalt and brighter sediment accounts for some of the streaks and splotches that appear in association with many craters.

In summary, Mars appears to be intermediate in both size and state of geologic evolution between the Earth, and the moon and Mercury. The ancient cratered surface so evident and well-preserved on the moon and Mercury, is preserved to a lesser extent on Mars and eradicated on Earth. In part, this preservation is due to the lack of an atmosphere on the moon and Mercury. The atmosphere of Mars, while allowing for eolian modification up to the present, has not supported a continuous hydrologic cycle such as that which dominates the surface of Earth. An additional factor in the preservation of the ancient cratered terrain is the lack of extensive compressional deformation, associated with plate tectonics and continental drift on Earth. On the moon, extensive surface volcanic activity ceased about three billion years ago, and the youngest plains units on Mercury appear of similar age (Murray et al., 1975). On Mars, volcanic activity continued past that time, perhaps extending close to the present.

REFERENCES

Hartmann, W. K. 1973b. Martian surface and crust: Review and synthesis. *Icarus* **19**:550–575.

Jones, K. L. 1974. Evidence for an episode of crater obliteration intermediate in Martian history. *J. Geophys. Res.* **79**:3917–3931.

Murray, B. C., Strom, R. G., Trask, N. J., and Gault, D. E. 1975. Surface history of Mercury: Implications for terrestrial planets. *J. Geophys. Res.* **80**:2508–2514.

Phillips, R. J., and Saunders, R. S. 1975. The isostatic state of Martian topography. *J. Geophys. Res.* **80**:2893–2898.

Phillips, R. J., Saunders, R. S., and Conel, J. F. 1973. Mars, crustal structure inferred from Bouguer gravity anomalies *J. Geophys. Res.* **78**:4815–4820.

Soderblom, L. A., Kreidler, T. J., and Masursky, H. 1973b. Latitudinal distribution of a debris mantle on the Martian surface. *J. Geophys. Res.* **78**:4117–4122.

Part VIII

GEOPHYSICS AND INTERNAL STRUCTURE

Editor's Comments on Papers 25 and 26

25 PHILLIPS and SAUNDERS
The Isostatic State of Martian Topography

26 JOHNSTON and TOKSÖZ
Internal Structure and Properties of Mars

Prior to the Viking landings, the only information about the interior of Mars came from the mean density, moment of inertia, topography and gravity measurements. The mean density of 3.96 g/cm^2 suggests a total iron content of about 25 percent by mass, as compared with 33 percent for the Earth (Anderson 1972). The amount of inertia factor ($C/MR^2 = 0.365$) indicates a concentration of mass toward the center and therefore provides evidence for the presence of a dense core. Detection of a weak dipole magnetic field by the Soviet Mars probe offers additional support for a core (Dolginov et al. 1973, 1975). The composition and size of the core are important in determining the internal density distribution and the thermal history. The density and core size depend strongly on the assumed composition. The chemistry of the Solar System condensation points to a FeS–Fe core (Lewis 1972). The interplay between core size, composition, and mantle density has been explored in papers by Anderson (1972), Johnston et al. (1974), and Johnston and Toksöz (Paper 25).

Paper 25 represents an updated version of an earlier article by Johnston and Toksöz. A major difference from the previous paper is the reduced value of the moment of inertia factor from 0.377 to 0.365. As a consequence, the estimated mantle densities drop from around 3.7 g/cm^3 to 3.4–3.5 g/cm^3, thus leading to an increase in the possible size of the core. Another consequence is a higher forsterite composition for the martian mantle (Fo_{75} vs. Fo_{56}). This new value (Fo_{75}) is nevertheless lower than the Fo_{90} deduced for the Earth's upper mantle (Birch 1969). Thus the martian mantle appears enriched in FeO relative to the Earth, as expected from the condensation model (Lewis 1972). However, Johnston and Toksöz find that FeO enrichment in the martian mantle is inconsistent with an FeS core and thus conclude that some metallic iron must also be present.

Topography and gravity also provide clues about the internal structure of Mars. Doppler tracking data from Mariner 9 have yielded gravity information that suggests Mars, like the Earth, has regions of thicker, less-dense crust underlying elevated topography (e.g., Tharsis) and thinner, denser crust beneath lowlands and basins (Amazonis, Libya Basin, and Oxia Palus) (Phillips et al. 1973).

Further analysis of gravity data by R. J. Phillips and R. S. Saunders (Paper 26) reveals two distinct topographic groups: an older region in isostatic equilibrium, and a younger, only partially compensated terrain that displays a strong correlation between gravity and topography. Since all the anomalous topographic features lie along the same trend, the authors believe that they were formed by a single process, which led to the uplift of the Tharsis ridge and concurrent subsidence of Chryse and Amazonis. It may be significant that the area of chaotic terrain is confined to the Chryse trough, where crustal thinning has occurred. Blasius et al. (1977) propose that tectonic collapse generated chaotic terrain, which would be consistent with the gravity data. There is some evidence for structural control of chaos (Wilson et al. 1973).

Crustal thickness has been estimated by Blasius and Cutts (1976), using gravity data from Phillips and Saunders (1975) and a simple two-layer model. They conclude that the crust beneath the highest part of the Tharsis uplift is 24 km thicker than under Olympus Mons, with a corresponding thinning of the lower lithosphere (upper mantle) under the Tharsis crest. Estimated crustal thicknesses range from 50 km below Olympus Mons to 74 km below Arsia Mons. It becomes apparent from the preceding papers that the crustal layer must be quite variable in thickness, but the calculations yield nonunique solutions unless they can be related to a direct measurement of crustal thickness by seismic techniques. This is one reason why the Viking seismology experiment was so important. Unfortunately, the Viking 1 seismometer failed to uncage, and useful data were obtained only from the Viking 2 seismometer.

The results of the seismology experiment were reported by Anderson et al. (1977). Although the experiment held great potential, the tentative findings to date did not justify including the paper here. Anderson et al. detected only one likely marsquake

(Richter magnitude 2.8) at an epicentral distance of 110 km during the first 146 sols (martian days) of operation and none subsequent to that (priv. comm.). The event displayed similar amplitudes to minor Californian earthquakes recorded by the same instrument. Assuming similar physical properties as on Earth, a discontinuity (crust-mantle boundary?) at 16 km depth was deduced. The relatively short duration of the experiment and scarcity of natural "events" prevent further conclusions at this time.

25

Reprinted from *J. Geophys. Res.* **80**:2893–2898 (1975)

The Isostatic State of Martian Topography

R. J. Phillips and R. S. Saunders

Jet Propulsion Laboratory, California Institute of Technology, Pasadena, California 91103

As is evidenced by the gravity data, the regional topography of Mars falls into two distinct age groups. The older group is isostatically compensated at relatively shallow depth. All of our analyses indicate that the younger group is only partly compensated and that the depth of partial compensation is also shallow. The young topography is composed primarily of the Tharsis plateau and the adjacent low areas of Chryse and Amazonis. They appear to have all undergone the same amount of partial compensation, suggesting a contemporaneous origin. All of the topography is consistent with a model of viscous relaxation in an environment of approximate thermal steady state. The minimum viscosity necessary to explain the partial compensation is 10^{27} P, a value typical of a number of lunar studies. Alternatively, the strength required to support the uncompensated fraction of the topography is similar to strengths determined for the earth and moon.

Introduction

It is well established that the regional topography of the earth is in approximate isostatic balance. It has also been advocated that the regional topography of the moon, apart from the near-side multiringed basins, is also in equilibrium [*O'Keefe*, 1968].

With the availability of Doppler gravity data and elevation data from the Mariner 9 Mars mission, it is now possible to study the question of isostasy on a third planet. In this paper we examine the state of isostasy of the Martian topography to the resolution of an eighth-order and eighth-degree spherical harmonic model. Applied to the earth, this resolution reveals the gross shape of the major continents. It must be kept in mind then that our discussions regarding Mars are concerned with regional features analogous, in relative scale at least, to the continents of the earth.

It is concluded herein that large areas of Mars are only partly compensated. In fact, at the resolution considered here, the gravity and topography of Mars are explained by two distinct populations of regional topography. One is distinctly older and largely compensated; the other is younger and only partly compensated. The younger features are contemporaneous and probably related in origin. All of the topography is consistent with a model involving a simple creep response of the interior under approximately steady state thermal conditions.

The concept of isostasy—that all crustal columns exert equal pressure at some depth—may be investigated globally by studying a scatter diagram relating observed gravity to the gravity predicted from topography (Figure 1). Each point on the diagram represents a specific geographic location. The ordinate is the free air gravity at the surface. The abscissa is the gravity that the topography alone should give if there were no lateral variations in density. This latter is equivalent to the complete Bouguer correction.

In the absence of significant gravity effects other than those resulting from topography and its elements of compensating mass, the scatter diagram has a simple interpretation. If most of the points fall along the abscissa, the topography is compensated at shallow depth, and isostasy prevails. In this situation there may be large variations in topography, but the mass of a unit column of rock down to the relatively shallow compensation depth remains nearly constant over the compensated regions, and the resulting variations in free air gravity are small.

However, if the points fall along a slope s, $0 < s \leq 1$, then the topography may be partly to completely compensated, depending upon the depth of compensation. That is, for a given slope there is a continuum of solutions ranging from partial compensation at shallow depth to complete compensation at greater depth. As the depth is increased, the compensating mass must increase because of the geometrical falloff of attraction. In the limit, the compensating mass is sufficient to achieve the balance required for isostasy, but the diminished attraction of the compensating mass is much less than that of the topography. This results in the possibility of compensation at great depth but a finite slope of the points on the scatter diagram. As the slope approaches unity, the amount of partial compensation must decrease, and the depth to complete compensation must increase. At the limiting slope of unity, the only real solution is zero compensation.

In Figure 1 (top) we have plotted such a diagram for the earth, sampled at 10° latitude and longitude increments between ±40° latitude, from eighth-order spherical harmonic representations of gravity [*Gaposchkin and Lambeck*, 1971] and topography [*Balmino et al.*, 1973]. The data, scattered about the x axis, indicate lack of correlation between topography and gravity. The spread of data is about 10 times smaller along the y axis, indicating at least 90% compensation.

In Figure 1 (bottom) we have plotted a similar diagram for Mars, based on the work of *Sjogren et al.* [1975] and *Christensen* [1975]. Here the 80° latitude swath is centered about the Mariner 9 periapsis latitude (~20°S), so that the gravity data are of high statistical confidence, corresponding (approximately) to the band of lowest spacecraft altitudes. The diagram shows two distinct trends, one lying along the x axis indicating regions of compensation, at relatively shallow depth, and the other lying along a slope of 0.4, representing regions having a correlation between gravity and topography.

The Martian geographic names used in this paper correspond to traditional names applied to albedo features on early maps of Mars. These names have been applied to specific topographic features by the International Astronomical Union. For purposes of describing regional features, we have adopted the following nomenclature. The topographically high region, occupied by the Tharsis Montes, between 80° and 130°W and 30°S to 30°N, is referred to as the Tharsis plateau. A separate lobe at 80° to 105°W and 45° to 25°S is the

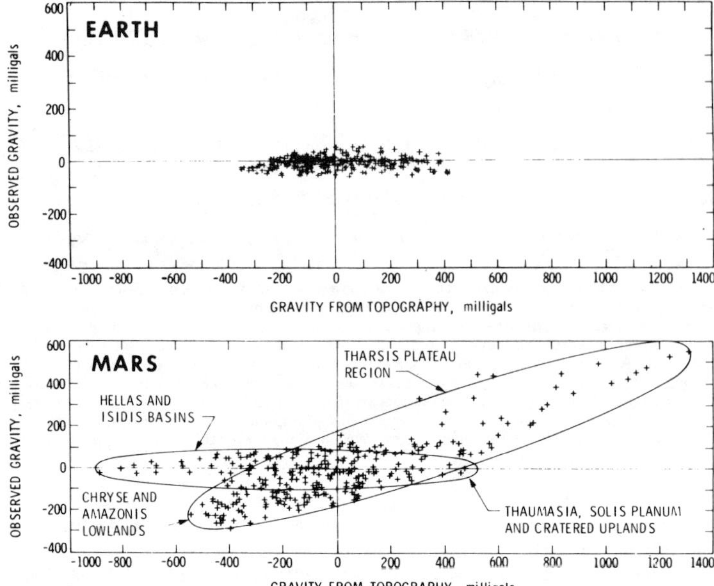

Fig. 1. Scatter diagrams for earth and Mars of observed gravity versus the gravitational equivalent of topography. Data are based on eighth-order spherical harmonic expansions with the J_0, J_1, and J_2 harmonics removed. Ellipses indicate trends of the two populations of Mars. Topographic densities for earth and Mars are 2.67 g/cm³ and 3.0 g/cm³, respectively.

Thaumasia plateau. Flanking the Tharsis plateau on the west and east are the Amazonis and Chryse lowlands. The Amazonis lowland is not well defined but lies between 150° and 180°W and 0° to 40°N. The Chryse lowland is a north-south trending trough extending from the Argyre Planitia (40°W, 50°S) northward in a north-northeasterly direction to the equator at 30°W and continuing in a north-northwesterly direction to its lowest point in Chryse Planitia at 45°W, 25°N. The trough is approximately 30° to 50° wide. The 'cratered uplands' occupy a region 20° to 30° to either side of a ridge that extends in a northeasterly direction from 0°W, 60°S, midway between Argyre Planitia and Hellas Planitia, to 300°W, 30°N. This region contains the highest concentration of large ancient craters. Official names associated with the largest presumed impact features on Mars refer to the plains surfaces within them, for example, Hellas Planitia. We will refer to the entire feature, including the mountainous rim deposits, as a basin. Thus the feature associated with the Isidis Planitia is the Isidis basin.

The interpretation of Martian topography and gravity requires some estimate of the age of various terrains. A wide range of ages has been proposed by various authors [*Soderblom et al.*, 1974; *Jones*, 1974; *Chapman*, 1974; *Hartmann*, 1973; *Carr*, 1973]. As the rate of crater production at Mars is in dispute, accepted rates ranging from about equal to the lunar rate to as much as 25 times the lunar rate, absolute ages of Martian terrains have not been established. We adopt the following ages but consider the older extreme to be more likely than the younger: most ancient features, Hellas, Isidis, and Argyre basins, 3×10^9 to 4×10^9 years; major volcanism forming volcanic plains of the Tharsis plateau, 10^8 to 10^9 years; formation of major shield volcanoes, 3×10^7 to 3×10^8 years. The older plains of the Tharsis and Thaumasia plateau region (pc and pm of *Carr et al.* [1973]) are considered to be of volcanic origin and of the order of 10^9 years old. The uplift and fracturing of the Tharsis plateau predate the oldest volcanics and must be somewhat in excess of 10^9 years old and perhaps as old as 3×10^9 years [*Chapman*, 1974].

The compensated topography of Mars is associated with the more ancient terrain of the planet, principally the Hellas and Isidis basins, the Thaumasia plateau, and the cratered uplands. The population showing the high correlation between gravity and topography is associated with the Tharsis plateau and the adjacent lowlands of Chryse and Amazonis. These are relatively young tectonic features that are associated with extensive volcanism, the bulk of which probably exceeds 100 m.y. in age. Were it not for the basin anomalies of opposite sign, we might conclude that the volcanic cover on the Tharsis plateau is sufficient excess mass and therefore the most reasonable explanation for the anomaly there. Analogous to the lunar circular mare fill, the volcanic flows may be superisostatic. However, there is no complementary explanation for the basin anomalies. The Chryse plains proper (40°W, 20°N) could be underlain by thick accumulations of low-density sediments that might contribute to the negative anomalies. However, the major portion of the area of negative anomalies is in extensively cratered terrain that cannot have a significant thickness of low-density cover.

As these topographic features lie along the same linear trend on the scatter diagram, it is suggested that these anomalous high and low regions of Mars have similar depths of (partial?) compensation, that they are manifestations of the same process, and that they have undergone the same amount of isostatic adjustment. We conclude that the anomalies coincide in time with the uplift of the Tharsis plateau and the contemporaneous subsidence of the Chryse and Amazonis lowlands. The cause is internal redistribution of material by an unknown process.

In summary, the topographic history of Mars involves an early period when topographic features were compensated at shallow depth followed by a later period for which the crustal behavior satisfies a suite of possible models ranging between the case where compensation is shallow and is only partly complete in at least 1 aeon and the other extreme where compensation occurs at great depth and to completion. Below we investigate these two end-members.

ISOSTATIC DEVIATION

The most generally applied test for isostatic equilibrium of a region is the 'isostatic anomaly.' As ordinarily used, this procedure requires some a priori information about subsurface structure and density. A model is set up, with known parameters, and the gravity effects are computed. The model usually involves a root/antiroot system, that is, an Airy model. The relief on the root is approximately $\rho_1/\Delta\rho_D \cdot h$, where ρ_1 is the crustal density, $\Delta\rho_D$ is the density contrast of the root, and h is the height of the topography. The isostatic anomaly is free air gravity minus topographic gravity minus root gravity or equivalently Bouguer gravity minus root gravity.

Trial approaches were made for Mars, assuming a uniform 'crustal' density, by varying the crustal mean thickness and density and attempting to find a combination that minimized the isostatic anomalies in various regions of the planet. It is clear that under the Tharsis plateau, for example, a compensated solution must be one in which the shape of the root coincides with the perturbation shape that satisfies the Bouguer gravity. No such solution exists [e.g., *Phillips et al.*, 1973], and in order to avoid the need to explicitly specify the shape of the compensating element (i.e., the mirror image root), we adopted the following approach. Because of the high correlation between topography and gravity over much of the planet, we assume that the Bouguer gravity results from a perturbation surface that represents partial to complete isostatic compensation of the topography. It is then possible to calculate the amount of topography that is in isostatic balance with this surface. The difference between the actual topography and the balanced fraction is defined as the isostatic deviation and is given by the following:

$$T^{ID}(\theta, \phi) = R \sum_{n,m} \xi_{nm}{}^{ID} S_{nm}(\theta, \phi) \qquad (1)$$

where R is the mean planetary radius, $S_{nm}(\theta, \phi)$ is a surface spherical harmonic, and the coefficient $\xi_{nm}{}^{ID}$ is given by

$$\xi_{nm}{}^{ID} = \xi_{nm}{}^{t} + \alpha_n \frac{\bar{\rho}}{\rho_1} \beta^{n+2} \left[1 + \frac{\rho_1}{\bar{\rho}} (\beta^{-3} - 1)\right] \cdot \left[\xi_{nm}{}^{g} - \frac{\rho_1}{\bar{\rho}} \frac{1}{\alpha_n} \xi_{nm}{}^{t}\right] \qquad (2)$$

where $\xi_{nm}{}^{t}$ and $\xi_{nm}{}^{g}$ are spherical harmonic coefficients for topography and gravity, respectively, $\bar{\rho}$ is the mean planetary density, ρ_1 is the density of the topography, $\alpha_n = (2n + 1)/3$, and $\beta = R/R_D$, where R_D is the radius to the compensating layer.

We realize that because of the two populations on the scatter diagram, there can be no one surface at a single depth that balances both the young and the old terrain. Therefore the task is twofold: to search in shallow depth regions to find a bound on the maximum depth of compensation for the ancient terrain and to conduct a general search for a depth of compensation for the young terrain.

Using (1), we have computed values for $T^{ID}(\theta, \phi)$ over the planet for a variety of values for $D = (R - R_D)$ and ρ_1. For D up to 100 km the isostatic deviation vanishes for the ancient terrain, while for values of D up to approximately 150 km the isostatic deviation of the young terrain remains constant, of significant value, and well correlated with topography. Beyond a transition region of 150–250 km, the isostatic deviation over the young terrain becomes large, shows amplification of the higher harmonics, and is uncorrelated with topography. We found no combination of values of D and ρ_1 that showed the young terrain to be even approximately compensated. A typical result for the isostatic deviation is shown in Figure 2 for $D = 50$ km and $\rho_1 = 3.0$ g/cm^3.

The behavior of the higher harmonics with increasing depth D follows from the fact that the harmonic contribution of the perturbation surface must fall off as β^{-n}, relative to the topography. Thus as D is increased, the perturbation surface has an increasing inability to cancel the higher topographic harmonics, leading to their relative amplification in the isostatic deviation. It is thus clear that only shallow partly compensated solutions are permissible in the isostatic deviation. Alternatively, the fact that the Bouguer gravity has a similar harmonic content to the Bouguer gravity correction (topographic gravity) supports the contention that internal density variations are relatively shallow. Further, were this not so, there would not be a linear relationship between gravity and topography on the scatter diagram for points representing the relatively young topography.

The harmonics $\xi_{22}{}^{g}$, $\xi_{32}{}^{g}$, $\xi_{33}{}^{g}$, and $\xi_{31}{}^{g}$ contribute most to the Tharsis-Chryse-Amazonis gravity anomaly and they are in fact the dominant harmonics in the gravitational field [*Sjogren et al.*, 1975]. The first three coefficients are correlated with the corresponding topographic coefficients. The $\xi_{31}{}^{g}$ coefficient is not correlated because $\xi_{31}{}^{t}$ receives contributions from both the young and the old terrain. It could be argued that only $\xi_{22}{}^{g}$, $\xi_{32}{}^{g}$, and $\xi_{33}{}^{g}$ should be tested for compensation and that this approach would circumvent the problem of amplification of the higher harmonics. We have done this by using (2) and for various values of ρ_1 have solved for R_D such that $\xi_{nm}{}^{ID}$ is zero. That is, we have solved for the depth to the surface that simultaneously balances the topography and satisfies the Bouguer gravity. These two constraints are satisfied by values of β that satisfy the following polynominal equation:

$$\beta^{n+2} + \frac{\rho_1}{\Delta\rho} \beta^{n-1} - \frac{\bar{\rho}}{\Delta\rho}\left[1 - \frac{\bar{\rho}}{\rho_1} \alpha_n \frac{\xi_{nm}{}^{g}}{\xi_{nm}{}^{t}}\right]^{-1} = 0 \qquad (3)$$

where $\Delta\rho = \bar{\rho} - \rho_1$. It is appropriate to use the magnitudes J_{nm} of the spherical harmonic coefficients in (3) if the phase angle difference $\Delta\lambda$ between the gravitational and topographic coefficients is small, indicative of correlation. The coefficient ξ_{nm} represents either a cosine or a sine term, C_{nm} or S_{nm}, respectively, in the usual spherical harmonic representation and

$$J_{nm} \cos [m(\lambda - \lambda_{nm})] = C_{nm} \cos (m\lambda) + S_{nm} \sin (m\lambda) \qquad (4)$$

and $\Delta\lambda = |\lambda_{nm}{}^{g} - \lambda_{nm}{}^{t}|$.

The results of solving (3) with J_{22}, J_{32}, and J_{33} are shown in Table 1. The second- and third-order harmonics yield quite different solutions, and there appears to be no single depth of compensation for the dominant harmonics.

On the basis of the arguments in this section, we conclude that the young topography on Mars is not compensated, at least in terms of classical concepts of isostasy.

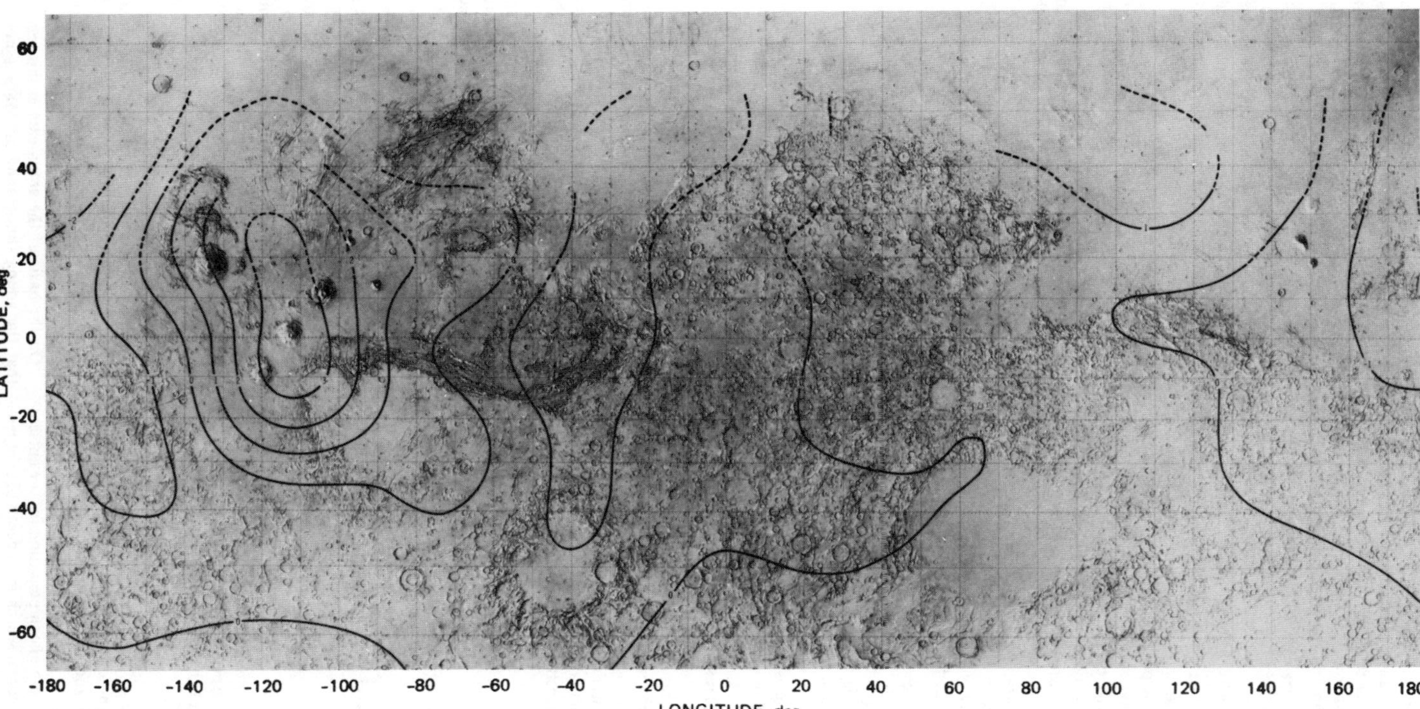

Fig. 2. Isostatic deviation for Mars with $\rho_1 = 3.0$ g/cm³ and a depth to compensation of 50 km. Contours are in kilometers. Solid contours indicate a standard deviation on the estimate of less than 500 m; dashed lines, 500–750 m; blank, greater than 750 m. We have not interpreted the anomaly centered at 110°E, 40°N, since it is a small lobe of an extensive anomaly in a region of very large standard deviation.

DISCUSSION

From the above analyses, we conclude that the depth of complete compensation of the old terrain is no greater than 100 km and that the depth of partial compensation of the younger terrain is no greater than about 150 km. It is reasonable to hypothesize that compensation of both Martian terrain types presently occurs at the same depth, which is no greater than about 100 km. If our simple model is correct, this surface of density discontinuity may be the crust-mantle boundary. For purposes of considering the response of the interior to surface loads, we also assume that this surface defines the base of the Martian lithosphere.

If the lithosphere has responded to surface loads by a Newtonian viscous creep, rheological histories that are consistent with the ages of major topographic features and the amount of relaxation require that shallow viscosity remain constant or increase with time. A history of decreasing viscosity is not probable. The thermal models of *Johnston et al.* [1974] also suggest constant or modestly increasing viscosity with time.

A viscosity of about 10^{27} P is a minimum value consistent with the amount of relaxation of Tharsis, for an assumed age of one billion years. Although this value is high compared to those obtained for the terrestrial interior, it is not unlike viscosity values obtained from a number of lunar studies [*Arkani-Hamed*, 1973; *Baldwin*, 1970, 1971; *Phillips et al.*, 1972].

Alternatively, we might consider that a threshold stress exists below which solid state creep will not occur. Stresses below the threshold are in elastic equilibrium and can be supported over long periods of geologic time. The threshold stress can be considered to be the strength of the Martian interior. The minimum strength of the terrestrial interior required to support the nonhydrostatic part of the flattening is 30 bars, which is also 'about the same as stresses associated with regional anomalies of the earth's gravity field' [*Caputo*, 1965]. *Chinnery* [1964] concludes that the strength of the earth's crust is about 10 bars, while *Baldwin* [1968] derives a value of 30–50 bars for the outer layers of the moon.

If the depth of partial to complete compensation is small compared to the wavelength of the harmonics of interest, then we need only inquire about the stresses associated with the uncompensated fraction of the topography as represented by $\xi_{nm}{}^{ID}$. The dominant uncompensated harmonic on Mars is $\xi_{22}{}^{ID}$, which imparts a stress difference of a maximum of 20 bars at the surface to 50 bars at the center for a homogeneous planet [*Jeffreys*, 1970]. As stated above, the earth and moon appear to support stresses of this magnitude, so it is possible that the uncompensated topography on Mars is also mechanically supported.

In conclusion, we feel that the regional topography of Mars formed at two distinct times. The uplift of the Tharsis plateau and the formation of the Chryse and Amazonis lowlands are contemporaneous. We find that these regions of younger topography are only partly compensated at shallow depth. Our criteria are, however, based upon classical concepts of equal weight in vertical columns; other types of equilibrium are not ruled out [*Artyushkov*, 1974]. Further, because of the gravity-topography correlation, we have assigned all of the anomalous gravity field to the topography and the compensating elements of the interior. The gravity field is consistent with a viscosity that is constant or increasing with time. The amount of relaxation of the Tharsis plateau corresponds to a shallow viscosity of 10^{27} P (minimum) over the last billion years, a number that is also representative of the moon. Alternatively, by terrestrial and lunar analogy, it is entirely possible that the uncompensated fraction of the topography is mechanically supported at present. If this is the case, it cannot have always been so. The compensation of Hellas argues against a threshold strength early in Martian history. In either case, dynamic support of Tharsis by convection is not necessary, nor is it likely.

TABLE 1. Phase Angle Difference and Depths of Compensation for Selected Harmonics

Harmonic Coefficient	$\Delta\lambda$/Fraction of Longitudinal Wavelength	Depth to Compensation, km		
		$\rho_1 = 3.0$	$\rho_1 = 3.3$	$\rho_1 = 3.6$
J_{22}	4.1°/0.02	1067	1077	1144
J_{32}	16.8°/0.09	588	574	575
J_{33}	9.3°/0.08	584	570	571

Values for ρ_1 are given in grams per cubic centimeter.

Acknowledgments. We thank J. E. Conel, F. Fanale, J. Lorell, W. J. Sjogren, and E. Christensen for useful discussions. John Morton did an outstanding job on the computer programing. Harold Masursky, David H. Scott, and Ralph Baldwin criticized the manuscript and made numerous helpful suggestions. This paper presents the results of one phase of research carried out at the Jet Propulsion Laboratory, California Institute of Technology, under contract NAS 7-100, sponsored by the Planetology Program Office, Office of Space Science, National Aeronautics and Space Administration.

REFERENCES

Arkani-Hamed, J., Viscosity of the moon, 1, After mare formation, *Moon, 6,* 100–111, 1973.
Artyushkov, E. V., Can the earth's crust be in a state of isostasy?, *J. Geophys. Res., 79,* 741–752, 1974.
Baldwin, R. B., A determination of the elastic limit of the outer layers of the moon, *Icarus, 9,* 401–404, 1968.
Baldwin, R. B., Absolute ages of the lunar maria and large craters, 2, The viscosity of the moon's outer layers, *Icarus, 13,* 215–225, 1970.
Baldwin, R. B., The question of isostasy on the moon, *Phys. Earth Planet. Interiors, 4,* 167–179, 1971.
Balmino, G., K. Lambeck, and W. M. Kaula, A spherical harmonic analysis of the earth's topography, *J. Geophys. Res., 78,* 478–481, 1973.
Caputo, M., The minimum strength of the earth, *J. Geophys. Res., 70,* 955–963, 1965.
Carr, M. H., Volcanism on Mars, *J. Geophys. Res., 78,* 4049–4062, 1973.
Carr, M. H., H. Masursky, and R. S. Saunders, A generalized geologic map of Mars, *J. Geophys. Res., 78,* 4031–4036, 1973.
Chapman, C. R., Cratering on Mars, 1, Cratering and obliteration history, *Icarus, 22,* 272–291, 1974.
Chinnery, M. A., The strength of the earth's crust under horizontal shear stress, *J. Geophys. Res., 69,* 2085–2089, 1964.
Christensen, E. J., Martian topography derived from occultation, radar, and spectral experiments, *J. Geophys. Res., 80,* this issue, 1975.
Gaposchkin, E. M., and K. Lambeck, Earth's gravity field to sixteenth degree and station coordinates from satellite and terrestrial data, *J. Geophys. Res., 75,* 4855–4883, 1971.
Hartmann, W. K., Martian cratering, 4, Mariner 9 initial analysis of cratering chronology, *J. Geophys. Res., 78,* 4096–4116, 1973.
Jeffreys, H., *The Earth,* 5th ed., p. 479, Cambridge University Press, London, 1970.
Johnston, D. H., T. R. McGetchin, and M. N. Toksöz, The thermal state and internal structure of Mars, *J. Geophys. Res., 79,* 3959–3971, 1974.
Jones, K. L., Evidence for an episode of crater obliteration intermediate in Martian history, *J. Geophys. Res., 79,* 3917–3931, 1974.
O'Keefe, J. A., Isostasy on the moon, *Science, 162,* 1405–1407, 1968.

Phillips, R. J., J. E. Conel, E. A. Abbott, W. L. Sjogren, and J. B. Morton, Mascons: Progress toward a unique solution for mass distribution, *J. Geophys. Res., 77,* 7106–7114, 1972.

Phillips, R. J., R. S. Saunders, and J. E. Conel, Mars: Crustal structure inferred from Bouguer gravity anomalies, *J. Geophys. Res., 78,* 4815–4820, 1973.

Sjogren, W. L., J. Lorell, L. Wong, and W. Downs, Mars gravity field based on a short-arc technique, *J. Geophys. Res., 80,* this issue, 1975.

Soderblom, L. A., C. D. Condit, R. A. West, B. M. Herman, and T. J. Kreidler, Martian planetwide crater distributions: Implications for geologic history and surface processes, *Icarus, 22,* 239–263, 1974.

(Received July 29, 1974;
revised February 27, 1975;
accepted March 3, 1975.)

26

Reprinted from *Icarus* **32**:73–84 (1977)

Internal Structure and Properties of Mars

DAVID H. JOHNSTON AND M. NAFI TOKSÖZ

Department of Earth and Planetary Sciences, Massachusetts Institute of Technology, Cambridge, Massachusetts 02139

Received August 4, 1976; revised January 3, 1977

Theoretical physical models of the Martian interior are presented in the light of new and revised data and constraints. These models include thermal evolution, densities, and seismic wave velocities. The interior of Mars appears to be Earth-like in many respects. Although thermal models indicate that Mars has passed its peak of evolution it may still have an asthenosphere and may be moderately active tectonically. Mars has an Fe–FeS core with a radius of 1500–2000 km. The mantle is enriched in FeO with an olivine composition of about Fo_{75}. Theoretically determined seismic wave velocities are relatively well constrained in the mantle with upper-mantle P_n velocities ranging from 7.64 to 7.80 km/sec. However, there are wide variations in V_P in the core dependent on composition. The shadow zone due to the core is larger than the Earth's.

In this paper we present a set of physical models of the Martian interior based on available data. Some of the results presented are extensions of work published in Johnston *et al.* (1974; hereafter referred to as Paper 1). However, improved theoretical techniques, revised constraints, and additional data that have become available since 1974 warrant another study of the structure and properties of the interior of Mars.

Interest in the structure of Mars has a long and involved history. Early studies of the thermal evolution include Urey (1951, 1952), MacDonald (1962), Kopal (1962), Lee (1968), Anderson and Phinney (1967), Hanks and Anderson (1969), and Reynolds and Summers (1969). More recent papers (Binder, 1969; Binder and Davis, 1973; Anderson, 1972; Ringwood and Clark, 1971; Toksöz and Johnston, 1974, 1976; Johnston *et al.*, 1974) have calculated density and compositional models including a metallic core. The debate over whether or not Mars has a core still continues, with Strangway *et al.* (1976) postulating a planet with an undifferentiated deep interior.

In this paper, thermal models are briefly presented followed by density and seismic velocity models. The final section summarizes the main features of the Martian structure.

DATA ON THE MARTIAN INTERIOR

Theoretical calculations of planetary interior models require constraints based on observed data to limit the problem to realistic models. Although Mars is not as strongly constrained as the Earth or Moon, data primarily from Mariner 9 have enabled us to revise and improve earlier thermal and density models. Many of these data are concerned with the interpretation of Mariner 9 photographs which yield evidence of surface evolution history, volcanism, and differentiation of the planet (Masursky, 1973; Soderblom *et al.*, 1974; Arvidson, 1974; Chapman, 1974; Jones, 1974; Malin, 1976; Murray *et al.*, 1972; Mutch and Head, 1975). These types of data are particularly useful in constraining thermal history models. A possible evolution history proposed by Masursky (1973) and its relation to thermal models is discussed in Paper 1. It appears that Mars, like the Moon (probably the Earth, Venus, and Mercury also), suffered through a period of intense bombardment about 4 b.y. ago (Tera and Wasserburg, 1976) shortly after origin. This implies that a crust had formed prior to this time, requiring relatively high initial temperatures. The later stages of Martian evolution show episodes of tectonic and volcanic activity. Tharsis volcanism probably occurred about a billion years ago (Soderblom *et al.*, 1974) and is a major constraint in the thermal evolution and temperature models of Mars.

Perhaps the most important data concerning the present structure of Mars are the mass, M, and moment-of-inertia factor, C/Ma^2. These roughly determine the density variation within the planet and indicate the presence of a high-density core. While the mass of Mars is well determined, the moment-of-inertia factor is not. The interpretation of the second zonal harmonic of the gravitational field (J_2) in terms of the hydrostatic and nonhydrostatic contributions depends critically on assumptions of crustal structure (Binder and Davis, 1973; Weir, 1975; Reasenberg, 1977). Binder and Davis (1973), assuming an isostatically compensated equatorial bulge of 8 km, estimate C/Ma^2 ranging from about 0.374 to 0.370.

Analysis of the gravity field and topography of Mars (Phillips and Saunders, 1975) indicates that the surface may be divided into two distinct groups: older crust isostatically compensated at a relatively shallow depth; and the Tharsis plateau, Chryse, and Amazonis lowlands, which are only partially compensated. The Tharsis plateau is the most prominent feature associated with an uncompensated gravity anomaly. These data suggest a description of Mars as a spheroid nearly in isostatic equilibrium plus an extra mass represented by the Tharsis region. Reasenberg (1977) uses this model to calculate the nonhydrostatic contribution of Tharsis to J_2. Thus, assuming that the rigidity supporting Tharsis could be relaxed, Reasenberg finds that the body, now in hydrostatic equilibrium, would have a moment-of-inertia factor of 0.3654 ± 0.001. This model is further justified in that the optical and dynamic flattening calculated are in much better agreement than those for other models. We therefore adopt for this paper the value of C/Ma^2 obtained by Reasenberg. This value is significantly lower than the 0.377 used in Paper 1. It alters the density models presented there in such a way as to emphasize an increased density contrast between the shallow layers and deep interior of Mars, possibly implying a larger core radius. Other physical parameters for Mars which are used in the thermal and density calculations are presented in Table I. A detailed discussion of these parameters may be found in Paper 1.

TABLE I

PARAMETERS USED IN MARS MODELS

Parameter	Value
Radius (km)	3389
Mean density (g/cm³)	3.96
C/Ma^2	0.3654
Heat of fusion (mantle) (J/g)	400
Cp (mantle) (J/g°C)	1.2
Surface temperature (°C)	−40
U concentration (ppb)	15
Th/U ratio	4
K/U ratio	50,000

THERMAL MODELS

The implied evolution history and presence of a core discussed in the preceding section place strong constraints on the thermal history of Mars. The presence of a weak magnetic field (Dolginov *et al.*, 1972) also seems to imply that present-day internal temperatures should produce a

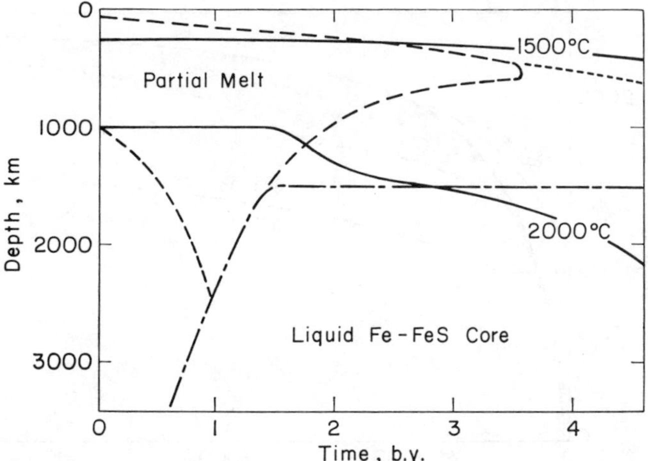

FIG. 1. Thermal evolution of the Martian interior as a function of depth and time based on the convection temperature calculations of Hsui *et al.* (1976). Regions of mantle partial melting and core formation are shown. The short dashed line represents the top of local partial melt associated with the convective currents.

molten or partially molten core. Condensation models of the solar nebula (Grossman and Larimer, 1974) suggest that the Martian core is enriched in FeS, and thus may contain small amounts of the heat-producing radioactive isotope ^{40}K (Lewis, 1971; Goettel, 1972).

Theoretical conduction thermal models consistent with the known constraints concerning the evolution of Mars and core formation are presented in Paper 1. Hsui *et al.* (1976) have calculated thermal models for Mars including the effect of solid-state convection by simultaneously solving the equation of motion and the time-dependent energy equation in spherical geometry. Since the process of core formation is not well understood, its effect upon convective structure is difficult to determine. Thus, prior to core formation, the calculations are carried out in the manner described in Paper 1, with similar initial conditions and physical parameters (Table I). After the separation of the core is completed, the core temperatures are assumed to follow an adiabatic gradient determined by the temperature at the core–mantle boundary. As in Paper 1, the presence of ^{40}K in the core is allowed.

The evolutionary history of Mars from Hsui *et al.* is shown in Fig. 1. It is assumed that core formation takes place before 1.5 b.y. after origin. The planet continues to heat up slightly after core formation but then starts to cool. Large tensile features of fractured plains (Carr, 1974) may have been associated with the heating-up period. A partially molten region in the upper mantle exists up to 1 b.y. ago. The thick lithosphere, however, resists lateral motions explaining the absence of plate tectonics. The short dashed line shows the top of local partial melt associated with the upwelling of convective currents. If water is present near the Martian surface, the solidus will be suppressed and partially molten regions may exist at a depth of 250 km at the present time. This could produce some present-day tectonic activity. There is a region of solid-state convection at the lower part of the mantle. The core is completely molten.

The present-day temperature profiles from both the conduction and convection

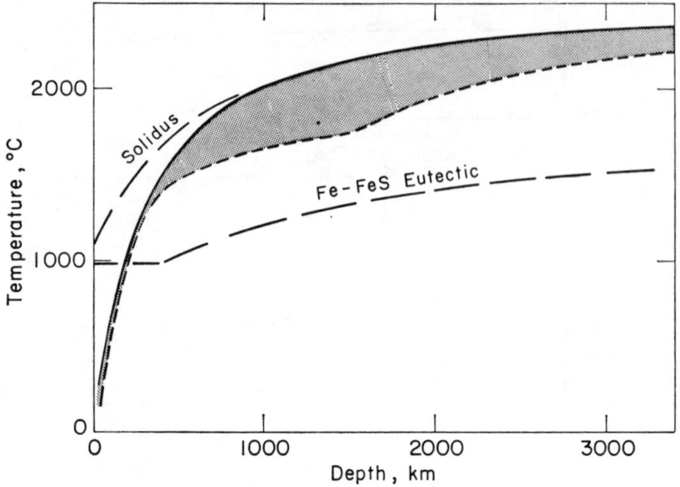

FIG. 2. Range of the present-day temperatures within Mars (shaded area), bounded by the conduction model of Paper 1 and the convection model of Hsui *et al.* (1976). Also shown are the dry-mantle solidus and the Fe–FeS eutectic temperatures.

models are interpreted to be bounds on the temperature distribution within Mars. These are used in the determination of density models and are shown in Fig. 2. Also shown are the dry-mantle solidus (extrapolated peridotite; Green and Ringwood, 1969) and an extrapolated Fe–FeS eutectic melting curve calculated by Usselman (1975a,b). In that convection most likely occurs within planetary interiors, we used the convective temperature profile in the determination of density and seismic wave velocities.

TABLE II

PARAMETERS USED IN DENSITY MODELS

Parameter	Value
Upper mantle (Fo_{75})	
K_0 (mbar)	1.255
dK/dT (kbar/°K)	−0.174
$\alpha(T = 20°C)$ ($°C^{-1}$)	2.4×10^{-5}
Lower mantle (spinel Fo_{75})	
K_0 (mbar)	1.893
Core	
K_0 (mbar)	1.2
$\alpha(T = 20°C)$ ($°C^{-1}$)	7.3×10^{-5}

DENSITY MODELS

Theoretical density models for Mars can be constructed using the mass and moment-of-inertia factor as constraints if a knowledge of the mantle and core compositions is assumed. These models are calculated for a warm, compressible, layered planet following the technique described in Paper 1 and in Solomon and Toksöz (1973). Since the core composition is not well determined, we present a range of possible density profiles.

The density models are derived by numerically integrating the equation of hydrostatic equilibrium where the density, ρ, is related to the pressure, P, by an equation of state, assumed to be the isothermal Birch–Murnaghan equation (Birch, 1938, 1952; Murnaghan, 1937):

$$P = \tfrac{3}{2}K_0[(\rho/\rho_0)^{7/2} - (\rho/\rho_0)^{5/2}]. \quad (1)$$

K_0 and ρ_0 are the isothermal bulk modulus and density at $P = 0$ and are computed at the temperature, T, at depths given by the thermal models shown in Fig. 2. The thermal expansion coefficient may be temperature dependent and is based on extrapolated data from experimental measurements for the appropriate material. Values for these parameters are listed in Table II.

Each model assumes a 50-km crust with a density of 3.0 g/cm^3 (Phillips *et al.*, 1973) which amounts to about 3% of the total mass of Mars. The bulk modulus and thermal expansion used are typical of terrestrial crustal rocks (Clark, 1966, p. 94).

As a starting model for the mantle, we assume a composition of Fo_{75}. Values of K_0 and dK/dT taken are given by Chung (1971). The use of Fo_{75} as a mantle model was based on preliminary density calculations in order to roughly estimate the STP mantle density, and thus its equivalent olivine composition.

A phase change from olivine to spinel phase may occur where the temperature profile crosses the P–T transition (Anderson, 1967a), as shown in Fig. 3. For the conduction thermal model, this occurs at a depth of about 1500 km and for the convection model at a depth of 1200 km. The standard temperature and pressure spinel-phase density is constrained to be 10% higher than the STP olivine-phase density. The elastic parameters for this layer are again taken from Chung (1971) and are listed in Table II. Other possible mantle phase changes would have smaller effects and are not considered in this paper.

The core composition (and therefore density) is considered to be a free parameter to be determined by the calculation. Core compositions considered range from $Fe_{85}S$ to FeS. An FeS composition has been proposed by Lewis (1972) for the Martian core, and $Fe_{85}S$ is similar to compositions proposed for the Earth (Usselman, 1975b; King and Ahrens, 1973).

Usselman (1975a) has shown from experimental determinations of the Fe–FeS eutectic temperature that a solid–solid phase transformation occurs in FeS at about 52 kbar. This is consistent with the high-pressure phase (hpp) evident from shock compression studies of pyrrhotite reported by King and Ahrens (1973). Assuming no further transitions, this is the phase present in the Martian core. Recent experimental work suggests a value of K_0 for the hpp of 1.2 mbar and a zero-pressure density increase of 15% over the low-

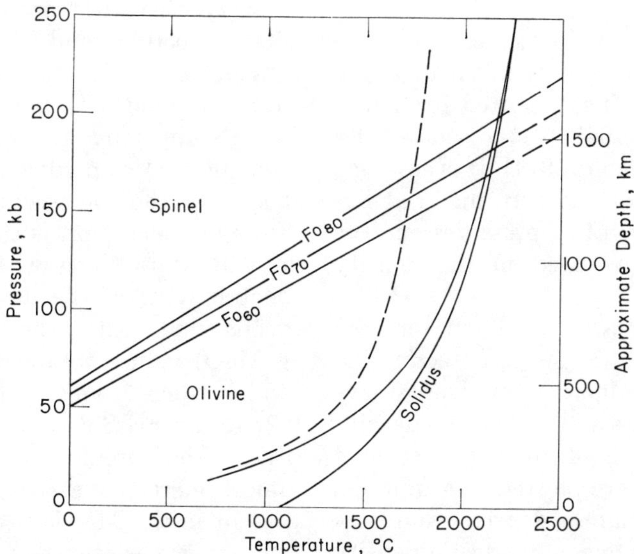

FIG. 3. Present-day temperature profiles from the conduction thermal model (solid line) and the convection model (dashed line) superimposed on the stability fields of olivine and spinel phase.

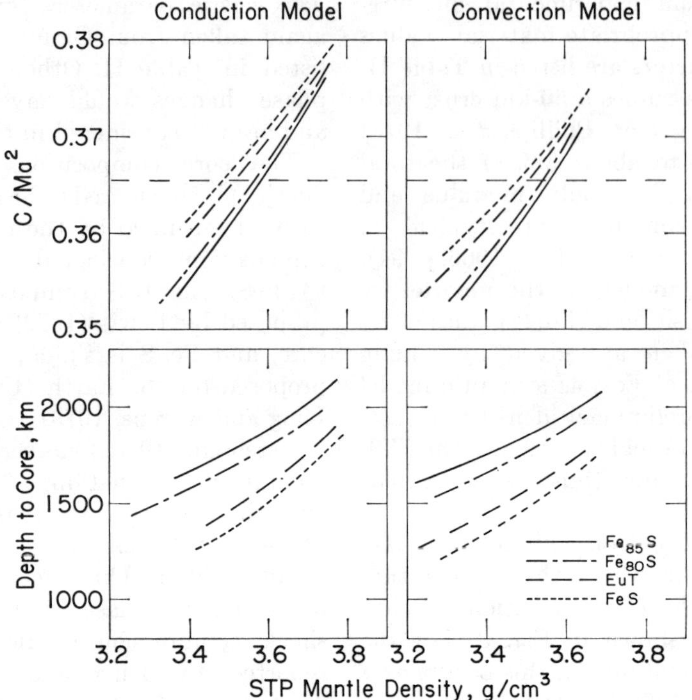

FIG. 4. Standard temperature and pressure upper-mantle density shown as a function of moment of inertia and depth to the core for varying core compositions (lower right corner) and for both temperature models. The horizontal dashed line is the moment-of-inertia factor used in this paper.

pressure phase (Ahrens, 1976; Pichulo *et al.*, 1976).

Thermal expansion in the core is temperature dependent and is taken to be an extrapolation of values reported for γ iron (Basinski *et al.*, 1955), as discussed by Siegfried and Solomon (1974). It is assumed that the melting of the core material has little effect at elevated pressures on the physical parameters used in the density calculations.

Density models are calculated for core compositions of $Fe_{85}S$, $Fe_{80}S$, eutectic, and FeS for both conductive and convective temperature profiles and fixed mean density of the planet. In Fig. 4, models of varying core compositions are plotted as a function of C/Ma^2, STP mantle density, and core depth. From these one may select a density model that best fits the moment-of-inertia factor. Valid density models satisfying both C/Ma^2 and the mean density are obtainable in every case with core radii varying from 1500 to 2050 km and STP mantle densities ranging from 3.47 to 3.58 g/cm³. The Martian mantle is more dense than the Earth's upper mantle, implying an enrichment in FeO as predicted by condensation models of the solar nebula (Lewis, 1972). However, such models that predict an FeS core in Mars along with extreme enrichment in FeO of the mantle appear to be inconsistent with the data. FeO/(FeO + MgO) values obtained from the density models range from 0.19 for an FeS core to 0.27 for an $Fe_{85}S$ core. Lewis (1972), on the other hand, predicts a value of 0.5. It is almost inescapable that some free iron is present in the Martian core.

Further variations in density and core radius may be obtained from Fig. 4 for different assumed values of C/Ma^2.

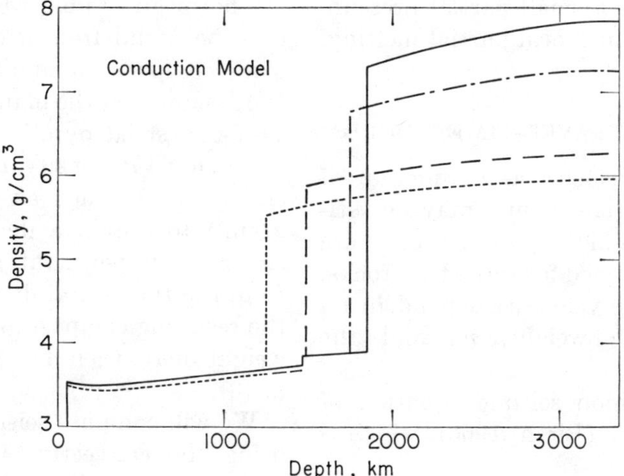

FIG. 5. Density profiles satisfying both the mean density and moment of inertia for varying core compositions (see Fig. 4) using temperatures from the conduction model.

Density profiles for valid models satisfying $C/Ma^2 = 0.3654$ are shown in Figs. 5 and 6 for the conduction and convection temperatures, respectively. Comparison of these figures shows the range of density variations due to extreme models of temperature and composition considered. We prefer models based on convective profiles (Fig. 6) since they represent more realistic thermal processes. Several features stand out. First, the change in core density is reflected primarily in core radius, with relatively small changes in the STP mantle density. This is important because it allows us to set rather specific bounds on the mantle composition and seismic velocities that will be discussed later. Second, close examination of these figures reveals that in the upper mantle, thermal expansion is dominant over compression resulting in a slight density inversion of about 2% at a depth of about 250–300 km. This is the

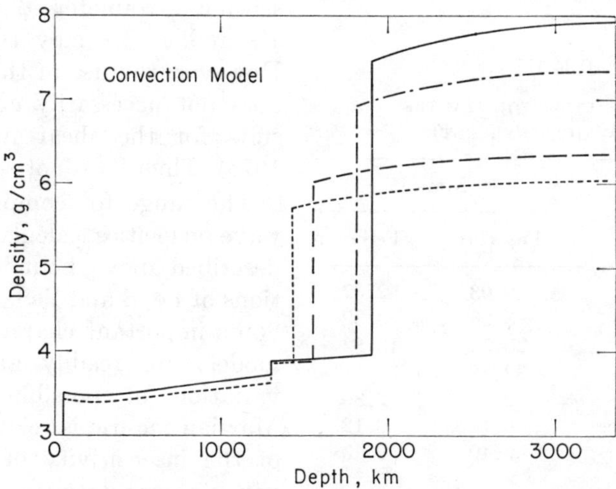

FIG. 6. Density profiles using temperatures from the convection model.

same region where a small partial pressure of water could induce local partial melting, as discussed earlier.

VELOCITY AND TRAVEL-TIME MODELS

Seismic wave velocity as a function of depth in the Martian mantle may be estimated using seismic equations of state empirically determined for terrestrial rocks. Most of these have velocities dependent on the mean atomic weight, $\bar{m}$, and the density, ρ.

The most common seismic equation of state is Birch's law (Birch, 1966), which is of the form

$$V = a(\bar{m}) + b\rho. \quad (2)$$

Simmons (1964) proposed a generalized form of the equation for P waves that will be used in this paper:

$$V_P = -0.98 + 0.7(21 - \bar{m}) + 4.60[\text{CaO}] + 2.76\rho, \quad (3)$$

reflecting the fact that Ca-rich rocks have a higher V than other rocks with similar $\bar{m}$.

Wave velocities in the core are determined from the calculated bulk modulus based on composition and density models. Since the core is molten (the shear modulus, $\mu = 0$),

$$V_{P_{core}} = (K/\rho)^{1/2}. \quad (4)$$

TABLE III

CHEMICAL ANALYSES FOR MANTLE MODELS (WEIGHT PERCENT)

Oxide	Pyrolite	Mars	
		Fe_{85} core	FeS core
SiO_2	45.16	39.03	42.32
Al_2O_3	3.54	3.06	3.32
FeO	8.04	20.53	13.82
MgO	37.47	32.38	35.11
CaO	3.08	2.66	2.89
K_2O	0.13	0.11	0.12
Fe_2O_3	0.46	0.40	0.43
Mean atomic weight		22.68	22.12

The mean atomic weight of the mantle may be found from a consideration of its oxide composition as discussed in Paper 1. We assume that the mantle may be modeled as a terrestrial pyrolite (Ringwood, 1966) with an STP density of 3.38 g/cm^3 with enough FeO added as wustite (ρ = 5.745 g/cm^3) to obtain a density equivalent to the STP upper-mantle density obtained from the theoretical density models. Given the resulting composition, the mean atomic weight may then be determined for use in (3).

We will compute seismic velocity models using the convective temperature profile and two compositions based on $Fe_{85}S$ and FeS cores. These represent two extremes in core radius and density. Table III lists the upper-mantle oxide compositions and mean atomic weights for these models. The lower mantle is assumed to have an $\bar{m}$ equal to that of the upper mantle. Furthermore, we propose a two-layered crust with an upper layer of 20 km and V_P = 6 km/sec and a 30-km lower layer with V_P = 7 km/sec, although such an assumption is not necessary for the models.

Shear wave velocities, V_S, may be found by adopting a reasonable value for Poisson's ratio, $\sigma = f(V_P, V_S)$, or by using Anderson's (1967b) equation of state using the seismic parameter $\phi \approx (21\rho/\bar{m})^3$. Given V_P and ϕ, V_S may then be determined. However, the use of this equation of state does not necessarily give independent results for the shear modulus (Anderson, 1976). Thus V_S is not strongly constrained.

The range for compressional and shear wave velocities calculated in the manner described above, bounded by core compositions of $Fe_{85}S$ and FeS, are shown in Fig. 7. Some important characteristics of velocity models are readily apparent. First, the variation of possible velocities in the Martian mantle is relatively small because of the insensitivity of mantle density to varying core density. While mantle densities are lower for the FeS core compared to

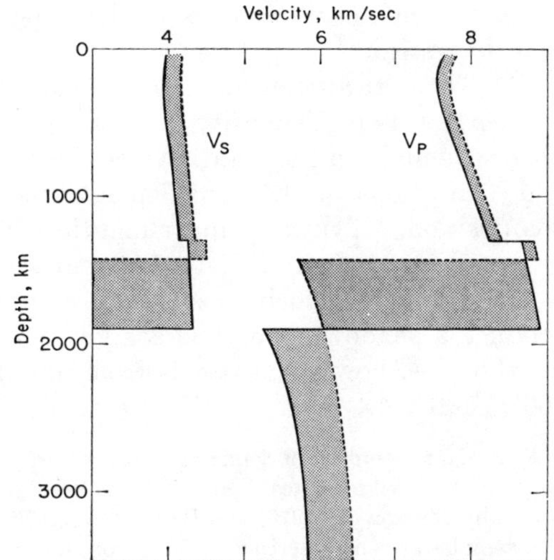

FIG. 7. Range of seismic wave velocities as a function of depth for both P and S waves (shaded areas), calculated using the densities shown in Fig. 6. The range is bounded by models with core compositions of FeS (dashed line) and $Fe_{85}S$ (solid line.

the $Fe_{85}S$ core, the mean atomic weight is also lower and thus the velocities are higher. Second, a large velocity contrast exists at the core-mantle boundary and this produces a seismic shadow zone which is more prominent than that which exists in the Earth. Upper-mantle (P_n) velocities range from 7.64 to 7.80 km/sec. A low-velocity zone associated with the density inversion occurs at about 250 km. Velocity increases at the base of the mantle are due to the olivine–spinel phase change. In reality this boundary may be smeared out and less sharp than shown on the model.

Poisson's ratio generally varies from about 0.30 at the base of the crust to 0.35 near the core–mantle boundary. The high value reflects the enrichment of FeO in the Martian mantle.

To demonstrate the effects of these models on seismic wave propagation, travel times were calculated for the FeS core composition. *P*–wave ray paths for a surface source (Marsquake or meteorite impact) shown in Fig. 8. Travel time versus distance for P waves is shown in the same figure. In the mantle, seismic ray propagation and the travel-time curve are well behaved, except for a reverse branch due to the velocity increase at the bottom. There is no shadow zone from the upper-mantle low-velocity zone due to the compensating effect of curvature, i.e., r/V is monotonically decreasing with depth. However, relatively lower amplitudes might be expected from 15 to 45°. The most striking feature of these models, of course, is the shadow zone due to the core. For an FeS core the range is from 90 to 152°, while for an $Fe_{85}S$ core this zone extends from 115 to 170°. The shadow zone for Mars is larger than the

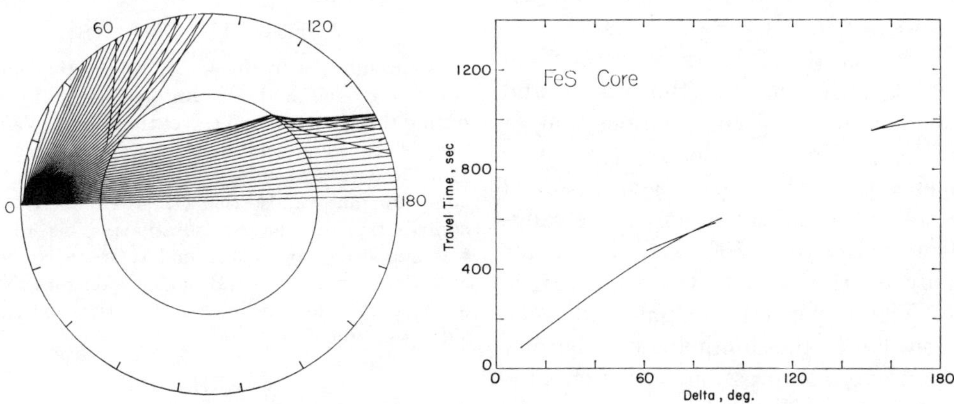

FIG. 8. P-wave ray paths and travel time versus distance curves calculated from the FeS core velocity model shown in Fig. 7.

Earth's due to a greater velocity contrast between the mantle and core.

SUMMARY AND CONCLUSIONS

We have presented a suite of physically realistic models of the interior of Mars that satisfy the available constraints. What has emerged is a picture of the Martian interior that is similar to the Earth's in many respects. Some inferences that can be made on the basis of these models are:

(1) Thermal history models indicate that Mars has passed its peak of evolution. However, it has an asthenosphere (about 250 km deep) and, most likely, a convecting lower mantle. The planet may still be moderately active tectonically.

(2) Mars has a core with a radius of about 1500–2000 km with a composition on the Fe-rich side of the Fe–FeS zero-pressure eutectic composition. More sulfur-rich cores do not satisfy the constraint of the moment of inertia while simultaneously providing an FeO-rich mantle. The presence of a molten core is consistent with an internally generated magnetic field, however, the low observed field strength may be due to either stratification within the core or less conductive material due to a higher sulfur content relative to the Earth's core. Alternate models of magnetic field generation infer distinctly different models (Strangway *et al.*, 1976).

(3) The mantle of Mars is enriched in FeO relative to that of the Earth and corresponds to an olivine composition of about Fo_{75}. The mantle is essentially homogeneous with the olivine-to-spinel phase transition occurring at depths ranging from 1200 to 1500 km dependent primarily on the present-day temperature profile. The upper mantle has a density inversion due to the dominance of thermal expansion over compression. This temperature increase could produce an asthenosphere at a depth of about 250 km. Local partial melting could occur in this zone if small amounts of water exist in the upper-mantle system.

(4) Theoretical seismic wave velocity models based on the density models exhibit narrow bounds on the mantle velocities but wide variations in the core dependent on composition. Typical upper-mantle P_n velocities range from 7.64 to 7.80 km/sec. A mantle low-velocity zone does not produce a shadow zone. The shadow zone for the core, however, extends from about 100 to 160°.

Note added in proof. Inorganic analyses of surface samples from Viking sites 1 and 2 have been reported by Clark *et al.* (1976) and Baird *et al.* (1976). The samples are characterized by high iron, moderate magnesium, calcium, and sulfur, and low aluminum abundances. A comparison of these values with the mantle models proposed in this paper (Table III) shows reasonable agreement, especially for the enriched Fe. A simple ascending differentiation of the mantle material will result in a substantial depletion of MgO (Maderazzo, 1977; Carmichael *et al.*, 1974), giving a surface olivine composition of nearly Fa_{100}.

Although sulfur is not included in our mantle model, the whole planet is enriched in sulfur relative to Earth abundances when the core is included. Density considerations preclude a 6–10% SO_3 abundance, as observed in the surface samples for the mantle.

The relatively high concentration of sulfur may be due to a process proposed by Huguenin (1976) in which sulfates and other salts formed by photochemical weathering precipitate from a brine and are enriched in a thin surface layer.

ACKNOWLEDGMENTS

We would like to thank Dr. Albert Hsui for the thermal models and Dr. Anton Dainty for computing the ray paths and travel-time curve. Valuable discussions were held with Marc Maderazzo and Drs. Robert Reasenberg, Robert Huguenin, and John Longhi. We also thank Joan Gomberg for her contribution to the computations. This research was supported by NASA under Grant NGS-7081 and Subcontract 66437B under Contract NAS1-9703, partly in preparation for the analysis of Viking seismic data.

REFERENCES

Ahrens, T. J. (1976). Shock wave data for mantle materials (abstract). *Eos, Trans. Amer. Geophys. Union* **57**, 326.

ANDERSON, D. L. (1967a). Phase changes in the upper mantle. *Science* **157**, 1165–1173.

ANDERSON, D. L. (1967b). A seismic equation of state. *Geophys. J.* **13**, 9–30.

ANDERSON, D. L. (1972). Internal constitution of Mars. *J. Geophys. Res.* **77**, 789–795.

ANDERSON, D. L. (1976). The lower mantle (abstract). *Eos, Trans. Amer. Geophys. Union* **57**, 327.

ANDERSON, D. L., AND PHINNEY, R. A. (1967). Early thermal history of the terrestrial planets. In *Mantles of the Earth and Terrestrial Planets* (S. K. Runcorn, Ed.), pp. 113–126. Interscience, London.

ARVIDSON, R. E. (1974). Morphologic classification of Martian craters and some implications. *Icarus* **22**, 264–271.

BAIRD, A. K., TOULMIN, P., III, CLARK, B. C., ROSE, H. J., JR., KEIL, K., CHRISTIAN, R. P., AND GOODLING, J. L. (1976). Mineralogic/petrologic implications of Viking geochemical results from Mars: Interim report. *Science* **174**, 1288–1293.

BASINSKI, Z. S., HUME-ROTHERY, W., AND SUTTON, A. L. (1955). The lattice expansion of iron. *Proc. Roy. Soc. London Ser. A* **229**, 459–467.

BINDER, A. B. (1969). Internal structure of Mars. *J. Geophys. Res.* **74**, 3110–3117.

BINDER, A. B., AND DAVIS, D. R. (1973). Internal structure of Mars. *Phys. Earth Planet. Int.* **7**, 477–485.

BIRCH, F. (1938). The effect of pressure upon the elastic properties of isotropic solids, according to Murnaghan's theory of finite strain. *J. Appl. Phys.* **9**, 279–288.

BIRCH, F. (1952). Elasticity and constitution of the Earth's interior. *J. Geophys. Res.* **57**, 277–286.

BIRCH, F. (1961). The velocity of compressional waves in rocks to 10 kilobars, Part 2. *J. Geophys. Res.* **66**, 2199–2224.

CARMICHAEL, I. S. E., TURNER, F. J., AND VERHOOGEN, J. (1974). *Igneous Petrology*, pp. 487–527, 603–620. McGraw-Hill, New York.

CARR, M. H. (1974). Techtonism and volcanism of the Tharsis region of Mars. *J. Geophys. Res.* **79**, 3943–3949.

CHAPMAN, C. R. (1974). Cratering on Mars, I. Cratering and obliteration history. *Icarus* **22**, 272–291.

CHUNG, D. H. (1971). Elasticity and equations of state of olivines in the Mg_2SiO_4–Fe_2SiO_4 system. *Geophys. J. Roy. Astron. Soc.* **25**, 511–538.

CLARK, B. C., BAIRD, A. K., ROSE, H. J., JR., TOULMIN, P., III, KEIL, K., CASTRO, A. J., KELLIHER, W. C., ROWE, C. D., AND EVANS, P. H. (1976). Inorganic analyses of Martian surface samples at the Viking landing sites. *Science* **194**, 1283–1288.

CLARK, S. P., JR. (1966). *Handbook of Physical Constants*. Geol. Soc. Amer. Memoir.

DOLGINOV, S. S., YEROSHENKO, Y. G., AND ZHUZGOV, L. N. (1972). The magnetic field in the very close neighborhood of Mars according to data from the Mars 2 and Mars 3 spacecraft. Goddard Space Flight Center Report X-690-72-434.

GOETTEL, K. A. (1972). Partitioning of potassium between silicates and sulphide melts: Experiments relevant to the Earth's core. *Phys. Earth Planet. Int.* **6**, 161–166.

GREEN, D. H., AND RINGWOOD, A. E. (1969). Mineralogy of peridotitic compositions under mantle conditions. *Phys. Earth Planet. Int.* **3**, 359–371.

GROSSMAN, L., AND LARIMER, J. W. (1974). Early chemical history of the solar system. *Rev. Geophys. Space Phys.* **12**, 71–102.

HANKS, T. C., AND ANDERSON, D. L. (1969). The early thermal history of the Earth. *Phys. Earth Planet. Int.* **2**, 19–29.

HSUI, A. T., TOKSÖZ, M. N., AND JOHNSTON, D. H. (1976). Thermal evolution of the Moon, Mercury and Mars (abstract). *Eos, Trans. Amer. Geophys. Union* **57**, 271.

HUGUENIN, R. (1976). Chemical evolution of the Martian atmosphere by surface weathering. In *Proc. NASA Colloquium on Water in Planetary Regoliths*. Dartmouth College.

JOHNSTON, D. H., MCGETCHIN, T. R., AND TOKSÖZ, M. N. (1974). Thermal state and internal structure of Mars. *J. Geophys. Res.* **79**, 3959–3971.

JONES, K. L. (1974). Evidence for an episode of crater obliteration intermediate in Martian history. *J. Geophys. Res.* **79**, 3917–3932.

KING, D. A., AND AHRENS, T. J. (1973). Shock compression of iron sulphide and possible sulphur content of the Earth's core. *Nature Phys. Sci.* **243**, 82–84.

KOPAL, Z. (1962). Thermal history of the Moon and of the terrestrial planets: Numerical results. JPL Technical Report No. 32-225.

LEE, W. H. K. (1968). Effects of selective fusion on the thermal history of the Moon, Mars and Venus. *Earth Planet. Sci. Lett.* **4**, 277–283.

LEWIS, J. S. (1971). Consequences of the presence of sulfur in the core of the Earth. *Earth Planet. Sci. Lett.* **11**, 130–134.

LEWIS, J. S. (1972). Metal/silicate fractionation in the solar system. *Earth Planet. Sci. Lett.* **15**, 286–290.

MACDONALD, G. J. F. (1962). On the internal constitution of the inner planets. *J. Geophys. Res.* **67**, 2945–2974.

MADERAZZO, M. (1977). The Viking 1 inorganic analysis experiment: Interpretation for petrologic information. M.S. Thesis. MIT, Cambridge, Mass.

Malin, M. C. (1976). Ph.D. Thesis, Calif. Inst. of Tech., Pasadena, Calif.

Masursky, H. (1973). An overview of geological results from Mariner 9. *J. Geophys. Res.* **78**, 4009–4030.

Murnaghan, F. D. (1937). Finite deformations of an elastic solid. *Amer. J. Math.* **59**, 235–260.

Murray, B. C., Soderblom, L. A., Cutts, J. A., Sharp, R. P., Milton, D. J., and Leighton, R. B. (1972). Geological framework of the south polar region of Mars. *Icarus* **17**, 328–345.

Mutch, T. A., and Head, J. W. (1975). The geology of Mars: A brief review of some recent results. *Rev. Geophys. Space Phys.* **13**, 411–416.

Phillips, R. J., Saunders, R. S., and Conel, J. E. (1973). Mars; Crustal structure inferred from Bouguer gravity anomalies. *J. Geophys. Res.* **78**, 4815–4820.

Phillips, R. J., and Saunders, R. S. (1975). The isostatic state of Martian topography. *J. Geophys. Res.* **80**, 2893–2898.

Pichulo, R. O., Takahashi, T., and Weaver, J. S. (1976). High pressure polymorphs of FeS (abstract). *Eos, Trans. Amer. Geophys. Union* **57**, 340.

Reasenberg, R. (1977). The moment of inertia and isostasy of Mars. *J. Geophys. Res.* **82**, 369–375.

Reynolds, R. T., and Summers, A. L. (1969). Calculations on the composition of the terrestrial planets. *J. Geophys. Res.* **74**, 2494–2511.

Ringwood, A. E. (1966). Mineralogy of the mantle. In *Advances in Earth Science* (P. M. Hurley, Ed.), pp. 287–356. MIT Press, Cambridge, Mass.

Ringwood, A. E., and Clark, S. P. (1971). Internal constitution of Mars. *Nature* **234**, 89–92.

Siegfried, R. W., II, and Solomon, S. C. (1974). Mercury: Internal structure and thermal evolution. *Icarus* **23**, 192–205.

Simmons, G. (1964). Velocity of compressional waves in various minerals at pressures to 10 kilobars. *J. Geophys. Res.* **69**, 1117–1121.

Soderblom, L. A., Condit, C. D., West, R. A., Herman, B. M., and Kreidler, T. J. (1974). Martian planetwide crater distributions: Implications for geologic history and surface processes, *Icarus* **22**, 239–263.

Solomon, S. C., and Toksöz, M. N. (1973). Internal constitution and evolution of the Moon. *Phys. Earth Planet. Int.* **7**, 15–38.

Strangway, D. W., Sharpe, H. N., and Peltier, W. R. (1976). Planetary magnetism—primordial or dynamo (abstract). *Lunar Science VII*, 842–844.

Tera, F., and Wasserburg, G. J. (1976). Lunar ball games and other sports (abstract). *Lunar Science VII*, 858–860.

Toksöz, M. N., and Johnston, D. H. (1974). The evolution of the Moon. *Icarus* **21**, 389–414.

Toksöz, M. N., and Johnston, D. H. (1976). The evolution of the Moon and the terrestrial planets. In *Proceedings of the Soviet–American Conference on the Cosmochemistry of the Moon and Planets*. In press.

Urey, H. C. (1951). The origin and development of the Earth and other terrestrial planets. *Geochim. Cosmochim. Acta* **1**, 209–277.

Urey, H. C. (1952). *The Planets, Their Origin and Development*. Yale University Press, New Haven, Conn.

Usselman, T. M. (1975a). Experimental approach to the state of the core. I. The liquidus relations of the Fe-rich portion of the Fe–Ni–S system from 30 to 100 kb. *Amer. J. Sci.* **275**, 278–290.

Usselman, T. M. (1975b). Experimental approach to the state of the core. II. Composition and thermal regime. *Amer. J. Sci.* **275**, 278–290.

Weir, S. (1975). Comment on Martian density models and the moment of inertia (abstract). *Eos, Trans. Amer. Geophys. Union* **56**, 387.

Part IX

CRATERING

Editor's Comments on Papers 27 and 28

27 **SODERBLOM et al.**
Martian Planetwide Crater Distributions: Implications for Geologic History and Surface Processes

28 **NEUKUM and WISE**
Mars: A Standard Crater Curve and Possible New Time Scale

In the absence of returned rock samples, which could be precisely dated by radiometric techniques in the laboratory, the chronology of events on Mars must be inferred by measurements of crater density. Calculation of absolute ages from crater frequencies involves uncertainties in knowledge of the influx of impacting objects in the Solar System throughout time. One way to overcome this problem is to relate the martian crater frequencies to the Moon, since the lunar cratering history has been accurately established from radiometric dates on Apollo and Luna samples (Wasserburg et al. 1972; Papanastassiou et al. 1971). The papers considered in this section approach the problem somewhat differently, and the results place constraints on the geological history of Mars.

W. K. Hartmann (1973) originally assumed that because Mars was located near the asteroid belt, it would be struck by ten times the number of particles hitting the Moon, although the actual mass distribution would be identical. The number of craters of a particular diameter, however, would be only six times that of the Moon since the impact velocity is lower by a factor of 1/1.4 (due to differences in orbital and gravitational factors). Hartmann's cratering rates lead to age estimates of several hundred million years for the Tharsis volcanoes.

The assumption that Mars is struck by 10 times the number of objects as the Moon has been challenged by Chapman (1974) and Soderblom et al. (Paper 27). Soderblom et al. conclude that Mars and the Moon have experienced a parallel cratering history. First,

Mars and Moon can both be divided into two distinct physiographic provinces: heavily cratered highlands and sparsely cratered lowlands, which Soderblom et al. relate to an exponential decline in the cratering rate by analogy to the Moon. Furthermore, comparable densities for 4–10 km craters occur on both lunar maria and martian cratered plains. Similar densities of "fresh" bowl-shaped craters on both the martian heavily cratered highlands and cratered plains indicate that only a short time elapsed between the end of the intense bombardment, erosion, and formation of the oldest cratered plains unit, which Soderblom et al. estimate to have occurred about 4 billion years ago, again by analogy to the Moon.

The largely circumstantial arguments of Soderblom et al. have been strengthened by data compiled by G. Neukum and D. U. Wise (Paper 28). They find that the martian crater frequency-diameter curve matches the standard lunar curve, if lunar diameters are shifted downward by 1.5x, which corresponds to the relative difference in impacting velocities on both bodies. The good fit for both crater curves implies that the Moon and Mars have had a similar impact flux, in agreement with Soderblom et al., but in contrast to Hartmann. Near saturation crater densities on Phobos, with values close to the lunar highlands, support the conclusion that lunar and martian crater fluxes did not differ by more than a factor of 2 (Veverka and Duxbury 1977). Neukum and Wise concur with Soderblom et al. that the erosional episode occurred before or during the time of deposition of the oldest plains unit, some 4 billion years ago, but determine the age of the Tharsis volcanoes to be around 2.6 billion years ago. (Neukum and Wise point out an error in Soderblom's estimate of $\sim$1 billion years for the Tharsis volcanoes, thus accounting for the age discrepancy).

Crater production rates estimated by Hartmann (1977) and Shoemaker and Helin (1977) show that the relative rates for Mars and the Moon do not differ by more than a factor of 2 or 3, which lies well within the range of uncertainty.

27

Reprinted from *Icarus* **22**:239–263 (1974)

Martian Planetwide Crater Distributions: Implications for Geologic History and Surface Processes[1,2]

L. A. SODERBLOM, C. D. CONDIT, R. A. WEST, B. M. HERMAN, AND T. J. KREIDLER

Center of Astrogeological Studies, U.S. Geological Survey, Flagstaff, Arizona 86001, and Division of Geological and Planetary Science, California Institute of Technology, Pasadena, California 91109

Received February 19 1974; revised April 12 1974

Population-density maps of craters in three size ranges (0.6 to 1.2 km, 4 to 10 km, and >20 km in diameter) were compiled for most of Mars from Mariner 9 imagery. These data provide: historical records of the eolian processes (0.6 to 1.2 km craters); stratigraphic, relative, and absolute timescales (4 to 10 km craters); and a history of the early postaccretional evolution of the uplands (>20 km craters).

Based on the distribution of *large craters* (>20 km diameters), Mars is divisible into two general classes of terrain, densely cratered and very lightly cratered—a division remarkably like the uplands–maria dichotomy of the moon. It is probable that this bimodal character in the density distribution of large craters arose from an abrupt transition in the impact flux rate from an early intense period associated with the tailing off of accretion to an extended quiescent epoch, *not* from a void in geological activity during much of Mars' history. Radio-isotope studies of Apollo lunar samples show that this transition occurred on the moon in a short time.

The *intermediate-sized* craters (4 to 10 km diameter) and the *small-sized* craters (0.6 to 1.2 km diameter) appear to be genetically related. The smaller ones are apparently secondary impact craters generated by the former. Most of the craters in the larger of these two size classes appear fresh and uneroded, although many are partly buried by dust mantles. Poleward of the 40° parallels the small fresh craters are notably absent owing to these mantles. The density of small craters is highest in an irregular band centered at 20° S. This band coincides closely with (1) the zone of permanent low-albedo markings; (2) the "wind equator" (the latitude of zero net north or south transport at the surface); and (3) a band that includes a majority of the small dendritic channels. Situated in the southernmost part of the equatorial unmantled terrain which extends from about 40° N to 40°S, this band is apparently devoid of even a thin mantle. Because this belt is also coincident with the latitutde of maximum solar insolation (periapsis occurs near summer solstice), we suggest that this band arises from the asymmetrical global wind patterns at the surface and that the band probably follows the latitude of maximum heating which migrates north and south from 25°N to 25°S within the unmantled terrain on a 50,000 year timescale.

The population of intermediate-sized craters (4–10 km diameter) appears unaffected by the eolian mantles, at least within the ±45° latitudes. Hence the local density of these craters is probably a valid indicator of the relative age of surfaces generated during the period since the uplands were intensely bombarded and eroded. It now appears that the impact fluxes at Mars and the moon have been roughly the same over the last 4 b.y. because the oldest postaccretional, mare-like surfaces on Mars and the moon display about the same crater density. If so, the nearness of Mars to the asteroid belt has not generated a flux 10 to 25 times greater

[1] Publication authorized by the Director of the United States Geological Survey.

[2] Contribution No. 2450, Division of Geological and Planetary Science, California Institute of Technology, 1201 E. California Blvd., Pasadena, California 91109.

than the lunar flux. Whereas the lunar maria show a variation of about a factor of three in crater density from the oldest to the youngest major units, analogous surfaces on Mars show a variation between 30 and 50. This implies that periods of active eolian erosion, tectonic evolution, volcanic eruption, and possibly fluvial modification have been scattered throughout Martian history since the formation and degradation of the martian uplands and not confined to small, ancient or recent, epochs. These processes are surely active on the planet today.

Introduction

The first close-up pictures of Mars, obtained by Mariners 4, 6 and 7, showed a surface dominated by large craters (Leighton *et al.*, 1965; Hartmann, 1966; Chapman *et al.*, 1969; Murray *et al.*, 1971). This early spacecraft photography revealed that large regions of the southern hemisphere and south polar region are densely populated by highly degraded, large craters (tens to hundreds of kilometers in diameter). Unfortunately, because these pictures provided a highly biased sample of the Martian surface, it was erroneously concluded that *most* of Mars was densely cratered. Our ideas about the geologic evolution of the planet became distorted. Only a small percentage of the images revealed younger terrains devoid of large craters (Sharp *et al.*, 1971a, 1971b).

During 1971 and 1972, Mariner 9 photographed the entire globe of Mars at much better resolution than did any of the previous missions. To the surprise of all, the new data revealed that most of the northern hemisphere is devoid of large craters and is characterized by a variety of smooth plains, immense shield volcanoes, numerous channels, huge canyons, and extensive regions of "chaotic" terrain (Masursky, 1973). These terrains, all of which are younger than the densely cratered southern uplands, record the evolution of the impact flux at Mars and permit a comparison with the impact flux history of the moon.

The data which form the basis for much of this paper are the planet-wide distributions of craters in three different diameter size ranges, each separated by about an order of magnitude: *small craters* (0.6 to 1.2km) *intermediate-sized craters* (4 to 10km), and *large craters* (>20km). Data for the first category were derived from Mariner 9 narrow angle B-camera images and for the latter two from wide angle Mariner 9 A-camera pictures.

The purpose of the following discussions are ultimately to: (1) establish the effects of eolian processes in modifying craters in the different size ranges; (2) describe the genetic relationships among these three size ranges of craters, (3) separate observables related to relative age of geologic provinces from those related to geographic variations in eolian transport and deposition; (4) compare the lunar and Martian cratering histories by studying the *rate of change* of Martian impact flux implied in the geologic record provided by Mariner 9, assuming uniformatarianism in the geologic evolution of Mars; and (5) apply this analysis to establish relative and absolute time scales for the geological evolution of Mars.

A Summary of Earlier Work—The Questions

Although the Mariner 4, 6, 7 images were not representative of Mars as a whole, they did provide an important data base which provoked a decade's research on the relative and absolute time scales of Martian evolution. In the analysis of Mariner 4 data, Leighton *et al.* (1965) were the first to draw an analogy between the Martian crătered terrain and the lunar uplands. Based on estimates of the antiquity of the lunar uplands by Shoemaker *et al.* (1962) and Hartmann (1965) they suggested that the large Martian craters were 2 to 5b.y. old. Soon thereafter, others (Witting *et al.*, 1965; Anders and Arnold, 1965; Baldwin, 1965) reevaluated the analogy drawn by Leighton *et al.* (1965) and concluded that the large Martian craters were a few hundred m.y. old rather than a few billion. These later analyses had two things in common: (1) the Martian impact flux was estimated to be roughly an order of magnitude higher than the lunar flux, and (2) the Martian crater density was found to be substantially lower than that of the lunar uplands. Taken together, these two observations indicated that a relatively short time was required to generate the Martian craters.

The following year Hartmann (1966) and Binder (1966) revised the age estimates back to the several billion year range pointing out that the Martian crater density should be compared with that of the old lunar maria, not with the uplands where the crater populations are saturated and, therefore, the net accumulation indeterminant. The old lunar maria have about an order of magnitude fewer craters than the Martian cratered terrains. Consequently, even if the Martian flux was 25 times higher than the lunar flux, the Martian craters would still require several billion years to form. Additionally, Öpik (1966) and Hartmann (1971) showed that the current population of Mars-crossing asteroids is so low that it would probably take billions of years to generate the Martian craters.

A related problem addressed by several authors was the degradation and retention of Martian craters. Hartmann (1966), Binder (1966), and Öpik (1966) noticed that the size frequency distribution of Martian craters showed a break in slope somewhere between 20 and 50 km. They attributed this change to depletion of smaller craters by some erosional process. Further, Chapman *et al.* (1969) pointed out that a few large craters (diameter >20 km) appeared much less degraded than the rest of the craters in the Mariner 4 pictures. They suggested that the erosion continued "until close to or after crater formation ended" and "crater formation and erosion appear closely tied together in time." Chapman *et al.* (1969) also noted that craters smaller than about 20 km appeared remarkably fresh compared to larger ones. They suggested that the degraded smaller craters of this size were simply invisible in the Mariner IV pictures and that the break in slope at several tens of kilometers, which others attributed to depletion of small craters, was in fact due to incomplete crater counts. From the Mariner 6 and 7 images Murray *et al.* (1971) noted that most of the small craters (<20 km diameter) had sharp rims and bowl-shaped interiors and therefore were very fresh. Thus the change in slope of the size frequency distribution of craters between 20 and 50 km, as noted by Hartmann (1966), Binder (1966), and Öpik (1966), was real and reflected the loss of most of the old craters smaller than about 15 km. McGill and Wise (1972) observed that not all of the small craters were "sharp and pristine," and that a few ghost-like remnants could be seen in the Mariner 6 pictures. Although these ghosts constitute a statistically small sample, their existence shows that the last episodes of catastrophic erosion did not completely erase all pre-existing small craters.

Murray *et al.* (1971) drew a crude parallel between the cratering histories of Mars and the moon and raised several unanswered questions regarding the similarities and differences in the two histories. Foremost among these questions was the nature and time of the intense modification that degraded the large Martian craters. Was the modification contemporaneous with the period of torrential impact flux while the last remnants of the accretional debris were being swept up, or did the obliteration occur during subsequent aeons of Martian history *after* the large craters had formed? Because of the uniform appearance of degraded large craters in the areas photographed by Mariners 6 and 7, Murray *et al.* (1971) concluded that "substantial large crater modification on Mars took place in the last 3 or 4 b.y....," *after* the large craters had formed.

In summary, prior to the Mariner 9 mission, a general consensus had been reached on a number of points. First, the large Martian craters have existed for a substantial fraction of Martian history; their age is measured in billions of years. Second, these large craters have been intensely eroded by some process which had obliterated most craters smaller than 15 km. Third, a new population of small fresh craters had formed since the last stage of obliteration.

The principal question remained: What was the nature of the obliteration of the large craters and when did it take place? Data of Chapman *et al.* (1969) from Mariner 4 suggested that formation and obliteration may have been synchronous. The data base was scanty—only two craters >20 km were identified as class 1 (the freshest). The morphology of large craters in the areas photographed by Mariner 6 and 7 appeared to be so uniform that Murray *et al.* (1971) concluded that most of the obliteration occurred after these craters formed. The difference between these two ideas has profound implications for the geologic history of Mars. The time at which obliteration ended provides a maximum age for the younger system of terrains discovered by Mariner 9—they could be extremely old or extremely young. One of the keys of the absolute timescale developed in this paper is to establish which alternative is correct.

Distribution of Large Martian and Lunar Craters—Global Dichotomies

Perhaps the most important observation made from the Mariner 9 data is that not all of Mars is characterized by ancient cratered terrains as was previously thought. The northern hemisphere is distinctly different from the southern hemisphere in that it contains a variety of smooth plains in which large craters are rare (McCauley *et al.*, 1972, Hartmann, 1973). Further, these young plains of the northern hemisphere show a variety of markedly fresh details in contrast to the degraded cratered terrains to the south (Masursky, 1973). Surfaces intermediate between these two types are rare. This extreme dichotomy in *apparent* surface age creates the impression that terrains on Mars are either ancient, dating back to the early phases of Martian history, or extremely young, perhaps developed in the latest stages of Martian evolution. Several geophysical models enhanced this impression in that they suggest melting and endogenic evolution of the Martian crust probably will not occur until billions of years after accretion (Anderson, 1969; Anderson and Phinney, 1967; Hanks and Anderson, 1969; MacDonald, 1962). This dichotomy in apparent age of Martian surfaces is remarkably like that of the Moon. The lunar uplands are densely cratered, ancient by any standard; the maria by contrast are very lightly cratered, retaining morphological details of their volcanic origin. The reason for this dramatic contrast in the cratering style of these two principal lunar terrains was predicted as resulting from a rapid decline in the early postaccretional impact flux (Shoemaker *et al.*, 1962, Hartmann, 1965). Before presenting and discussing the new planetwide crater data for Mars, it is appropriate to first summarize the impact flux history of the moon—which has only recently moved from the hypothetical to the observed.

In Fig. 1 lunar crater densities are plotted against the radiogenic ages of the material derived from Apollo 11, 12, 14, 15, and 16 missions. The dichotomy in the abundance of large craters between lunar uplands and maria regions can be understood directly from Fig. 1. Before about four b.y. ago the fluxes were torrential, being great enough to saturate the surface within a period of tens-of-millions of years. Exponential decay rates of the population of impacting debris of about 10^7 to 10^8 yr are implied for this early period. These decay rates are consistent with collision lifetimes of accretional debris and probably represent the tailend of the accretional sweep-up. The change in the impacting flux from this early surge to the quiescent cratering rate occurred over a very brief period between 3.5 and 4 b.y. ago.

The shape of the curve in Fig. 1 shows why lunar surfaces with populations of large craters intermediate between the maria and the uplands are rare—the time interval during which they would have had to originate was very brief. Hence, at least in the case of the moon, the uplands–maria dichotomy does not require an exotic evolutionary history with an extensive void in geologic activity. Rather, a continuous generation of new surfaces

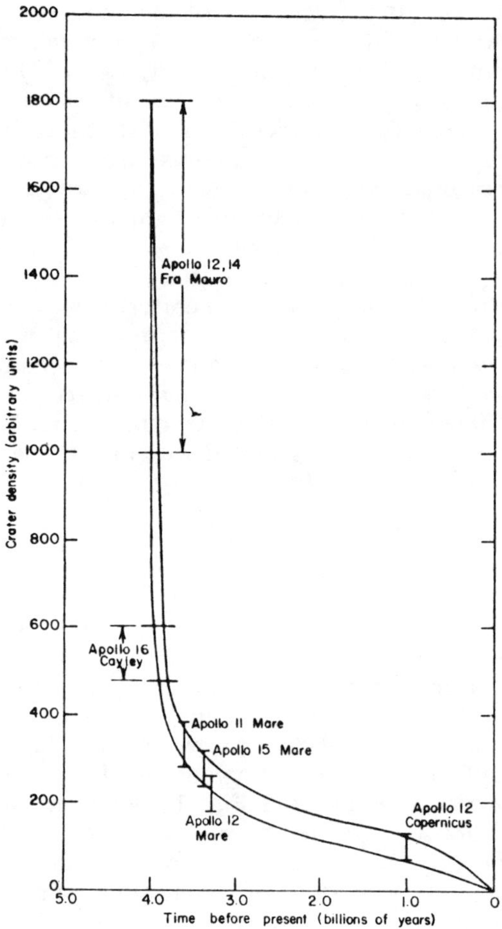

FIG. 1. Total accumulation of impact craters on lunar surfaces as a function of age. Shown here are the ages of various lunar volcanic and impact units, determined from isotopic age analyses of Apollo samples, plotted against the total amount of cratering developed on those surfaces. The isotopic age analyses are taken from Wasserburg *et al.* (1972a, 1972b); Papanastassiou *et al.* (1971a, 1971b); and Silver (1971). The integrated impact flux measurements are those of Soderblom and Lebofsky, 1972, and Soderblom and Boyce, 1972. The approximately linear increase in degree of cratering from the present back to about 3.5 b.y. implies a fairly constant flux for that period. Before about 3.5 b.y. the fluxes were dominated by a torrential but rapidly declining flux probably generated as the last remnants of the accretional debris distributions were swept up.

throughout geologic history would create this dichotomy as a result of the abrupt termination of early intense cratering.

Figure 2 is a map of the areal density of *large* Martian craters (>20 km diameter). The Martian global dichotomy is quite evident both in this illustration and a similar map prepared by Hartmann, 1973. The contrasting cratered terrains and smooth plains divide Mars approximately into two hemispheres separated by a great circle tilted about 20° at the equator. The differing characteristics between these two global terrains on Mars are remarkably similar to the morphological and crater density differences which distinguish the lunar maria–uplands units. On both Mars and the Moon terrains displaying intermediate morphological forms or crater density are rare. The overall similarities of these two global dichotomies is extremely important in testing models for Martian flux history.

THE NATURE AND PERIOD OF DEGRADATION OF LARGE MARTIAN CRATERS

The data of Fig. 2 illustrates that the cratered uplands on Mars are only 20 to 40% saturated by large craters, in contrast to the lunar uplands which are 50 to 75% saturated. As noted above, previous authors concluded that the lower density and severely degraded appearance of the large Martian craters was a result of either an early period of intense erosion roughly contemporaneous with crater formation (Chapman *et al.*, 1969) or that it was caused by major horizontal redistribution of material which occurred after most of the large craters formed (Murray *et al.*, 1971). Figure 2 shows tremendous variability in crater retention throughout the Martian uplands. It has been suggested by Wilhelms (1973) that the morphological form and uniform appearance of large craters in areas like that photographed by Mariner 6 (c.f. 6N19 and 6N21) is consistent with episodes of subsequent burial, probably to a large extent by volcanic processes. On the other hand, in other areas photographed by Mariner 9 (Fig. 3), large craters show a wide and gradational range of morphology from fresh and pristine to highly degraded and nearly invisible. We

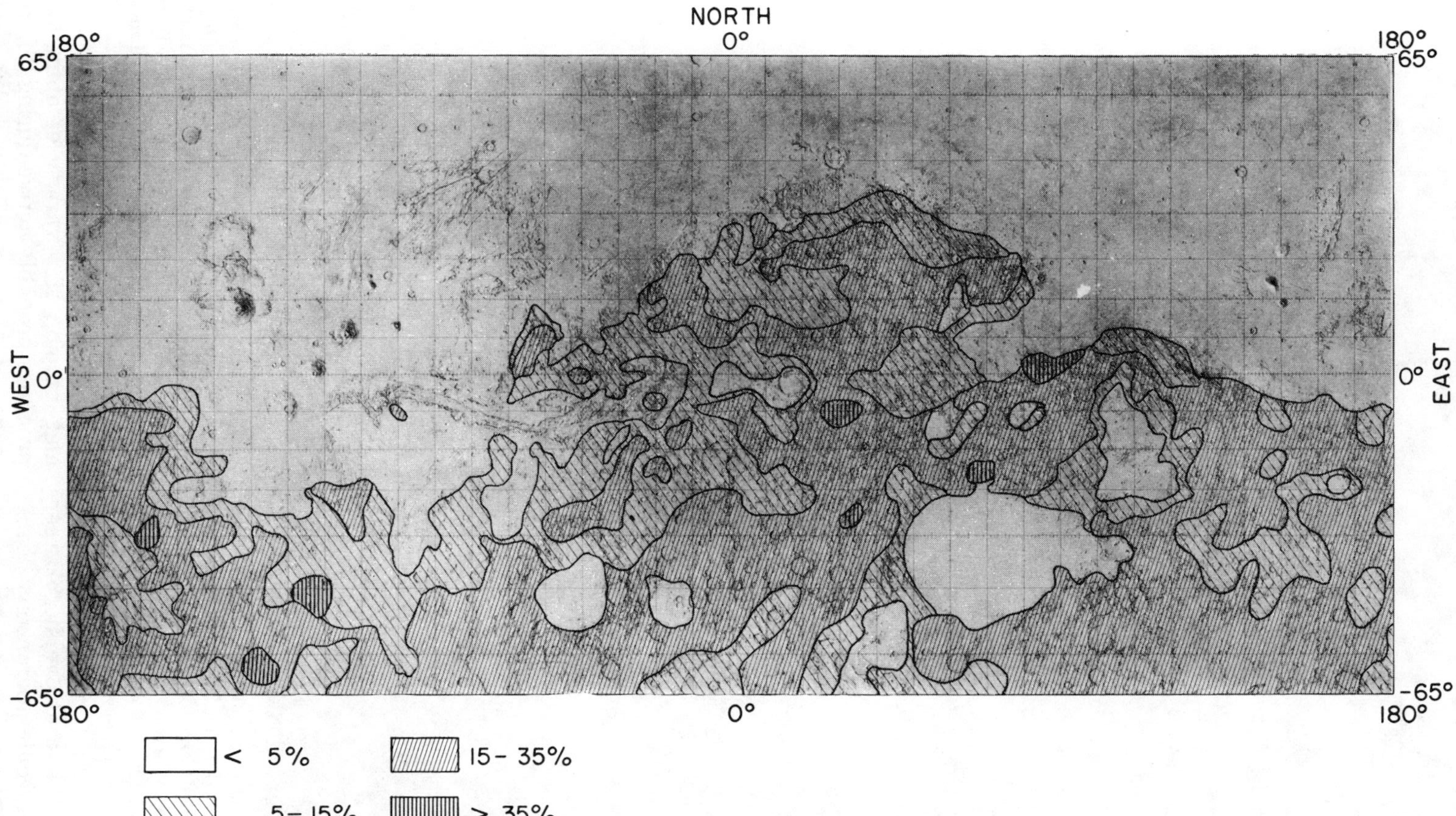

FIG. 2. Density map of large craters (>20 km diameter) on Mars. The four categories refer to the percentage of the surface area occupied by the interior walls and floors of large craters within Mariner 9 wide-angle frames. These data were averaged, smoothed, and contoured. The boundary between the sparsely cratered northern regions and densely cratered southern regions is approximately a great circle, tilted about 20° to the equator, which divides the Martian globe into two dramatically different kinds of terrain.

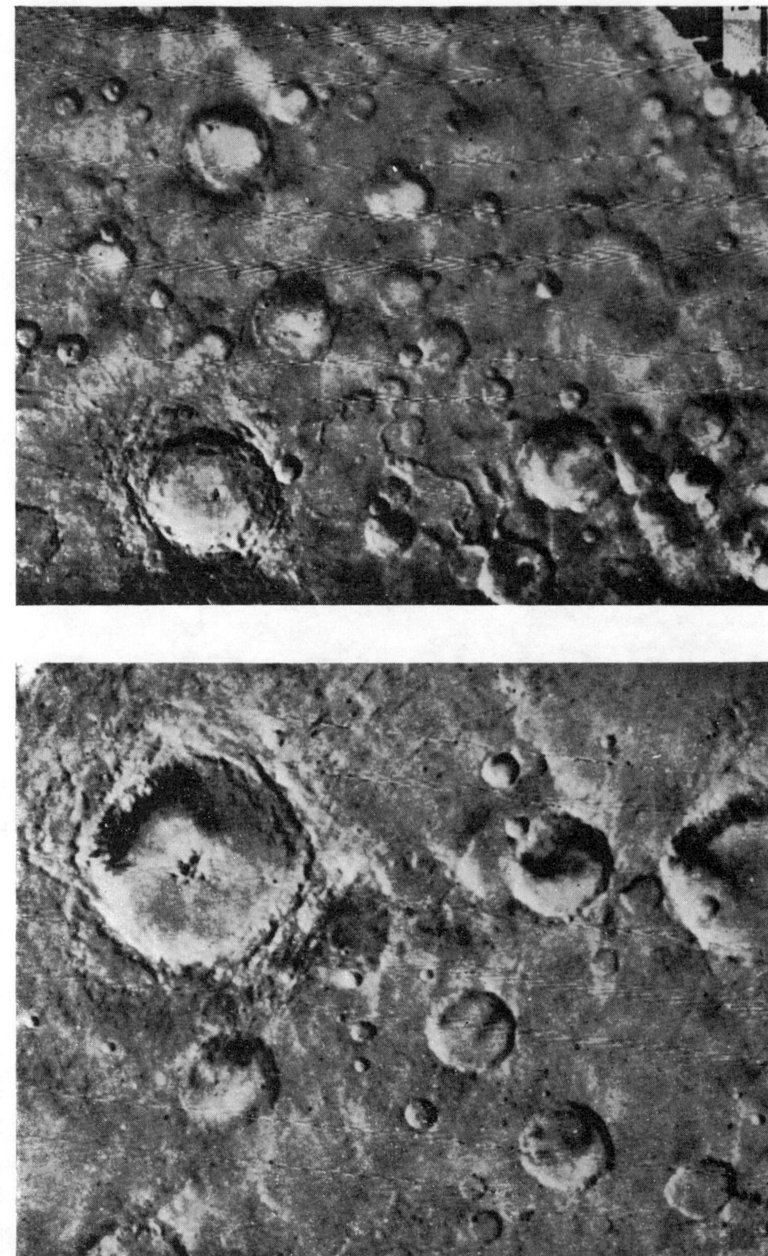

FIG. 3. Variation in the degree of preservation of large craters. These Mariner 9 frames show, contrary to earlier thinking based on Mariner 6 and 7 photographs, that some areas in the ancient cratered terrain display a wide range of crater degradation. This implies that the intense modification of these populations was roughly contemporaneous with their formation. The two Mariner 9 A-frames (Upper figure DAS 7290743 and lower figure DAS 8047028) cover approximately 300 × 400 km and were acquired with a sun elevation angle of about 25°.

conclude that the controversy between subsequent and contemporaneous degradation arises because of oversimplification and lumping of a variety of obliterative processes probably active during different stages of Martian evolution. We divide these stages of modification into (1) an early intense erosional epoch (possibly including deposition) which was contemporaneous with crater formation recorded (as suggested by Chapman *et al.*, 1969), and (2) multiple stages of volcanic, eolian, fluviatile, and other ad hoc depositional processes which created the uniform buried appearance of large craters and smooth intercrater surfaces, in areas such as those photographed by Mariner 6. We know a variety of plains-forming processes generated extensive plains in the northern hemisphere; it is not surprising that similar burial processes also occurred in the Martian uplands. In unravelling the flux history for Mars, the first stage. the contemporaneous obilteration, is most important.

Figure 4 shows one of the oldest Martian plains located in Lunae Planum. Termed *crater plains* by McCauley *et al.* (1972) and

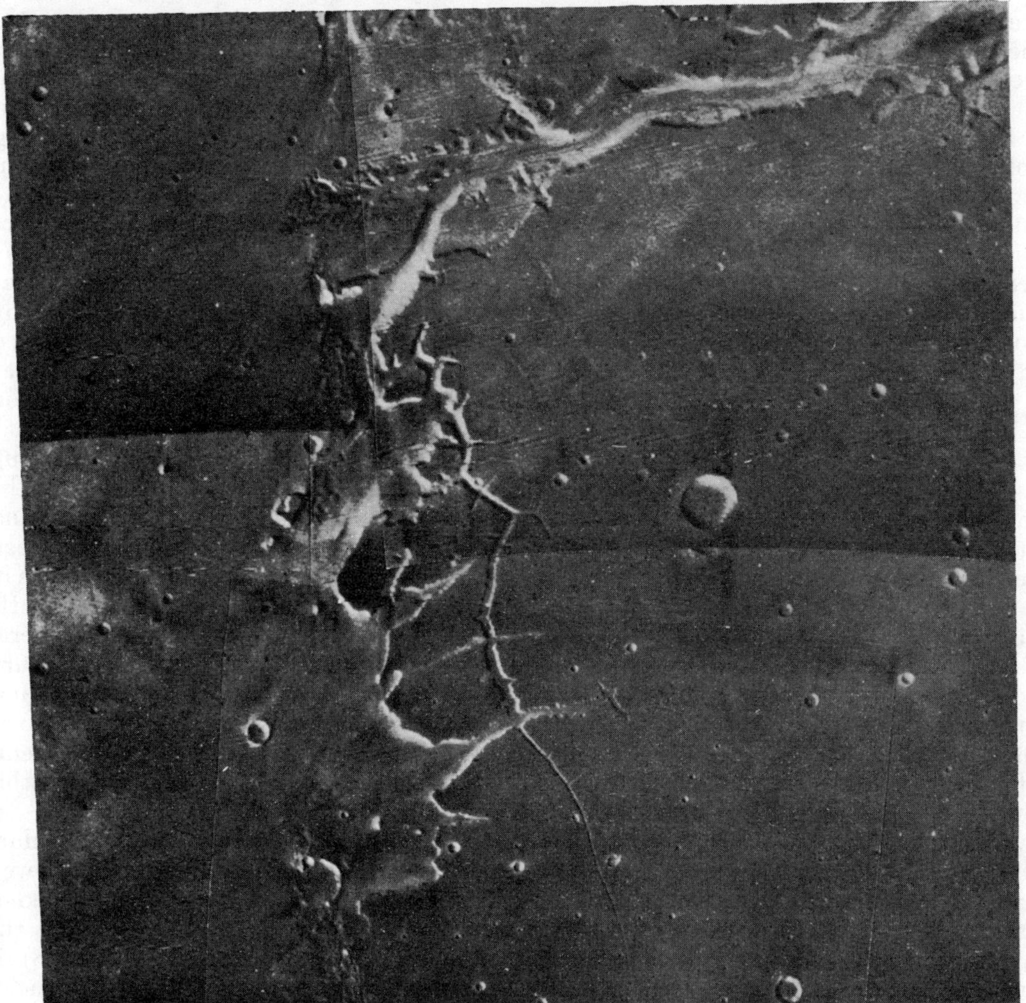

FIG. 4. Oldest "post-uplands" terrain. In the eastern half of the mosaic is the Lunae Planum plateau (60°W, 10°N) which has been down-cut on the east and north by complex young channel systems. These cratered plains have a relatively high density of "small bowl-shaped" craters whereas large craters (>20 km diameter) are rare. Because of the very fresh appearance of the craters on such cratered plains, the intense levelling, characteristic of large craters, must have ceased before the plains were formed. The mosaic which covers an area roughly 1,000 km on a side, includes wide angle frames: DAS 7399733, DAS 7399383, DAS 7471693, and DAS 7471343.

by Carr *et al.* (1973), this unit is characterized by a rarity of large craters (>20 km diameter) but also by a high density of smaller bowl-shaped craters (diameter <20 km). Most of the craters formed on this surface appear relatively fresh indicating that the obliterative epoch contemporaneous with formation of large craters had ended before these old plains were generated. Further as will be shown later, the density of small bowlshaped craters on the uplands and these old plains differ by less than 30% suggesting that the time between the cessation of obliteration and emplacement of the old plains was a small period of Martian history.

In summary we conclude that (1) major obliteration of large craters was contemporaneous with the early period of intense bombardment; (2) in later periods large craters were further modified by processes of burial; and (3) the oldest surfaces among the lightly cratered terrains must have been generated soon after the early intense cratering ceased.

Distribution of Small Craters (0.6 to 1.2 km)—Influence of Eolian System

The standard technique for establishing the sequence of emplacement of the lunar maria has been to measure the population density of small fresh superposed craters (about 300 m to 3 km diameter). As most lunar craters in this size range are essentially pristine, it can be assumed that the total population which has accumulated on the more surfaces is still preserved. Therefore, the older the surface the greater the density of craters. A natural approach is to apply the same technique to Mars. On Mars, however, it must be used with caution because the interaction between the eolian regime and small craters is complex. We will first assess those effects.

Figure 5 is a map of the density of craters 0.6 to 1.2 km in diameter over the lattitude range from 65°S to about 50°N. The data were obtained from counting crater densities on about 1800 narrow angle B-frames acquired after revolution 100 (approximately after the dust storm cleared). The lower limit of 0.6 km was chosen because it is well above the resolution cutoff and is roughly equivalent to ten picture elements (pixels) in the poorest resolution pictures used. After the B-frames were examined and their craters counted, the data were plotted and mapped as three density classes. The distribution in Fig. 5 has been smoothed and averaged to suppress statistical noise introduced by the low number of craters per frame.

The first-order observation to be made from Fig. 5 is that the density of these small craters is strongly correlated with latitude. For instance, the density changes abruptly from the highest to lowest at roughly 35°S latitude. No obvious changes occur in the regional morphology of the cratered terrain across that boundary. Such a strong latitudinal correlation in the distribution of these crater densities suggests that they are not good age indicators of regional terrains on Mars. Recent studies by Soderblom *et al.*, (1973a), have shown that both polar regions are surrounded by young debris mantles, apparently derived from the erosion and dispersion of the polar sediments. Figure 6 shows the distribution of those debris mantles mapped in that study. As can be seen by comparing Figs. 5 and 6, the boundary between the highest and lowest density units around 35°S coincides with the boundary between the mantled and the unmantled terrains in the southern hemisphere. The mantle thickness must be on the order of several tens of meters thick in order to bury craters 0.5–1 km in diameter. Very few small craters are found on top of the mantling blanket, suggesting very recent, perhaps continuing, reworking of this material.

Figure 7 is an albedo map of Mars derived from photographs by the Lowell Observatory, Flagstaff, Arizona. In comparing Figs. 5–7, it can be seen that the region of high crater density (20°S) is approximately coincident with the zone of permanent dark areas in the southern hemisphere. Observations in the northern hemisphere (Soderblom *et al.*, 1973a) have shown that the mobile materials being transported out of the polar regions appear as bright streaks on underlying terrain. The correlation of the dark areas with the

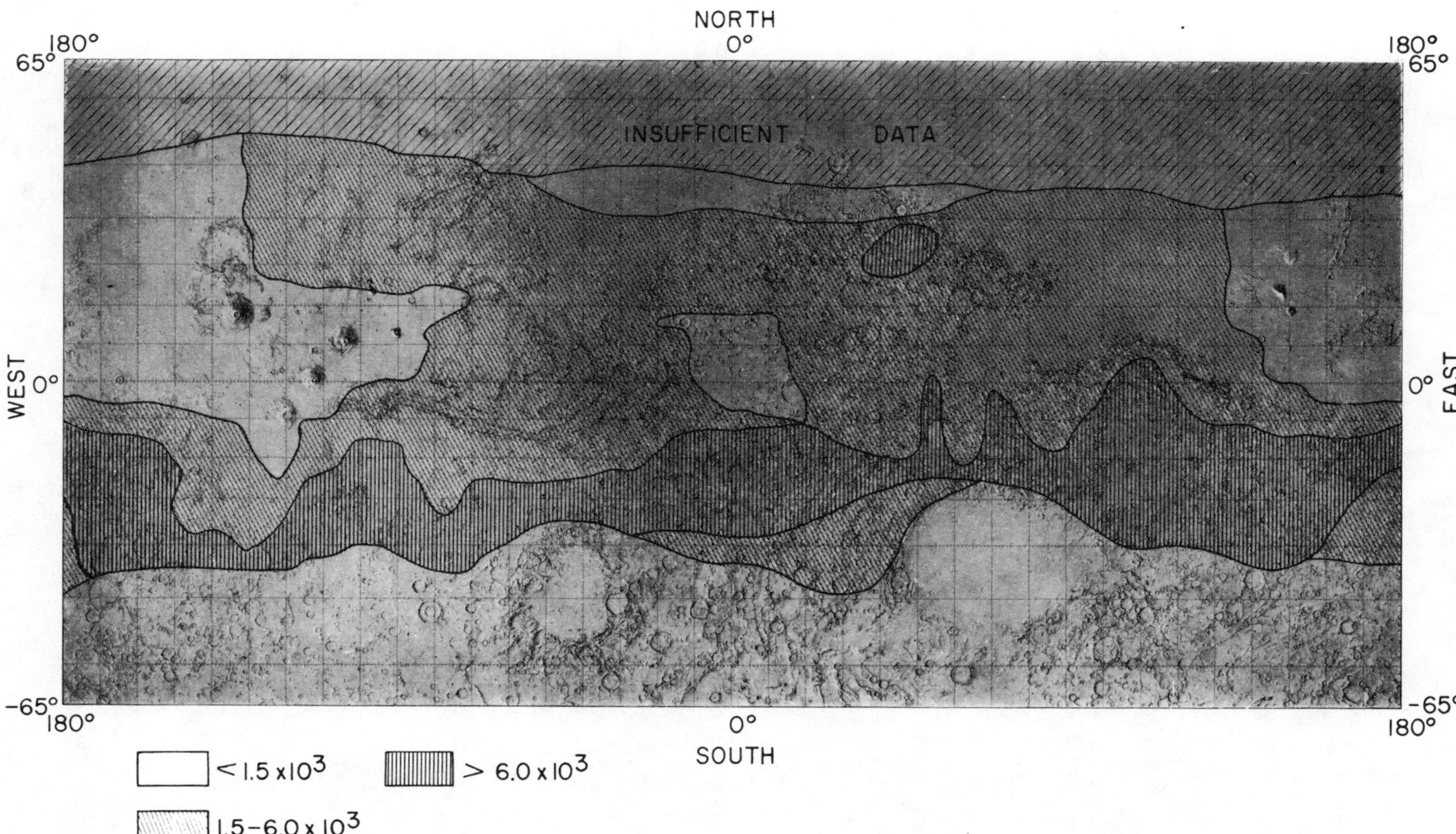

FIG. 5. Density map of small craters (0.6–1.2 km diameter), compiled from crater densities measured on about 1,800 Mariner 9 narrow-angle frames. The densities shown represent the number of craters per $10^6 km^2$. Within the broad units shown, there are many irregularities which are not mappable because the high-resolution frames are so scattered. The map distribution shown has been statistically smoothed to prevent aliasing these high frequency variations into broad patterns.

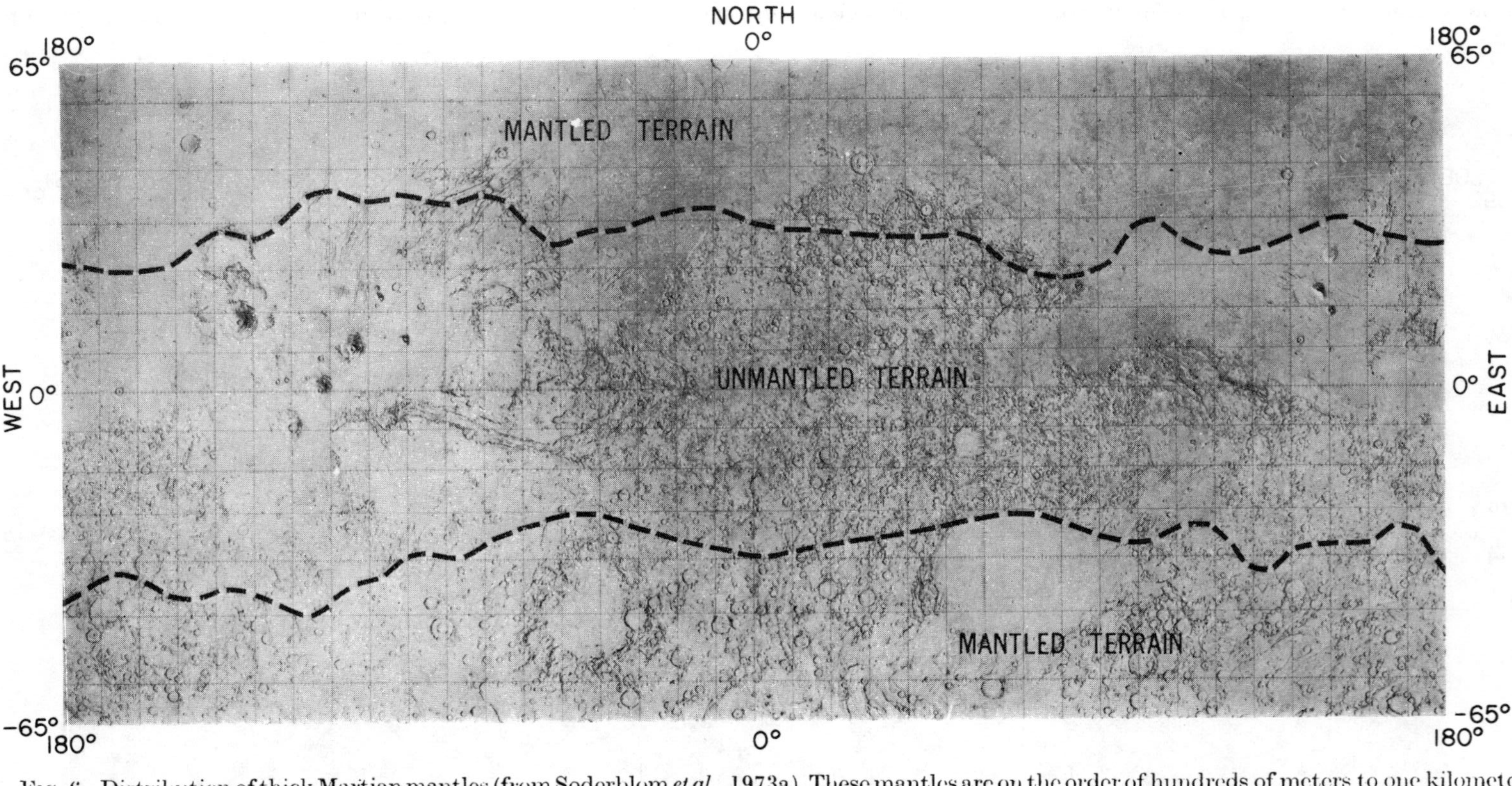

FIG. 6. Distribution of thick Martian mantles (from Soderblom *et al.*, 1973a). These mantles are on the order of hundreds of meters to one kilometer thick. They are easily visible in Mariner 9 narrow-angle frames as they are draped over small craters, causing them to appear shallow with sharp rims.

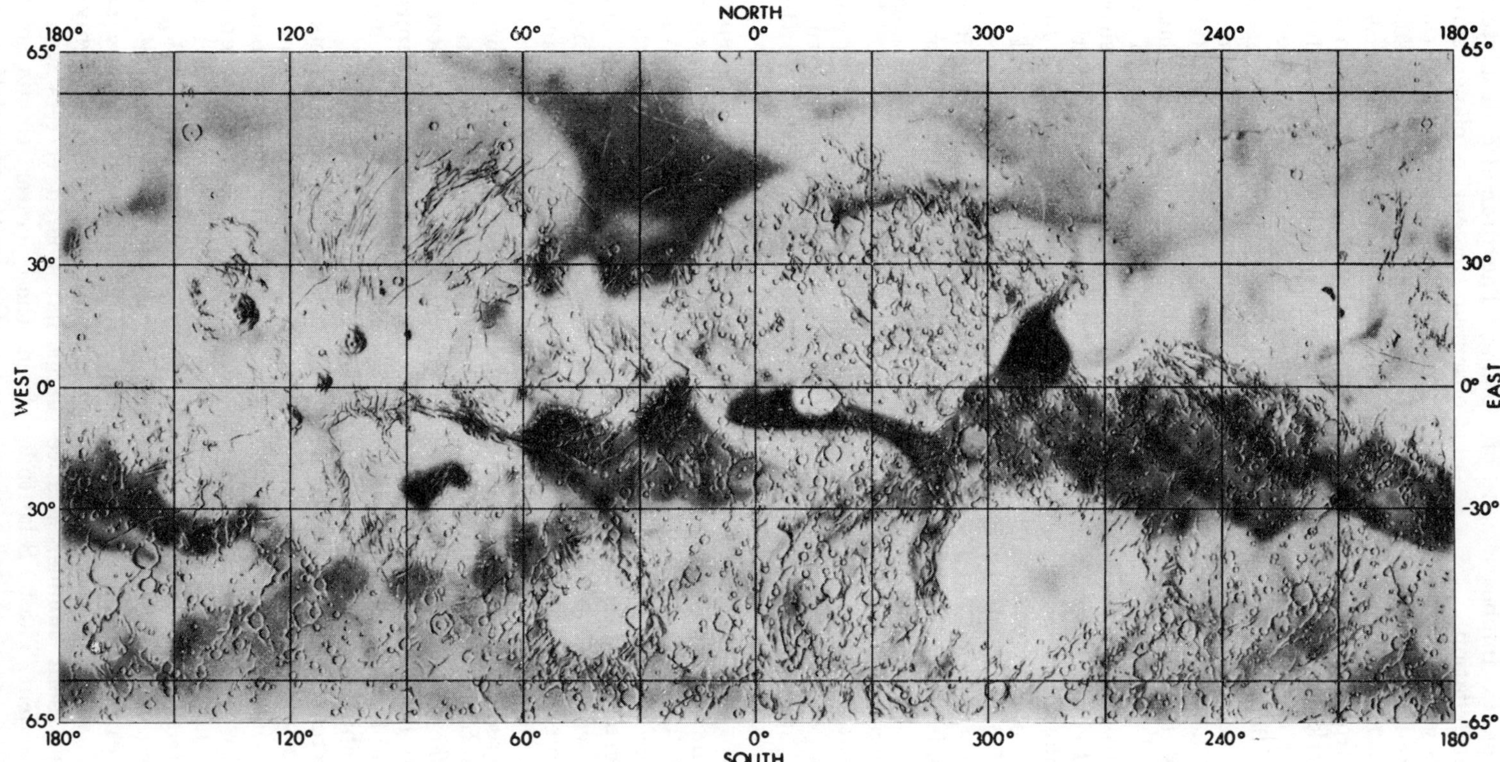

Fig. 7. Global patterns of permanent and recurrent albedo on Mars (from International Planetary Patrol Photography, compiled at Lowell Observatory, Flagstaff, Arizona). The major zone of low albedo on Mars is located between the equator and about 25°S and coincides with the belt of highest density of small craters (Fig. 5).

band of high density of small craters and absence of the mantle implies a zone of non-deposition or deflation.

Figure 8 shows yet another kind of data which is correlated with the distributions shown in Figs. 5–7. Shown here are wind vectors derived from surface markings of both light and dark streaks cast downwind from topographic obstacles (Sagan *et al.*, 1973). The wind field maps derived from IRIS data show the same pattern (Conrath *et al.*, 1973). The surface winds blow southward across the equator; the "wind equator;" defined here as the latitude of zero net north to south flow, is actually centered in the 15° to 20°S zone. Figure 9 shows the distribution of small dendritic channels (averaging 4 km in width) derived by D. Pieri (1974). The density of small dendritic channels, formed on top of the levelled uplands, is strongly peaked in the low albedo, high crater density band about 20°S—again suggesting an area where mantles are absent.

Thus from a variety of evidence it appears that (1) north and south of the 40° latitudes thick young eolian mantles (about 100 m) have buried small craters (0.6–1.2 km diameter) and other small features, and (2) the "unmantled" equatorial zone between 40°N and 40°S described by Soderblom *et al.*, 1973a, is actually composed of two units: the northern two-thirds is thinly mantled (about 50 m) so that the density of 0.6–1.2 km craters is reduced but the mantle is not thick enough to visibly affect craters a few km in diameter; the southern one-third is apprently free of debris so that the low albedo and subtle details (small shallow channels) are visible.

Two questions remain. First, why are the demarcations between the thick mantle and the equatorial unmantled zone so abrupt? Second, why is the dark zone not centered within the equatorial unmantled zone but situated in the southernmost part? The answer to both of these questions probably involves the global asymmetry in the surface wind patterns. Today periapsis and southern summer solstice are roughly coincident. The 20°S latitude band receives the greatest solar input during summer than elsewhere on the planet. The dramatic dust storm which obscured the entire planet early in the Mariner 9 mission began near the southern summer solstice when the sub-solar latitude was about 20°S. As Mars precesses on its axis and periapsis revolves in the orbital plane, the latitude of greatest heating moves north and south between ±25° with a 50 000 yr cycle (Murray *et al.*, 1973, and Ward, 1973). We present the hypothesis that the dark band of high small crater density follows this latitude and migrates north and south with the 90 000-year period. Thus the sharp demarcation between thickly mantled terrains and unmantled terrains at the 40° latitudes arises because the dark band continually sweeps the equatorial unmantled zone. Today the dark zone is near its southernmost limit. Hence its southern boundary coincides, today, with the edge of the thick mantle.

Two anomalies in small crater density (Fig. 5) appear in the northern equatorial zone; one centered at 35°N, 315°W, and the other at 0°N, 0°W. The topographic map based on ultraviolet spectrometer data (Hord *et al.*, 1972) shows that the high density anomaly (35°N, 315°W) is situated on a high plateau rising above the surrounding terrains. The small craters in that zone have not been buried by eolian debris. The zone near 0°N, 0°W of low crater density is coincident with a low saddle and probably represents a trap of eolian debris, a trap of volcanic flows or a trap controlling some other depositional process. Other areas of low density in the northern hemisphere are generally correlated with young volcanic plains and mantled areas.

To understand the effects of these mantles on craters of different diameters, the crater frequency distributions were measured for eight areas, four north and four south of the mantled–unmantled boundary near 220°W, 40°S (Fig. 10). This region was chosen because of the uniformity in regional morphology across the mantle boundary (Fig. 6). Highly degraded large flat floored craters were excluded from these crater counts as we were interested in the populations developed

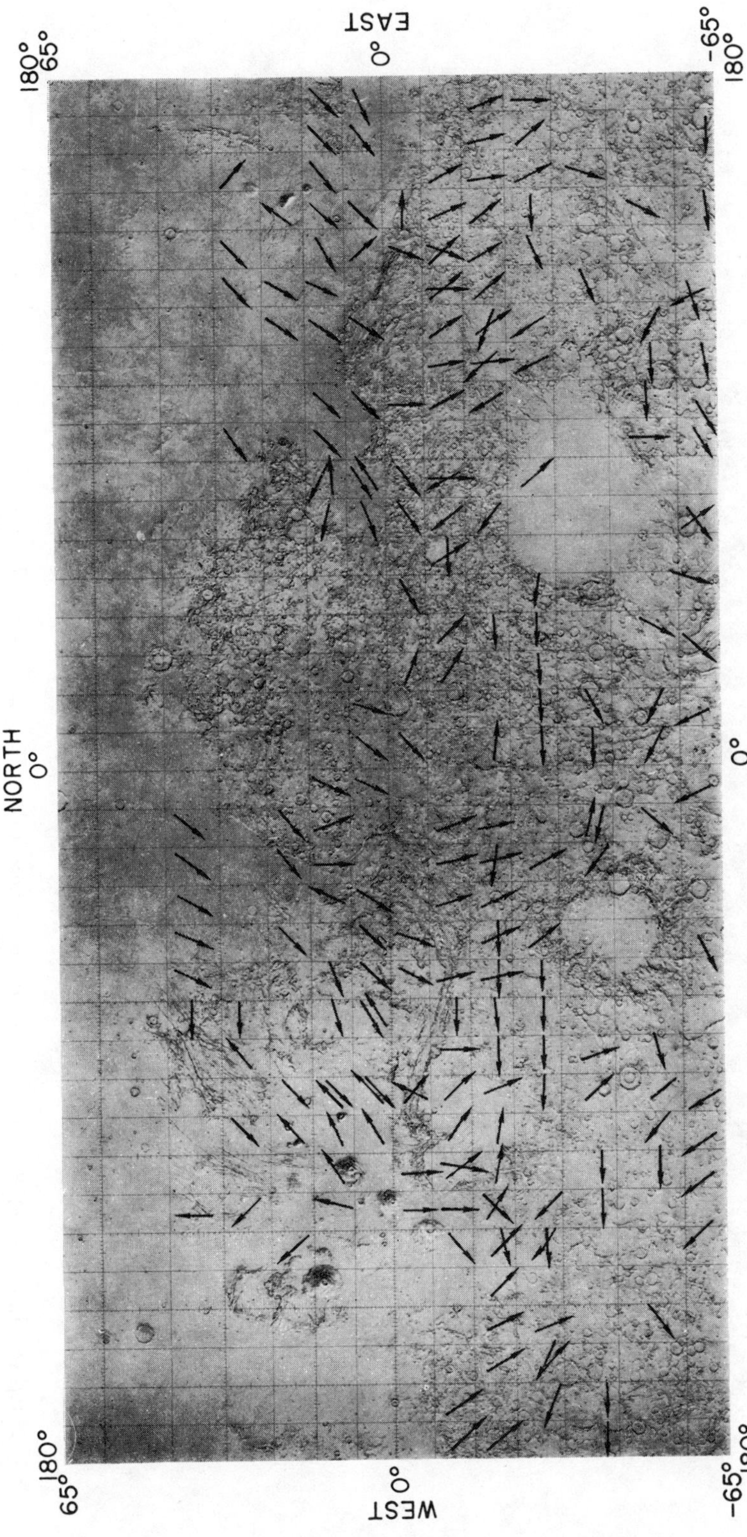

FIG. 8. Direction of surface winds during the 1971 global dust storm (from Sagan *et al.*, 1973). The arrows show the wind directions indicated by light and dark streaks which formed downwind from topographic features during the Mariner 9 dust storm. The data have been averaged over 10° × 10° squares. Notice that the "wind equator" (zone of zero north south net flow) is displaced roughly 20° south of the equator—again coincident with the zone of high small crater density and low albedo.

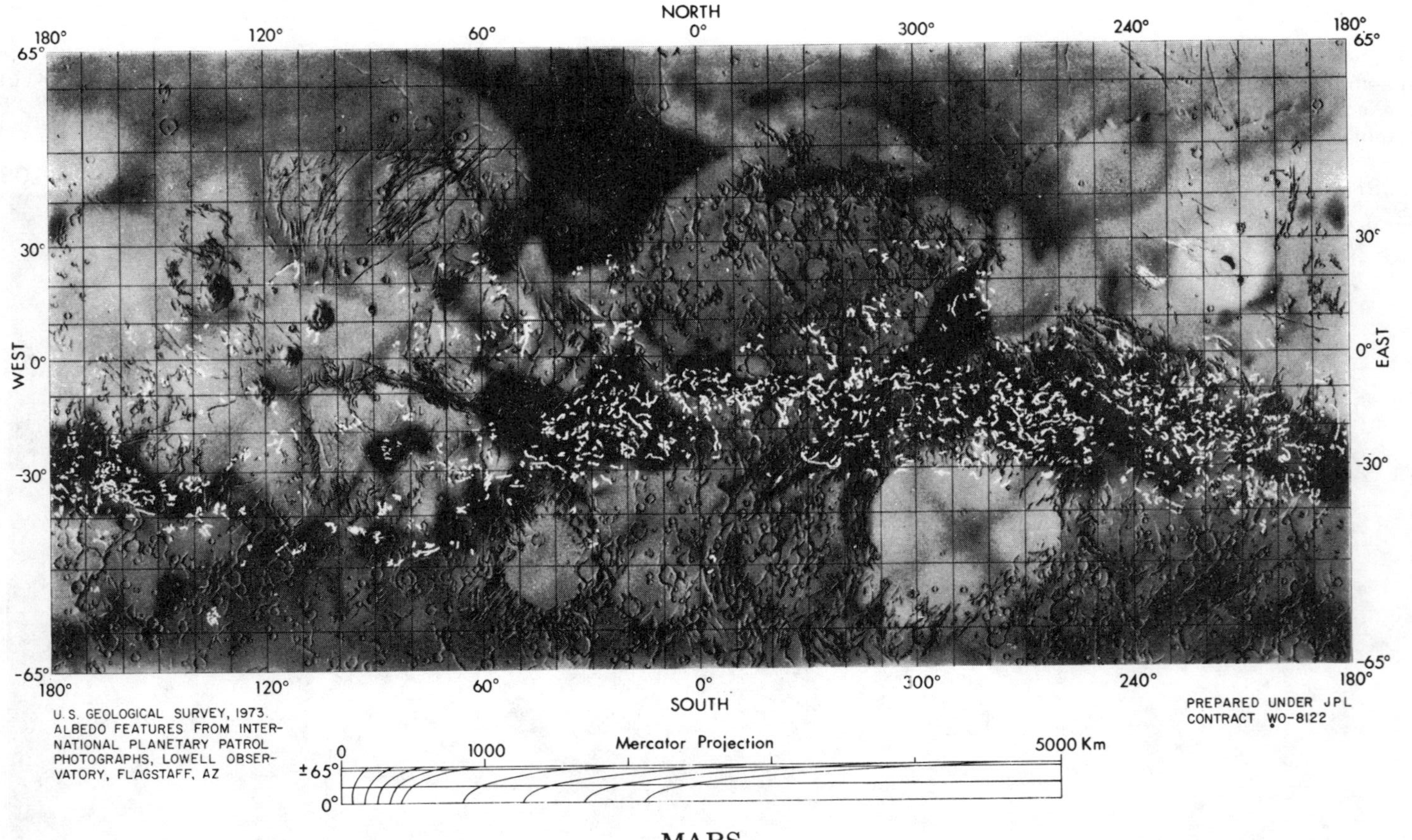

FIG. 9. Distribution of small filamental channels (from Pieri, 1974). They are small (1–7 km wide) U-shaped, and often dendritic. The channels are represented in white; their widths are exaggerated for display purposes. These small dendritic channels are apparently confined to the cratered, levelled uplands. Notice the high density centered on a band roughly 20°S latitude. These small channels are morphologically distinct from the presumably much younger mammoth cliff-rimmed canyons and channels in the equatorial zone.

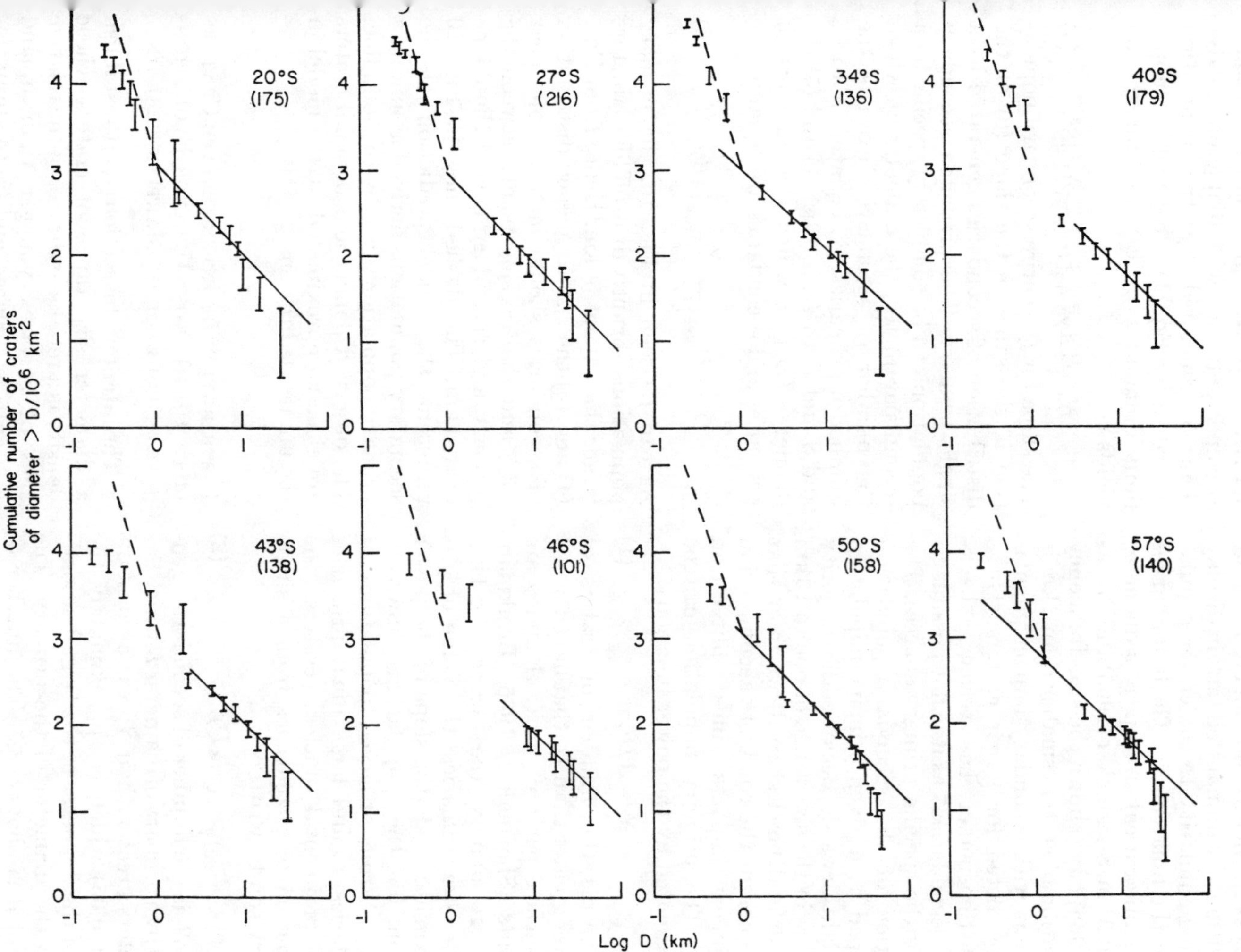

FIG. 10. Cumulative crater frequency curves for small fresh craters north and south of the southern mantle boundary, measured in eight areas along 220°W longitude where the regional morphology is relatively uniform. The mantle boundary is about 40°S latitutde in that area. The solid lines represent an assumed primary population; the dashed lines represent the model prediction of secondary populations that such primary populations would generate. It can be seen that the mantle severely affects the density of craters of diameters less than about 1 km.

after the intense early erosion. As can be seen from Fig. 10, there is a marked change in the form of the size–frequency distribution across the mantle boundary. The four distributions for unmantled terrain display a striking similarity to the crater populations in the lunar maria. The lunar curves show a similar break in slope at a diameter of about 3 km. Shoemaker (1965) and Brinkman (1966) have shown the size–frequency distribution of craters smaller than 3 km is consistent with secondary impact populations predicted for larger craters. These craters Shoemaker has termed "background secondaries" as it is not possible to identify the specific primary sources. The derivation of the secondary population developed by a given primary population is straightforward. Because the majority of craters in the size range of one to 15 km are fresh and bowl shaped, we infer that they represent the complete accumulation of primary impacts since obliteration ceased. The primary population can be approximated by the power function:

$$N = AD^{\gamma}, \tag{1}$$

where N is the number of craters of diameter greater than D; that have formed on a surface per unit area; A and γ are constants (Shoemaker, 1965; Brinkman, 1966). Assuming the craters >1 km diameter are primaries, the value of γ is approximately −1 (the slope of the power function for the population shown in Fig. 10). From observed populations of secondaries around large lunar primaries and experimental nuclear craters, the distribution of secondaries from a single primary can be written

$$N_s = (S/\alpha D)^{\beta}, \tag{2}$$

where N_s is the number of secondaries generated by a primary of diameter D which has diameters greater than S; α is a constant (about 0.05) which is approximately the ratio of the diameter of the largest secondary to the diameter of the primary. The value of β is about −3.5. The number of secondaries generated by the total population of primaries can be written

$$N_s = -\int_{S/\alpha}^{D_{\max}} \left(\frac{S}{\alpha D}\right)^{\beta} A\gamma D^{\gamma-1}\, dD, \tag{3}$$

where $D_{\max}$ is the largest primary crater which contributes; S/α is the smallest primary capable of generating a secondary of diameter S; N_s is the number of secondaries per unit area with diameters greater than S generated by the population described by Eq. (1). Realizing the upper limit dominates in Eq. (3), integration yields

$$N_s = [A\gamma D_{\max}^{\gamma-\beta}/(\beta-\gamma)\alpha^{\beta}]S^{\beta}. \tag{4}$$

The slope for the entire secondary population is the same as the slope (β) of the distribution of secondaries generated by a single primary. The predicted secondary population can be completely specific by a second parameter, the diameter at which the number of secondaries exceeds the number of primaries. By equating N to N_s and S and D to D_c in Eqs. (1) and (5) the diameter (D_c) at which the numbers of primaries and secondaries are equal is

$$D_c = D_{\max}\left[\frac{\gamma}{(\gamma-\beta)\alpha^{\beta}}\right]1/(\gamma-\beta). \tag{5}$$

As can be seen in Fig. 10, the primary populations abruptly drop off for diameters larger than about 50 km. Hence from Eqs. (4) and (5) the size frequency distribution of secondaries should have a slope near −3.5 and the crossing point between the primary and secondary curves should be near 1 km. The dashed lines in Fig. 10 represent the model prediction of the secondary populations derived assuming a primary population given by the solid line. The observed populations of small craters are closely approximated by the model in the unmantled regions.

Distribution of Intermediate Craters (4 to 10 km): Post-Uplands Evolutionary Sequence of Martian Terrain

The intent in this section is (1) to establish a diameter size range of craters whose density is a reliable indicator of the relative ages of surfaces younger than upland obliteration surfaces, and (2) to apply this tool to establish the relative ages of regional units formed *after* intense cratering ceased. This size range must (1) include only craters formed after the

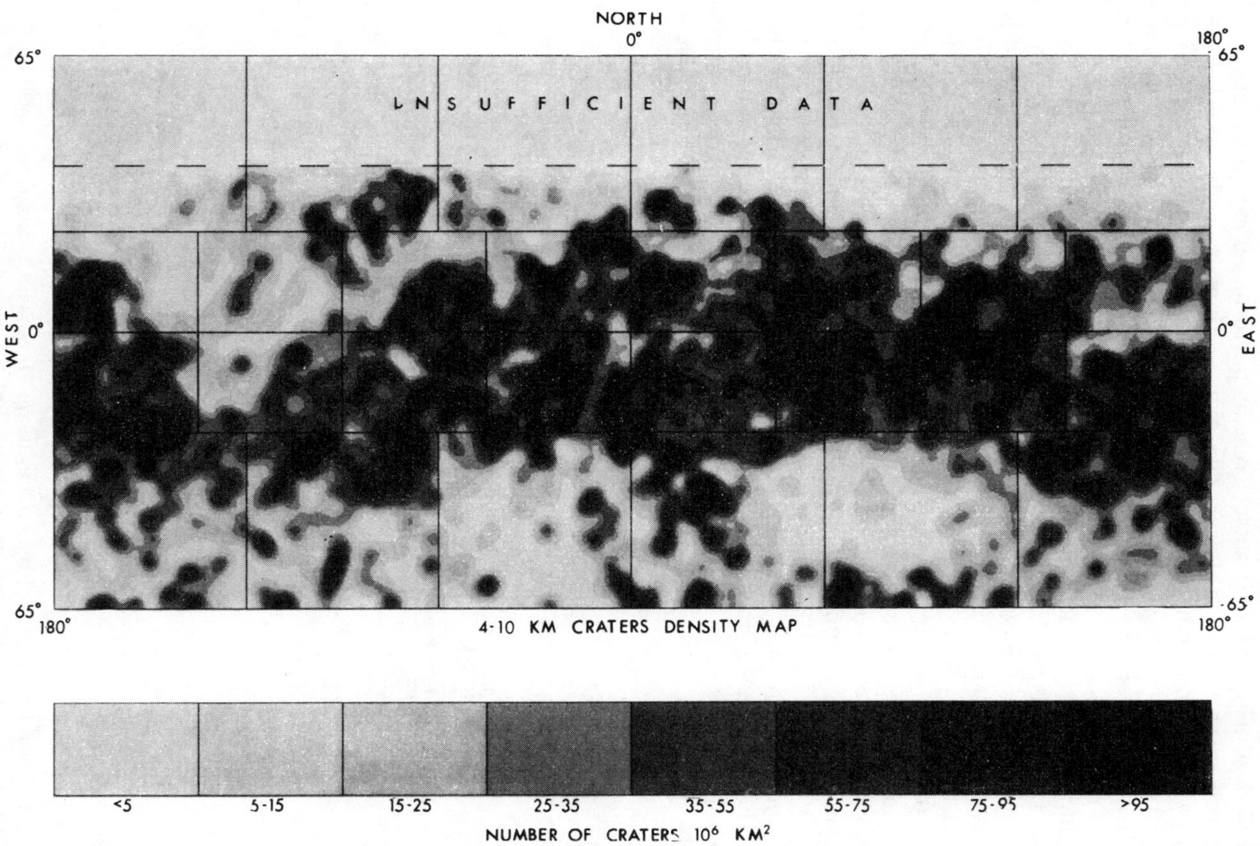

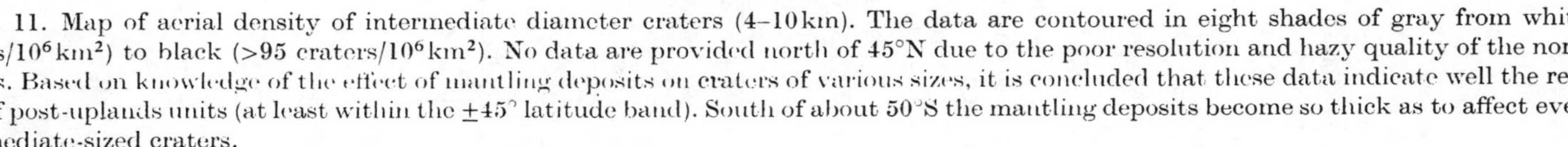

FIG. 11. Map of aerial density of intermediate diameter craters (4–10 km). The data are contoured in eight shades of gray from white (<5 craters/10^6 km^2) to black (>95 craters/10^6 km^2). No data are provided north of 45°N due to the poor resolution and hazy quality of the northern images. Based on knowledge of the effect of mantling deposits on craters of various sizes, it is concluded that these data indicate well the relative ages of post-uplands units (at least within the ±45° latitude band). South of about 50°S the mantling deposits become so thick as to affect even the intermediate-sized craters.

intense erosion of the uplands, and (2) include craters whose density is insensitive to the debris mantles. The first constraint places an upper limit on the size range (craters must be small enough to have been erased during the intense levelling). The second provides a lower limit (craters large enough not to have been buried by the mantles). Murray *et al.* (1971) noticed that craters larger than 10 to 15 km show severe degradation and that smaller "bowl-shaped" craters a few km in diameter appear fresh and pristine. Hartmann (1973) concluded that most craters larger than 10–20 km are separated in age from smaller fresh craters by some erosional epoch. Hence 10 km represents a good *upper limit* as almost all the smaller craters are clearly younger than upland degradation event. The lower limit can be obtained directly from Fig. 10. Craters smaller than about 1 km are strongly affected within the mantle. A conservative *lower limit* of 4 km was chosen to minimize the effects of the mantles.

Figure 11 is a map of the density of 4 to 10 km craters in the latitude band from 45°N to 65°S. These data were obtained by mapping the crater densities on 1600 Mariner 9 wide-angle frames by using the reseau patterns as sampling grids. These data were then compiled on 1:5 000 000 scale photomosaics and averaged over a 1.5° grid. These maps were digitized and smoothed with Gaussian filters until the random statistical noise disappeared. The area north of 45°N latitude was excluded because photography was too low in resolution and obscured by atmospheric hazes. Throughout the equatorial zone the changes in crater density coincide with changes in the geomorphic styles of density units as shown in Fig. 11. The southern mantle boundary is faintly visible in that the crater density in the mantled zone is very irregular. Figure 12 is a histogram showing the percentage of the area which has a given 4–10 km crater density. The spike at zero crater density reflects the strong control of the mantle in the Hellas

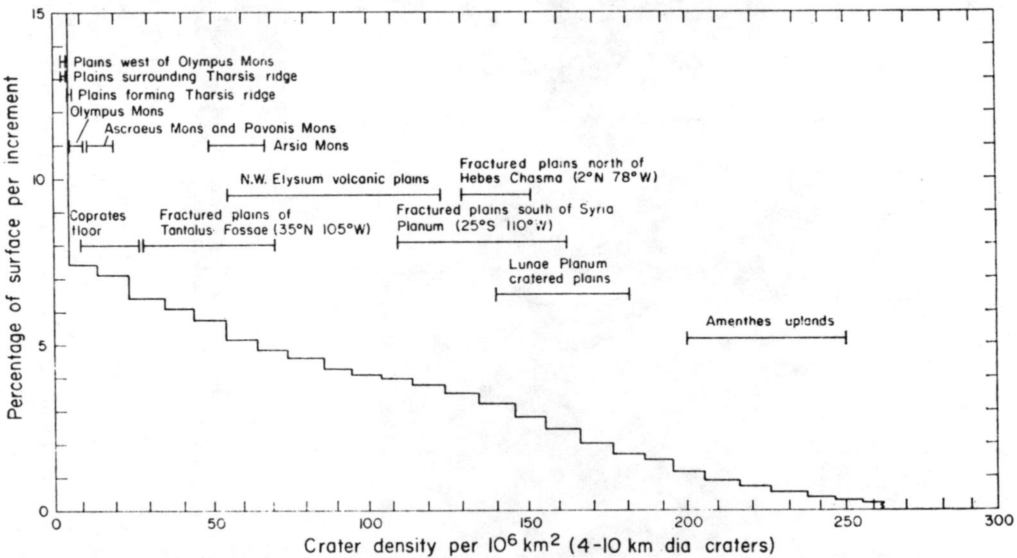

FIG. 12. Histogram of relative ages of regional units on Mars. This histogram, based on the data of Fig. 11, shows the percentage of area (within $\pm 40°$ Lat.) per increment of crater density. The crater density is the number of 4 to 10 km craters per 10^6km^2. In comparison, the lunar maria range between about 100 and 300 on this scale. Also shown (as horizontal bars) are the ranges of crater density measured for a variety of post-uplands plains and features. These bars are referenced horizontally to the crater density scale; their vertical positions were chosen to simplify the display and carry no specific meaning.

basin and area south of about 50°S latitude. It appears that the mantles there are locally thick enough to bury craters as large as a few kilometers in diameter. Hence use of this kind of data to establish stratigraphic sequences south of about 40°S is of dubious value. For this reason the mantled areas (poleward of ±40° latitude) are avoided in the global stratigraphy described below.

The following is a brief discussion of the sequence of emplacement of the post-uplands surfaces, based on the 4 to 10 km crater distributions. Figure 13 is a reference chart showing the locations of the areas described.

The most heavily cratered area on Mars in the intermediate crater range (4–10 km) is found in the Amenthes uplands in a strip between 300°W, 10°S and, 260°W, 5°N. This is also the area most heavily populated by large (>20 km diameter) craters (Fig. 2 and Hartmann, 1973, Fig. 5) displaying a wide range of preservation. This area probably most closely resembles the primordial crust just as it appeared after the early intense erosion and bombardment.

The next oldest major plains units, mentioned earlier, are found throughout the northern hemisphere. These cratered plains are typified by the Lunae Planum plateau (Fig. 4). The large (>20 km diameter) craters are notably absent on these surfaces whereas the populations of intermediate (4–10 km diameter) craters are only 20–30% lower than the Amenthes uplands. The Lunae Planum cratered plains show the same density as the north polar mottled cratered plains (Soderblom *et al.*, 1973b). Similar crater densities are found in many fractured plains (see Fig. 12, plains near Hebes Chasma and Syria Planum). We suggest that these cratered and fractured plains with a 4–10 km crater density between 130 and 180 craters 10^6 km^2 can be classified as a global plains system, the first developed after the uplands degradation. Because they often show complex wrinkle ridges like those found in the lunar maria, it is inferred that many of these plains have a volcanic origin.

The Elysium Fossae volcanic units (Carr, 1973) constitute the second major regional volcanic system. The large error bar for this system in Fig. 12 probably indicates a family of units rather than poor statistics. Volcanic units of similar age can be found throughout the equatorial region east of Elysium an example being the southwest of the Coprates Canyon near 15°S, 85°W.

The Tharsis system is the youngest major volcanic system recognized here. Arsia Mons volcano (South Spot) and the fractured plains of Tantalus Fossae to the north (35°N, 105°W) and to the northwest of Ceraunius Fossae (15°N, 110°W) are about the same age and in fact may be part of the same complex. The two large volcanoes north of Arsia Mons on the Tharsis ridge, Ascraeus Mons and Pavonis Mons (Middle and North Spot) both show crater densities about one-third that on Arsia Mons. The youngest stratigraphic unit in the Tharsis system extends from the eastern equatorial area in Elysium through Olympus Mons (Nix Olympica) and surrounds the three older volcanoes to the east. This volcanic plains unit buries an erosional scarp (left central part of Fig. 4) west of the Lunae Planum plateau. This scarp is the western wall of the erosional canyon cutting through western Lunae Planum to the north.

In summary, the global stratigraphy can be subdivided on the basis of 4–10 km crater populations as follows:

A. The intensely "levelled" uplands surface is the oldest with a density of 200–250 craters/10^6 km^2.

B. Next the cratered plains of the equatorial and north polar regions mottled cratered plains (Soderblom *et al.*, 1973b) were emplaced; some of these were subsequently fractured. These show densities between 130 and 180 craters/10^6 km^2.

C. The Elysium Fossae volcanics were the next major system of units with densities between 60 and 120 craters/10^6 km^2.

D. The last major system is the Tharsis ridge complex with a variety of volcanic and fractured plains and four major volcanoes with crater densities between 10 and 70 craters/10^6 km^2.

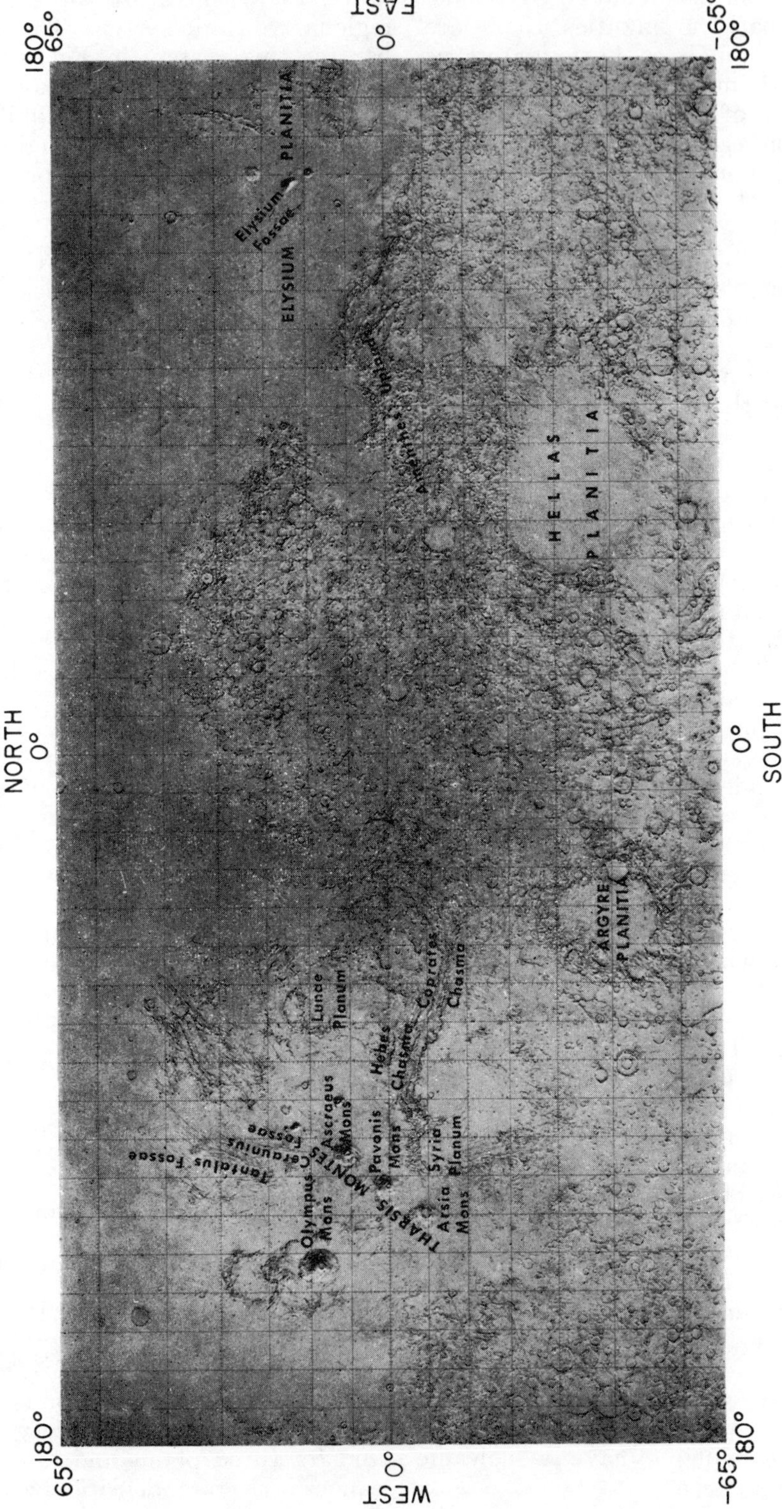

FIG. 13. Locations of regions named and described in the text and figures.

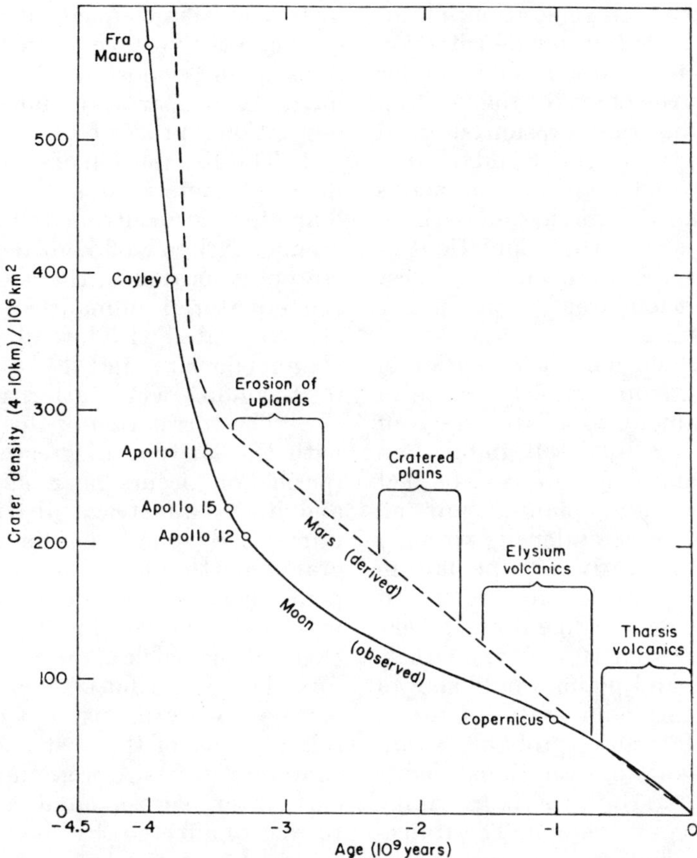

FIG. 14. Comparison of impact flux history derived for Mars with that observed for the moon. The lunar data are the same as in Fig. 1. The derivation of the Martian curve is discussed in the text.

DISCUSSION: AN ABSOLUTE TIME SCALE FOR MARTIAN EVOLUTION

Several fundamental observations about the Martian crater populations and their similarities to lunar populations have led us to the induction that the impact flux histories for Mars and the Moon have been about the same. Our approach differs from others mentioned earlier, in that we do not require a knowledge of the ratio of impacting flux at Mars or the moon, past or present. Further, we compared the entire spectrum of craters on surfaces of all ages of Mars and the Moon. Hartmann (1966) pointed out the fallacy of comparing the Martian populations with those of the uplands of the Moon because the lunar uplands are saturated, thus their net accumulation is indeterminate. We add to that concept that is it fallacious to compare the lunar maria with the densely cratered uplands on Mars which were probably once saturated and since have suffered depletion by erosion. The Martian uplands usefulness in recording flux history is therefore even less than that of the lunar uplands. Following this reasoning we choose to compare the crater density spectra for the lunar maria and the Martian cratered plains where the craters are undegraded and the statistics complete.

There are four key observations that support our induction. First, Mars and the moon show strongly bimodal crater populations—surfaces show either a few sparse craters or show remnants of a saturated population of very large craters. Second, an epoch of intense erosion of the large Martian craters appears to have been contemporaneous with their generation. Third, the small bowl-shaped crater densities on the old eroded Martian uplands are similar to the density on the oldest of the Martian plains. This means that erosion stopped before the oldest plains formed, and that the time between the end of erosion and plains formation was short. Fourth, and perhaps most imporant, is that the populations of small bowl-shaped craters on the oldest lunar maria and the oldest Martian plains are about the same.

The overall sequence in the evolution of the flux at Mars, from an early period of intense bombardment to a later stage of quiescent cratering by itself implies an early abrupt change in the flux. The fact that the uplands and plains show an abrupt transition in crater density strongly suggests to us an early abrupt fall in cratering rate, one which can arise only if the early intense bombardment had a very short half life (approx. 10^7–10^8 yr). Otherwise the uplands and plains would show a gradational transition in cratering style, which they do not. Mars probably swept up the last remnants of accretional debris during its first 1–1.5 b.y., perhaps taking slightly longer than the moon. The transition in the flux history inferred by the uplands–maria dichotomy probably occurred between 3 and 4 b.y. ago. This further implies the same ages (3–4 b.y.) for the oldest plains and maria on both Mars and the moon. The fact that the crater densities on such old plains on both planets are the same leads us to believe that the fluxes have been about the same at Mars and the moon during the last 4 b.y.

Conclusions

The basic observations and conclusions based on the global crater studies presented here can be summarized as follows:

1. Mars and the moon both show marked global dichotomies in cratering style (densely cratered uplands and smooth plains).

2. In the most primitive Martian uplands a large variation in morphological preservation implies an early period of intense erosion coupled with the tailing off of the early postaccretion bombardment. Subsequently most of the Martian areas have been partially buried by other depositional processes.

3. The distribution of small craters a few hundred meters to a few kilometers in diameter is controlled by the eolian regime. A band of high density of these craters is located in the southern part of the equatorial unmantled zone—a zone also correlated with low albedo, high small channel density, and the belt of symmetry for the global wind patterns.

4. The correlation of this unusual band with the latitude of greatest solar input (perihelion occurs near summer solstice) and its asymmetrical distribution in the unmantled zone suggests this band migrates north and south on a 50 000-yr period between the boundaries of the thick mantles and continually registers the edges of the thick mantles.

5. The distribution of intermediate craters (4–10 km) is a measure of the relative ages of the post-uplands surfaces. Craters of this size were destroyed by the early intense erosion and are large enough to be unaffected by mantles (at least within ±45° latitudes).

6. Four main stratigraphic systems are recognizable from the density of intermediate craters: (a) ancient eroded uplands; (b) volcanic cratered plains (e.g., Lunae Planum); (c) Elysium Fossae volcanics; and (d) Tharsis volcanics.

7. The similarity in the upland–mare dichotomy on Mars and the moon, the association of Martian upland erosion with the intense bombardment, the similarity of the Martian small crater populations on intensely eroded uplands and on cratered plains, and the similarity in crater density on old Martian and lunar "maria" lead to the inference that the Martian and lunar flux histories are nearly the same. More simply, similar cratering flux histories have led to similar global dichotomies and crater densities.

8. If the lunar and Martian flux histories are about the same, Mars has not had prolonged epochs of inactivity but rather the geologic evolution of the planet

has been a gradual process spread throughout the last 4 b.y.

Acknowledgments

Bruce C. Murray helped initiate and support much of the work presented in this study. We are grateful for careful review of the manuscript by him, D. E. Wilhelms, D. H. Scott, G. W. Colton, C. R. Chapman and W. K. Hartmann.

This work was conducted at the Center of Astrogeology, U.S. Geological Survey, Flagstaff, Arizona, and at the California Institute of Technology and the Jet Propulsion Laboratory, Pasadena, California, under contracts W13,204; W12,872; WO-8122, and NAS 7-100.

References

Anders, E., and Arnold, J. (1965). Age of craters on Mars. *Science* **149**, 1494–1496.

Anderson, D. L. (1969). The evolution of terrestrial-type planets. *Appl. Opt.* **8**, 1271–1277.

Anderson, D. L., and Phinney, R. A. (1967). Early thermal history of the terrestrial planets, *In* "Mantles of the Earth and Terrestrial Planets," (S. K. Runcorn, ed.) p. 113–126. Interscience, New York.

Baldwin, R. B. (1965). Mars: An estimate of the age of its surface. *Science* **149**, 1498–1499.

Binder, A. B. (1966). Mariner IV: Analysis of preliminary photographs. *Science* **152**, 1053–1055.

Brinkman, R. T. (1966). Lunar crater distributions from Ranger VI photographs. *J. Geophys. Res.* **71**, 340–342.

Carr, M. H., Masursky, H., and Saunders, R. S. (1973). A generalized geologic map of Mars. *J. Geophys. Res.* **78**, 4031–4036.

Chapman, C. R., Pollack, J. B., and Sagan, C. (1969). An analysis of the Mariner-4 cratering statistics. *Astron. J.* **74**, 1039–1051.

Conrath, R., Curran, R., Hanel, R., Kunde, V., Maquire, W., Pearl, J., Pirraglia, J., Walker, J., and Burke, T. (1973). Atmospheric and surface properties of Mars obtained by infrared spectroscopy on Mariner 9. *J. Geophys. Res.* **78**, 4267–4278.

Hanks, T. C., and Anderson, D. L. (1969). The early thermal history of the earth, *Phys. Earth Planet. Interiors* **2**, 19–29.

Hartmann, W. K. (1965). Secular changes in meteoritic flux through the history of the solar system. *Icarus* **4**, 207–213.

Hartmann, W. K. (1966). Martian cratering. *Icarus* **5**, 565–576.

Hartmann, W. K. (1970). Lunar cratering chronology. *Icarus* **13**, 299–301.

Hartmann, W. K. (1971). Martian cratering. III. Theory of crater obliteration. *Icarus* **15**, 410–428.

Hartmann, W. K. (1973). Martian cratering. 4. Mariner 9 initial analysis of cratering chronology. *J. Geophys. Res.* **78**, 4096–4116.

Hord, C. W., Barth, C. A., Steward, A. I., and Lane, A. L. (1972). Mariner 9 ultraviolet spectrometer experiment: photometry and topography of Mars. *Icarus* **7**, 443–456.

Leighton, R. B., Murray, B. C., Sharp, R. P., Allen, J. D., and Sloan, R. K. (1965). Mariner 4 photography of Mars: initial results. *Science* **149**, 627–630.

Masursky, Harold. (1973). An overview of geological results from Mariner 9. *J. Geophys. Res.* **78**, 4009–4030.

McCauley, J. F., Carr, M. H., Cutts, J. A., Hartmann, W. K., Masursky, H., Milton, D. J., Sharpe, R. P., and Wilhelms, D. E. (1972). Preliminary Mariner 9 report on the geology of Mars. *Icarus* **17**, 289–327.

MacDonald, G. J. F. (1962). On the interval constitution of the inner planets. *J. Geophys. Res.* **67**, 2945–2974.

McGill, G. E., and Wise, D. U. (1972). Regional variations in degradation and density of Martian craters. *J. Geophys. Res.* **77**, 2433–2441.

Murray, B. C., Soderblom, L. A., Sharp, R. P., and Cutts, J. A. (1971). The surface of Mars. 1. Cratered Terrains. *J. Geophys. Res.* **76**, 313.

Murray, B. C., Ward, W. R., Young, S. C. (1973). Periodic insolation variations on Mars. *Science* **180**, 638–640.

Öpik, E. J. (1966). The Martian surface. *Science* **153**, 255–265.

Papanastassiou, D. A., and Wasserburg, G. J. (1971a). Lunar chronology and evolution from Rb–Sr studies of Apollo 11 and 12 samples. *Earth Planet. Sci. Let.* **11**, 37–62.

Papanastassiou, D. A., and Wasserburg, G. J. (1971b). Rb–Sr ages of igneous rocks from the Apollo 14 mission and the age of the Fra Mauro Formation. *Earth Planet. Sci. Let.* **12**, 36–48.

Pieri, D. C. (1974) Distribution of small channels on the Martian surface. *Icarus*, to be submitted.

Sagan, Carl, Veverka, J., Fox, P. Dubisch, R., French, R., Gierasch, P., Quam, L., Lederberg, J., Levinthal, E., Tucker, R., Eross, B., and Pollack, J. B. (1973). Variable features on Mars, 2, Mariner 9 global results. *J. Geophys. Res.* **78**. 4163–4196.

Sharp, R. P., Soderblom, L. A., Murray, B. C., and Cutts, J. A. (1971a). The surface of Mars. 2. Uncratered terrains. *J. Geophys. Res.* **76**, 331–342.

Sharp, R. P., Murray, B. C., Leighton, R. B., Soderblom, L. A., and Cutts, J. A. (1971b). The surface of Mars. 4. The south polar cap. *J. Geophys. Res.* **76**, 357–368.

Shoemaker, E. M., Hackman, R. J., and Eggleton, R. C. (1962). Interplanetary correlation of geologic time. *Advances Astronaut. Sci.* **8**, 70–89.

Shoemaker, E. M. (1965). Preliminary analysis of the fine structure of the lunar surface in Mare Cognitum. *In* "The Nature of the Lunar Surface," pp. 23–78. Johns Hopkins Press, Baltimore.

Silver, L. T. (1971). U–Th–Pb isotopic systems in Apollo 11 and 12 regolith materials and a possible age for the Copernicus event. Abstract, *Trans. Am. Geophys. Union* **52**, 534.

Soderblom, L. A., and Boyce, J. M. (1972). Relative ages of some near-side and far-side terra plains based on Apollo 16 metric photography. *Apollo 16 Preliminary Science Report*, 29–3 to 29–6.

Soderblom, L. A., and Lebofsky, L. A. (1972). Technique for rapid determination of relative ages of lunar areas from orbital photography, *J. Geophys. Res.* **77**, 279–296.

Soderblom, L. A., Kreidler, T. J., and Masursky, H. (1973a). Latitudinal distribution of a debris mantle on the Martian surface. *J. Geophys. Res.* **78**, 4117–4122.

Soderblom, L. A., Malin, M. C., Cutts, J. A., and Murray, B. C. (1973b). Mariner 9 observations of the surface of Mars in the north polar region. *J. Geophys. Res.* **78**, 4197–4210.

Ward, W. R. (1973). Large scale obliquity variations on Mars. *Science* **181**, 260–262.

Wasserburg, G. J., Huneke, J. C., Papanastassiou, D. A., and Podosek, F. A., and Turner, G. (1972a). Age determinations on samples from the Apollo 14 landing site. *Space Res.*, (XII), **1**, 39–42.

Wasserburg, G. J., and Papanastassiou, D. A. (1972b) Age of an Apollo 15 mare basalt; lunar crust and mantle evolution. *Earth Planet. Sci. Let.* **13**, 97–104.

Wilhelms, D. E. (1973). Cratered terrains. Presentation at the International Colloquium on Mars, California Institute of Technology.

Witting, J., Marin, F., and Stone, C. A. (1965). Mars: age of its craters. *Science* **149, 1496–1498.**

28

Reprinted from *Science* 194:1381–1387 (1976)

Mars: A Standard Crater Curve and Possible New Time Scale

Cratering links to lunar time suggest that Mars died long ago.

G. Neukum and D. U. Wise

One major goal in planetary science is to determine the chronology of development of the surfaces of the terrestrial planets, especially our neighbor Mars. Whereas for the moon we have rock specimens whose ages have been determined radiometrically, we will not have any way to analyze the ages of martian rocks in the near future. Nevertheless, the surface of Mars has been mapped extensively by the Mariner 9 and recent Viking missions. These pictures allow some qualitative classification of old or young features according to their stratigraphic relations and apparent degree of erosion. Fortunately—for the purpose of age determination from photographs—Mars is impact-cratered. Differences in impact crater frequencies at different sites reflect differences in age. Recently, two attempts have been made to determine absolute ages for Mars from its measured crater frequencies, based on extrapolations from the cratering chronology of the lunar surface (*1*, *2*). Unfortunately, a straightforward comparison of martian and lunar crater frequencies does not necessarily yield true ages: relative impact rates and the time dependence of the martian cratering rate are not known; and it is not certain whether the same meteoroid population bombarded both planets. Some of these questions are discussed by Hartmann (*1*) and Soderblom *et al.* (*2*).

No martian cratering time scale is better than the lunar-terrestrial one it refers to. Thus, before comparing the impact conditions on Mars with those in the earth-moon system, we will briefly review present knowledge of the cratering chronology in the earth-moon system.

Cratering Chronology in the Earth-Moon System

Four independent determinations of the cratering chronology in the earth-moon system may be found in the literature (*1*–*5*). We have compiled these data in Fig. 1, where crater frequencies per square kilometer are plotted against exposure ages of sites where the craters accumulated until today. The time dependence of crater frequency in the data sets of Baldwin (*3*), Hartmann (*1*, *4*), and Neukum *et al.* (*5*) in Fig. 1 appears similar. The absolute data agree within a factor of 2 to 3. In the first billion years after the formation of the lunar crust 4.4 to 4.5 billion years ago (*6*, *7*), the impact rate or number of craters as a function of time (dN/dt) dropped sharply. There are no crater frequency data for age-dated surfaces in the span 1 to 3 billion years. For the last 1 billion years, lunar crater frequencies have been measured (*8*) for structures whose ages have been very accurately determined (*9*). The terrestrial impact crater frequency data shown in Fig. 1 for the Canadian Shield area and other parts of North America are reduced to lunar impact conditions for gravitative acceleration and are compatible with the lunar data within a factor of 2 to 3. The terrestrial data are much more uncertain than the lunar data, as reflected by the large error bars. The data for the last 1 billion years together with the lunar measurements for an age of ≈ 3.2 billion years strongly suggest an impact rate varying at the most by a factor of 3 to 4 during that time. A nearly constant flux is indicated, especially in the last 1 billion years (*10*).

The crater frequency data of Soderblom *et al.* (*2*) in Fig. 1 do not seem to agree with those of the other authors for ages $\lesssim 3.5 \times 10^9$ years, differing by a factor of ≈ 5 from the data of Baldwin (*3*) and Hartmann (*1*, *4*) and a factor of ≈ 10 from those of Neukum *et al.* (*11*). In addition, there is a difference in radiometric ages for some of the same Apollo landing sites because of progress in age measurements and differences in interpretation of lunar rock ages (*12*).

The minor differences between the frequency data of Baldwin and Hartmann and those of Neukum *et al.* can be partly explained as follows. Crater frequency data are commonly obtained for different size ranges. To intercompare these data, it is necessary to apply a size distribution law. The lunar impact crater production size-frequency distribution (the undisturbed image of the meteoroid mass-velocity distribution) has been determined recently (*13*) and is displayed in Fig. 2 (*14*–*16*). All the data of Neukum *et al.* in Fig. 1 have been reduced by using this "calibration" or standard distribution to project the frequency measurements to the diameter D = 1 km intercept. Baldwin (*3*) and Hartmann and Wood (*16*) used $N \sim D^{-1.8}$ and $N \sim D^{-2}$, respectively, as calibration lines, which seem to be good approximations at larger sizes but are too flat in the 1-km size range. They usually reduced their frequency data to sizes larger than 3 km. For the 1-km size range they have included some data reported by others, especially for the

Gerhard Neukum is research scientist at the Max-Planck-Institut für Kernphysik, 6900 Heidelberg, West Germany. Donald U. Wise is professor of geology at the University of Massachusetts, Amherst 01003.

younger Apollo sites. Applications of the slopes of −1.8 or −2 to those data in the kilometer size range or extrapolation from larger sizes ($D \geq 3$ km) would commonly result in values higher than those derived with the standard curve by a factor of about 2 or 3.

The overall picture we derive from Fig. 1 is the following. For ages of around 4 billion years all authors' data agree fairly well. For younger ages the discrepancies are as great as one order of magnitude. The data of Soderblom *et al.* differ the most, and we believe that this difference may reflect an error in the conversion (*11*) of their measurements to the published frequencies of craters 4 to 10 km in diameter. If the inconsistency between figures 1 and 14 in (*2*) is removed, their values fall close to those of the other authors.

On the grounds stated above we have elected to use the curves drawn as solid lines in Figs. 1 and 2 as the representation of lunar impact cratering. Choice of these curves has the added advantage for our martian correlations that they are derived in the same size range as the martian curves, by use of the same equipment and data reduction methods.

Martian and Lunar Production Size-Frequency Distributions

Before discussing the Mars models derived from different lunar data sets (*1–5*) and comparing them with a time scale derived from the lunar data (*17*), we will discuss the importance of a precise knowledge of the martian impact crater production size-frequency distribution in comparison to the lunar one shown in Fig. 2.

The lunar curve shows a characteristic steepening around a crater diameter of 2 km. This steepening was detected when the Surveyor and Lunar Orbiter photographs were investigated and was widely interpreted as due to an admixture of secondary craters produced by the larger primary ones (*2*). We do not agree with this interpretation. In our investigations (*8, 13*) we found that the lunar crater size distribution curve remained the same for crater counts on ejecta blankets, floors, or terrace deposits of younger (Eratosthenian and Copernican) lunar craters as well as on homogeneously cratered mare areas and a few light plains areas with ages ranging to more than 4 billion years. We excluded obvious secondary craters by applying criteria developed by Oberbeck and Morrison (*18*). The resulting general curve is the calibration distribution of Fig. 2.

The constancy in shape of the lunar curve regardless of distance to large craters (outside the discontinuous ejecta blanket) capable of producing secondary craters, and on the continuous ejecta blankets of large craters too young to have secondaries from other sources superimposed on them (such as Tycho), argues that the curve does not include significant numbers of secondary craters but represents the primary-production population. The constancy with time of the slope and inflection point of the standard lunar curve suggests that the size and velocity distribution of bodies impacting the lunar surface has not changed significantly over the last 4 billion years.

Whatever view is favored for the steep branch of the lunar crater size distribution, this part of the curve is very important for the derivation of a martian crater time scale based on the lunar one, because most of the lunar data have been obtained in this size range, for crater diameters between ≈ 500 m and ≈ 5

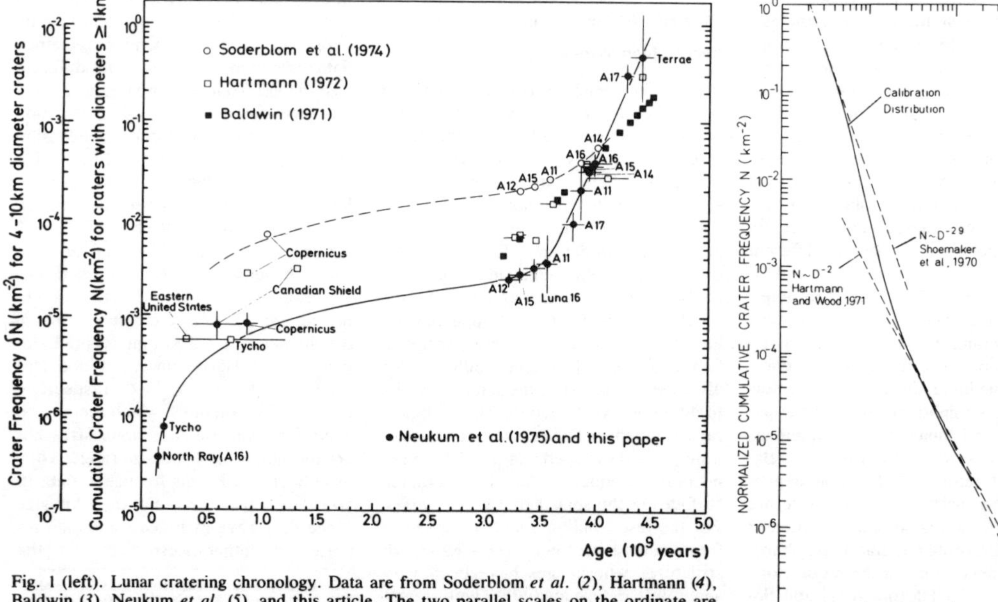

Fig. 1 (left). Lunar cratering chronology. Data are from Soderblom *et al.* (*2*), Hartmann (*4*), Baldwin (*3*), Neukum *et al.* (*5*), and this article. The two parallel scales on the ordinate are cumulative crater frequency [the number of craters equal to or larger than a certain diameter D (here $D = 1$ km) per unit area; compare Fig. 2] and frequency (the number of craters with diameters between 4 and 10 km per unit area). These scales are proportional (*35*). The lunar ages are those measured radiometrically for the Apollo lunar rocks (*12*), except ages > 4 billion years from Baldwin, which are inferred from morphological criteria. The terrestrial crater frequency data are reduced to lunar conditions, the ages from the terrestrial data (*4, 5*) are based on geologic estimates and radiometric measurements. Terrae are lunar highlands; the prefix A means Apollo. **Fig. 2 (right). Lunar calibration curve, or production size–frequency distribution. These data (*13*) reconcile former linear distribution laws (also given) which partly contradicted each other. The steepening at small sizes (*15*) is confirmed and precisely determined.**

km. To compare martian with lunar crater frequency data, we have to know about the shape of the martian impact crater size distribution in this size range.

The martian crater size distribution is known to follow power law $N \sim D^{-2}$ in the size range from several kilometers to about 100 km (*1, 2, 19*), practically identical with the lunar case. At a diameter of about 2 km a steepening similar to that in the lunar case was detected (*2, 20*). These measurements were not sufficient to determine the martian production size distribution in this size range, although they indicated that it was not too dissimilar from the lunar one.

Before reporting new measurements, we wish to state what we can learn from an exact knowledge of the martian distribution curve in the size range where it steepens. Figure 3 shows such a steepening distribution in a somewhat simplified and exaggerated form. Different impact conditions on different planets will result in characteristic effects on a distribution of this form. Age effects—that is, similar ratios of exposure times of the surfaces to cratering—will result in similar ratios of crater frequencies at any diameter ($\Delta \log N$ = constant for the right half of Fig. 3). The same is true for different cross-section effects, in which slower meteorites are more easily deflected by a planet's gravity. This is equivalent to a difference in flux. These effects do not change the shape of the curve—that is, the crater frequency at some diameter in the steep part of the curve relative to the value at some diameter in the flat part of the curve. In other words, age differences and cross-section effects are such that the curves can be shifted vertically and should coincide. The shape of the curve, in terms of the diameter at which the inflection point occurs in Fig. 3, does not remain constant if the impact velocity is different on different planets, or if target properties have an effect on the sizes of the craters produced. Thus, even with the same meteoroid mass distribution, different (average) impact velocities on Mars and on the moon or different target properties will result in different crater sizes for the same projectile mass, and the shapes or inflection points of the curves for the moon and Mars will differ. The ratio of crater frequencies at two fixed diameter values in the steep and flat parts of the curve will be different, as seen in the left half of Fig. 3 ($\Delta \log N_1 \neq \Delta \log N_2$ for a diameter shift by a constant factor).

To derive a martian time scale by comparing martian crater frequencies with lunar ones, we have investigated the shape of the martian production size frequency curve in comparison with the lunar calibration distribution of Fig. 2 (*17*). The ideal conditions for measuring a standard curve are: a homogeneously cratered area of a large enough extent; adequate resolution of imagery; a single, well-defined time of surface origin; and the absence of resurfacing by processes other than impact cratering. In that this ideal cannot be met with the Mariner 9 imagery, the curve has been pieced together from a variety of areas and imagery scales.

The most homogeneous areas of high-resolution Mariner 9 imagery (B frames) were selected, and the corresponding lower-resolution A frames covering a large area surrounding the B frames were checked for quality of imagery and homogeneity of larger craters. Most areas proved to be inadequate for our purpose because they had experienced extensive resurfacing resulting in irregularities (bumps) in the distributions. These resurfacing processes have been extensively documented (*19, 21*). Details of our own measurements are given in (*17*).

The most homogeneous "test" areas we found with both high-resolution and low-resolution Mariner 9 coverage are Elysium Planitia (240°W, 30°N) and the Alba Patera volcano central cone area (110°W, 40°N). These measurements are displayed in Fig. 4, a and b, where the solid lines through the data represent a computer least-squares fit. This curve (*22*) through the Alba and Elysium points is the one we consider the best derivable

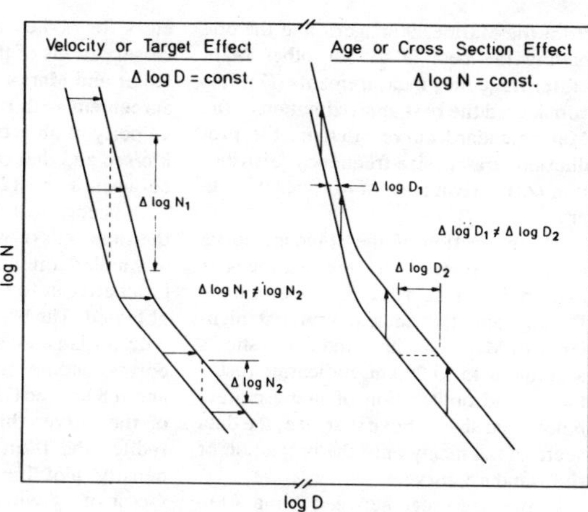

Fig. 3. Effects of different impact conditions on a size distribution curve whose slope is different at different crater sizes; N is the cumulative number of craters with diameters greater than a given diameter D. Velocity or target effects cause the curve to shift left or right while maintaining a constant change in log D. The bend in the curve produces an inconstant ΔN for different diameters. For age or cross-section effects the curve shifts vertically and the reverse relations obtain.

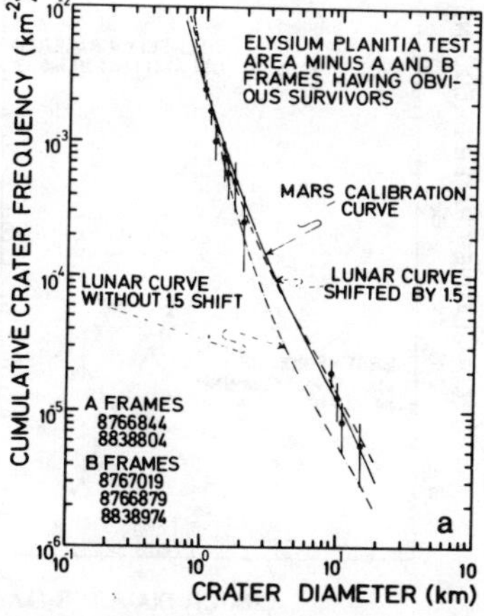

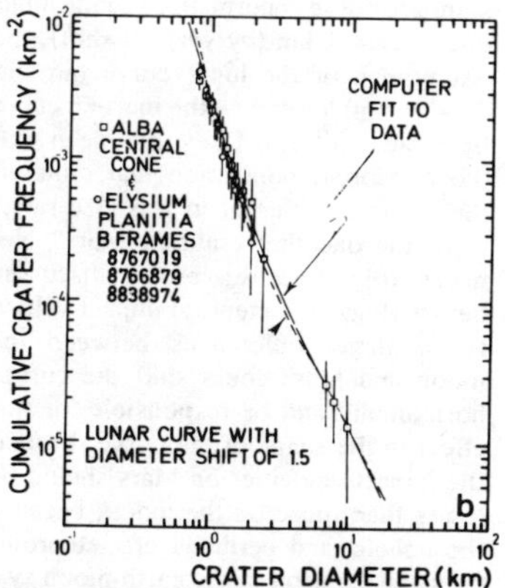

Fig. 4. (a) Test plot for Elysium Planitia after frames showing obvious survivor craters were eliminated. The lunar standard curve is included with and without a diameter shift by a factor of 1.5. In the latter case, the curve is vertically "age-shifted" to fit approximately at small crater sizes. (b) Standard curve for Alba central cone. Thin edges of the cone were excluded from the counting area. Included are B-frame counts from the second-best test area, Elysium Planitia.

from the Mariner 9 imagery and the one having the best fit to our other Mars crater frequency measurements (17). It is considered the best approximation of the Mars standard curve—that is, the production crater size-frequency distribution (23) derivable from Mariner 9 imagery.

The projection of the standard curve to larger craters and older surfaces is explored in Fig. 5 for the older Lunae Planum and the ancient cratered highlands of Mars. The highland curve shows a break at 15 to 20 km, indicating resurfacing and obliteration of many craters below that size. Above that size, the data seem to fall nicely onto the projection of the standard curve.

In the size range between 3 and 7 km the statistical limitations of the Mariner data are severe. It is possible that the curve has a slight bump in this region, but this seems unlikely because (i) a few exceptionally homogeneous areas show no or almost no bump, (ii) the size of the bump is variable and related to numbers of obvious survivor craters (those not extinguished by resurfacing), and (iii) by analogy, the lunar curve shows no obvious bump near this size range. Detailed work on the Viking imagery, when it becomes generally available, should resolve the issue.

As seen in Fig. 4, simple superposition of the lunar and martian standard curves (taking out age or cross-section differences by shifting vertically, as discussed in connection with Fig. 3) shows that they are not identical. If the ratio of martian and lunar crater frequencies at $D = 10$ km is compared with that at $D = 1$ km, a difference of a factor of 2.5 is found. In other words, if the martian standard curve is normalized to the lunar one at $D = 1$ km (by vertical shift), the flat branch of the lunar curve (around $D = 10$ km) lies below the martian curve by a factor of 2.5. Does this mean that the meteoroid population that cratered the moon is different in this size range from the one that cratered Mars? Not necessarily, because, as we discussed before (Fig. 3), meteoroid impact velocity or target differences between the moon and Mars could shift the curves horizontally and be responsible for this effect in the shape of the curve. In fact, the impact velocities on Mars should be lower than those on the moon, because the aphelia and perihelia of meteoroids crossing the orbit of the earth-moon system are judged to be smaller on the average than those of meteoroids crossing the orbit of Mars. Lower velocities on Mars would result in smaller crater diameters for bodies with the same masses. Observations of the velocities of Apollo-Amor and Mars-crossing asteroids are in agreement with this: the average impact velocity of objects hitting the moon is 15 km/sec and that of bodies hitting Mars is between 8 and 12 km/sec (24).

In accord with this velocity argument, the lunar curve was shifted horizontally to smaller crater diameters to test whether a better fit to the Mars curve could be obtained. The best fit of the two curves is with a diameter shift by a factor of 1.5, corresponding to a Mars velocity of about 8 km/sec (25). Because of the slope of the curve, this diameter shift would reduce the 1-km intercept of the crater density plot for a Mars surface by a factor of 4.8 with respect to an identically aged lunar surface. If the cross-sectional effect for more effective gravitational capture of slower bodies is included, the factor is reduced to 4.5.

A lunar curve with a diameter shift of a factor of 1.5 is contrasted with the standard Mars curves in Fig. 4, a and b, to show the high degree of similarity between them and to illustrate the need for a diameter shift if coincidence is to be obtained. Whether the best value of the diameter shift is precisely 1.5 awaits refinement of the standard curve with Viking imagery.

A diameter shift could also be produced by target effects (Fig. 3). At present, we have no detailed information on martian surface composition and its effect on possible differences in crater sizes. Such an effect is considered small though, since martian surface rock composition is supposed to be not too different from the lunar one.

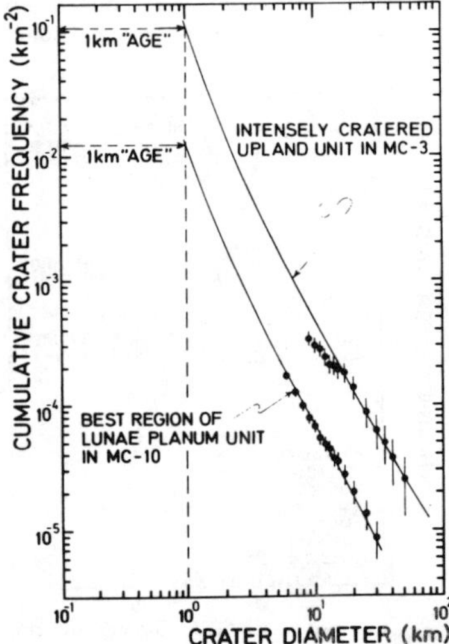

Fig. 5. Comparison of the projection of the standard Mars curve (slope ~ −1.8 at $D \geqslant 10$ km) with two older cratered areas. MC-3 and MC-10 are map quadrangle numbers.

In any case, we have found that the meteoroid populations cratering Mars and the moon in their middle to late history are probably not two populations with different mass distributions, but the same family of objects in the size range investigated. A simple diameter shift by a factor of 1.5 removes all differences in the shapes of the two standard curves within the error limitation. It is not necessary to assume an additional target effect, because the diameter shift corresponds to a reasonable impact velocity difference of 15 km/sec on the moon and 8 km/sec on Mars (25).

The correlation by diameter shift strengthens our interpretation that the steep parts of the standard curves at $D \lesssim 2$ km are not due to an admixture of secondary craters but represent the original primary production size-frequency distribution.

Absolute Age Correlations

The reasonably good fit of the lunar standard crater curve to cratered lunar surfaces less than 4 billion years old implies that the production population of lunar impacting bodies has not changed significantly in that time. Similarly, the close fit of the standard Mars curve to a wide range of ages of martian surfaces (17) suggests that the bodies impacting Mars also have not changed over a very long (but not radiometrically determined) period of time. The likelihood that one family of objects is involved argues for the same or at least a very similar time dependence of impact flux on both Mars and the moon. This is important, because at present we have no other means to determine the time dependence for Mars.

Hartmann (1) and Soderblom *et al.* (2) have also considered some of the problems in linking the martian and lunar cratering histories. They assumed that the time dependence of lunar and martian impact flux was the same. Hartmann also took into account the velocity effect by considering the present-day asteroid velocities but assumed a constant martian size distribution with a slope of −2 for all sizes. In this way he arrived at a crater frequency difference of a factor of 1.5 for all sizes, which is in accord with our calculation for the flat part of the standard curves ($D \gtrsim 5$ km) but differen from our values at smaller sizes. Soderblom did not account for impact velocity differences on Mars and the moon.

Both Hartmann and Soderblom *et al.* considered a possible difference in impact flux at Mars because of its greater proximity to the asteroid belt. Soderblom *et al.* made the ad hoc assumption that the cratering flux at Mars was about the same as that at the moon (*26*), and Hartmann presented various arguments in favor of an impact flux at Mars ten times higher than that at the moon, which gives a six times higher cratering rate on Mars (because of the velocity effect). Both assumptions have been criticized (*19, 21*). Our work discussed above and in (*17*) places some additional constraints on these assumptions about the relative lunar and martian impact fluxes.

As pointed out by Soderblom *et al.* (*2*), the highlands of Mars and the moon were cratered at the time of the early intense bombardment, between 4.5 and 4 billion years ago. We assume that the martian highlands, like the lunar highlands, show the cratering record from about 4.4 billion years until the present. A frequency of 1-km craters of 480,000 per 10^6 km^2 has been determined for the 4.4-billion-year-old (*6, 7*) lunar highlands by using the craters in the 100-km size class and the lunar calibration curve to project to the 1-km size (Fig. 1). For the martian highlands the frequency of 1-km craters, from Fig. 5 and from a second area in quadrangle MC-3 (*17*), is 100,000 per 10^6 km^2. This result would argue for about the same flux at Mars as at the moon because martian crater frequencies at $D = 1$ km should be a factor of 4.5 lower than lunar ones for identically aged surfaces because of the velocity effect.

Lunar highland crater populations have commonly been interpreted (*16, 27*) to be in saturation (or a state of equilibrium) with respect to the crater frequencies (that is, preexisting craters are so densely packed that they are extinguished by subsequent impacts, keeping the crater densities constant at all times). We agree with this for craters smaller than ≈ 50 km but not for those larger than 50 to 70 km (for the moon), since the slope of the crater size distribution for this range is not –2 as expected for saturated surfaces. Instead, the slope has been measured as –2.4 to –2.6 (*5, 19, 28*). The martian highlands are clearly undersaturated (*1, 4*) at all sizes visible at Mariner 9 A-frame resolution; that is, we deal with production populations to which our martian standard curve can be directly applied.

There is some uncertainty in the lunar highland point in Fig. 1. Considering the Terrae point together with the Apollo 17 point at 4.3 billion years, and taking the average behavior of crater frequency as a function of age (solid line in Fig. 1), we regard this uncertainty to be smaller than a factor of 2.

Finding similar lunar and martian absolute fluxes could be coincidental. If the martian highlands were 100 million years younger than 4.4 billion years the martian flux could be about a factor of 2 higher. However, another correlation between crater frequency and age can be derived from measurements of crater densities on Mars' satellite Phobos (*29*). Using the crater densities on Phobos and the standard Mars curve for craters 2 to 5 km in diameter, we obtain a 1-km crater frequency of 350,000 per 10^6 km^2. [Since the craters have no ejecta blankets they can be measured without significant superposition even for these small sizes (*29*).] This crater frequency value would be expected for the surface of Phobos if it was exposed for 4.5 to 4.6 billion years and experienced approximately the same flux as the moon, with the velocity effect taken into account. An age of 4.5 to 4.6 billion years for the solidification of Phobos is cosmologically very likely (*30*). This supports our assumption of an age of about 4.4 billion years for the solidification of the martian crust. We conclude that the flux at Mars was the same as that at the moon early in the histories of these planets, with an uncertainty of about a factor of 2. This result disagrees with Hartmann's assumption (*1*) that the martian flux is ten times higher than the lunar one, but roughly agrees with the assumption by Soderblom *et al.* (*2*) of about the same cratering rate (*26*), which would correspond to about the same flux (a factor of 2 higher) for objects forming craters in the size range 4 to 10 km.

These considerations have given the constraints necessary to derive a martian time scale on the basis of our lunar results in Fig. 1.

1) The flux at Mars is the same as that at the moon.

2) The 1-km crater frequency values for Mars are a factor of 4.5 lower than the lunar ones for identically aged surfaces.

3) The time dependence of impact flux is the same for both planets.

Following these arguments, we have shifted the lunar cratering chronology curve of Fig. 1 to 1-km crater frequency values, a factor of 4.5 smaller, to reduce them to martian conditions. This curve is displayed in Fig. 6, together with Soderblom's and Hartmann's relationships, which differ markedly from ours. This difference reflects several factors:

1) In shifting their lunar curve, Soderblom *et al.* made the reasonable but ad hoc assumption of a higher cratering

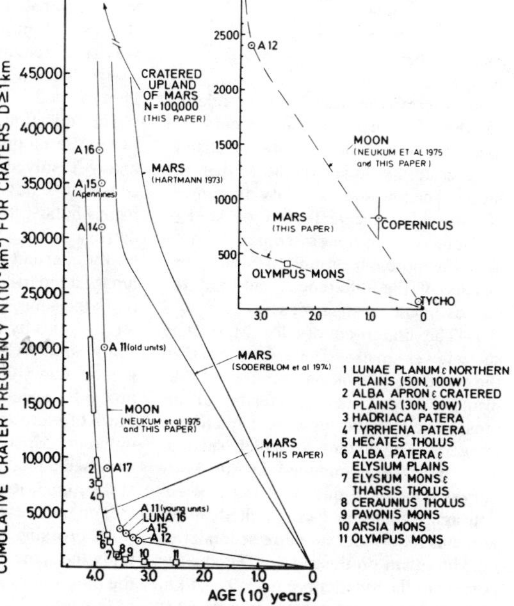

Fig. 6. Cumulative crater frequencies plotted against age for the moon and Mars. The older Mars curves of Hartmann (*1*) and Soderblom *et al.* (*2*) are given for contrast to indicate the radical shift backward in time of martian events. The prefix A means Apollo.

rate on Mars (factor of ≈ 1.5) and did not include the velocity effect. Thus, their curve is shifted with respect to the lunar curve in the opposite direction from ours.

2) The lunar counts on which the martian cratering chronology of Soderblom *et al.* is based lie much higher than any other values (Fig. 1) for ages < 4 billion years.

3) Although Hartmann's lunar crater frequency–time curve is higher than ours by only a factor of 2 to 3 (Fig. 1) but is lower than Soderblom's by a factor of 5 for ages ≲ 3.5 billion years, their Mars curves appear similar in Fig. 6. This is a coincidental agreement, due to Hartmann's assumption of a higher impact flux at Mars (*31*). If Hartmann had assumed equal martian and lunar fluxes, his martian curve would fall near our lunar curve.

The new curve leads to greatly increased estimates of the ages of martian events. The Lunae Planum lava flows are now 3.9 billion years old, the older cratered plains near 30°N, 90°W 3.8 billion years old, and Alba and the Elysium volcanics about 3.7 billion years old. The various volcanic landforms have been interpreted to be as young as 100 to 500 million years (*1, 2*), but are now estimated to be 2.5 to 3.9 billion years old [based on Carr's crater counts (*20*)]. The bulk of the volcanic cone activity was in the time span 3.4 to 3.8 billion years. Olympus Mons, the last great volcanic construct, is 2.5 billion years old.

Uncertainties

Another way of describing the relation between the lunar and martian curves is to say that the frequency ratio of small to large craters is lower for Mars than the moon. The observed velocity differences of asteroids indicate that some kind of diameter shift in the crater size distributions is probably required, but other causes of the differences between the curves might be suggested.

1) The crater counts for Mars may show fewer small secondary craters than the counts for the moon because of lower primary impact energy, later burial, atmospheric slowing, or poorer quality of imagery. The extent to which secondaries have been included in our Mars curves is open to question, but it is encouraging that the factors cited above argue as follows for no more secondaries on Mars than on the moon. The steeper portion of the lunar curve near $D = 1$ km (Fig. 2) must represent primary impacts where measured, for example, on the apron and terraces of the extremely young (≈ 100 million years) Tycho (*9*), where secondary overprints from other large impacts are highly unlikely. If the primary bodies impacting Mars are not from a markedly different family of objects, we would expect a similar steeping for the small sizes in the martian crater distribution, and any secondaries would make the curve steeper rather than shallower.

2) A higher proportion of very large bodies may be included in the family of objects impacting Mars. We know of no way to evaluate this factor quantitatively at present.

3) Differences between Mars and the moon in target characteristics (strength and shock-wave velocity) and in gravity effects could affect the ability of impacting bodies to lift out material or influence crater enlargement by slumping (*32*). These factors, like the velocity differences, could lead to diameter shifts, but in an unknown direction. We suspect that these are relatively modest effects, but await the general availability of the Viking data for better information on target behavior. Our use of Phobos, with its essentially zero gravity, is debatable. Experiments by Johnson *et al.* (*33*) indicate that gravity is a significant factor in determining crater size for small explosion craters in cohesionless sand. However, these authors caution against using this result for cohesive materials and for higher explosive energies. Since the gravitational stresses in kilometer-sized craters are small compared to the tensile strength of most rocks, we think the gravity effect on crater diameter would be small for a cohesive rock mass like Phobos.

The question of relative fluxes between the moon and Mars has not been finally resolved, but relatively narrow constraints have been given. The data from Phobos, to the extent that its crater density can be linked to the conditions on the martian surface, suggest that the lunar and martian highlands are of essentially the same age. Uncertainties in crater densities by factors of more than 2 or 3 are unlikely. Hence, age differences greater than 100 to 200 million years or flux differences between the moon and Mars of more than a factor of 2 or 3 are unlikely. Most previous workers have assumed a much higher flux at Mars than at the moon. Recent work by Shoemaker (*34*) on the present-day Apollo-Amor and Mars-crossing asteroids suggests essentially the same cratering rate at Mars and the moon (for crater diameters > 10 km) in fair accord with the conclusions from our work.

Conclusions

At Mariner 9 resolution, the impact crater production size-frequency distribution of Mars is generally similar to that of the moon for crater diameters in the range 0.8 to 50 km, and it appears to have been relatively stable through time. The lunar and martian crater curves can be brought into near coincidence by a diameter shift appropriate to reasonable impact velocity differences between bodies hitting Mars and the moon. This indicates that a common population of bodies impacted both planets and suggests the same or a very similar time dependence of impact flux. Constraints on relative lunar and martian fluxes can be obtained by comparing crater frequency data for the lunar and martian highlands and for Mars' satellite Phobos.

These cratering constraints provide the basis for a tentative martian time scale derived from lunar data. Previous time scales have painted a picture of a disorderly planetary evolution of Mars, punctuated by a strange pulse of Tharsis Ridge tectonic and volcanic activity late in geologic history. The new scale suggests a much more orderly evolution, with Mars, like the moon, winding down most of its major planetary tectonic and volcanic disturbances in the first 1.5 billion years of its history. By 2.5 billion years ago the volcanic-tectonic circus on Mars had folded.

References and Notes

1. W. K. Hartmann, *J. Geophys. Res.* **78**, 4096 (1973).
2. L. A. Soderblom, C. D. Condit, R. A. West, B. M. Herman, T. J. Kreidler, *Icarus* **22**, 239 (1974).
3. R. B. Baldwin, *ibid.* **14**, 36 (1971).
4. W. K. Hartmann, *Astrophys. Space Sci.* **17**, 48 (1972).
5. G. Neukum, B. König, H. Fechtig, D. Storzer, *Proc. 6th Lunar Sci. Conf.* (1975), p. 2597.
6. The probable date of ≈ 4.4 billion years ago for the final differentiation of the lunar crust and the ability to have recorded the early cratering has been discussed [for example, see D. A. Papanastassiou and G. J. Wasserburg, *Lunar Sci.* **7**, 665 (1976); Baldwin (7)]. The early lunar age determinations yielded a strong age peak around 4 billion years and only few greater ages; this was interpreted as due to an extraordinary peak in cratering rate at that time ("lunar cataclysm") [F. Tera, D. A. Papanastassiou, G. J. Wasserburg, *Earth Planet. Sci. Lett.* **22**, 1 (1974)]. Such a peak in cratering rate is not supported by studies of the lunar crater populations and is probably a misconception [Baldwin (7); W. K. Hartmann, *Icarus* **24**, 181 (1975); Neukum *et al.* (5)].
7. R. B. Baldwin, *Icarus* **23**, 157 (1974).
8. G. Neukum and B. König, *Proc. 7th Lunar Sci. Conf.*, in press.
9. R. Arvidson, G. Crozaz, R. J. Drozd, C. M. Hohenberg, C. J. Morgan, *Moon* **13**, 259 (1975); R. Arvidson, R. J. Drozd, E. Guinness, C. M. Hohenberg, C. Morgan, R. Morrison, V. Oberbeck, *Lunar Sci.* **7**, 25 (1976).
10. On the basis of the terrestrial Canadian Shield impact crater data, Baldwin (3) and Neukum *et al.* (5) suggested a peak in cratering rate 400 to 600 million years ago. Recent lunar data (8) indicate that a great fluctuation is unlikely. At the most, the recent flux rate has varied within a factor of 2.
11. The crater frequency data of Soderblom *et al.* were not in terms of crater numbers originally.

They determined the erosional state of craters in the size range 100 m to kilometers and converted the erosion data to crater frequencies, possibly erroneously [figures 1 and 14 in (*2*) are inconsistent]. This is discussed by G. Neukum (*Moon*, in press).

12. The meaning of the Apollo and Luna radiometric ages for the lunar cratering record and of terrestrial age data for the North American cratering record is discussed by Neukum *et al.* (*5*) and G. Neukum and P. Horn (*Moon*, in press).
13. G. Neukum, B. König, J. Arkani-Hamed, *Moon* **12**, 201 (1975).
14. The constant power law distributions also shown in Fig. 2 for comparison are taken from Baldwin (*3*), Hartmann and Wood (*16*), and Shoemaker *et al.* (*15*).
15. E. M. Shoemaker *et al.*, *NASA Spec. Publ. SP-325* (1970).
16. W. K. Hartmann and C. A. Wood, *Moon* **3**, 3 (1971).
17. G. Neukum and D. W. Wise, *NASA Tech. Memo. TM X 74316* (1976).
18. V. R. Oberbeck and R. H. Morrison, *Moon* **9**, 415 (1974).
19. C. R. Chapman, *Icarus* **22**, 272 (1974).
20. M. H. Carr, *NASA Tech. Memo. TM X 3364* (1976), p. 152.
21. K. L. Jones, *J. Geophys. Res.* **79**, 3917 (1974).
22. The polynomial fit of the standard curve is (for the age of the Alba cone)

$$[\log N = a_0 + a_1 \log D + a_2(\log D)^2 + a_3(\log D)^3 + a_4(\log D)^4]$$

with $a_0 = -2.605$, $a_1 = -2.998$, $a_2 = 0.578$, $a_3 = 0.637$, $a_4 = -0.498$, N = cumulative crater frequency per square kilometer, and D = crater diameter in kilometers.

23. We have not distinguished in our counts between single impacts and crater doublets produced by broken-up projectiles [V. R. Overbeck and M. Aoyagi, *J. Geophys. Res.* **77**, 2419 (1972)]. The effect of doublets on the size-frequency curve is considered minor because of their relative rarity (*19*).
24. The asteroids with perihelions inside the earth's orbit are called Apollo asteroids, and those with perihelions between 1 and 1.3 A.U. Amor asteroids. Those with perihelions (crossing Mars' orbit) greater than 1.3 A.U. are called Mars crossers. The relative impact velocities have been given (*9*).
25. Shock theory would suggest the law $D \propto v^{2/3}$ for constant mass, where D is the crater diameter and v is the impact velocity. Chemical explosion data [M. D. Nordyke, *J. Geophys. Res.* **66**, 3439 (1961)] show a variation of crater diameter as the 1/3.4 power of energy expended. Use of this relationship gives essentially the same results as use of the 1/3 power.
26. Soderblom *et al.* use impact flux (mass-related) as synonymous with cratering rate (diameter-related), and thus do not account for any velocity differences. In fact, they do not set the cratering rate at Mars equal to that at the moon over the whole past. In their crater chronology diagram, the martian crater frequency (time integral of impact rate) is about a factor of 1.5 higher than the lunar frequency for ages > 2 billion years and about the same as the lunar frequency for ages < 1 billion years.
27. D. E. Gault, *Radio Sci.* **5**, 272 (1970).
28. C. R. Chapman and R. R. Haefner, *J. Geophys. Res.* **72**, 549 (1967).
29. J. B. Pollack *et al.*, *Icarus* **17**, 394 (1972). These authors considered the Phobos crater population saturated. We do not agree with respect to the larger craters ($D > 2$ km) because they lie far apart from each other.
30. The crystallization ages of meteorites also fall in this age span. It is quite possible that some of the meteorites stem from asteroid-type bodies that are similar in size to Phobos. The meteorite ages reflect the time of solidification of these bodies. This is a corroborating argument for solidification of Phobos 4.5 to 4.6 billion years ago.
31. We have not corrected Hartmann's original martian cratering frequency data given for a crater size distribution law to the −2 power in our reduction to D = 1 km. The velocity effect in going from large crater sizes to 1 km would lower Hartmann's values by a factor of about 2.
32. H. J. Moore, *U.S. Geol. Surv. Prof. Pap. 812-B* (1976).
33. S. W. Johnson, J. A. Smith, E. G. Franklin, L. K. Moraski, D. J. Teal, *J. Geophys. Res.* **74**, 4838 (1969).
34. E. M. Shoemaker, paper presented at the Planetary Program Principal Investigators meeting, Flagstaff, Ariz. (1976).
35. For a distribution law $N = f(D)$, $N \propto \delta N$. The data of Soderblom *et al.* in Fig. 1 are the original data in terms of δN, and those of Neukum *et al.* are the original data in terms of N. The relation between N and δN used here is derived from the calibration distribution of Fig. 2 (*13*): $N(D = 1$ km$) = 96\ \delta N$. Baldwin originally gave his data as cumulative crater frequencies N for craters ≥ 161 km in diameter. We used his distribution law $N \propto D^{-1.8}$ for reduction to $N(D = 5$ km$)$ and the calibration distribution for reduction to $N(D = 1$ km$)$. Hartmann gave his crater frequency data originally "in arbitrary units relative to an average mare frequency." We used his Apollo 14 (Fra Mauro) counts for absolute reduction to $N(D = 1$ km$)$ in applying our calibration distribution. In this way, all data have been reduced by application of a consistent method and are directly comparable.
36. This work was supported by the Deutsche Forschungsgemeinschaft, and the Planetology Office of the National Aeronautics and Space Administration, and was completed while D. U. Wise was a guest scientist at the Max-Planck-Institut für Kernphysik, Heidelberg. B. König was most helpful in the data reduction. Discussions with R. Arvidson, K. Blasius, M. Carr, J. Cutts, R. Greeley, J. Guest, H. Moore, R. W. Shorthill, and J. Veverka helped clarify our ideas.

Editor's Comments on Paper 29

29 **CINTALA, HEAD, and MUTCH**
Characteristics of Fresh Martian Craters as a Function of Diameter: Comparison with the Moon and Mercury

Although martian craters may superficially resemble their lunar and mercurian counterparts, a more detailed examination reveals numerous differences caused by the enhanced effectiveness of erosion on Mars and gravity variations among the terrestrial planets. Hartmann (1972) proposed that the diameters at which craters undergo a change in morphology depends chiefly on the planet's gravity. Ejecta blankets and secondary craters accumulate closer to the crater rims on Mercury than on the Moon, as would be expected from its increased gravity (Murray et al. 1974). (The gravitational acceleration on Mercury is 0.37 that of Earth and on the Moon, 0.17; ballistic deposition is related to $R = \mu \sin 2\gamma/g$ where γ is the angle of ejection, μ the particle velocity, and R the trajectory arange). On the other hand, crater depth-diameter relations appear quite similar on both bodies (Malin and Dzurisin 1977, 1978). Depth-diameter values for degraded lunar and martian craters are also quite similar, thereby suggesting that much of the crater degradation on Mars might not have been caused by eolian erosion (Cintala et al. 1976a).

A more systematic survey of interplanetary cratering has been conducted by E. I. Smith (1976b) and M. J. Cintala, J. W. Head, and T. Mutch (Paper 29). The onset of crater wall terracing occurs at about the same crater diameters for Mars and the Moon. However, mercurian craters develop terraces at a *smaller* diameter, which is inconsistent with gravity effects alone, since Mercury and Mars have nearly the same gravity. Smith attributes the discrepancy to greater erosion on Mars, which has destroyed proportionally more terraces in smaller craters, while Cintala et al. suggest differences in subsurface strengths and meteorite impact velocities. Some support for the influence of subsurface materials on crater parameters comes from lunar and planetary studies (Head 1976; Pike 1976; Cintala et al. 1977).

Smith shows that central peaks appear at approximately the same crater diameter on Mercury and Mars, but at a much higher diameter on the Moon, which is the sequence expected on the basis of gravity. However, a lower percentage of the larger martian craters have central peaks than do mercurian ones, which may reflect the greater degree of erosion on Mars. Cintala et al. find that although the onset of central peaks occurs between 10–20 km on all three bodies, the fraction of craters with central peaks at a given diameter is much less for Mars than either Mercury or the Moon, which again may possibly be an effect of erosion. However, the martian curves described by Cintala et al. differ considerably from Smith's. Some of this difference may be accounted for by sampling bias: Smith only sampled 193 craters from a limited portion of Mars, whereas Cintala et al. accumulated a data base of 33,810 "fresh" craters from a total of 46,559 martian craters of all degrees of degradation from Mariner 9.

Preliminary analysis of Viking imagery suggests that onset of terracing and central peak formation occurs at lower diameters than previously observed from Mariner 9, presumably because of the improved resolution (Carr et al. 1977; Wood et al. 1978). The data tentatively indicate that the statistics for diameter of onset and frequency of terracing and central peak development differ from one terrain type to another (Smith and Hartnell 1977; Wood et al. 1978). One possibility is that substrate strengths and perhaps even density of impacting objects may largely offset the effects of gravity.

29

Reprinted from *Geophys. Res. Lett.* **3**:117–120 (1976)

CHARACTERISTICS OF FRESH MARTIAN CRATERS AS A FUNCTION OF DIAMETER: COMPARISON WITH THE MOON AND MERCURY

Mark J. Cintala
James W. Head
Thomas A. Mutch

Department of Geological Sciences
Brown University
Providence, Rhode Island 02912

Abstract. Martian craters defined as fresh on the basis of morphologic parameters have been analyzed for the presence and abundance of various morphologic features as a function of size. Bowl-shaped craters dominate the fresh crater population below about 15 km. The onset of central peaks occurs at about 5 km. Craters above about 15 km often have terraced walls, central peaks, and hummocky floors; at diameters of 40 km and greater, these features dominate fresh martian crater morphology. Central peak onset occurs at smaller diameters on Mars and the Moon than on Mercury, and terrace onset occurs at similar diameters on the Moon and Mars, but at larger diameters than on Mercury. Since Mars and Mercury have a similar surface gravitational acceleration (greater than twice that of the Moon), gravity-controlled crater features should appear at similar diameters on the two planets. However, the differences in onset and abundances of central peaks and terraces on Mars and Mercury indicate that processes other than gravitational effects may also be important.

Introduction

Recent space flights to Mars and Mercury have shown that cratering is a significant process on these planets, as well as on the Moon. While fresh craters on the three planetary bodies and the Earth show similarities, differences have also been noted. For instance, the first appearance of central peaks (Gault et al., 1975; Smith and Sanchez, 1973) occurs at smaller diameters on the Earth than on the Moon, Mars, or Mercury (Hartmann, 1973). Causes for variation in fresh crater morphology may be related to gravitational differences from planet to planet (Hartmann, 1972) and their effects during crater excavation and modification (Gault et al., 1975). Other variations, such as impact velocity (Wetherill, 1975) or substrate variations (Head, 1976), may also be important. Preliminary information on the morphology of fresh craters has been presented for the Moon (Smith and Sanchez, 1973) and Mercury (Gault et al., 1975), based on relatively small samples of the total fresh crater population. Preliminary observations have also been made on martian crater morphology and morphometry, but are also restricted to small samples of the total fresh crater population (Hartmann, 1972, 1973; Pike, 1971; Murray et al., 1971). The purpose of this paper is to present a more comprehensive analysis of the morphologies of fresh craters on Mars and to compare the characteristics and size distribution of morphologic features of martian craters with those on the Moon and Mercury.

Based on criteria presented below, 33,810 fresh craters were chosen from a population of 46,559 martian craters of all sizes and morphologies. The data base consists of a catalog of all craters visible on Mariner 9 A frames. Crater location, rim crest diameter, and a series of morphologic features, as well as several other parameters, are recorded for each of the craters. The procedures for recording the data and the details of the morphologic classification scheme are presented elsewhere (Arvidson et al., 1974; Arvidson, 1974). Craters classified as volcanic on the basis of these morphologic parameters have not been included in this study.

Definition of Fresh Craters

Fresh lunar craters over several kilometers in diameter have a crisp appearance and are characterized by rays, satellitic craters, radial rim facies, a sharp rim crest, and terraces or a bowl shape depending on size (Pohn and Offield, 1970; Arthur et al., 1963; Wood, 1972; Smith and Sanchez, 1973; Head, 1974; Howard, 1974; Pike, 1974). Relative youth of craters is determined both by crispness and stratigraphic relationships. Fresh craters are of Copernican age (with rays) or Erathosthenian age (fresh morphologies without rays) (Wilhelms and McCauley, 1971) and span an age range of about the last 3 billion years (Head, 1974, Fig. 3). Large Imbrian-age craters show only slightly higher levels of degradation, but pre-Imbrian craters are usually heavily degraded (Head, 1974). The freshest lunar crater class of Arthur et al. (1963), Class I, is characterized by craters with sharp rim crests, while Class II craters have blurred or possibly broken rims. Class I craters correlate with craters of Copernican, Erathosthenian, and even Imbrian age (C. A. Wood, personal communication). Fresh craters on Mercury show similar characteristics and correspond to Class I lunar craters (Gault et al., 1975).

Using similar criteria, fresh martian craters are defined as unmodified or nearly unmodified structures which appear in one of three categories (Fig. 1): 1) deep, flat-floored, with terraced walls; 2) deep, bowl-shaped, with terraced walls; and 3) deep, bowl-shaped, with non-terraced walls.

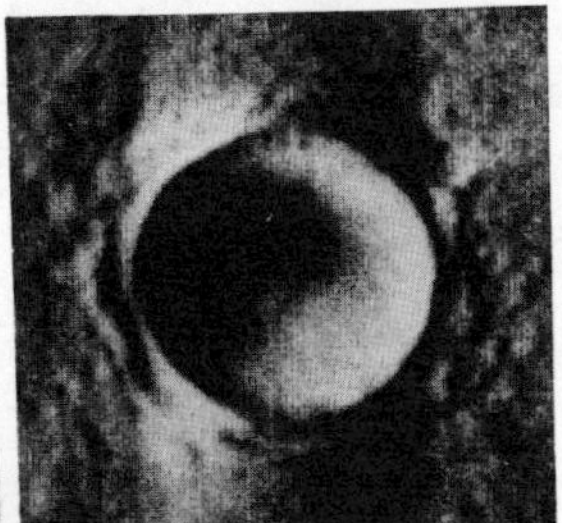

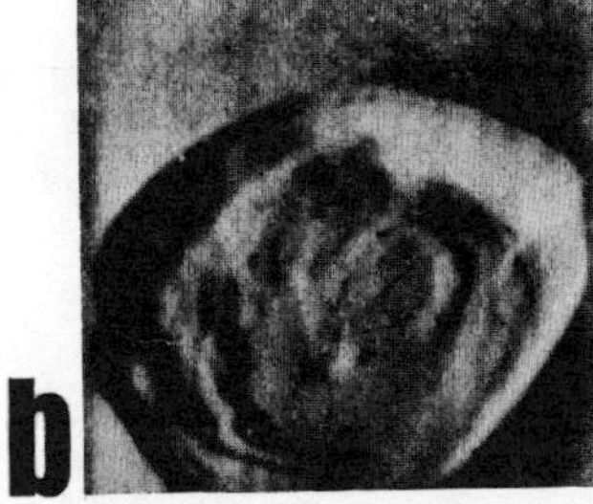

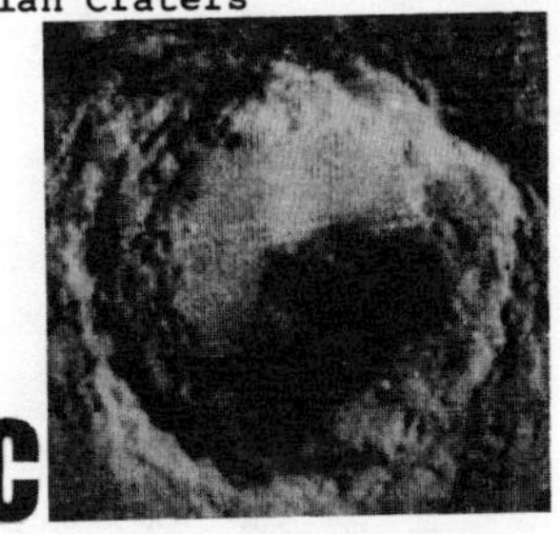

Figure 1. Fresh martian craters characteristic of the three types described in the text; a) a 9 km diameter, deep, bowl-shaped crater with non-terraced walls, (DAS 08838839); b) an 11 km diameter, deep, bowl-shaped crater with terraced walls (DAS 08297879); c) a 59 km diameter, deep, flat-floored crater with terraced walls (DAS 07506488).

The deep category is a qualitative expression for relative crater depth and compares to other categories such as shallow and extremely shallow. Details of modified martian craters are described elsewhere (Jones, 1974; Arvidson, 1974). Of the fresh crater population thus defined, over 98 percent have raised rims, while this feature is seen in less than 25 percent of the remaining population of non-fresh craters. In addition to the morphologic crispness of these craters, a relatively young age is supported by superposition relationships. Less than one percent of the fresh craters defined on morphologic grounds have breached rims or superposed craters, compared to 25 percent for the more highly degraded craters.

In a morphologic sense, the martian fresh crater class is approximately equal to Class I lunar craters of Arthur et al. (1963). Less than one percent of the population have the blurred and broken rims characteristic of Classes II and III. Detailed correlation with the Pohn and Offield (1970) lunar scale is more difficult because of the lack of information on the characteristics of the fine-textured exterior deposits on Mars. This is a result of resolution limitations of Mariner 9 A frames, and some eolian modification of exterior crater deposits as exhibited in B frames (Soderblom et al., 1973). However, fresh martian craters appear to be as fresh or fresher than about 4.2 to 4.4 on the Pohn and Offield scale, based on characteristics of the crater interiors. On the Moon, this value lies close to the beginning of the Imbrian Period (Head, 1974).

Crater Morphologic Characteristics and Their Variation with Size

Bowl-Shaped Craters. The vast majority of fresh martian craters are bowl shaped. The classification of craters as bowl shaped at small diameters is partly controlled by the resolution limits of Mariner A frames. Craters with diameters less than 1 km cannot be resolved. This same resolution effect accounts for a decrease in apparent abundance of bowl-shaped craters at diameters less than 3 km (Fig. 3). It is likely that many of the bowl-shaped craters here classified as fresh would show some effects of degradation in higher-resolution pictures, but a reclassification of some bowl-shaped craters from fresh to degraded would not influence our present conclusions. Fig. 2a shows the number of fresh bowl-shaped craters as a function of size and Fig. 2b shows the percentage of fresh craters which are bowl shaped, as a function of size. Almost 100 percent of all fresh craters less than 20 km in diameter are bowl shaped. Over 98 percent of the total bowl-shaped crater population are less than 20 km in diameter.

Flat-floored craters. Fig. 2c illustrates the percentage of the fresh crater population which has flat floors, as a function of crater size. Less than 1 percent of the fresh craters below 20 km diameter are characterized by flat-floors. The occurrence of flat floors increases markedly above about 20 km diameter. Between 20 and 30 km, about 20 percent of the fresh craters have flat floors, and above 50 km all fresh craters contain this feature.

Central peaks. Central peaks are seen in 685 fresh craters. No central peaks are observed in fresh craters less than 5 km in diameter, and an extremely small percentage of the total population between 5 and 10 km contain central peaks (Fig. 2d). Because of the large population, these are not visible in Fig. 2d. However, their presence is indicated in Fig. 4b. The percentage of fresh craters showing central peaks steadily increases above 10 km until over half of the fresh craters between 60 and 100 km contain central peaks.

Wall terraces. These features are seen in 483 fresh craters but are not found at diameters below 10 km (Fig. 2e). Less than 5 percent of the fresh craters between 10 and 20 km display terraces but the proportion of terraced craters increases with increasing crater size until 100 percent of the craters over 50 km diameter have terraces.

Floor hummocks. In the lunar case floor hummocks are associated with fresh Copernican-and Eratosthenian-age craters and are seen predominantly in flat-floored craters over 15-20 km in diameter, although they do occur at lower diameters (Pohn and Offield, 1970; Smith and Sanchez, 1973; Head, 1974; Cintala and Head, 1976). Lunar floor hummocks are generally less than a kilometer in diameter, which is approximately the resolution of Mariner 9 A frames. On Mars, floor hummocks are seen in only a small percentage of fresh craters (Fig. 2f). This distribution may be due to resolution effects and small amounts of eolian infilling.

Comparison with the Moon and Mercury

A comparison of the frequency of occurrence

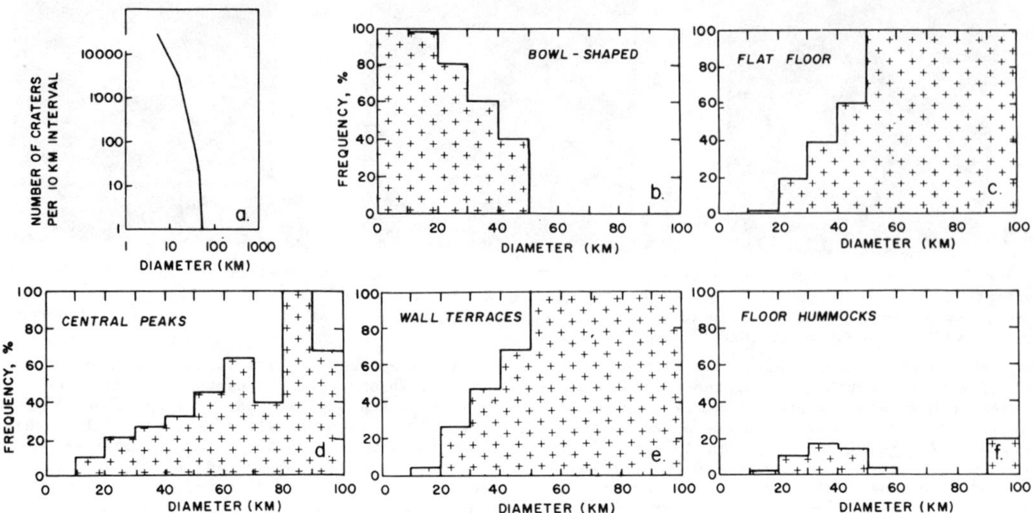

Figure 2. Morphologic characteristics of the fresh martian crater population as a function of diameter; a) number of fresh bowl-shaped craters as a function of size; b-f) percentage of fresh craters with a particular feature as a function of crater size (e.g. - 98.5% of the fresh craters between 10-20 km diameter are bowl-shaped). b) bowl-shaped; c) flat-floored; d) central peaks; e) wall terraces; f) floor hummocks.

of terraces and central peaks as a function of diameter for Mars, Mercury, and the Moon is shown in Fig. 4a, b. The onset of terraces occurs at about the same diameter on the Moon and Mars. However, the percentage of martian craters with terraces in each size class falls consistently below values for both the Moon and Mercury below 50 km. A small percentage of central peaks are seen at diameters of less than 10 km on Mars. On the basis of data presented by Gault et al., (1975) and Smith and Sanchez (1973), central peaks do not occur until 10-20 km diameter on the Moon and Mercury (Fig. 4b). However, a

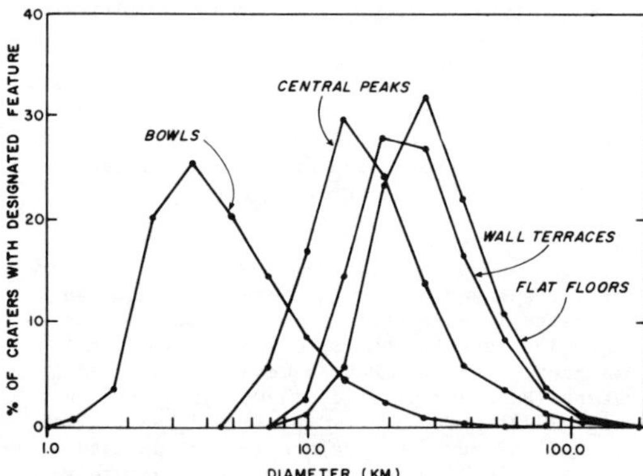

Figure 3. Summary of fresh crater morphologic variations with crater diameter. Points indicate what percentage of craters with specific feature occurs in each diameter interval (e.g. 20% of all bowl-shaped craters occur between 2 and 3 km diameter). The area under each curve represents total population of each feature. Modified from Mutch et al., 1976.

recent compilation of lunar crater data shows that, within a 13 million square kilometer area on the nearside, 5 percent of the Class I craters between 5 and 10 km diameter have central peaks (C. A. Wood, personal communication). Martian central peaks show a slower rate of increase at higher diameters than do lunar and mercurian craters. Small amounts of eolian infilling on Mars may serve to obscure or bury the central peaks in many cases, while not changing the general fresh-crater appearance.

Previous investigators have attributed the differences in central peak and terrace frequencies between the Moon and Mars (Hartmann, 1972; 1973) and the Moon and Mercury (Gault et al., 1975) to dissimilar surface gravitational field strengths. The Moon and Mercury differ in surface gravitational accleration by over a factor of 2 (Moon, 0.16 relative to Earth=1; Mercury, 0.37). The gravitational acceleration at the surface of Mercury is approximately the same (about 5 percent less) as that of Mars (0.38). Since Mars and Mercury have similar surface gravities, features such as terraces which are thought to be gravity-controlled (Gault et al., 1975) should appear at similar diameters if gravity is the dominant factor. However, terraces appear at smaller diameters on Mercury than on Mars and the Moon (Fig. 4a). The differences between these planets strongly suggest that, in addition to gravitational effects, other factors such as varying impact velocities and dissimilar substrate characteristics may also be important. Compilation of a complete data base for the Moon and Mercury will allow more detailed comparisons to be made.

Acknowledgements. This work was performed under NASA grants NGR-40-002-008 and NGR-40-002-116 which are gratefully acknowledged. Thanks

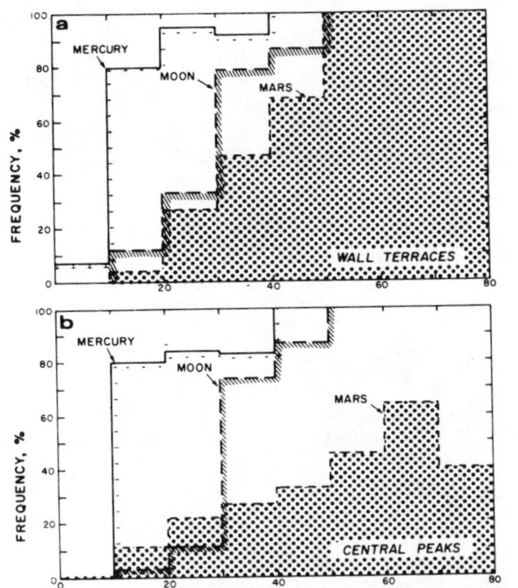

Figure 4. Comparison of Mars data with the Moon and Mercury for the frequency of occurrence of terraces (a) and central peaks (b) as a function of crater diameter (Mercury data from Gault et al., 1975; lunar data from Smith and Sanchez, 1973, as modified by Gault et al., 1975). In Gault et al. Lunar wall terrace data below 10 km is incorrectly plotted but is corrected here. Martian central peak data below 10 km are enhanced for visibility.

are extended to Ray Arvidson and Ken Jones for the compilation of the Mars data bank, to R. Stockman, R. D'Alli, S. Grenander, and C. A. Wood for discussion and to J. Ralske for help in preparation of the manuscript, and to D. E. Gault and K. A. Howard for their critical reviews of the manuscript.

References

Arthur, D. W. G., A. P. Agnieray, R. A. Horvath, C. A. Wood, and C. R. Chapman, Comm. of the Lunar Planetary Lab., p. 71, The University of Arizona, Tucson, AZ, 1963.

Arvidson, R. A., T. A. Mutch and K. L. Jones, Craters and Associated Aeolian Features on Mariner 9 Photographs: An Automated Data Gathering and Handling System and Some Preliminary Results, Moon, 9, 105-114, 1974.

Arvidson, R. A., Morphologic Classification of Martian Craters and Some Implications, Icarus, 22, 264-271, 1974.

Cintala, M. J., J. W. Head, and T. A. Mutch, Depth/Diameter Relationships for Martian and Lunar Craters, EOS, 56, 389, 1975.

Cintala, M. J., J. W. Head, and T. A. Mutch, Depth/Diameter Relationships for Craters on Mars, the Moon, and Mercury, submitted to Lunar Science VII, Lunar Science Institute, Houston, TX, 1976.

Cintala, M. J. and J. W. Head, The Relationship of Morphology and Morphometry in Fresh Lunar Craters, submitted to Moon, 1976.

Gault, D. E., J. E. Guest, J. B. Murray, D. Dzurisin, and M. C. Malin, Some Comparisons of Impact Craters of Mercury and the Moon, J. Geophys. Res., 80, 2444-2460, 1975.

Hartmann, W. K., Interplanet Variations in Scale of Crater Morphology--Earth, Mars, Moon, Icarus, 17, 707-713, 1972.

Hartmann, W. K., Martian Cratering, 4, Mariner 9 Initial Analysis of Cratering Chronology, J. Geophys. Res., 78, 4096-4116, 1973.

Head, J. W., Processes of Lunar Crater Degradation: Changes in Style with Geologic Time, Moon, 12, 299-329, 1974.

Head, J. W., The Significance of Substrate Characteristics in Determining Morphology and Morphometry of Fresh Lunar Craters, submitted to Lunar Science VII, Lunar Science Institute, Houston, TX, 1976.

Howard, K. A., Fresh Lunar Impact Craters: Review of Variations with Size, Proc. 5th Lunar Sci. Conf., Geochim. Cosmochim. Acta, 1, 61-69, 1974.

Jones, K. L., Evidence for an Episode of Crater Obliteration Intermediate in Martian History, J. Geophys. Res., 79, 3917-3931, 1974.

Murray, B. C., L. A. Soderblom, R. P. Sharp, and J. A. Cutts, The Surface of Mars I, Cratered Terrains, J. Geophys. Res., 76, 313-330, 1971.

Mutch, T. A., R. A. Arvidson, K. L. Jones, J. W. Head, and R. S. Saunders, Geology of Mars, Princeton Press, in press, 1976.

Pike, R. J., Depth/Diameter Relations of Fresh Lunar Craters: Revision from Spacecraft Data, Geophys. Res. Lett., 1, 291-294, 1974.

Pike, R. J., Genetic Implications of the Shapes of Martian and Lunar Craters, Icarus, 15, 384-395, 1971.

Pohn, H. A. and T. W. Offield, Lunar Crater Morphology and Relative-Age Determination of Lunar Geologic Units--Part 1. Classification, Geol Surv. Res., 153-162, 1970.

Smith, E. I. and A. G. Sanchez, Fresh Lunar Craters: Morphology as a Function of Diameter, A Possible Criterion for Crater Origin, Mod. Geol., 4, 51-59, 1973.

Soderblom, L. A., T. Kreidler, and H. Masursky, Latitudinal Distribution of a Debris Mantle on the Martian Surface, J. Geophys. Res., 78, 4117-4122, 1973.

Wetherill, G. W., Late Heavy Bombardment of the Moon and Terrestrial Planets: Proc. 6th Lunar Sci. Conf., Geochim. Cosmochim. Acta, Suppl. 6, 2, 1539-1561, 1975.

Wilhelms, D. W. and J. F. McCauley, Geologic Map of the Near Side of the Moon, Geologic Atlas of the Moon, 1971.

Wood, C. A., The System of Lunar Craters, Revised, Moon, 3, 408-411, 1972.

Part X

THE MARTIAN ATMOSPHERE

Editor's Comments on Paper 30

30 OWEN et al.
The Composition of the Atmosphere at the Surface of Mars

The presence of CO_2 as the major constituent of the martian atmosphere, at a mean surface pressure of 6–7 mb, had already been established by the Mariner 4 mission (Kliore et al. 1965). Traces of water vapor, carbon monoxide, and molecular oxygen had been detected by telescopes (Spinrad et al. 1963; Kaplan et al. 1969; Carleton and Traub 1972). The Mariner 9 IR spectrometer generally confirmed the earlier findings (Conrath et al. 1973). In addition, the UV spectrometer measured seasonally varying ozone (Barth et al. 1973). The failure to detect nitrogen was attributed to either very efficient eddy mixing (Dalgarno and McElroy 1970) or loss by photodissociation (Brinkman 1971). Gainduk (1971) found traces of NO_2 from earth-based observations.

The Viking mass spectrometer verified that CO_2 comprised 95 percent of the atmosphere. Small amounts of N_2, A, Xe, Kr, and Ne were detected for the first time (Owen and Biemann 1976; Nier et al. 1976; Biemann et al. 1976; Owen et al. 1976). The discovery of N_2 was considered significant, as a necessary ingredient for potential martian life and as an indicator of atmospheric evolution. The observed concentration of Ar (1.6 percent) fell far below earlier Soviet estimates—now known to have been in error (Istomin and Grechnev 1976; Istomin et al. 1975). A summary of the atmospheric composition has been provided by T. Owen, K. Biemann, and others (Paper 30).

Other important results are the noble gas abundances and isotopic ratios. The noble gases show the same relative pattern on Mars as on the Earth and in chondrites, but the absolute abundances are one hundred times less. This fact, among others, led Anders and Owen (1977) to conclude that the level of outgassing on Mars had been considerably lower than on Earth. A similar conclusion could be drawn from the argon isotopic ratios. The new data led Owen et al. to conclude that Mars may have once had a denser atmosphere; yet the development of the martian atmosphere may have followed a different path from that on Earth.

30

Reprinted from *J. Geophys. Res.* **82**:4635–4639 (1977)

The Composition of the Atmosphere at the Surface of Mars

TOBIAS OWEN,[1] K. BIEMANN,[2] D. R. RUSHNECK,[3] J. E. BILLER,[2]
D. W. HOWARTH,[4] AND A. L. LAFLEUR[2]

We have confirmed the discovery of N_2 and ^{40}Ar by the Entry Science Team, and we have also detected Ne, Kr, Xe, and the primordial isotopes of Ar. The noble gases exhibit an abundance pattern similar to that found in the terrestrial atmosphere and the primordial component of meteoritic gases. Xenon appears to be underabundant in comparison to the meteoritic ratio, as it is on earth. The isotopic ratios $^{15}N/^{14}N$, $^{40}Ar/^{36}Ar$, and $^{129}Xe/^{132}Xe$ are distinctly different from the terrestrial values, implying different evolutionary histories for volatiles on the two planets. The noble gas abundances indicate that at least 10 times the present atmospheric amount of N_2 and 20 times the CO_2 abundance were released by the planet during geologic time; the outgassing of a large amount of water must also have taken place. There is thus an explanation for the high surface pressure and abundance of water required at some early epoch to cut the dendritic channels observed on the Martian surface.

This report is a summary of the current status of our investigations of the composition of the Martian atmosphere. We have been using the mass spectrometers that function as the analytical components of the molecular analysis experiments on the two Viking landers. The basic concepts underlying the experiment and its instrumentation have been discussed by *Anderson et al.* [1972]; a detailed description of the flight instruments is currently being prepared for publication (D. R. Rushneck, private communication, 1977).

The first accounts of our analyses of the Martian atmosphere have appeared in three previous papers [*Owen and Biemann*, 1976; *Biemann et al.*, 1976*a*; *Owen et al.*, 1976]. Since these publications, we have obtained additional data from the two landers, and we have refined some of our earlier conclusions. This process is continuing, special emphasis being placed on the acquisition of supplementary calibrations using the Engineering Breadboard (EBB) version of the instrument and the spare flight-configured model. The instruments on the spacecraft were turned off on March 13, 1977 (Viking Lander 1), and April 5, 1977 (Viking Lander 2), to avoid potential hazards to other experiments from problems that developed with the high-voltage supplies.

In order to discuss the atmospheric data we must briefly review the manner in which they were acquired. Although the molecular analysis instrument was designed primarily for the detection of organic compounds in the gas chromatographic mode [*Biemann*, 1974], the mass spectrometer's high sensitivity (dynamic range, 6–7 orders of magnitude), high mass range (m/e, 12–200), and resolution (1:200 at m/e = 200; better at lower mass) were used to advantage in determining the composition of the atmosphere, particularly its minor constituents. The penalty that one pays for resolution and sensitivity is a certain loss of accuracy, mainly because the residual background in the instrument becomes more significant and the long-term reproducibility of the fragmentation pattern is lowered.

Prior to the Viking missions the detection of even traces of N_2 in the Martian atmosphere was deemed to be extremely important because of its relevance both for the history of volatiles on Mars and for possible Martian biology. Previous data suggested that it must be a minor component or could be almost completely absent [*Barth et al.*, 1969; *Dalgarno and McElroy*, 1970; *McElroy*, 1972]. One of the major problems in a mass spectrometric determination of N_2 in the Martian environment is the interference of CO^+ (from CO or CO_2) with the ion current of N_2^+ at m/e = 28. For this reason the gas reservoir of the instrument is coupled via separate valves to two cavities, one containing Ag_2O and LiOH for the oxidation of CO to CO_2 and the absorption of all CO_2 and the other containing $Mg(ClO_4)_2$ for the removal of the resulting water (Figure 1).

We thus have three distinct options for analyzing the Martian atmosphere: (1) we can admit the atmosphere directly to the mass spectrometer (an unaltered or unfiltered sample), (2) we can use the chemical scrubbers to reduce CO and CO_2, and (3) we can repeat the scrubbing procedure, progressively admitting fresh samples to the gas reservoir and thereby building up the partial pressure of the trace gases.

During the fourth and fifth days after the landing of Viking Lander 1 (July 20, 1976) a total of six atmospheric analyses were performed at approximately 6-hour intervals. In the first four of these analyses, CO and CO_2 were removed; in the last two analyses, samples of unaltered atmosphere were used. During the third analysis the spectrometer shut down temporarily, leaving us with a total of five sets of mass spectral scans. Analysis of these spectra confirmed the presence of nitrogen and argon in the Martian atmosphere as reported by *Nier et al.* [1976*a*] and led to the discovery of ^{36}Ar [*Owen and Biemann*, 1976]. The isotope ratio $^{36}Ar/^{40}Ar$ was found to be 3.34×10^{-4}, approximately one-tenth the terrestrial value (3.43×10^{-3}), while the total abundances of ^{36}Ar and ^{40}Ar in the Martian atmosphere relative to the planet's mass were 0.0075 and 0.08, respectively, of the comparable terrestrial values.

In these first analyses we also detected molecular oxygen and (of course!) carbon dioxide, previously known from ground-based observations. We observed considerable scatter in the oxygen measurements (a factor of ±2), which we have attributed to instrumental causes. No correlation with external effects (time of day, operation of other instruments, etc.) was noted, nor have we detected a seasonal variation in the abundance of any atmospheric component during the lifetime of the mission.

[1] Department of Earth and Space Sciences, State University of New York, Stony Brook, New York 11794.
[2] Department of Chemistry, Massachusetts Institute of Technology, Cambridge, Massachusetts 02139.
[3] Interface, Inc., Fort Collins, Colorado 80522.
[4] Guidance and Control Systems Division, Litton Industries, Woodland Hills, California 91364.

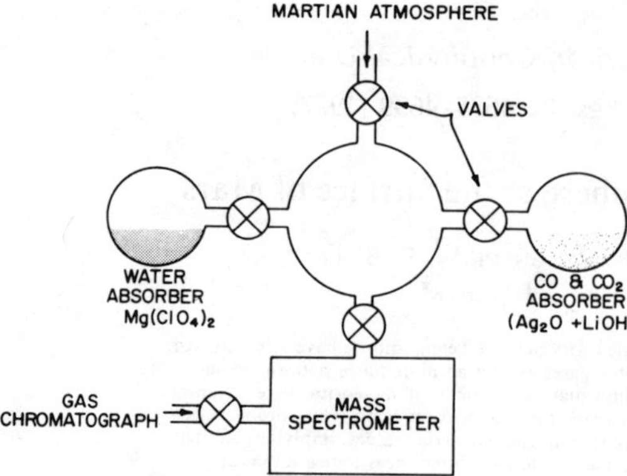

Fig. 1. Schematic diagram of the atmospheric inlet system of the molecular analysis experiment. The valves can be cycled independently. (Diagram is not to scale.)

Variable amounts of water vapor, 0.07% carbon monoxide, and up to 0.03 ppm of ozone (also variable) have also been detected by ground-based or spacecraft observations [*Barth*, 1974; *Owen*, 1974*a*; *Young and Young*, 1977]. The mass spectrometric data on H_2O are rather meaningless because of adsorption on the relatively large surface area of the interior of the atmospheric inlet system and of the ion source (which is not heated in the atmospheric analysis mode). Low concentrations (less than a few percent) of CO cannot be detected with our system because of the interference by the large amount of CO_2 in analyses of the unmodified atmospheric sample and the concomitant removal of CO along with CO_2 when the sample was exposed to Ag_2O in the other type of analysis. The abundance of ozone is below our detection limit.

Subsequent analysis of enriched samples on Viking Lander 1 (VL 1) led to the establishment of ratios for the abundant isotopes, including confirmation of the report by *Nier et al.* [1976*b*] that the abundance of the naturally occurring heavy isotope of nitrogen (^{15}N) is enhanced in the Martian atmosphere by a factor of approximately 1.7 over the terrestrial value [*Biemann et al.*, 1976*a*]. Other isotope ratios measured during this sequence ($^{13}C/^{12}C$, $^{18}O/^{16}O$, and $^{36}Ar/^{38}Ar$) appear to exhibit terrestrial values within the errors of measurement (±10%).

The successful deployment of Viking Lander 2 (VL 2) on Mars (September 3, 1976) afforded us the desired opportunity to increase the sensitivity of our experiment by further enrichment of atmospheric samples. We had established from tests of the two instruments carried out during the cruise from earth to Mars that the background in the mass spectrometer on VL 2 was much lower than that in the instrument on the first lander, making it superior for the detection of trace amounts of atmospheric gases. It was also clear from experience gained with the operation of VL 1 that the best way to maintain this low instrumental background would be to perform the atmospheric analyses prior to any analyses of the Martian soil. A delay in obtaining the first soil sample for our instrument with VL 2 provided the opportunity to design an optimized enrichment sequence for detecting trace gases in the atmosphere.

Using chemical scrubbing sequences of 5 and 10 cycles, we obtained enrichments of 4 and 6.3 times the yield from a single cycle. Mass spectra of the ×6.3 enriched sample showed an indication of krypton, but the identification was not conclusive. After evaluating the performance of the instrument we changed the internal timing of the sequence and obtained a ninefold enrichment with 15 cycles. This sample was analyzed 16 times by the mass spectrometer with the electron multiplier gain increased by a factor of 5.3 over its nominal value. These spectra gave clear evidence of the presence of krypton. The analysis was subsequently repeated after an additional 15 cycles with a multiplier gain of 28 times nominal.

The results are shown in Figure 2, which indicates the appearance of the averaged mass spectrum in the vicinity of the krypton and xenon isotopes. The characteristic isotopic pattern of krypton is clearly evident; the broad peak at $m/e = 80$ is an artifact caused by the high partial pressure of argon in the instrument. In the case of xenon, ^{129}Xe is much more abundant, in relation to the other xenon isotopes, in the Martian atmosphere than in the terrestrial atmosphere. The absence of high molecular weight organic compounds in the soil [*Biemann et al.*, 1976*b*], the absence of spectroscopically active high molecular weight compounds in the atmosphere [*Barth*, 1974; *Owen*, 1974*a*], and the clean instrumental background in this mass range make us quite confident that the peaks shown in this region of Figure 2 are chiefly caused by xenon on Mars.

It is not yet possible to compute accurate abundances for

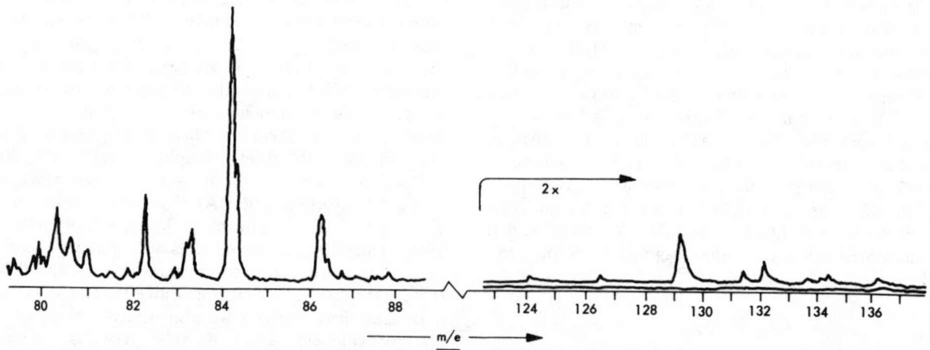

Fig. 2. Mass spectra of enriched samples of the Martian atmosphere in the region of krypton and xenon. The spectra are averages of nine scans, the lower lines are averages of three background scans. The vertical scale is linear; it has been increased by a factor of 2 for xenon.

these two gases or even to give precise values for the ratios of their isotopes. At this low level of detection the instrumental response is noisy, nonlinear, and distorted by memory effects (degassing from the pump). Thus the small peaks at m/e = 124 and 126, while appearing to be anomalously high for xenon, are probably caused by a very weak hydrocarbon background from the ion pump, not by Martian gases. There is, however, no question that ^{129}Xe is much more abundant than ^{132}Xe and ^{131}Xe rather than almost equal, as is the case on earth. To make further progress, the mass spectrometer must be recalibrated in a manner that duplicates experimental conditions on Mars. Such tests are now being conducted with the third flight unit that was built for this mission as a spare instrument.

This completed the studies of the atmosphere carried out prior to solar conjunction (November 15, 1976). The prime objective for postconjunction work was a search for neon. The difficulty posed by this gas is that each of the abundant isotopes is overlapped in the mass spectrum by doubly ionized species that are much more abundant than neon on Mars: $^{20}Ne^+$ by $^{40}Ar^{++}$ and $^{22}Ne^+$ by $^{44}CO_2^{++}$. By using a lower ionizing energy (~40 eV versus 70 eV nominal) we could reduce the contribution from the doubly ionized species, but we had already found that this was inadequate to reveal the presence of the neon isotopes with our standard procedures. There was no way in which we could affect the argon concentration, but we felt that we might be able to reduce the residual CO_2 abundance by exposing the enriched sample to the LiOH capsule during the entire solar conjunction period (~2 months). This indeed proved to be the case. When we analyzed the mixture at low ionizing energy on sol 118, the residual concentration of CO_2 was found to be reduced from 0.33 to 0.015% of its abundance in the free atmosphere.

At this level we found that $^{44}CO_2^{++}$ contributed less than one third of the intensity of the observed peak at m/e = 22 at low ionizing energy. This evaluation was made by using preflight calibration data for the CO_2 cracking pattern that we confirmed by subsequent analyses of an unaltered sample of the Martian atmosphere (sol 191). The resulting abundance for ^{22}Ne in the Martian atmosphere is 0.25 ppm. To determine the total neon abundance, we have to make an assumption for the value of $^{20}Ne/^{22}Ne$ on Mars. Both a 'planetary' value of 8.2 ± 0.4 and a 'solar' value of 12.5 ± 0.4 have been identified in meteorites (see the work of *Heymann* [1971] for a review); in the earth's atmosphere the value is 9.8, and the value at the surface of the sun is probably close to 13 [e.g., *Geiss et al.*, 1970]. In view of this potential range of values we adopt $^{20}Ne/^{22}Ne$ = 10 ± 3 for Mars. Including the other uncertainties in this determination, we find a neon abundance of $2.5^{+3.5}_{-1.5}$ ppm for the Martian atmosphere.

TABLE 1. Composition of the Lower Atmosphere

Gas	Proportion
Carbon dioxide (CO_2)	95.32%
Nitrogen (N_2)*	2.7%
Argon (Ar)*	1.6%
Oxygen (O_2)	0.13%
Carbon monoxide (CO)	0.07%
Water vapor (H_2O)	0.03%†
Neon (Ne)*	2.5 ppm
Krypton (Kr)*	0.3 ppm
Xenon (Xe)	0.08 ppm
Ozone (O_3)	0.03 ppm†

* Discovered by Viking experiments.
† Variable.

TABLE 2. Isotope Ratios in Atmospheric Gases

Ratio	Earth	Mars
$^{12}C/^{13}C$	89	90
$^{16}O/^{18}O$	499	500
$^{14}N/^{15}N$	277	165
$^{40}Ar/^{36}Ar$	292	3000
$^{129}Xe/^{132}Xe$	0.97	2.5

Uncertainties in the Mars values are presently ±10% except for Ar and Xe (see text).

A summary of our current best estimate for the composition of the atmosphere near the surface of Mars is given in Table 1; isotope ratios are summarized in Table 2. The uncertainties in abundances not specifically discussed in the text are of the order of ±20% of the values. We expect to be able to improve this precision after we have completed the additional calibrations that are currently under way. A detailed interpretation of the results must obviously await these improvements, but a few general comments can be made at this time.

In our previous papers we have used the relative abundances of the noble gases in the terrestrial and Martian atmospheres as a guide to the total volatile inventories on both planets [*Owen and Biemann*, 1976; *Owen et al.*, 1976]. The discovery of neon lends additional weight to this approach, since it increases the similarity between the noble gas abundance pattern on Mars and that found in the earth's atmosphere and in the primordial meteoritic component (Figure 3). The difference in the relative abundances of ^{36}Ar and ^{40}Ar on the two planets (Table 2) may be interpreted in one of several ways:

1. Mars has a volatile inventory identical to that of the earth, but either (1) an early catastrophic event removed over 90% of its atmosphere, or (2) Mars never degassed as much as the earth.

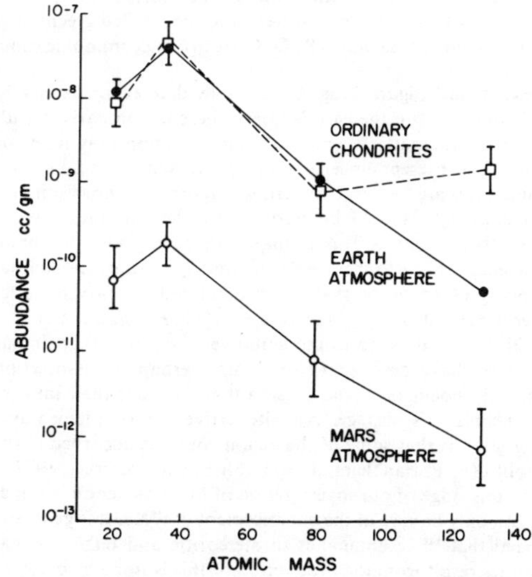

Fig. 3. Abundances of noble gases in ordinary chondrites and in the atmospheres of the earth and Mars. Error bars are only approximate, since systematic errors remain to be assessed. Meteoritic abundances are from *Signer* [1964]. Abundances are in cubic centimeters per gram (at STP) of planet (or meteorite).

2. Mars is deficient in the most volatile elements in comparison to the earth, and degassing is less complete.

The choice among these alternatives is not obvious. In our previous papers we favored part 2 of alternative 1, but in collaboration with E. Anders we are developing a strong case for alternative 2. Catastrophic loss seems unlikely, as will be shown below.

The end result of either of these approaches is the same: they lead to the prediction that at least 10 times more nitrogen and 20 times more carbon dioxide were outgassed on Mars than the atmosphere now contains. The equivalent of a layer of water about 10 m deep over the entire planet would also have been produced. (The basic argument involves a comparison of the Martian atmosphere with the earth's volatile inventory, using the noble gases as a guide. It goes back to the work of *Rubey* [1951, 1955], *Brown* [1952], and *Holland* [1964]; it has been developed for Mars by *Owen* [1966, 1974*a*, *b*, 1976], *Fanale* [1971, 1976], and *Levine* [1976] (and work cited in these papers); its first application to Viking data was in our three previous papers. Subsequent treatments by *Rasool and Le Sergeant* [1977] and by *Anders and Owen* [1977] lead to similar conclusions regarding the estimates of total outgassing.) This prediction for nitrogen agrees with the minimum value deduced from the enhancement of $^{15}N/^{14}N$ by *McElroy et al.* [1976], and the water abundance agrees with the 10-m layer required by the 'normalcy' of $^{16}O/^{18}O$ (M. B. McElroy, private communication, 1977). This agreement may be fortuitous; preferential degassing of water could produce a 40-m layer, and some of the degassed water may be bound in rocks and thus not exchangeable with the atmospheric CO_2. In any event, even the smaller value for the outgassed water abundance is adequate to explain the carving of the fluvial features observed on the planet's surface [*Milton,* 1973] at a time when the ancient atmosphere exhibited the higher surface pressure implied by the missing CO_2. We must emphasize, however, that there is still no adequate model for climatic change on Mars that permits the planet's surface to be warm enough for this to occur. A water vapor controlled greenhouse is an attractive possibility (R. D. Cess, private communication, 1977).

Examining Figure 3 again, we note that xenon is clearly underabundant in the earth's atmosphere in comparison with the meteoritic pattern, and the same deviation may hold for Mars. The present uncertainties in the xenon and krypton abundances are too large to permit a rigorous comparison, but this ambiguity should be resolved by the calibration work currently in progress. It is an important point, since the xenon deficiency on earth has been attributed to the preferential trapping of xenon in shales and other sedimentary material after it was outgassed [*Canalas et al.*, 1968; *Fanale and Cannon,* 1972]. One is thus led to the tentative conclusion that similar processes have been active on Mars, perhaps in association with the epochs of fluvial erosion that have left their imprint on the planet's surface. An alternative (or supplementary) suggestion is that some of the xenon could be adsorbed in the regolith (F. P. Fanale et al., unpublished manuscript, 1977).

At this stage of our investigation of Martian xenon isotopes we can only be sure of the enhancement of ^{129}Xe. It is generally agreed that ^{129}Xe anomalies in meteoritic and terrestrial gas samples result from the production of this isotope by decay of extinct ^{129}I [*Reynolds,* 1960, 1963]. We find that the ratio $^{129}Xe/^{132}Xe$ is 2.5^{+2}_{-1}; the terrestrial atmospheric value is 0.97, and in the carbonaceous and ordinary chondrites, values as high as 4.5 and 9.6 have been reported [*Signer,* 1964; *Mazor et al.*, 1970; *Pepin,* 1964]. There are several possible interpretations of this anomaly. Catastrophic loss of an early atmosphere at just the right number of ^{129}I half-lives could produce the observed abundance of ^{129}Xe, but such an event would have to 'shut off' very abruptly. A gradual decrease in the intensity of a T-Tauri solar wind (for example) would seem to require that a point would eventually be reached at which the lighter noble gases would be preferentially swept away from the upper atmosphere. This in turn would lead to a fractionation from the pattern observed in the terrestrial atmosphere and the meteorites which is apparently not the case (Figure 3). But a quantitative model should be developed before this alternative is completely abandoned.

A more satisfactory explanation may be found by postulating a difference in the manner in which a volatile-rich veneer was accumulated by the earth and Mars. This is one aspect of a model that attempts to account for the volatile abundances on all of the inner planets in terms of late-accreting veneers with a composition resembling that of the C-3V meteorites [*Anders and Owen,* 1977].

There is clearly much more to be done in the way of improving the precision of the existing data and refining their interpretation, but the following points seem well established:

1. The present atmosphere on Mars represents only a small fraction of the total amount of volatiles outgassed by the planet. The potential for a high surface pressure (>100 mbar) and abundant water at some past epoch is therefore available.

2. The noble gases in the Martian atmosphere exhibit a relative abundance pattern similar to that in the earth's atmosphere and (except for Xe) to that in the primordial component of the meteorites. The existence of a 'planetary component' is thus proven, supporting the arguments of those who favor a fractionation of noble gases prior to the formation of the planets.

3. Despite this basic similarity between the volatile inventories on earth and Mars the histories of the two atmospheres, perhaps even the preformation histories, have been very different. The isotopic ratios of nitrogen, argon, and xenon provide ample data for this assertion.

Acknowledgments. This work was supported by NASA research contracts NAS 1-10493 and NAS 1-9684. We thank E. Anders, B. Clark, O. A. Schaeffer, and P. Toulmin III for helpful discussions.

REFERENCES

Anders, E., and T. Owen, Origin and abundance of volatiles on the earth and Mars, *Science,* in press, 1977.

Anderson, D. M., K. Biemann, L. E. Orgel, J. Oro, T. Owen, G. P. Shulman, P. Toulmin III; and H. C. Urey, Mass spectrometric analysis of organic compounds, water, and volatile constituents in the atmosphere and surface of Mars, *Icarus, 16,* 111, 1972.

Barth, C. A., The atmosphere of Mars, *Annu. Rev. Earth Planet. Sci., 2,* 333–368, 1974.

Barth, C. A., W. G. Fastie, C. W. Hord, J. B. Pearce, K. K. Kelley, A. I. Stewart, G. E. Thomas, G. P. Anderson, and O. F. Raper, Mariner 6: Ultraviolet spectrum of Mars' upper atmosphere, *Science, 165,* 1004, 1969.

Biemann, K., Test results on the Viking gas chromatograph-mass spectrometer experiment, *Origins Life, 5,* 417, 1974.

Biemann, K., T. Owen, D. R. Rushneck, A. L. Lafleur, and D. W. Howarth, The atmosphere of Mars near the surface: Isotope ratios and upper limits on noble gases, *Science, 194,* 76–78. 1976*a*.

Biemann, K., J. Oro, P. Toulmin III, L. E. Orgel, A. O. Nier, D. M. Anderson, P. G. Simmonds, D. Flory, A. V. Diaz, D. R. Rushneck, and J. E. Biller, Search for organic and volatile inorganic compounds in two surface samples from the Chryse Planitia region of Mars, *Science, 194,* 72–76, 1976*b*.

Brown, H., Rare gases and the formation of the earth's atmosphere, in *The Atmospheres of the Earth and Planets*, edited by G. P. Kuiper, chap. IX, University of Chicago Press, Chicago, Ill., 1952.

Canalas, R. A., E. C. Alexander, Jr., and O. K. Manuel, Terrestrial abundance of noble gases, *J. Geophys. Res., 73*, 3331–3334, 1968.

Dalgarno, A., and M. B. McElroy, Mars: Is nitrogen present?, *Science, 170*, 168, 1970.

Fanale, F. P., History of Martian volatiles: Implications for organic synthesis, *Icarus, 15*, 279–303, 1971.

Fanale, F. P., Martian volatiles: Their degassing history and geochemical fate, *Icarus, 28*, 179–202, 1976.

Fanale, F. P., and W. A. Cannon, Origin of planetary primordial rare gas: The possible role of adsorption, *Geochim. Cosmochim. Acta, 36*, 319–328, 1972.

Geiss, J., P. Eberhardt, F. Bühler, and J. Meister, Apollo 11 and 12 solar wind composition experiments: Fluxes of He and Ne isotopes, *J. Geophys. Res., 75*, 5972, 1970.

Heymann, D., The inert gases, in *Handbook of Elemental Abundances in Meteorites*, edited by B. Mason, pp. 29–66, Gordon and Breach, New York, 1971.

Holland, H. D., On the chemical evolution of the terrestrial and cytherean atmospheres, in *The Origin and Evolution of Atmospheres and Oceans*, edited by P. J. Brancazio and A. G. W. Cameron, pp. 86–101, John Wiley, New York, 1964.

Levine, J. S., A new estimate of volatile outgassing on Mars, *Icarus, 28*, 165–169, 1976.

Mazor, E., D. Heymann, and E. Anders, Noble gases in carbonaceous chondrites, *Geochim. Cosmochim. Acta, 34*, 781–824, 1970.

McElroy, M. B., Mars: An evolving atmosphere, *Science, 175*, 443, 1972.

McElroy, M. B., Y. L. Yung, and A. O. Nier, Isotopic composition of nitrogen: Implications for the past history of Mars' atmosphere, *Science, 194*, 70–72, 1976.

Milton, D. J., Water and processes of degradation in the Martian landscape, *J. Geophys. Res., 78*, 4037–4048, 1973.

Nier, A. O., W. B. Hanson, A. Seiff, M. B. McElroy, N. W. Spencer, R. J. Duckett, T. C. D. Knight, and W. S. Cook, Composition and structure of the Martian atmosphere: Preliminary results from Viking 1, *Science, 193*, 786–788, 1976*a*.

Nier, A. O., M. B. McElroy, and Y. L. Yung, Isotopic composition of the Martian atmosphere, *Science, 194*, 68–70, 1976*b*.

Owen, T., The composition and surface pressure of the Martian atmosphere: Results from the 1965 opposition, *Astrophys. J., 146*, 257–270, 1966.

Owen, T., What else is present in the Martian atmosphere?, *Comments Astrophys. Space Phys., 5*, 175, 1974*a*.

Owen, T., Martian climate: An empirical test of possible gross variations, *Science, 183*, 763, 1974*b*.

Owen, T., Volatile inventories on Mars, *Icarus, 28*, 171–177, 1976.

Owen, T., and K. Biemann, Composition of the atmosphere at the surface of Mars: Detection of argon-36 and preliminary analysis, *Science, 193*, 801–803, 1976.

Owen, T., K. Biemann, D. R. Rushneck, J. E. Biller, D. W. Howarth, and A. L. Lafleur, The atmosphere of Mars: Detection of krypton and xenon, *Science, 194*, 1293–1295, 1976.

Pepin, R. O., Isotopic analyses of xenon, in *The Origin and Evolution of Atmospheres and Oceans*, edited by P. J. Brancazio and A. G. W. Cameron, pp. 191–234, John Wiley, New York, 1964.

Rasool, S. I., and L. Le Sergeant, Implications of the Viking results for volatile outgassing from earth and Mars, *Nature, 266*, 822–823, 1977.

Reynolds, J. H., Determination of the age of the elements, *Phys. Rev. Lett., 4*, 8–10, 1960.

Reynolds, J. H., Xenology, *J. Geophys. Res., 68*, 2939, 1963.

Rubey, W. W., Geologic history of seawater, *Geol. Soc. Amer. Bull., 62*, 1111, 1951.

Rubey, W. W., Development of the hydrosphere and atmosphere, with special reference to the probable composition of the early atmosphere, *Geol. Soc. Amer. Spec. Pap., 62*, 631–650, 1955.

Signer, P., Primordial rare gases in meteorites, in *The Origin and Evolution of Atmospheres and Oceans*, edited by P. J. Brancazio and A. G. W. Cameron, pp. 183–190, John Wiley, New York, 1964.

Young, L. D. G., and A. T. Young, Interpretation of high-resolution spectra of Mars, IV, New calculations of the CO abundance, *Icarus, 30*, 75–79, 1977.

Editor's Comments on Paper 31

31 McELROY, YUNG, and NIER
Isotopic Composition of Nitrogen: Implications for the Past History of Mars' Atmosphere

The Viking entry and landing site experiments established for the first time the presence of a few percent N_2 in the martian atmosphere (Nier et al. 1976a; Owen and Biemann 1976). However, N^{15} was found to exhibit a 75 percent enrichment relative to the terrestrial isotopic N^{15}/N^{14} ratio, unlike the O^{18}/O^{16} and C^{13}/C^{12} ratios that were the same as on Earth (Nier et al. 1976b; Biemann et al. 1976). The enrichment in N^{15} has been attributed to preferential loss of N^{14} due to dissociative reactions in the upper atmosphere. According to models developed by M. B. McElroy, Y. L. Yung, and A. D. Nier (Paper 31), the current nitrogen isotopic ratio suggests that Mars may once have had a much higher initial abundance of N_2 at a pressure of at least several mb, and perhaps as much as 30 mb. If CO_2 were present in the same proportion to N_2 as on Earth, and using the upper limit of the initial N_2 pressure, the atmospheric pressure may have been as high as 1800 mb.

On the other hand, the Ar^{40}/Ar^{36} isotopic ratio apparently shows that much less volatiles were evolved on Mars than that inferred from the N isotopes (Owen and Biemann 1976; Anders and Owen 1977). The Ar^{40}/Ar^{36} ratio is ten times the terrestrial value. However, the significance of the Ar^{40} concentration is complicated by its link to radioactive decay of K^{40} and its gradual release to the surface over time. Therefore, Ar^{36} is considered to be a better index of atmospheric evolution. The amounts of Ar^{36}, Kr, and Xe are all 100 times less abundant in the martian than in the terrestrial atmosphere, which implies a considerably lower level of martian outgassing than on Earth. Anders and Owen (1977) find that Mars has only released 27 percent of its Ar^{36} and H_2O to the surface, as compared with 74 percent for the Earth.

In summary, the new Viking data suggest that Mars may have once had an atmosphere 10 to 100 times denser than at present, with the lower range given by Ar^{36} and rare gas abundances and the upper bound implied by N^{15} enrichment.

31

Reprinted from *Science* **194**:70–72 (1976)

ISOTOPIC COMPOSITION OF NITROGEN: IMPLICATIONS FOR THE PAST HISTORY OF MARS' ATMOSPHERE

Michael B. McElroy, Yuk Ling Yung, and Alfred O. Nier

Abstract. *Models are presented for the past history of nitrogen on Mars based on Viking measurements showing that the atmosphere is enriched in ^{15}N. The enrichment is attributed to selective escape, with fast atoms formed in the exosphere by electron impact dissociation of N_2 and by dissociative recombination of N_2^+. The initial partial pressure of N_2 should have been at least as large as several millibars and could have been as large as 30 millibars if surface processes were to represent an important sink for atmospheric HNO_2 and HNO_3.*

Nitrogen accounts for about 2.5 percent of the present martian atmosphere (*1*). It is clear, however, that the concentration of nitrogen on Mars must have been much higher in the past. The heavy isotope, ^{15}N, is enriched in the present atmosphere by about 75 percent relative to a terrestrial standard (*2*). In contrast, the relative abundances of oxygen and carbon isotopes on Mars appear to be similar to values observed for the earth (*1, 2*). It is hard to escape the conclusion that Mars must have lost an appreciable amount of N_2 to space over the past 4.5×10^9 years. The initial abundance of N_2 may have been large enough to provide an atmospheric partial pressure of at least several millibars (*2*).

Escape of nitrogen from Mars may proceed by production of fast atoms in the exosphere, by either dissociative recombination of N_2^+ (*3*)

$$N_2^+ + e \rightarrow N + N \qquad (1)$$

or electron impact dissociation of N_2 (*4*)

$$e + N_2 \rightarrow e + N + N \qquad (2)$$

or predissociation of N_2 (*5*)

$$h\nu + N_2 \rightarrow N + N \qquad (3)$$

It appears that the source should be dominated by a combination of reactions 1 and 2. The escape flux due to reaction 1 will be a fairly sensitive function of the value assumed for the partial pressure of CO_2, since N_2^+ ions are removed mainly by charge transfer to CO_2. Loss of nitrogen due to reaction 2 is proportional to the mixing ratio for N_2, at least for small values of this parameter.

According to Michels (*6*), reaction 1 involves production of both $N(^4S)$ and $N(^2D)$ in approximately similar amounts. If we assume that ion and electron temperatures in Mars' exosphere should have values near 400°K, we may compute an average speed for atoms formed by reaction 1 of magnitude 4.96 km sec^{-1}. The initial velocities for ^{15}N should be adjusted by a factor of 0.95 in order to allow for the heavier mass of the less abundant isotope. The velocity required to escape Mars' gravitational field from an altitude of about 210 km above the planetary surface is 4.68 km sec^{-1}. It would appear that the escape rate for reaction 1 should have magnitude approximately equal to the integrated rate for recombination of N_2^+ above the exobase (*7*).

We expect that the dominant path for electron impact dissociation of N_2 should also involve production of $N(^2D)$ and $N(^4S)$. Production of atoms with sufficient energy to escape Mars would require that the target molecules be raised to internal states with energies at least 15.6 ev above the molecular ground state. A significant yield for fast N would not be expected if the molecules were excited to internal states with energies above 19 ev: autoionization should represent the dominant reaction path in this case (*8*). We are therefore concerned primarily with excitation of N_2 to internal states with energies in the range 15.6 to 19 ev. Strobel (*9*), using oscillator strengths measured by Wight *et al.* (*8*), estimates that approximately 16 percent of all dissociative reactions (reaction 2) might result in production of atoms with velocities sufficient for escape. His result is adopted as a guide for the analysis presented below.

We first consider a model in which the surface is taken to be passive—that is, a model in which we explicitly ignore the possibility that appreciable quantities of nitrogen might be removed from the atmosphere by irreversible heterogeneous processes at the surface (*4*). We assume that the concentration of atmospheric CO_2 remains constant with time and use the model described by McElroy and Yung (*10*) to evaluate the past evolutionary history of N_2. Results are shown in Fig. 1 for η, the efficiency with which escaping atoms may be formed by reaction 2, equal to 0.16, 0.032, and 0.016; that is, with the efficiency taken equal to the value given by Strobel (*9*) and reduced by factors of 5 and 10, respective-

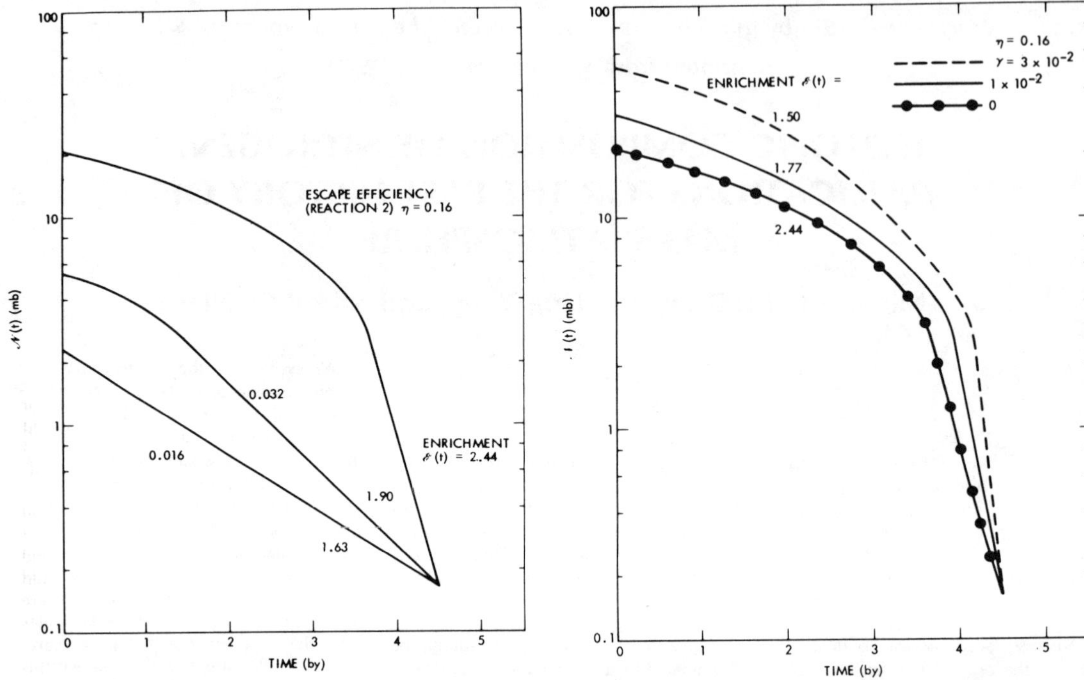

Fig. 1 (left). Abundance of N_2 (in millibars) as a function of time (in units of 10^9 years). The escape efficiencies for reaction 2 were assumed to be 0.16, 0.032, and 0.016, respectively. The corresponding enrichment factors at time t = 4.5 billion years are 2.44, 1.90, and 1.63, respectively, for an eddy diffusion coefficient $K = 3 \times 10^8$ cm^2 sec^{-1}. Fig. 2 (right). Abundance of N_2 (in millibars) as a function of time (in units of 10^9 years). The escape efficiency for reaction 2 was assumed to be 0.16. Results are shown for heterogeneous reactivity coefficients (for HNO_2 and HNO_3) $\gamma = 3 \times 10^{-2}$, 1×10^{-2}, and 0. Corresponding enrichment factors at time t = 4.5 billion years are 1.50, 1.77, and 2.44, respectively, for K = 3×10^8 cm^2 sec^{-1}.

ly. This leads to predicted values for the enrichment of ^{15}N equal to 144, 90, and 63 percent, respectively; that is, the ratio $^{15}N/^{14}N$ exceeds the terrestrial value by factors of 2.44, 1.90, and 1.63, respectively. A model with an efficiency of 0.03 would be consistent with the Viking measurements. The initial concentration of N_2 would correspond to a partial pressure of about 5 mbar in this case. The contributions to the escape flux due to reactions 1 and 2 are comparable for an efficiency of 0.032.

Figure 2 shows results obtained with a model in which we allowed for incorporation of HNO_2 and HNO_3 into surface minerals, with the heterogeneous reaction coefficient, γ, taken equal to 3×10^{-2}, 1×10^{-2}, and 0 (*4*). For all cases shown in Fig. 2 we used the efficiency factor η given by Strobel (*9*). A reaction coefficient γ equal to 10^{-2} provides a satisfactory fit to the Viking data, and would imply an initial partial pressure of N_2 equal to about 30 mbar. The model requires a net integrated deposition of nitrogen in minerals of magnitude 2.5×10^{23} atoms (N) per square centimeter averaged over the martian surface. The es-

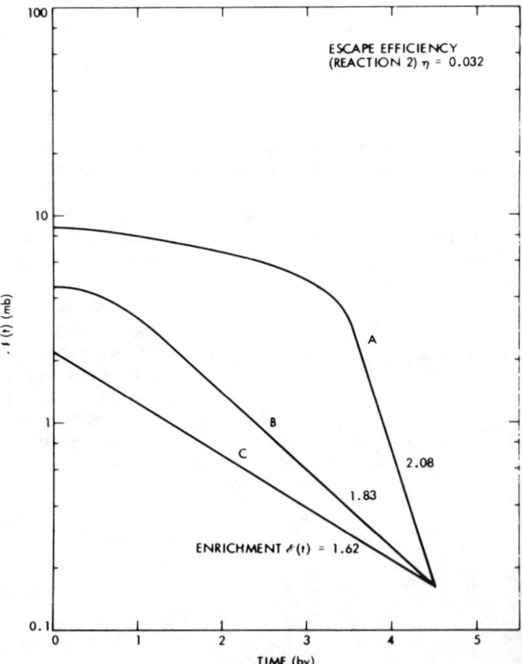

Fig. 3. Abundance of N_2 (in millibars) as a function of time (in units of 10^9 years). The escape efficiency for reaction 2 was assumed to be 0.032. Curve A represents results for the case in which the atmospheric CO_2 pressure was assumed to fall below 0.65 mbar for one-third of a climatic cycle. Curves B and C represent results for the cases in which the CO_2 pressure rises to 65 mbar for one-third and one-tenth of a climatic cycle, respectively. The enrichment factors at time t = 4.5 billion years for curves A, B, and C are 2.08, 1.83, and 1.62, respectively, for $K = 3 \times 10^8$ cm^2 sec^{-1}.

cape rate for nitrogen is dominated by reaction 2 for all models shown here.

Figure 3 shows results obtained with a model in which we allow for a secular variation in the partial pressure of atmospheric CO_2, which might arise because of changes in the obliquity and orbital eccentricity of Mars if the atmospheric pressure were controlled by an equilibrium with the polar ice cap, as discussed for example by Ward *et al.* (*11*). Curve A is based on the assumption that the partial pressure of CO_2 was equal to 10 percent of the present value for approximately 30 percent of the time over the past 4.5×10^9 years. Curves B and C model the case for which the pressure might have been much higher in the past: the partial pressure of CO_2 is taken as 65 mbar, for 33 percent of the time in curve B and for as much as 10 percent of the time in curve C (*12*). For the model in Fig. 3 we assumed a passive surface, $\gamma = 0$, and $\eta = 0.032$. The initial partial pressure of N_2 for all three cases lies in the range 2 to 9 mbar.

The measurement of enriched ^{15}N on Mars provides a powerful diagnostic tool which may be used to derive invaluable data on the past history of Mars. The uncertainty indicated by the spread of results shown in Figs. 1 to 3 may be narrowed considerably by careful laboratory studies of the velocity distribution of atoms formed by electron impact dissociation of N_2. An appropriate laboratory investigation would clarify the role that heterogeneous processes might play in the chemistry of nitrogen on Mars and would permit a quantitative estimate of the rate at which nitrogen may have been incorporated in surface minerals over the past 4.5×10^9 years. It is difficult, in any event, to escape the conclusion that the abundance of atmospheric nitrogen on Mars must have been much higher in the past: the partial pressure of the gas may have been as high as 30 mbar, and was surely no less than about 2 mbar.

References and Notes

1. A. O. Nier, W. B Hanson, A. Seiff, M. B. McElroy, N. W. Spencer, R. J. Duckett, T. C. D. Knight, W. S. Cook, *Science* **193**, 786 (1976); T. Owen and K. Biemann, *ibid*., p.801.
2. A. O. Nier, M. B. McElroy, Y. L. Yung, *ibid*., **194**, 68 (1976).
3. M. B. McElroy, *ibid*. **175**, 443 (1972).
4. Y. L. Yung, D. F. Strobel, T. Y. Kong, M. B. McElroy, *Icarus*, in press.
5. R. T. Brinkmann, *Science* **174**, 943 (1971).
6. H. Michels, private communication.
7. We assume here that atoms are released isotropically. Half of the atoms formed will be emitted in the upward hemisphere and may escape to space without suffering further collisions. We use a model atmosphere given in (*4*) to describe mean martian conditions. The model has an exobase at 210 km. It is interesting to note that reaction 1 could provide a selective mechanism for escape of ^{14}N, although for most models considered here nitrogen is lost by reaction 2 rather than reaction 1, and one would not expect the escape efficiency for reaction 2 to depend critically on isotopic composition. Moreover, the temperature assumed here for ions may be conservative on the low side, which would further reduce the dependence of the escape flux on the isotopic composition of N_2. We shall assume that ^{15}N and ^{14}N should escape with equal probability from the exobase. Clearly this matter should be reexamined as further data become available on the detailed kinetics for reactions 1 and 2, and as further measurements are made to define the temperature of Mars' high-altitude ionosphere. If we were to neglect the escape flux due to reaction 2 and if we were to further postulate that reaction 1 should favor escape of ^{14}N by a factor of 2, the initial pressure of N_2 would be 0.56 mbar and the enrichment for ^{15}N would be about 110 percent for a passive surface. If we were to reduce the escape efficiency for reaction 2 to a value $\eta = 0.008$, and again assume that reaction 1 favors reduction of fast ^{14}N by a factor of 2, the enrichment for ^{15}N would be about 140 percent, and the initial N_2 pressure would have a value of 1.1 mbar, for models similar to Fig. 1.
8. G. R. Wight, M. J. Van Der Wiel, C. E. Brion, *J. Phys. B* **9**, 675 (1976).
9. D. F. Strobel, private communication.
10. M. B. McElroy and Y. L. Yung, *Planet. Space Sci.*, in press.
11. W. R. Ward, B. C. Murray, N. C. Malin, *J. Geophys. Res.* **79**, 3387 (1974).
12. O. B. Toon drew our attention to the possibility that the CO_2 pressure might have been considerably higher in the past, before the formation of the Tharsis Ridge. Formation of the ridge may be expected to have altered the gravitational moment of Mars, resulting in a change in the insolation of the polar cap and appropriate perturbations of the atmospheric CO_2, if we assume that the concentration of the gas in the atmosphere is determined in large measure by the temperature of the polar ice.
13. Supported by NASA contracts NAS-1-9697 and NAS-1-10492 to the University of Minnesota and Harvard University, respectively.

2 September 1976

Part XI

CLIMATIC CHANGES ON MARS

Editor's Comments on Paper 32

32 WARD
Large-Scale Variations in the Obliquity of Mars

Two principal lines of geological evidence indicate that the climate of Mars must have differed in the past: (1) a network of dendritic runoff gullies, which formed during a milder period when rain could fall (Pieri 1976), and (2) the layered polar deposits, generally believed to represent eolian sediments, which record periodic changes in the ability of the atmosphere to transport dust (Cutts 1973a).

M. Milankovitch (1941) originally proposed that variations in orbital parameters of the planets could affect the amount of solar insolation and hence the climate. His hypothesis received strong support from data derived from deepsea cores (Hays et al. 1976). The mean annual insolation is $\bar{I} = S/\pi \sin i(1 - e^2)^{-1/2}$, where i = obliquity or tilt of the axis from the normal to the equator, and e is eccentricity. It can be readily seen that the mean insolation at higher latitudes depends strongly on i, but is fairly insensitive to e, except at the equator during perihelion. Although e ranges between 0.004 and 0.141 on Mars, the average polar insolation changes only by 1 percent according to Murray et al. (1973) and Ward (Paper 32).

Since obliquity dominates the insolation changes at higher latitudes, W. R. Ward (Paper 32) lays the groundwork for study of climatic changes by calculating the amplitude and periodicity of obliquity variation for Mars. The obliquity ranges between 15° and 35°, with the present value of 25.1°, over 120,000 and 1.2 million year cycles. In a subsequent paper, Ward et al. (1974) find that the extremes of obliquity can produce a 30°K difference in polar temperatures, corresponding to atmospheric pressures ranging between 0.3 and 30 mb—a factor of 100. (The pressures were derived assuming a permanent polar CO_2-ice reservoir. Since this is unlikely, exchange of adsorbed CO_2 in the regolith with the atmosphere could provide a source of CO_2 [Fanale and Cannon 1974].)

Tectonic changes, particularly redistribution of mass, can alter the planet's moment of inertia, which in turn can affect the precession frequency and ultimately the obliquity. Toon et al. (1977) conclude that the departure from isostatic equilibrium of the Tharsis volcanoes could change the principal moment of inertia by $\sim$6 percent, which would have reduced the mean obliquity from a former value of 34° to the present 25°. The higher polar insolation occurring during the period of increased obliquity would have increased the polar regolith temperatures—hence increased atmospheric pressures, larger atmospheric greenhouse effects, and an overall warmer climate, thereby permitting stabilization of liquid water.

While deposition of polar laminations could have conceivably been regulated by obliquity-induced pressure variations, the effects may not have sufficed to stabilize liquid water over much of the surface (Mutch et al 1976). Liquid water at the surface may have been confined to an early period of much denser atmosphere (Pieri 1976; McElroy et al. 1976). This question remains an active subject of research (e.g., Pollack 1977, unpubl. notes).

32

Reprinted from *Science* **181**:260–262 (1973)

LARGE-SCALE VARIATIONS IN THE OBLIQUITY OF MARS

William R. Ward

Geological and Planetary Sciences
California Institute of Technology

Large-Scale Variations in the Obliquity of Mars

Abstract. *Large-scale variations in the obliquity of the planet Mars are produced by a coupling between the motion of its orbit plane due to the gravitational perturbations of the other planets and the precession of its spin axis which results from the solar torque exerted on the equatorial bulge of the planet. The obliquity oscillates on a time scale of approximately* 1.2×10^5 *years. The amplitude of this oscillation itself varies periodically on a time scale of* 1.2×10^6 *years. The present-day obliquity is approximately 25.1 degrees. The maximum possible variation is from about 14.9 to 35.5 degrees. Significant climatic effects must be associated with the phenomenon.*

Since Leighton and Murray's article (*1*) on carbon dioxide and other volatiles on Mars, there has been considerable interest in acquiring an understanding of the atmospheric history and behavior of volatiles on Mars. Climatic changes stemming from alterations in the orbit of Mars are of prime importance to the atmospheric problem. Murray *et al.* (*2*) presented modulations in the yearly polar insolation produced by variations in the orbital eccentricity, *e*, of Mars as a result of the gravitational perturbations of the other planets. The average yearly insolation at the poles is equal to $(S/\pi) \times \sin\theta/(1-e^2)^{1/2}$, where S is the solar constant at the distance of Mars (1.52 A.U.) and θ is the planet's obliquity (the angle between the spin axis and the normal to the orbit plane). Changes in the polar insolation thus depend on the second power of the eccentricity. Although the eccentricity varies from .004 to .141 (*3*), the resulting alteration in the average polar insolation is only of the order of 1 percent. However, the polar insolation would be more sensitive to changes in the obliquity, were such changes to occur. Recent dynamical calculations indicate that, indeed, the obliquity of Mars has undergone large periodic variations—sometimes changing by as much as $\sim 17°$ in $\sim 6 \times 10^4$ years. The climatic effects of this phenomenon must be significant, and it would not be surprising if they were intimately associated with the shaping of many of the surface features.

Secular changes in the orbits of the planets due to their mutual gravitational perturbations were calculated to lowest order in the disturbing function by Brouwer and van Woerkom (*4*). These included changes in the orbital inclination, *i*, and the longitude of the ascending node, Ω, for each planet, as well as the changes in the eccentricity and argument of the perihelion employed by Murray *et al.* (*2*). In deriving the secular changes, Brouwer and van Woerkom used the auxiliary variables $p = \sin i \cos\Omega$ and $q = \sin i \sin\Omega$ for each planet in order to avoid complications that arise with vanishingly small inclinations. The motions of all the planets are coupled to each other and the problem reduces to a system of eight linear differential equations (Pluto was excluded in these calculations) which, when solved simultaneously, yield eight eigenvectors and eigenfrequencies, ω_k. The motion of any planet is given by expressions of the form

$$p = \sum_k N_k \cos(\omega_k t + \delta_k) \quad (1)$$

$$q = \sum_k N_k \sin(\omega_k t + \delta_k) \quad (2)$$

where the amplitudes, N_k, and phase constants, δ_k, are found from the values of p and q in the epoch 1900, and t is time. The longitudes and inclinations are referred to the ecliptic of 1950. The evolution of the orbit plane is most easily understood when its position is expressed relative to the invariable plane of the solar system. When this is done, one eigenvector (with $\omega = 0$) vanishes. Table 1 is a list of the amplitudes and phase constants for the remaining seven terms for Mars, along with the eigenfrequencies and their corresponding periods (P_k). The inclination to the invariable plane can be obtained from the relation

$$\sin i = (p^2 + q^2)^{1/2} \quad (3)$$

This quantity is the lower curve in Fig. 1, plotted for 5×10^6 years back in time. Two basic periodicities can be seen in the plot: a rapid oscillation of small amplitude ($\sim 1°$) superimposed on a slower one of large amplitude ($\sim 5°$). The first has a period of approximately 1.6×10^5 years, the second a period of the order of 1.2×10^6 years. From Table 1, we see that the solution is dominated by the third, fourth, and fifth terms. The large-

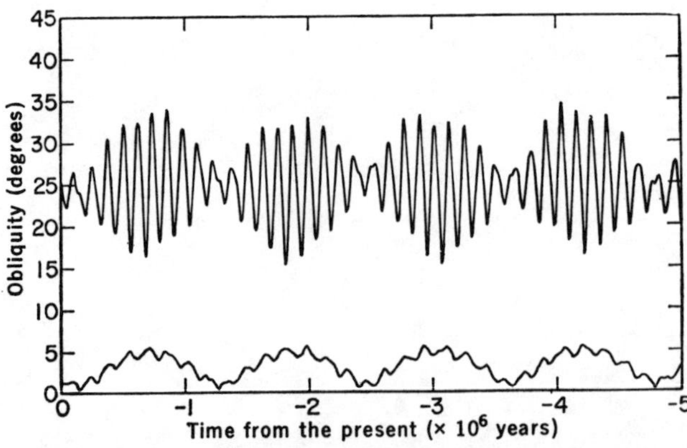

Fig. 1. Variations in the obliquity and orbital inclination of Mars for the past 5×10^6 years.

amplitude changes result from a beat between the two largest terms and have a frequency corresponding to $(\omega_3 - \omega_4)$. The small-amplitude modulations are due to the fifth term and have a frequency given by $(\omega_5 - \omega_4)$. The theoretical maximum of $\sin i$ is found from $\sum_k |N_k| = 0.102$ (*3*), which corresponds to an inclination of 5.87°. The largest and smallest values obtained in Fig. 1 are 5.60° and 0.23°, respectively. Theoretically the inclination can become zero, although this does not happen for the time interval shown. The rotation of the ascending node is also dominated by the third and fourth terms so that the period for a complete cycle is of the order of 7×10^4 years, comparable to P_3 and P_4. The nodes move in the opposite sense to the orbital motion.

Both a changing inclination and a rotating node are capable of producing a variation in the obliquity. If the spin axis of Mars were fixed in inertial space, the time evolution of the obliquity could be obtained from the motion of the orbit normal. However, the problem is complicated by the fact that the spin axis of Mars is also moving in inertial space as a result of the solar torque exerted on the equatorial bulge of the planet. This motion must be taken into account before the actual change in the obliquity can be ascertained. In fact, it has been shown (*5*) that if the precession frequency of the spin axis is much larger than the relevant frequencies describing the motion of the orbit normal, the obliquity will remain essentially constant. In such a case, the spin axis has the ability to "follow" the moving orbit normal. It is this effect which partially suppresses variations in the earth's obliquity during the motion of its orbit plane. However, for Mars the situation is considerably different.

The equations governing the motion of the spin axis observed from a noninertial reference frame attached to the moving orbit plane (with the x-axis in the instantaneous direction of the ascending node on the invariable plane and the z-axis in the direction of the normal to the orbit plane) take the form,

$$\frac{d\theta}{dt} = -\sin i \cos\phi\left(\frac{d\Omega}{dt}\right) + \sin\phi\left(\frac{di}{dt}\right) \quad (4)$$

$$\frac{d\phi}{dt} = -\alpha\cos\theta - \cos i\left(\frac{d\Omega}{dt}\right) + \sin i \sin\phi \cot\theta\left(\frac{d\Omega}{dt}\right) + \cos\phi \cot\theta\left(\frac{di}{dt}\right) \quad (5)$$

Table 1. Frequencies (ω_k), corresponding periods (P_k), phase constants (δ_k), amplitudes for the inclination (N_k), and amplitudes for the obliquity (N_k') of Mars.

k	ω_k (arc second/year)	P_k (years)	δ_k (degrees)	N_k	N_k'	N_k'/N_k
1	− 5.202	249,200	272.06	0.00180	− 0.00422	− 2.342
2	− 6.571	197,200	210.06	0.00180	− 0.01393	− 7.740
3	− 18.744	69,100	147.39	− 0.03589	− 0.05940	1.655
4	− 17.633	73,500	188.92	0.05025	0.08678	1.727
5	− 25.734	50,400	19.58	0.00965	0.01356	1.405
6	− 2.903	449,500	207.48	− 0.00126	0.00081	− 0.643
7	− 0.678	1,912,900	95.01	− 0.00123	0.00012	− 0.101

The azimuthal angle of the spin axis, ϕ, is measured from the instantaneous position of the ascending node. From measurements of the spin axis orientation (*6*), one can deduce a current value of $\phi(t=0) = \Phi = 356.3°$, and a present value of the obliquity of Mars of $\theta(t=0) = \Theta = 25.1°$. The quantity α is a measure of the frequency of spin axis precession. A recent determination of the oblateness of Mars yields a precession period, $P = 2\pi/\alpha\cos\Theta$, of 175,000 years (*7*), from which $\alpha\cos\Theta = 7.42$ seconds of arc per year. To solve for the obliquity to first order in the orbital inclination we first solve the zero-order equations with $i = 0$ to obtain

$$\theta = \Theta \quad (6)$$

$$\phi = \Phi - \alpha t\cos\Theta - \Omega + \Omega_0 \quad (7)$$

where Ω_0 is the current location of the ascending node. These are then substituted into the right-hand side of Eq. 4. With the aid of Eqs. 1 and 2, the resulting first-order equation can be solved, yielding

$$\theta = \Theta - \sum_k\left(\frac{\omega_k N_k}{\omega_k + \alpha\cos\Theta}\right)[\sin(\omega_k t + \alpha t\cos\Theta + \delta_k - \Phi - \Omega_0) - \sin(\delta_k - \Phi - \Omega_0)] \quad (8)$$

This expression well illustrates the effect of the spin precession on the variation in the obliquity. If $\alpha \rightarrow 0$, one obtains a first-order expression for the variation in the obliquity of a spin axis fixed in inertial space, which can also be computed directly from the relation $\cos\theta = (\hat{s} \cdot \hat{n})$, where $\hat{s}$ is a unit vector in the direction of the spin axis and $\hat{n}$ is the unit orbit normal. On the other hand, if $\alpha \rightarrow \infty$, $\theta \rightarrow \Theta$ and the obliquity remains constant in spite of the movement of the orbit normal. However, Eq. 8 also illustrates another possible behavior. Since $\omega_k < 0$, if for any k, $|\omega_k + \alpha\cos\Theta| < 1$, a resonance-type phenomenon occurs and the variation in the obliquity due to the kth eigenvector is enhanced. Table 1 shows $N_k' = \omega_k N_k/(\omega_k + \alpha\cos\Theta)$ for Mars. In all but the last two terms the amplitudes are enlarged. Hence, rather than suppressing obliquity changes, the precession of the spin axis magnifies them.

Equation 8 is plotted in the upper portion of Fig. 1. The dominance of the third and fourth terms can be recognized. There are rapid oscillations with a period of the order of 1.2×10^5 years as a result of the frequencies $(\omega_3 + \alpha\cos\Theta)$ and $(\omega_4 + \alpha\cos\Theta)$ of the largest terms. The amplitude of these oscillations varies periodically on a larger time scale of approximately 1.2×10^6 years, which corresponds to half a cycle of the beat frequency $\frac{1}{2}(\omega_3 - \omega_4)$. There is a significant correlation between the two plots of Fig. 1.

The variation in the obliquity is quite dramatic. From Eq. 8 we see that the maximum amplitude of the variation is $\sum |N_k'| = 0.179$, which is equivalent to $\sim 10.3°$. The maximum total variation is twice this value, or $\sim 20.6°$. The oscillations occur about the value,

$$\Theta + \sum_k N_k' \sin(\delta_k - \Phi - \Omega_0) \quad (9)$$

which is equal to 25.2°. Hence, Mars is very close to this position now. During its dynamical history, its obliquity has varied from $\sim 14.9°$ to $\sim 35.5°$.

Although Eq. 8 is only a first-order solution, we have numerically integrated Eqs. 4 and 5 and the agreement with Fig. 1 is quite good. We have also obtained an analytic solution correct to second order in the orbital inclination (*8*). However, insight into the physical nature of this phenomenon is furnished by the first-order solution presented here.

Recent Mariner 9 orbital photographs have revealed several remarkable features on the surface of Mars, including regions composed of unusual layered structures (designated laminated terrain), a vast system of canyons and channels, and a persistent residual south polar cap thought to be composed of water ice (*9*). It is an exciting prospect that the origin of some of these features may be linked to periodic climatic changes associated with the oscillating obliquity. No complete understanding of the history and behavior of volatiles on Mars is possible until this effect is taken into account (*10*).

References and Notes

1. R. B. Leighton and B. C. Murray, *Science* **153**, 136 (1966).
2. B. C. Murray, W. R. Ward, S. C. Yeung, *ibid.* **180**, 638 (1973).
3. D. Brouwer and G. M. Clemence, *Planets and Satellites* (Univ. of Chicago Press, Chicago, 1961), p. 50.
4. D. Brouwer and A. J. J. van Woerkom, *Astron. Pap. Amer. Ephemeris Naut. Alm.* **13** (part 2), 81 (1950).
5. W. R. Ward, in preparation.
6. J. Lorell *et al.*, *Science* **175**, 317 (1972).
7. ———, *Icarus*, in press.
8. W. R. Ward, in preparation.
9. B. C. Murray *et al.*, *Icarus* **17**, 328 (1972).
10. W. R. Ward, B. C. Murray, M. C. Malin, in preparation.

11. I acknowledge valuable discussions with Dr. Bruce C. Murray. Supported by NASA grant NGL 05-002-003. Contribution No. 2322 of the Division of Geological and Planetary Sciences, California Institute of Technology.

26 March 1973

Part XII

THE CANALS OF MARS

Editor's Comments on Paper 33

33 SAGAN and FOX
The Canals of Mars: An Assessment after Mariner 9

No aspect of the study of Mars has attracted as much popular interest as the notorious canals, nor has any phase of astronomical observation withstood the test of time so poorly. Although P. Lowell's notions of a dying advanced civilization on Mars represent an extreme viewpoint, several astronomers besides Lowell, ranging from Schiaparelli to Slipher and Pickering, claimed to have seen the canals. Other astronomers denied the reality of these features. To Antoniadi, the human eye connects discontinuous blotches into straight lines in a kind of mental data compression. A wide variety of geological explanations (sand dunes, dikes, faults, and so forth) have also been offered (Gifford 1964; Jamison 1965; Katterfeld and Hedervari 1969).

The final word comes from the Mariner 9 TV mapping experiment, which provided total surface coverage at 1 km resolution and nearly 10 percent coverage at 100 m resolution. C. Sagan and P. Fox (Paper 33) convincingly expose the illusion of the canals. Only in rare instances do the alleged canals line up with albedo markings, crater chains, or other topographic features. Such isolated correlations may be expected by chance alone. Only one classical canal (Agathodaemon) is real: Valles Marineris. The width of this vast canyon (~100km) lies within the resolution limits of Earth-based telescopes and thus should have been easily seen (and it was!). Sagan and Fox conclude: "The vast majority of the canals appear to be largely self-generated by the visual observers of the canal school, and stand as monuments to the imprecision of the human eye-brain-hand system under difficult observing conditions."

There may be another factor involved in the canal controversy: Our compulsive curiosity over the existence of life (especially intelligent life) on other planets, or elsewhere in the universe. Although the canals as artifacts can be consigned to the dustbin of disproven theories, the need to establish whether life (no matter how primitive) has evolved on Mars has provided the primary motivation behind the technically complex and successful Viking missions.

33

Reprinted from *Icarus* 25:602–612 (1975)

The Canals of Mars: An Assessment after Mariner 9

CARL SAGAN AND PAUL FOX

Laboratory for Planetary Studies, Cornell University, Ithaca, New York 14853

Received November 21, 1974; revised March 31, 1975

The Lowellian canal network has been compared with the results of Mariner 9 photography of Mars. A small number of canals may correspond to rift valleys, ridge systems, crater chains, and linear surface albedo markings. But the vast bulk of classical canals correspond neither to topographic nor to albedo features, and appear to have no relation to the real Martian surface.

The canals of Mars have been something of an embarrassment to planetary astronomers since attention was called to their existence by Schiaparelli in 1877. The reality of most of the canals, much less the processes producing them, has been the subject of heated controversy; and the initial hypothesis by Lowell that they were the constructs of intelligent beings on Mars has led to their general classification somewhere in the no-man's land between science and fiction. Since a variety of independent observers—including some, such as Trumpler, intensely skeptical of the hypothesis that the canals are artificial—have produced canal maps of tolerable mutual consistency, it cannot be claimed that there is no phenomenon to explain. On the other hand, by no means all of the canals displayed on the classical maps are seen by all or even most observers, and the observational problems in detecting features of intrinsically low contrast and small angular size under poor seeing conditions are severe. Also the term canal has been used to describe a variety of quite different features. For example, Agathodaemon, Cerberus, and Thoth-Nepenthes are dark markings of considerable extent which have been repeatedly photographed and about whose reality there can be no question. But they are hardly fine dark lines following great circles for thousands of kilometers; instead they are irregular, broad quasilinear features. On the other hand, the great bulk of very regular Lowellian canals have never been photographed, have not been seen by all or in some cases even by most observers, and are probably of dubious reality. There are no more than 10 or 20 so-called canals whose reality is likely; and no more than half a dozen whose reality is beyond question.

Probably no Martian observable has elicited so many varied and mutually exclusive hypotheses as the canals. A very incomplete list of these hypotheses follows.

(1) Canals are the waterways, or the vegetation along the banks of waterways, constructed by an advanced civilization on Mars (Lowell, 1906).

(2) Canals are natural valleys or surface cracks produced by a range of geological processes. (Wallace, 1907; Pickering, 1921; Tombaugh, 1950). The idea that the canals are ridges has also been suggested (Joly, 1898) and a variety of authors have proposed some combination of rifts or faults with ridges, dikes, or escarpments (Schiaparelli, 1893; Arrhenius, 1918; Wasiutynski, 1946; Katterfel'd, 1959; Hope, 1966). In detail many of these suggestions are physically naive or based upon now-outmoded geophysical models—for example the idea that the canals are ridges produced by tidal interaction with asteroids on near-grazing collisions, cracks produced during contraction from an initially molten state, etc.

(3) Canals are accidental alignments of disconnected fine surface detail (Maunder, 1894; Antoniadi, 1930). A specific version of this hypothesis attri-

butes some canals to chains of large craters (Sagan, 1966).

(4) The canals are crater rays as on the lunar surface (Plassman, 1901; Oncley and Fulmer, 1966). Why lunar rays are bright and Martian canals dark is not readily apparent in this explanation.

(5) The canals are rectilinear sand dunes (Gifford, 1964).

(6) A new hypothesis which might be suggested by the discovery of long linear windblown streaks on Mars (Sagan *et al.*, 1972, 1973): the canals are albedo features produced by a covering or uncovering of surface material by windblown particulates.

No topographic feature on Mars has large enough dimensions to cast a shadow detectable from Earth, as has been stressed to us by W. A. Baum. It follows that features reported by ground-based observers must, if real, arise from regional albedo or roughness differences, these can, however, be topography-associated (Sagan *et al.*, 1972, 1973).

Mariner 9 has achieved a complete mapping of the surface of Mars down to 1km resolution (A-frames) and spot photography of a few percent of the surface down to 100m resolution (B-frames). The best photographic resolution of Mars obtained before spacecraft reconnaissance was several hundred kilometers and the best imaginable visual resolution perhaps slightly better than 100km. The Mariner 9 photography has now been mosaicked, and it should therefore be possible to make some definitive conclusions about the nature of the Martian canals. In the following figures we have employed mosaics of Mariner 9 A-frame photography.

Since Mariner 9 was in orbit about Mars for more than half a Martian year, it is unlikely that the planet was viewed at a time when canals were not to be seen. While the early part of the mission was marked by the waning phases of a great dust storm, the later stages of the mission were remarkably clear. Variable-features information obtained during the mission seems to correlate well with other ground-based information (Sagan *et al.*, 1972, 1973). The mosaics are chiefly useful for their topographic information. Albedo information is contained within each Mariner 9 A-frame but no attempt has been made to preserve albedo information from frame to frame within the mosaic. In some of the mosaics "gores" between adjacent A-frames can be seen.

In Fig. 1 are displayed four representations of the same region of Mars in the region between Elysium Planitia and Hesperia Planum. The representations are Fig. 1a, Schiaparelli, 19th Century; Fig. 1b, Antoniadi, 1929; Fig. 1c, Mariner 9, photography, 1971–72; Fig. 1d, cartography based on Mariner 9 photography and Earth-based photography. The network of single and double canals of Schiaparelli is found by Antoniadi, under superior seeing conditions, to resolve into an array of disconnected fine detail (Hypothesis 3). Mariner 9 shows no evidence of any "canals" in this region. The vertical sequence of spots in the middle of Fig. 1b might possibly be an array of craters, just barely too small to have been given names, which may be seen in Fig. 1c. The largest distinct crater near the center of Fig. 1c is Gale about 150km across. To its southwest is the more subdued but larger crater, Herschel. It is barely possible that one of these craters corresponds to the dark semicircular outline toward the lower left of Fig. 1b, at the edge of Mare Cimmerium. But these are very dubious correlations.

In the remaining figures we have chosen to compare the Mariner 9 results with the canal cartography of Slipher (1962), the most recent representative of the Lowellian Canal School. In Slipher's as in Lowell's maps, cartographic errors of 5° in latitude and longitude are not uncommon and errors up to 10° or 15° exist. For example Hellas Planitia is shown by Slipher centered at 300°W rather than 290°W. Where common albedo features such as this exist we have distorted Slipher's cartography to bring it into agreement with ground-based or space-borne photographic data sources. A somewhat subjective interpolation was utilized to position features adjacent to displaced features.

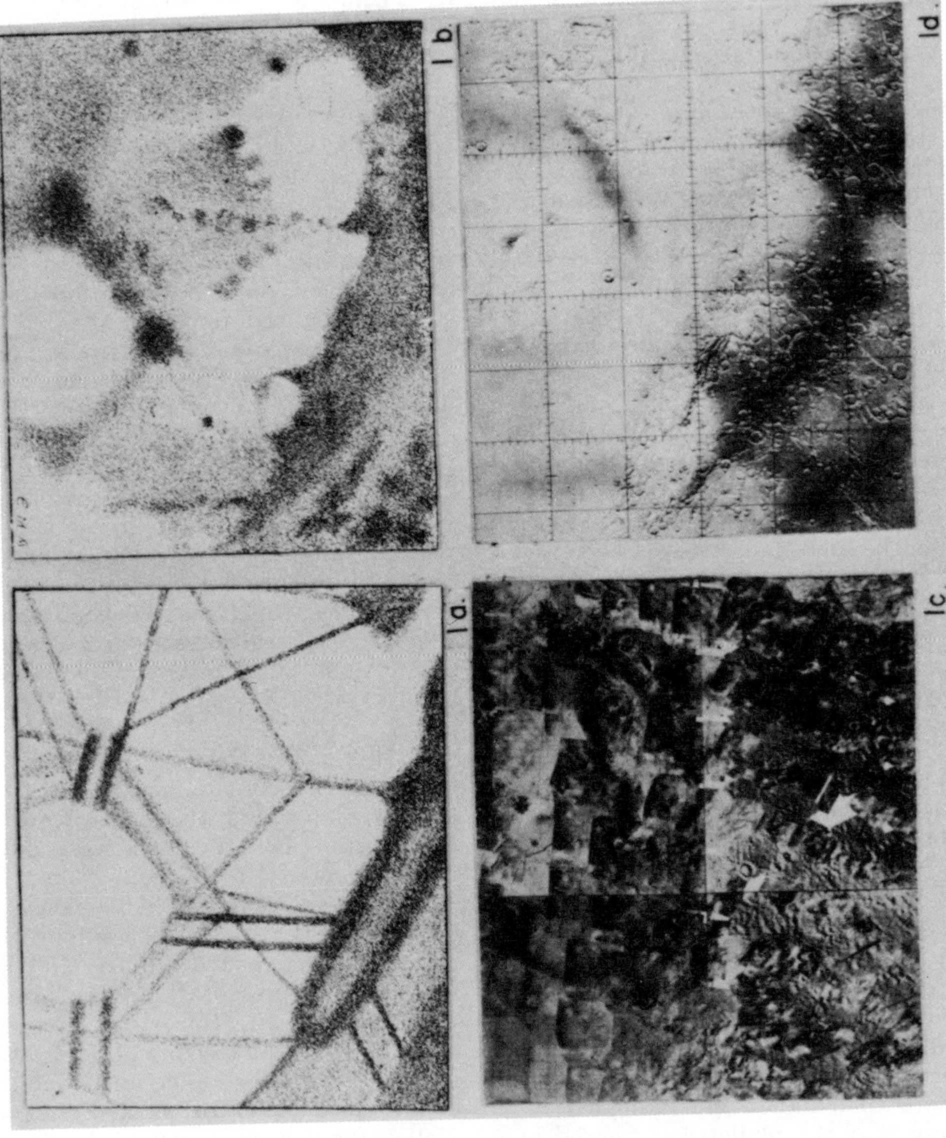

FIG. 1. The region between Elysium Planitia and Hesperia Planum according to Schiaparelli (1a), Antoniadi (1b), Mariner 9 photography (1c), and cartography based on Mariner 9 and Earth-based photography (1d). Figure (1c) is not corrected for the albedo and shading of adjacent con-

In making comparisons between ~100 straight lines and a surface which requires $\sim 10^{10}$ bits to characterize it, it is clear that there will be occasional correlations which are due merely to chance. Therefore the following conclusions represent an upper limit to any real correlation between true Martian surface features and classical canals. The subsequent figures were prepared by laying down the canals as opaque tape on a transparent overlay placed over the Mariner 9 results. Occasionally a faint shadow of the overlay can be discerned following one side of the "canal"; of course, these are not true surface albedo features. As mentioned above, the Mariner 9 imagery was not well controlled photometrically, and it is possible that faint albedo features which may cross many adjacent A-frames exist on Mars but are indetectable in these figures. But features with contrast $\gtrsim 5\%$ should be discernible. In many classical descriptions the canals are reported to have high contrast.

Figures 2, 3, and 4 display typical comparisons. Canals are shown as straight or straightish white thick lines; dark areas as drawn by Slipher are shown as broad pin-striped overlays. In Fig. 2 we see a striking connection between the canal Agathodaemon and the great rift valley, Valles Marineris. But this is a low-albedo feature, 5000km long, easily resolvable and of high contrast to ground-based observers. Elsewhere in Figs 2, 3, and 4, no correlations are apparent. Indeed there are many major topographic or albedo features which show no hint of having been observed by Slipher despite the fact that they have much larger two-dimensional angular extent than many of the linear features which were drawn by Lowell and his followers.

Figure 5 is our best effort at a superposition of Slipher's global canal map with the Mariner 9 cartography. The baseline map here used was prepared by J. Inge and the Lowell Observatory on the basis of Mariner 9 topographic data and photographable ground-based albedo data. In addition to Elysium and Hellas, a number of control points were employed, including Dawes' forked bay (5°S, 0°W), Aurorae Sinus (12°S, 49°W), Ascraeus Mons (11°N, 104°W), and Lunae Planum (15°N, 65°W). The canal Thoth-Nepenthes is drawn in the position given by Slipher and is obviously 5° displaced from the albedo feature of the same name. The correlation of canals with other topographic or albedo features is again seen to be extremely poor. For example, the large shield volcanoes Pavonis Mons, Arsia Mons, and Olympus Mons have not been noted at all, while Ascraeus Mons is shown as the nexus of seven canals. It is a discernible albedo feature—as, however, was Olympus Mons. It is possible that the convergence of three canals at coordinates (13°N, 124°W) is intended to correspond to the position of Olympus Mons.

In the light of these results, we now discuss each of the six canal hypotheses in turn.

(1) Mariner 9 photographed the surface of Mars with a resolution and surface coverage which would have detected a civilization of our own extent and level of development had it been there (Sagan and Wallace, 1971). No such civilization was uncovered. No features which strongly suggest intelligent life were discerned. The data counterindicate Hypothesis 1.

(2) There is at least one correspondence between a classical canal and a valley or ridge. By far the most striking case is the connection between Agathodaemon and Valles Marineris, whose width (~100km) corresponds well to the resolution limits of Earth-based photography and visual observations. It should have been seen and it was. It is approximately but not compulsively linear. A possible second case is Ceraunius, which had been deduced from preliminary doppler radar spectroscopy to be a ridge system (Sagan *et al.*, 1967; Sagan and Pollack, 1966). The feature is near Ceraunius Fossae and may join with Mareotis Fossae, producing a system which is 25° in latitude or about 1500km long. It is in many places hundreds of kilometers wide. Like Valles Marineris it may be a region where bright or dark material is temporarily trapped. It is just possible that similar cases include correlation of the canal Styx with the Phlegra Montes. On the other hand there are a great many other Martian features designated Valles, Fossae,

FIG. 2. Connection of Slipher's canal cartography with Mariner 9 imagery near Solis Lacus. Dark areas are shown by pinstripes, canals by straight ... [illegible] ...ted from this Mariner 9 mosaic.

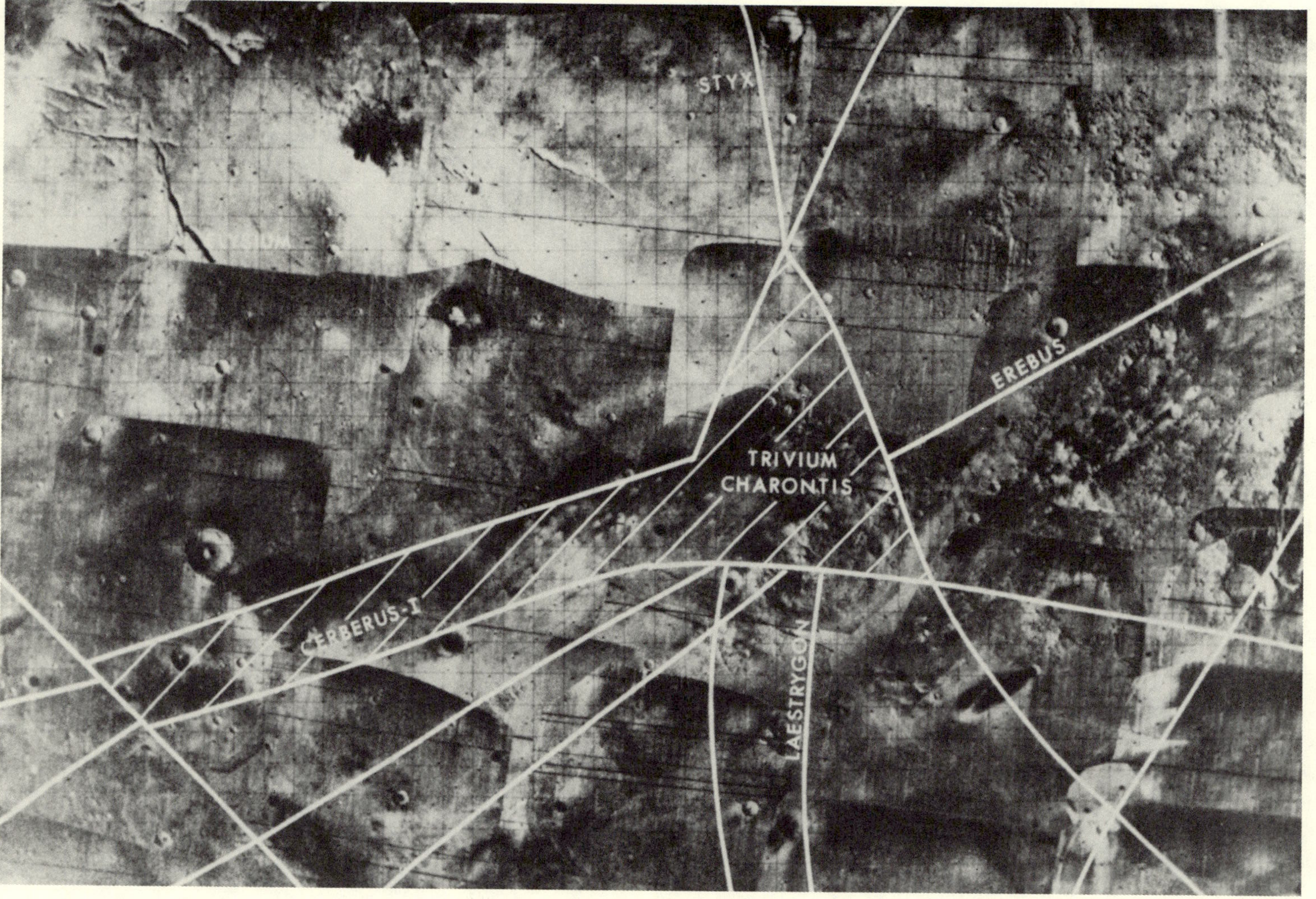

FIG. 3. Connection of Slipher's canal cartography with Mariner 9 imagery near Trivium Charontis. Dark areas are shown by pinstripes, canals by straight lines. The mosaic is from Mars chart 15. Only topographic information can be extracted from this Mariner 9 mosaic.

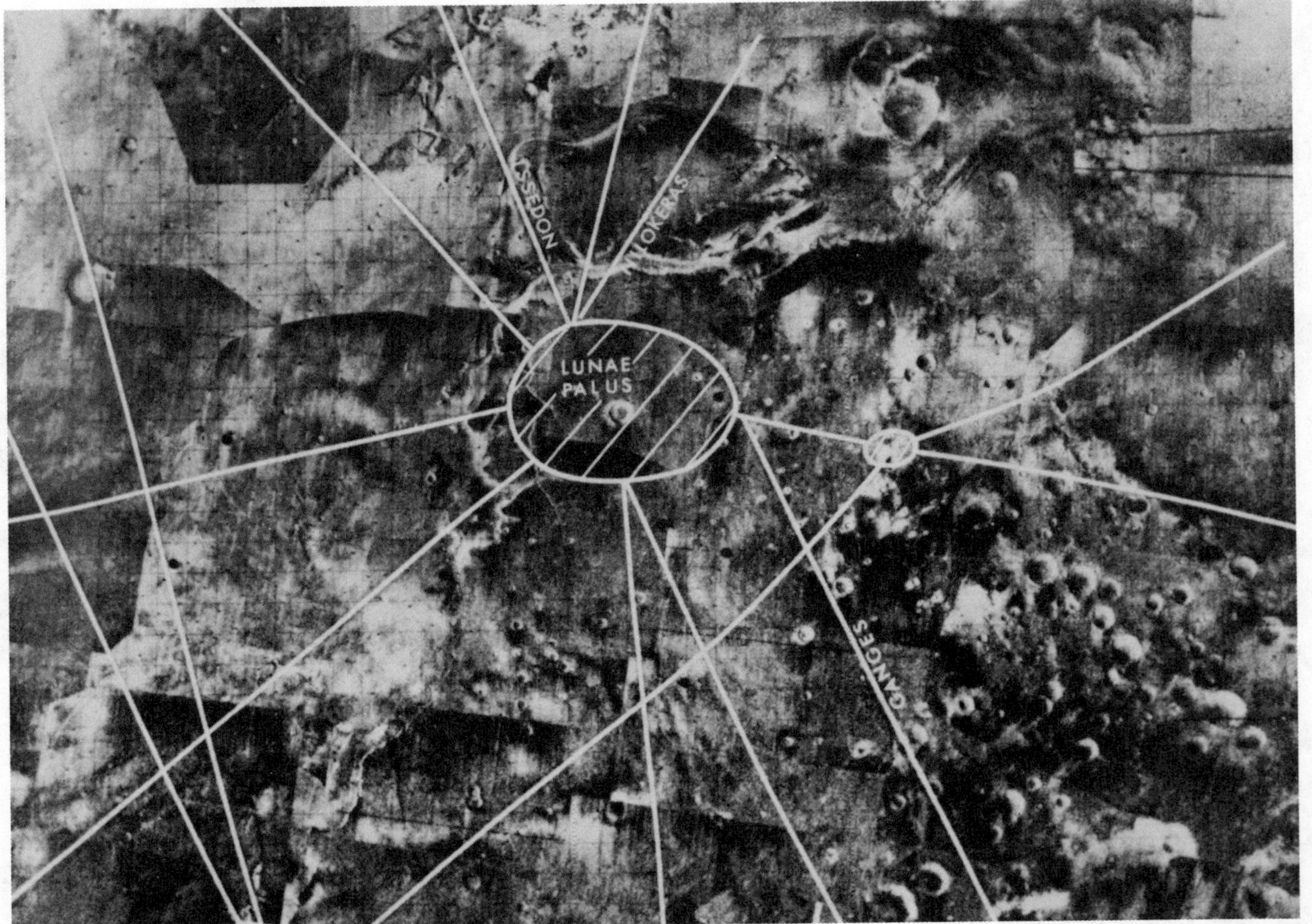

FIG. 4. Connection of Slipher's canal cartography with Mariner 9 imagery near Lunae Palus. Dark areas are shown by pinstripes, canals by … from this Mariner 9 mosaic.

and Montes which correspond closely to none of the classical canals. Most of these cases are either smaller in linear or lateral extent or are at higher latitudes than the features just mentioned.

(3) Sagan and Pollack (1966) asked why, if some subset of Martian canals is due to the psychophysiological alignment of disconnected fine detail, there are not similar psychophysiological linearities seen on other astronomical objects. It is possible that Mars is the only planet where there are a multitude of features of low contrast and marginal resolution which can be so strung up. In any case only a few examples of the Maunder/Antoniadi Hypothesis 3 could be found. The most striking is the chain of large craters proceeding from the northern tip of Margaritifer Sinus, and which includes the craters Trouvelot, Rutherford, Becquerel, Curie, and Sklodowska. It is possible that this was drawn as Oxia (Fig. 5) or as the unnamed canal immediately west of Oxia. Likewise striking is the crater chain which proceeds northeast from Argyre Planitia and includes the craters Galle, Wirtz, Helmholtz, and Lohse. Lohse may just possibly be a feature detectable from the Earth. It is apparently shown by Slipher as the conjunction of seven canals (Fig. 5). The canal proceeding through these craters (Fig. 5) is unnamed. The craters Lowell (52° S, 82° W) and Schiaparelli (2° S, 343° W) do not seem to be parts of any canal system, a deep but inadvertent irony visited on them by the International Astronomical Union's Committee on Martian Nomenclature. As in the case of other Martian craters observed directly by Mariner 9 (Sagan *et al.*, 1972) the floors of these craters may darken seasonally, due to the removal of windblown dust, and both the existence and seasonal variations of some canals can be understood in this way. A similar remark applies to linear topographic features discussed in the paragraph above.

(4) Only a handful of cases of crater rays were uncovered by Mariner 9. All cases are of such small contrast and angular extent as not possibly to be detectable from the Earth. On a planet with extensive windblown dust, impact crater rays should have very short lifetimes, and should not be expected to contribute to features visible from the Earth (cf. Sagan and Pollack, 1966).

(5) Dune fields have been discovered on Mars (Cutts and Smith, 1973; Sagan *et al.*, 1972). But there are no cases of individual dunes of sufficiently extensive size to be seen from the Earth. Also, in all cases where linear dunes are found, they are found collectively, which does not correspond to the classical canal description.

(6) A few Martian "canals," notably Cerberus and perhaps Thoth-Nepenthes, are real surface albedo features which are not strongly connected to topography. They may represent locales where, for meteorological reasons, fine bright dust is unlikely to be deposited; or where, for reason of geochemistry or surface roughness, there is a preferential abundance of lower albedo material. The longest dark-crater-associated streak uncovered by Mariner 9 is in the Gaea region of Mars, and is seen in Fig. 3 extending at right angles to the nearest canals. It is not impossible that some canals reported in the past are connected with time-variable features of this sort.

We conclude that while a small subset of the classical Lowellian canals corresponds to topographic or albedo features on Mars, the bulk of the canals do not. Indeed there are many canals where there are no real surface features, and there are many real surface features where there are no canals. Although we have not pursued the relevant statistical study, we have the impression that there exists an anticorrelation between the cartographic accuracy of a map and the number of canals it displays. The vast majority of the canals appear to be largely self-generated by the visual observers of the canal school, and stand as monuments to the imprecision of the human eye–brain–hand system under difficult observing conditions.

Acknowledgments

This research was supported by the Planetology Programs Office, and the Exobiology Programs Office, NASA Headquarters, under grants

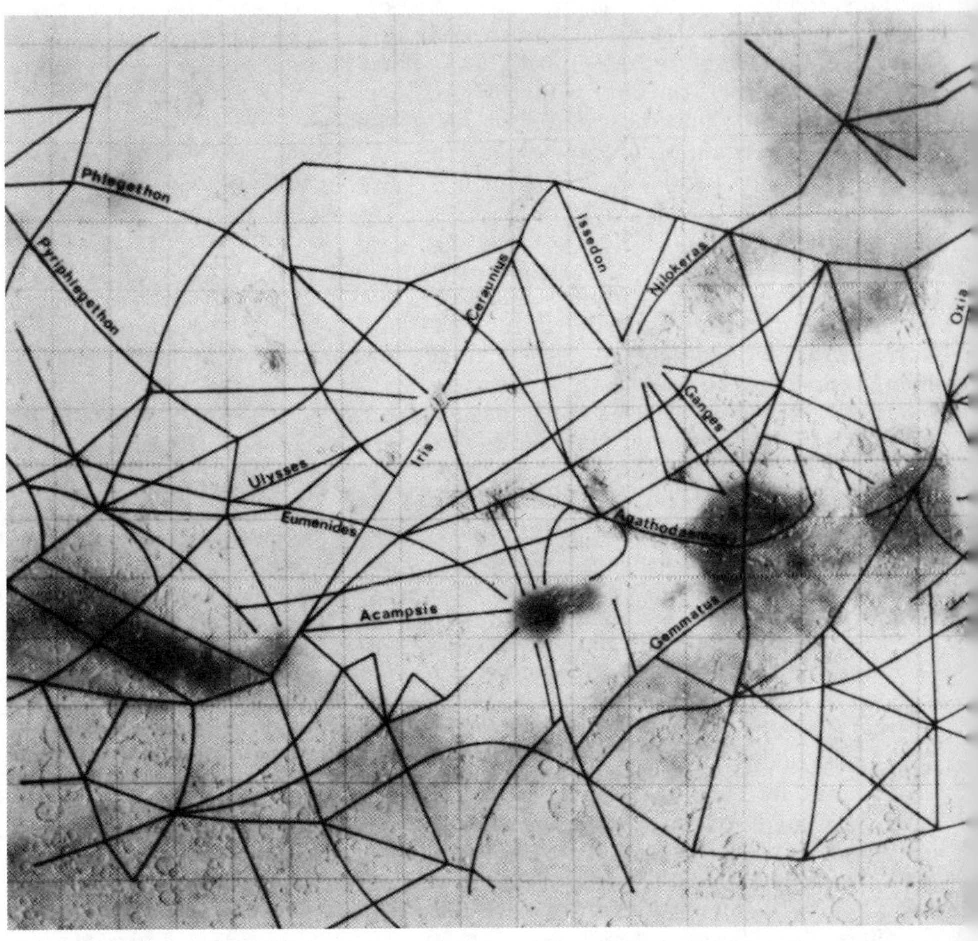

FIG. 5. The Lowellian canal network superposed c

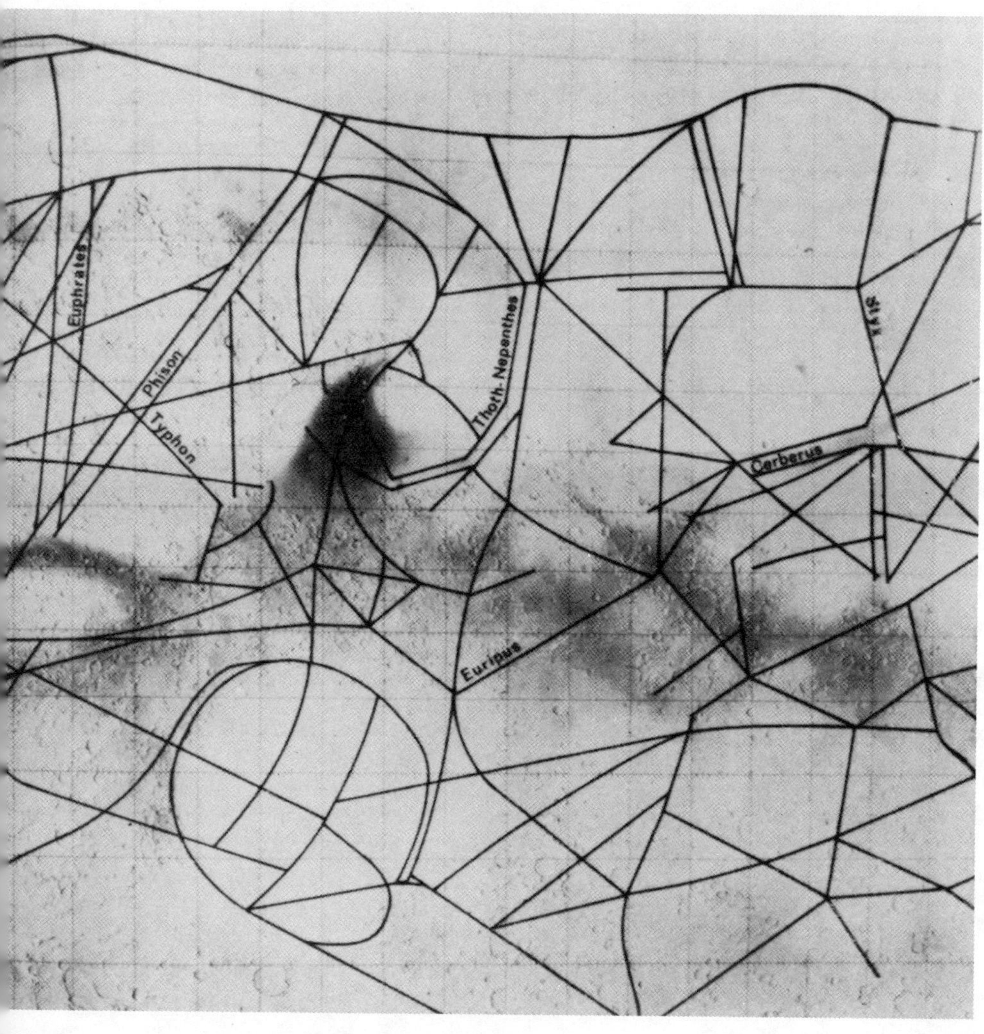

ner 9 Mars cartography and Earth-based albedo.

NGR 33-010-220 and NGR 33-010-101. We are grateful to W. A. Baum and J. Veverka for helpful comments.

References

ANTONIADI, E.-M. (1930). *La Planète Mars*. Hermann, Paris.

ARRHENIUS, S. (1918). *Destinies of the Stars*. Putnam, New York.

CUTTS, J. A., AND SMITH, R. S. U. (1973). Eolian deposits and dunes on Mars. *J. Geophys. Res.* **78**, 4139–4161.

GIFFORD, F. A. (1964). The Martian canals according to a purely aeolian hypothesis. *Icarus* **3**, 130–135.

HOPE, E. R. (1966). Transl. and comments on Katterfel'd, G. N. (1959). The question of the tectonic origin of linear formations on Mars. Defense Res. Board, Canada, Tech. Transl. T 445 R.

JOLY, J. (1898). On the origin of the canals of Mars. *Sci. Trans. Roy. Dublin Soc., Ser. II* **6**, 249–268.

KATTERFEL'D, G. N. (1959). The question of the tectonic origin of linear formations on Mars. *Izv. Vses. Geogr. Obschch.* **91**, 272.

LOWELL, P. (1906). *Mars and Its Canals*. Macmillan, London.

MAUNDER, E. W. (1894). The canals of Mars. *Knowledge* **17**, 249–252.

ONCLEY, P. B., AND FULMER, C. (1966). The Martian "canali" as meteoritic crater rays. *Trans. Amer. Geophys. Union* **47**, 482.

PICKERING, E. C. (1921). *Mars*. Badger, Boston.

PLASSMANN, J. (1901). *Ist Mars ein bewohnter Planet?* Frankfurt.

SAGAN, C. (1966). High-resolution planetary photography and the detection of life. Proc. Caltech–JPL Lunar and Planet. Colloq. (H. Brown *et al.*, Eds), JPL Tech. Memo. 33-266, 279–287.

SAGAN, C., AND POLLACK, J. B. (1966). On the nature of the canals of Mars. *Nature* **212**, 117–121.

SAGAN, C., POLLACK, J. B., AND GOLDSTEIN, R. M. (1967). Radar doppler spectroscopy of Mars. I. Elevation differences between bright and dark areas. *Astron. J.* **72**, 20–34.

SAGAN, C., AND WALLACE, D. (1971). A search for life on Earth at 100 meter resolution. *Icarus* **15**, 515–554.

SAGAN, C. VEVERKA, J., FOX, P., DUBISCH, R., LEDERBERG, J., LEVINTHAL, E., QUAM, L., TUCKER, R., POLLACK, J. G., AND SMITH, B. A. (1972). Variable features on Mars: Preliminary Mariner 9 television results. *Icarus* **17**, 346–372.

SAGAN, C., VEVERKA, J., FOX, P., DUBISCH, R., FRENCH, R., GIERASCH, P., QUAM, L., LEDERBERG, J., LEVINTHAL, E., TUCKER, R., EROSS, B., AND POLLACK, J. B. (1973). Variable features on Mars, 2. Mariner 9 global results. *J. Geophys. Res.* **78**, 4163–4196.

SCHIAPARELLI, G. (1893). The planet Mars. *Natura ed Arte*. Transl. by E. C. Pickering (1894). *Astron. Astrophys.* **13**, 635–640, 714–723.

SLIPHER, E. C. (1962). *Mars*. Sky Publishing, Cambridge, Mass.

TOMBAUGH, C. W. (1950). Geological interpretation of the markings on Mars. *Astron. J.* **55**, 184.

WALLACE, A. R. (1907). *Is Mars Habitable?* Macmillan, London.

WASIUTYŃSKI, J. (1946). Studies in hydrodynamics and structure of stars and planets. *Astrophys. Norvegica* **4**, 241–244.

Part XIII

BIOGEOCHEMISTRY

Editor's Comments on Papers 34 and 35

34 KLEIN
The Viking Biological Investigation: General Aspects

35 OYAMA, BERDAHL, and CARLE
Preliminary Findings of the Viking Gas Exchange Experiment and a Model for Martian Surface Chemistry

Three of the instruments on board the Viking Landers were designed to establish whether or not life exists on Mars. To the surprise of most scientists, all three experiments yielded positive results, thereby indicating at the very least, a peculiar chemical reactivity of the soil, with the ability to mimic Earth-like biological responses. During the months that followed the Viking landings, an intense debate ensued, as to whether the observed reactions were produced by organisms or chemical oxidants ("superoxides") in the martian soil. While at present, the consensus of opinion favors an inorganic chemical explanation, only the results of the labelled release experiment are still compatible with biology. But H. P. Klein, in summarizing the Viking biology experiments (Paper 34) believes the remaining ambiguities may remain unresolved.

Results from the pyrolytic release (carbon assimilation) experiment are the most difficult to interpret. The reproducibility of the experiment could be questioned because of gas leaks and variable incubation temperatures (Horowitz et al. 1976, 1977). While suppression of the PR response at the Utopia site and especially the "under-the-rock" sample was initially attributed to higher levels of soil moisture, which presumably prevented the reactions (Horowitz et al. 1976), additional experiments with moistened samples indicate that "water vapor does *not* inhibit the fixation reaction" (Horowitz et al. 1977). Now, the reduced response of the "under-the-rock sample" is attributed to protection from UV radiation.

The rates of release of O_2 and CO_2 in the gas exchange experiment suggest a chemical, rather than biological reaction (Oyama et al. 1977a). As anticipated from the superoxide theory, the level of O_2 release from the "under-the-rock" sample in the GEX experiment was significantly lower than for the surface soils. On the other hand, the evolution of CO_2 in the labelled release experiment was about the same for both surface and subsurface specimens. These results are puzzling. Photochemical reactions would not be expected to occur beneath the soil, yet the LR experiment apparently remains unaffected by the depth of the soil samples. Furthermore, the LR response is suppressed by sterilization at 160°C and is substantially reduced by "cold sterilization" at 50°C (Levin and Straat 1977). On the other hand, the kinetics of the reaction and consumption of only 10 percent of the nutrients cast doubt on biological interpretation.

V. I. Oyama, B. J. Berdahl, and G. C. Carle (Paper 35) have developed an inorganic explanation for the martian soil chemistry. Photochemical reactions, catalyzed by γ-Fe_2O_3 (maghamite) produce metal superoxides or peroxides, which can release O_2 upon addition of water as in the GEX experiment, or can oxidize organic nutrients, therby evolving CO_2, as in the labelled release experiments. Laboratory experiments have attempted to simulate the martian environment (Ponnamperuma et al. 1977). However, the laboratory reactions were conducted in UV light from 160 to 180 nm, which wavelengths had been filtered out in the actual Viking experiments, in order to prevent synthesis of organic matter from CO. Furthermore, Na_2O_2 and KO_2 were used as reagents, which are unlikely to exist on Mars, given the low Na and K content of the soil (Clark et al. 1977). Thus, further experiments will be required to provide a more realistic simulation of martian soil chemistry.

The failure of the gas chromatograph-mass spectrometer to detect organic matter, while not intended to be a direct test for the presence of life, places severe constraints on its existence (see Biemann et al. 1977). Although the gas exchange experiment and labelled release experiment results can be understood in terms of the presence of strong soil oxidants, the pyrolytic release experiment requires additional assumptions (Klein 1978). In fact, a major weakness of the nonbiological models is that at least three different oxidants are required in the martian soil in order to explain the observed reactions (Klein, 1978).

34

Reprinted from *J. Geophys. Res.* **82**:4677–4680 (1977)

The Viking Biological Investigation: General Aspects

H. P. Klein

NASA Ames Research Center, Moffett Field, California 94035

The Viking biological investigation has tested four different hypotheses regarding the possible nature of Martian organisms. While significant results were obtained for each of these, tests of three of the hypotheses appear to indicate the absence of biology in the samples used, while the fourth is consistent with a biological interpretation. The original assumptions for each experiment and the experimental procedures that were utilized to test these assumptions are reviewed.

Introduction

Before the two Vikings reached Mars, speculation about the prospects for life on that planet ranged from extreme pessimism to optimism. At that time, when the biology experiments were actually selected for final inclusion on Viking, our information about Mars was far from complete. Under these circumstances it is not surprising that different ideas emerged concerning what Martian organisms might be like and what procedures or techniques would best elicit evidence of their metabolism. As a result of these diverse ideas it was decided to incorporate several different biological experiments into the Viking payload in order to test a number of different, and sometimes conflicting, assumptions about the characteristics of Martian organisms [*Young*, 1976]. Indeed, had it been possible to fly additional biological experiments and thereby extend the number of different ideas and techniques that could be tested, this would have been a valuable thing to do. However, the biological portion of the payload was constrained to approximately 0.03 m^3 of volume and 15 kg of weight [*Klein*, 1976].

By supplying four different sets of environmental conditions within which to conduct incubations (Table 1), the three experiments that were finally selected for the Viking biology instrument actually tested four different hypotheses. One of these assumed that active Martian metabolism was limited by the availability of water. Another assumed that biological activity would best be seen under conditions approximating those on Mars. The remaining two tested for heterotrophic metabolism, one by using a very dilute aqueous solution of simple organic compounds, the other by utilizing a concentrated mixture of many organic compounds. All of these have now been tested, and the results from these are the subjects of the ensuing papers.

As of this writing the two Viking landers have been operating on the surface of Mars for 8½ and 7 months, respectively. All during this time the Viking biology instrument has performed exceptionally well, with few instances of instrument anomalies, and those that have occurred are sufficiently well understood so that they do not interfere with the interpretations of the data. Accordingly, for the papers that follow, one can safely assume that all of the information is based on statistically significant data with strong 'signal-to-noise' ratios and that we are not engaged here in describing or explaining artifacts produced by the instrumentation.

Initially, in the so-called 'nominal' mission we expected to perform a total of 13 or 14 separate experiments on the two spacecraft. With no a priori knowledge of the local Martian surface environments our overall initial strategy was essentially based on the concept that there might be local heterogeneity in the surface of Mars in the vicinity of the spacecraft. We planned therefore to perform the biological experiments in a 'survey' mode, testing sample after sample by using the same experimental sequences until any one of the experiments yielded a 'presumptive' positive result. (For a discussion of the criteria that were to be used to arrive at this judgment, see *Hubbard* [1976], *Levin and Straat* [1976*a*], and *Oyama et al.* [1976].) Once such a presumptive positive result was observed, our strategy called for repeating that particular experiment by using a heat-'sterilized' sample. This procedure, of using heated 'controls' to confirm biological processes, is deeply rooted in biological tradition. More important, it was based on hundreds of ground-based tests of the Viking concepts using terrestrial soil samples (from desert soils to rich garden soils), lunar samples, and pure cultures of microorganisms, the results of which enabled us accurately to determine the presence or absence of living organisms.

In point of fact, our intended nominal strategy was quickly discarded as the very first sample was analyzed. Two of the three experiments yielded presumptive positive results, while the third produced evidence for oxidizing surface material at that site [*Klein et al.*, 1976]. This combination of findings, together with the demonstrated lack of organic compounds [*Biemann et al.*, 1976], required substantial changes to the original experimental strategies and resulted in major departures from experiment sequences that we had anticipated using in these experiments. In all cases every element of flexibility inherent in the instruments was called into play in an effort to use the equipment both as biological and as chemical laboratories in order to discriminate between the two mechanisms that might be responsible for the presumptive positive results.

By now, considerable progress has been made in resolving the issues raised by the first set of analyses. To achieve this, we have exceeded the number of experiments initially planned, having carried out 26 experiments. A few additional experiments still remain to be performed, but on the basis of all the information now available, it is likely that the results of one of the two experiments that yielded presumptive positives, the Pyrolytic Release experiment [*Horowitz et al.*, 1976], are nonbiological in origin. The Labeled Release experiment, which also has consistently yielded presumptive positive biological results [*Levin and Straat*, 1976*b*], remains ambiguous.

In arriving at any final judgment on the fundamental question, Is there life on Mars?, we must carefully consider not only the actual experimental results that have been obtained but also the context within which these data were obtained. We must examine the assumptions on which each of the experimental techniques was based, the conditions under which the

TABLE 1. Environmental Parameters in the Viking Biology Investigation

Experiment	Nutrients Added	Water Added	Illumination
Pyrolytic Release	none*	none	light and dark
	none	trace§	light and dark
Gas Exchange	none	moist‖	dark
Labeled Release	dilute solution of simple organic compounds†	moist¶	dark
Gas Exchange	concentrated solution of organic and inorganic compounds‡	wet**	dark

*A mixture of CO_2 and CO was introduced into the incubation chamber.
†See *Levin and Straat* [1976*b*].
‡See *Oyama* [1972].
§Approximately 80 μg of H_2O was injected into an incubation chamber (2.6-cc volume) containing a 0.25-cc sample.
‖Approximately 0.5 cc of nutrient solution was added below the 1-cc sample in the chamber (8.7-cc volume).
¶Approximately 0.115 cc of nutrient solution was added to the 0.5-cc sample in the chamber (3.25-cc volume).
**Approximately 2.5 cc of nutrient solution was added to the 1-cc sample in the chamber (8.7-cc volume).

experiments were actually carried out, and the data themselves.

THE GAS EXCHANGE EXPERIMENT (HUMID NONNUTRIENT MODE)

As is indicated in Table 1, the Gas Exchange (GEX) experiment tested two different concepts of Martian biology. In one mode the fundamental assumption was that the sole limiting factor to growth of Martian organisms is water. Here it was assumed that nutrients, perhaps in the form of simple organic compounds formed photochemically as is described by *Hubbard et al.* [1971], already are present in the Martian surface and that organisms would be dormant in the dry Martian environment until enough moisture became available to stimulate the dormant organisms into metabolic activity, which was to be measured by analyzing the atmosphere above the incubating system with a gas chromatograph system.

Now let us examine some of the experimental conditions under which this assumption was actually tested. First, the Martian samples were incubated in the presence of Martian atmosphere to which additional carbon dioxide, krypton, and helium were added in order to bring the total pressure to approximately 200 mbar to facilitate subsequent gas sampling [*Edelson et al.*, 1975]. After introduction of approximately 0.5 cc of nutrient solution into the incubation cell, under conditions in which the nutrient does not come into contact with the samples, the atmosphere rapidly becomes saturated with water at the incubation temperatures of 8–15°C. After an incubation period for this phase of the experiment of approximately 7 days the experiment was terminated.

In this experiment, which was performed twice, once at each of the landing sites, the findings for both were essentially the same. While physical (desorption of some gases) and chemical (generation of oxygen) phenomena were noted, there was nothing in the data to suggest the presence of metabolic activity on the basis of the criteria that had been developed for this experiment, and therefore the results of this experiment can be said to be negative with regard to biology.

In terms of interpreting the results obtained, if we regard both the assumptions and the experimental conditions to be valid, we must conclude that the samples that were assayed did not contain metabolizing organisms. However, the original assumption may be incorrect in that some source of energy may be a requirement to stimulate metabolic activity of organisms on Mars. In this case a negative result would not preclude the existence of 'life' in the samples tested. Alternatively, it is possible that one or more of the experimental conditions employed during these tests prevented the accumulation of biological signals. For example, in this experiment, as in all of the biological experiments, incubation temperatures ran some tens of degrees warmer than ambient surface temperatures at the two sites [*Kieffer*, 1976]. Another issue is whether a total incubation period of 7 days was sufficient to demonstrate metabolic activity in view of tests with Antarctic soils that required months of incubation to produce presumptive positive results [*Oyama et al.*, 1976]. Other potential sources of possible inhibition of metabolism include the high pressure and alteration of the incubation atmosphere, as indicated above.

THE GAS EXCHANGE EXPERIMENT (WET NUTRIENT MODE)

In running the GEX experiment, in the presence of added organic compounds and inorganic salts the fundamental assumption made was that a significant fraction of the Martian 'biota' is composed of heterotrophic organisms [*Oyama et al.*, 1976]. Therefore the addition of organic compounds was deemed necessary to elicit metabolic response. Furthermore, it was assumed that this response would be expressed only in an aqueous environment, and finally, the presence of a large number of different organic and inorganic compounds [*Oyama*, 1972] was assumed not to be inhibitory to the expression of this metabolism.

This experiment was performed three times, for periods of 200 (Viking 1), 31 (Viking 2), and 116 (Viking 2) sols. However, Martian atmosphere was present for only a portion of these incubation periods, 13, 19, and 78 sols, respectively. For the remainder of the time the atmosphere consisted of carbon dioxide, krypton, and helium. Once again the total atmospheric pressure was approximately 200 mbar, and incubation temperatures were in the 8–15°C range.

While some gas changes were noted in the three trials of this experiment [*Oyama*, 1977], none of these fit the criteria for biological activity. Consequently, on the basis of the original assumptions for this experiment and provided that the conditions under which they were tested were adequate, we can conclude that no viable organisms were present in the samples. On the other hand, a negative finding for this experiment does not rule out the possible presence of autotrophs (i.e., chemosynthetic organisms) in the samples. In addition, the corollary assumptions concerning the nutrient mixture used, as well as the high water activity under which these tests were conducted, must also be assessed in arriving at the biological significance of the data from this experiment. Finally, even if the original assumptions are correct, some of the experimental conditions (temperature, pressure, and 'artifical' atmosphere) may have precluded positive biological findings in this experiment.

THE PYROLYTIC RELEASE EXPERIMENT

The fundamental assumption for this experiment stems from considerations of the characteristics of the planet Mars.

Since both carbon dioxide and carbon monoxide were known to be present in the atmosphere of Mars, it was assumed that organisms in the local ecology of that planet would have developed the capacity to assimilate one or both of these gases [*Horowitz et al.*, 1972] and convert these to organic matter. A basic tenet of this experiment is that metabolic activity would best be demonstrated under conditions approximating the ambient conditions on Mars as closely as possible.

The actual Pyrolytic Release (PR) experiments were conducted under conditions which, in many aspects, did approximate those on Mars, but in some they did not. Incubations were carried out either in the light or in the dark for 5-day periods. In those cases in which illumination was used, wavelengths below about 320 nm were filtered out in order to avoid false positives [*Hubbard*, 1976]. The incubation temperatures again were in the 10–18°C range.

As has been reported [*Horowitz et al.*, 1976], initial results of the PR experiment, as well as later attempts to repeat the original conditions, resulted in weak but significant presumptive positives. However, later experiments designed to elucidate the mechanisms yielding these results appear to rule out a biological explanation for these results.

As with the GEX experiments, it is possible that in the case of the PR experiment the basic assumptions are not correct. The photochemical synthesis of simple organic compounds has not been ruled out on the surface of Mars. If in the steady state this process were to supply organic matter to the surface, the need for autotrophic fixation in the Martian ecology may be obviated, and only heterotrophs may be present. In this case a negative result in this experiment would not preclude the existence of life in the samples tested.

In examining the experimental conditions under which the PR tests were conducted, the duration of the incubation periods, the incubation temperatures, and the fact that short-wavelength solar radiation was unavailable to the surface samples all raise questions regarding the adequacy of these experiments in providing the requisite conditions for measuring this type of metabolism even if Mars should have indigenous autotrophic organisms.

THE LABELED RELEASE EXPERIMENT

This experiment [*Levin*, 1972] makes the assumption that heterotrophic organisms are present on Mars and that these organisms would be capable of decomposing one or more simple organic compounds of the type reported to be produced from the so-called 'primitive reducing atmosphere' in laboratory simulations [*Miller and Urey*, 1959] and including some that have been found in carbonaceous chondrites [*Kvenvolden et al.*, 1970]. For this experiment, even more than for the others, the inclusion of a heated control was deemed absolutely necessary in order to avoid the possibility of obtaining false positive results. The conditions under which samples that were not heat sterilized were incubated, now four times, for periods of about 13, 52, and 90 days, were Marslike with the exception that incubation was carried out at around 10°C; and of course, there was the addition of a small volume of water containing the dilute solution of organic substrates [*Levin*, 1972]. As with the GEX experiment, the incubation cell was somewhat pressurized (to approximately 60 mbar) during incubation. Heat sterilization was carried out on three samples, first at approximately 160°C, and later at 50°C and 44°C.

What has been observed in all of the analyses performed on nonsterilized samples is an active initial decomposition of the nutrient, the reaction leveling off, in each case, at a time when

TABLE 2. Comparison of Data From the LR and GEX Experiments

Sample	Oxygen Released (GEX)*	Carbon Dioxide Produced (LR)*
Viking 1 (surface)	770	~30
Viking 2 (surface)	194	~30
Viking 2 (subrock)	70	~30

*Nanomoles per 1-cc sample.

about 95% of the added radioactivity still remained in the surface material. Prior heating of the Martian samples had substantial effects on this process, 160°C for 3 hours completely abolishing the reaction [*Levin and Straat*, 1976*b*].

On the basis of all of the experiments performed to date, the Labeled Release (LR) experiment, unlike the other biological experiments, yielded data which met the criteria originally developed for a positive. On this basis alone the conclusion would have to be drawn that metabolizing organisms were indeed present in all samples tested. Can we believe such a conclusion? Clearly, we must be wary of this in the face of information indicating that all of the samples tested yielded oxygen in the GEX experiment upon introduction of water. The evidence for strongly oxidizing chemicals in these samples is quite convincing. However, it should also be recognized that the two phenomena are not directly related (Table 2). From the relatively constant amount of decomposition that took place in all cases in the LR experiment, it would appear that all of the Mars samples contained excess oxidants. Thus it is reasonable to assume either that the factor limiting the LR reaction in each case was the depletion of some constituent in the mixture of substrates supplied in the nutrient or that the samples analyzed in these experiments contained at least two kinds of oxidants.

While a nonbiological (i.e., a chemical 'oxidant') theory may well explain the LR data, it does not seem likely that the ambiguity in interpreting this experiment will be resolved on Mars by the remaining Viking experiments.

SUMMARY

We have tested a number of different concepts about the nature of hypothetical Martian organisms over the course of the past few months. We have seen significant and quite reproducible results in each of our experiments. For each experiment, except for the LR experiment, we must conclude that there were no organisms present within the limits of detectability for these experiments and that all of the observed reactions for these were the result of nonbiological phenomena. Otherwise, we must question the fundamental assumptions made for these experiments and the conditions under which they were actually carried out. In the case of the LR experiment, from the very beginning of operations on Mars we recognized the possibility that the striking changes seen during incubation could be the result of nonbiological processes, but our attempts to discriminate between biological and nonbiological mechanisms by manipulating sterilization temperatures and the length of incubation have not resolved the issue.

Finally, we must not overlook the fact, in assessing the probabilities of life on Mars, that all of our experiments were conducted under conditions that deviated to varying extents from ambient Martian conditions, and while we have accumulated data, these and their underlying mechanisms may all be coincidental and not directly relevant to the issue of life on that planet.

REFERENCES

Biemann, K., J. Oro, P. Toulmin III, L. E. Orgel, A. O. Nier, D. M. Anderson, and P. G. Simmonds, Search for organic and volatile inorganic compounds in two surface samples from the Chryse Planitia region of Mars, *Science, 194,* 72–76, 1976.

Edelson, H. E., F. S. Brown, O. W. Clausen, A. J. Cole, J. T. Cragin, R. J. Day, C. H. Debenham, R. E. Fortney, R. I. Gilje, D. W. Harvey, F. A. Jackson, J. A. Katherler, J. L. Kropp, S. J. Loer, J. L. Logan, Jr., O. D. Minnick, E. M. Noneman, W. D. Potter, G. T. Rosiak, and J. S. Shapiro, The Viking lander biology instrument, *Rep. 21020-6003-RU-00,* TRW Syst. Group, Redondo Beach, Calif., 1975.

Horowitz, N. H., J. S. Hubbard, and G. L. Hobby, The carbon-assimilation experiment: The Viking Mars lander, *Icarus, 16,* 147–152, 1972.

Horowitz, N. H., G. L. Hobby, and J. S. Hubbard, The Viking carbon-assimilation experiments: Interim report, *Science, 194,* 1321–1322, 1976.

Hubbard, J. S., The pyrolytic release experiment: Measurement of carbon-assimilation, *Origins of Life, 7,* 281–292, 1976.

Hubbard, J. S., J. P. Hardy, and N. H. Horowitz, Photocatalytic production of organic compounds from CO and H_2O in a simulated Martian atmosphere, *Proc. Nat. Acad. Sci. U.S., 68,* 574–578, 1971.

Kieffer, H., Soil and surface temperatures at the Viking landing sites, *Science, 194,* 1344–1346, 1976.

Klein, H. P., General constraints on the Viking biology investigation, *Origins of Life, 7,* 273–279, 1976.

Klein, H. P., N. H. Horowitz, G. V. Levin, V. I. Oyama, J. Lederberg, A. Rich, J. S. Hubbard, G. L. Hobby, P. A. Straat, B. J. Berdahl, G. C. Carle, F. S. Brown, and R. D. Johnson, The Viking biological investigation: Preliminary results, *Science, 194,* 99–105, 1976.

Kvenvolden, K., J. Lawless, K. Pering, E. Peterson, J. Flores, C. A. Ponnamperuma, I. R. Kaplan, and C. Moore, Evidence for extraterrestrial amino-acids and hydrocarbons in the Murchison meteorite, *Nature, 228,* 923–926, 1970.

Levin, G. V., Detection of metabolically produced labeled gas: The Viking Mars lander, *Icarus, 16,* 153–166, 1972.

Levin, G. V., and P. A. Straat, Labeled release—An experiment in radiorespirometry, *Origins of Life, 7,* 293–311, 1976*a*.

Levin, G. V., and P. A. Straat, Viking labeled release biology experiment: Interim results, *Science, 194,* 1322–1329, 1976*b*.

Miller, S. L., and H. C. Urey, Organic compound synthesis on the primitive earth, *Science, 130,* 245–251, 1959.

Oyama, V. I., The gas exchange experiment for life detection: The Viking Mars lander, *Icarus, 16,* 167–184, 1972.

Oyama, V. I., Preliminary findings of the Viking gas exchange experiment in a model for Martian surface chemistry, *Nature, 265*(5590), 110–114, 1977.

Oyama, V. I., B. J. Berdahl, G. C. Carle, M. E. Lehwalt, and H. S. Ginoza, The search for life on Mars: Viking 1976 gas changes as indicators of biological activity, *Origins of Life, 7,* 313–333, 1976.

Young, R. S., The origin and evolution of the Viking mission to Mars, *Origins of Life, 7,* 271–272, 1976.

Reprinted from *Nature* **265**:110–114 (1977)

Preliminary findings of the Viking gas exchange experiment and a model for Martian surface chemistry

V. I. Oyama, B. J. Berdahl & G. C. Carle

National Aeronautics and Space Administration, Ames Research Center, Planetary Exploration Office, Moffett Field, CA 94035

Oxygen and CO_2 were evolved from humidified Martian soil in the gas exchange experiment on Viking Lander 1. Small changes in N_2 gas were recorded. A model of the morphology and a hypothesis of the mechanistics of the Martian surface are proposed.

THE early results from the gas exchange experiment (GEX), on Viking Lander 1 have already been reported but were limited to preliminary findings only[1]. This paper examines the findings and interprets them in terms of a model of the soil surface morphology and chemistry.

All major GEX activities that occurred after landing up to the end of the first incubation cycle are given in Table 1. In Table 2 are listed the sample processing and transport environments to which the soil was subjected up to the point when the first analysis was made; they may have relevance to the extent and nature of the gaseous responses.

Data on instrument performance will be reported at a later date. Descriptions of the concepts governing the design of the experiment, modes of operation and results obtained are described elsewhere[2–7].

Humid mode results

The first incubation cycle began with the addition of 1 cm^3 Martian soil to the 8.7 cm^3 test cell (0.465 $cm^3 \pm 0.035$ cm^3 solid volume, estimation based upon solid density range of 2.6–3.0 $g\,cm^{-3}$ for the soil in the test cell and a bulk density of 1.3 $g\,cm^{-3}$) (Viking Physical Properties Team). In the process of loading the soil and sealing the test cell, Martian atmosphere was trapped within the chamber at the prevailing pressure of 7.6 mbar (ref 8). A 0.132 cm^3 volume of starting gas at 8.394 atm consisting of 5.51% Kr, 2.84% CO_2, and 91.65% He and 0.567 cm^3 of M4 nutrient[3] were added to the lower cavity of the test cell, an amount insufficient to wet the soil but providing 100% relative humidity. The final pressure in the test cell after the above additions is estimated to be 150 mbar. Results of the first and subsequent gas chromatographic analyses of the headspace gases during the humid mode are shown in Table 3. After analyses of the headspace, the total gases in the test cell were estimated from Ostwald coefficients for aqueous solutions[9–14] to determine the contribution of the gases dissolved in the liquid phase. The results were corrected for the initial contributions of the original trapped Martian atmosphere[15], the starting Kr/CO_2/He gases and Ne introduced by the nutrient injection and losses from sampling the headspace gas. Solution coefficients were used for CO_2 to correct for the estimated basicity of the soil (Table 3). This technique was used only for CO_2 to approximate the magnitude of the CO_2 change.

The results show the CO_2, O_2, N_2 and Ar and/or CO (the Ar and CO are not resolved by this gas chromatograph) were evolved by the soil sample. The maximum amount of N_2, 83 nmol, appeared on Sol 11 (Sol is a Martian day, equal to 24 h, 39 min and 35 s) and decreased to 77 nmol by Sol 15. Oxygen, on the other hand, after reaching its maximum on Sol 11, appears to have reached a plateau. From the assumed total oxidation of ascorbate in the M4 nutrient[3], the net amount of O_2 produced equals 775 nmol (690 plus the 85 associated with the oxidation of ascorbate). The maximum amount of CO_2

Table 1 Major events in the first cycle of the gas exchange experiment

Event	Date (Sol)*	Elapsed time from landing (s)
Landing	0	0
Initiate chromatographic column rejuvenation	2	206,156
Soil sample added to test cell	8	682,426
Seal test cell	8	683,439
Add incubation gas, Kr/CO_2/He	9	759,219
Inject 0.57 ml nutrient (humid mode)	9	759,879
First headspace gas analysis (humid)	9	769,876
Second analysis	10	860,176
Third analysis	11	948,916
Fourth analysis	13	1,126,756
Fifth analysis	15	1,302,556
Inject 2.3 ml nutrient (wet mode)	16	1,383,339
First headspace gas analysis (wet)	16	1,392,916
Second analysis	17	1,481,656
Third analysis	18	1,570,456
Fourth analysis	20	1,744,456
Fifth analysis	25	2,189,320
Sixth analysis	28	2,446,105

The column rejuvenation cycle is the process by which the gas chromatograph columns, which have been opened to the vacuum of space, are cured of the need to be primed for oxygen in order to make quantitative measurements of oxygen in any given sample. The test cell had two positions on the carousel: the first to receive sample and the second to be sealed against the head-end. The latter contained all the attachments for adding gases and nutrient and for removing sample from the headspace. The soil was contained in a porous cup suspended over a sump that allows drainage, when required, and variable additions to either humidify or wet the soil. Test cell volume empty was 8.7 cm^3 and the sample loop volume was 0.1114 cm^3. The amount of gas lost by sampling varied as a function of the total pressure in the test cell, the relative sample loop and headspace volumes and the gas distributions which were a function of temperature and gaseous species.

*Sol (the Martian day) is 24 h, 39 min, 35 s long.

Table 2 Soil sample history

Location	Temperature (K)	Time duration (h)
Soil surface (0–5 cm deep)	190–208	N.A.
Collection head (collect then dump)	204–206	0.09
Arrival at Biology PDA (soil processor) through shuttle (dump from PDA to SDA)	239–280	0.26
Arrival at Biology SDA (soil distribution) through soil dump to test cell	280–282	0.05
Arrival of soil in test cell through seal of test cell	282	0.28
Seal of test cell through humidification (addition of 0.57 ml nutrient)	282	21.2
Humidification through first analysis	282	2.78

Soil temperature estimated, all other temperatures directly measured.

attained occured on Sol 13 and was approximately, 9,800 nmol. This decreased on Sol 15 to 9,400 nanomoles. No conclusion on the presence or change of CO can be made because of the low values of the combined Ar/CO peak. The values in Tables 3 and 4 of Ne and Kr demonstrate the consitency of the internal standards and the apparent precision for the gas analyses.

Wet mode results

On Sol 16, additional (estimated volume = 2.268 cm^3) M4 nutrient was injected. With the amount added earlier, the nutrient, now measuring 2.835 cm^3, wetted the soil. The results for the wet mode are shown in Table 4. The level of N_2 monotonically rose for an increase of 11 nmol whereas in the humid mode we had seen an increase followed by a decrease in the N_2 content.

Oxygen monotonically decreased from the plateau that was attained at the end of the humid mode and reached a minimum. The total amount of O_2 consumed equalised 325 nmol which is 78% of the ascorbate-oxygen equivalence added.

With water now making contact with the soil, the gaseous content of the interstitial water could be estimated. For the 1 cm^3 soil amount, an estimate of 5–50 μm particle sizes (personal communication from H. J. Moore, Viking Physical Properties Team) yielded an interstitial soil–water volume of about 35% (ref. 16). This value was used to estimate the total CO_2 in the solution contacting the soil, as separated from the CO_2 in the larger volume of water in the test cell sump that was treated separately as if it were unaffected by the soil. The interstitial soil solution and solution in the sump compartment shared a common headspace atmosphere from which the CO_2 was measured. The calculations show that CO_2 estimates are only meaningful to the extent that they represent a minimum.

During the entire first cycle, no H_2, NO or CH_4 was observed in the headspace.

Model for Martian surface chemistry

From very preliminary laboratory data (personal communication from Dr Edward Merek, Ames Research Center) using terrestrial volcanic ash, it was observed that some of the CO_2 adsorbed at Martian surface temperatures, 190–268 K, was desorbed at the higher incubation temperature of the test cell, 281–283 K, during the first 21.23 h that the soil was sealed in the test cell. However, when the same volcanic ash was then humidified, major desorption of the CO_2 occured in 2.78 h. When the amount of CO_2 desorbed was related to the Martian soil, the 9,800 nmol of CO_2 reduced to 0.33 mg CO_2 desorbed per g soil. This value of desorbed CO_2 which is assumed to be adsorbed by Martian soil is within the model proposed by Fanale and Cannon[17] and supports the method we used to estimate total CO_2. The CO_2 production in the wet mode was probably due to oxidation of nutrient organic compounds.

The unexpected amount of O_2 accompanying the desorption of CO_2, 9,800 nmol/775 nmol (13 : 1 CO_2/O_2) from the minimum 270 : 1 CO_2/O_2 ratio in the Martian atmosphere[15] represents an oxygen enrichment of approximately 20 times in the Martian soil. It is improbable that the excess amounts of O_2 found beyond that expected from adsorption of atmospheric O_2 comes from the direct or biological photocatalysis of water or hydroxyl groups, for the reasons that the critical temperature for O_2 is too low for adsorption in the quantities needed and the diffusion rate of O_2 at 7.6 mbar is too rapid to allow for enrichment to occur. A more likely source for the O_2 is chemically bound oxygen which is stable at GEX cell temperatures but allows displacement of oxygen by water. The surface of the grains of rock or mineral are good sources for this chemically bound oxygen. A simple estimate of the surface area of the grains can be made to assess whether all of the oxygen can be accommodated. The surface area of the grains are

Table 3 Gas composition (corrected) in gas exchange test cell (humid mode)

	Time from humidification (h)				
	2.78	27.86	52.51	101.91	150.74
	Mars Date				
	Sol 9	Sol 10	Sol 11	Sol 13	Sol 15
Gas	Quantity of gas* (nmol)				
N_2	73	76	83	79	77
O_2	500	650	690	690	690
CO_2	5,900	8,300	9,500	9,800	9,400
Ar†	8	7	13	9	8
Ne‡	20	19	18	20	21
Kr§	2,000	2,000	2,000	2,000	2,000

*The gas chromatograph detector data were sampled at 1 s intervals, digitised and fitted to a skewed gaussian distribution from which peak heights were obtained. The gas in the headspace was obtained from the ratio of the sample loop volume to the total headspace. The cumulative gas composition was corrected for sampling losses by referencing absolute changes in the krypton values for successive samples. Corrections were made for pressure sensitivity in this flight instrument caused by a partial restriction in the gas sampling system which prevented total evacuation of the sample loop to ambient pressure before filling (three times) from the test cell. The volume for krypton was corrected for pressure as follows:

$$\text{nmol Kr} = 37.77\,(P_c)\exp -0.118\,(V_p)\exp 1.016$$

where P_c is the test cell pressure in mbar, V_p is the peak height in volts. The value for each gas was corrected by the ratio of the term 37.77 (P_c)exp −0.118 to the similar Kr value from a pressure-insensitive instrument. P_c was determined from an inventory of all sources of pressure such as the enclosed atmosphere, the pressure of the initialisation gas injected and any additional injections of helium required to increase cell pressure. The gas composition as stated was corrected by removal of contributions from known sources (for example trace contaminants in injected gases) and for the amount dissolved in the liquid phase. An estimate of dissolved gases was made from literature values for the Ostwald coefficients. The effects of pH on the CO_2 distribution were included by estimating changes in apparent CO_2 levels upon nutrient injections in LR (second injection) and upon the wet-mode nutrient injection in gas exchange. The relationship used was

$$\frac{\text{(nanomoles, dissolved)}}{\text{(nanomoles, gas phase)}} = L\,\frac{\text{(volume, liquid)}}{\text{(volume, gas)}}$$

where the L values for CO_2 are as follows. Sol 9 and 10: $L = 21.4$; Sol 11–15: $L = 28.4$; Sol 16: $L = 40.4$; Sol 17–28: $L = 68.5$. The data shown in this table and in Table 4 are revised from those given in ref. 1 by better values for the atmospheric composition and the composition of the krypton/carbon dioxide gas. The correction, subtracting these contributions from the total, tend to increase the values, but do not alter the conclusions reached.

†Assumed to be argon (argon is not resolved from carbon monoxide on this column).

‡Mean value for neon = 20.02±1.09 (±5.44%).

§Mean value for krypton = 1,975±0 (±0%).

Table 4 Gas composition (corrected) in gas exchange test call (wet mode)

	Time from 2.3 ml nutrient injection (h)					
	2.66	27.31	51.98	100.31	223.88	295.21
	Mars date					
	Sol 16	Sol 17	Sol 18	Sol 20	Sol 25	Sol 28
Gas	Quantity of gas (nmol)					
N_2	76	78	78	79	86	87
O_2	670	600	550	510	460	450
CO_2	9,600	11,000	12,000	13,000	13,000	12,000
Ar	8	8	7	8	9	8
Ne	100	150	160	170	170	180
Kr	2,000	2,000	2,000	2,000	2,000	2,000

See Table 3 for details of test and calculations.

assumed to be 400–2,000 cm^2 per cm^3 of soil (personal communication from H. J. Moore for 5–50-μm particle sizes). Assuming 33.4 $Å^2$ coverage per molecule, a monolayer of O_2 covers 1,560 cm^2, and is accommodated by the available grain surface. The requirement that chemical absorption cannot extend beyond a monomolecular layer[26] is pertinent.

A model of surficial materials which can accommodate both oxygen and CO_2 is solid silicate particles covered with fine iron oxide particles. The carbon dioxide desorbed from a monolayer would be accommodated on 19,710 cm^2 of the iron oxide surface (assuming the 33.4 $Å^2$ for CO_2 also). The iron oxide cover would be approximately three particles deep, under the conservative assumption that each iron oxide particle is a cube that has six times the surface area of the rock surface it covers. The least surface area is obviously a sphere having four cross sectional diameters of surface for each area of the mineral it would cover. If the mean particle size in the atmosphere (2 μm diameter) (ref. 19) is indicative of the iron oxide diameters on the surface, the mean thickness of the three-layer iron oxide is 6 μm. This is thick enough to make the inner surface opaque to the solar ultraviolet, but also too thick to be consistent with $\leqslant 2$ μm thickness reported by the Inorganic Chemical Analysis Team (ICAT) on the limits of iron oxide coating allowed for measurement of underlying silicate surfaces[18]. If the value of the particle sizes were closer to the lower conservative estimate of 0.2 μm (ref. 20), the mean thickness would be compatible with the ICAT limit of detection.

These estimates of surface area are preliminary best estimates and may be revised at a future date. Lunar fines[25] have surface areas near 5×10^3 cm^2 g^{-1} while clays[16] have surface areas in the range of 10^5–10^7 cm^2 g^{-1}. The results of the GEX may ultimately yield an estimate of the surface area of the Martian surface materials.

The mechanical structuring of the oxygen affixed to the inner mineral surface instead of on the outer iron oxide coat is consistent with known chemicals, alkali and alkaline earth metal peroxides and superoxides that are capable of producing both O_2 and H_2O_2 in the presence of water vapour[21]. A candidate for one of the metal superoxides is calcium because of calcium's apparent abundance in the Martian soil[18]. For each superoxide hydrolysed by water, an equivalent alkaline site is produced that accounts for the resorption of the desorbed CO_2. But in the natural state, the basicity of the inner surface is not transferred to the iron oxide coat, because the mineral surface is mechanically separated at the molecular level from the iron oxide coating of several layers. The H_2O_2 formed from superoxide hydrolysis will, in the presence of abundant iron oxide catalyst, directly participate in the oxidation of the exact number of oxidisable equivalents such as formate in the Labelled release (LR) nutrient[22] or GEX M4 nutrient[3]. The alkali and alkaline earth metal superoxides are only one of many examples of the proposed chemistry on a stoichiometric basis that can well be applied to the ozonides, perchlorates, permanganates, and so on.

The iron oxide external surfaces provide at low temperature a monomolecular layer of adsorbed CO_2 which when activated by the solar ultraviolet flux splits off nascent oxygen and CO. The latter is free to return to the atmosphere. At higher temperatures, ultraviolet catalysis is limited by the quantity of CO_2 adsorbed. Other sources of nascent oxygen are photochemically formed ozone and hydrogen peroxide which are catalytically decomposed by the surficial iron oxide to form O_2 and O and H_2O and O, respectively. Then ascent oxygen, by simple diffusion (albeit an inefficient process, but time is not of the essence) passes through the porous iron-oxide matrix and attacks the inner mineral surface where it reacts with alkali and alkaline earth metal oxides to form the hydroxy-peroxides, superoxides, peroxyradicals and/or ozonides, if it does not first recombine with CO or react with another nascent oxygen. On the inner mineral surface, the products of the nascent oxygen reaction remain relatively stable but are potentially vigorous oxidisers.

If the oxygen–soil model is correct, it implies that the oxidiser source is distributed generally throughout the entire soil sample bed, that is, through the entire depth of the trench dug on the first sample taken. On the other hand, if the activity were associated only with the surface, there would have to be ten times the activity from just the surface material. If the red colour is associated with this activity, the colour photographic data substantiate that the entire trench is the same as the surface (personal communication from T. A. Mutch, Viking Lander Imaging Team). Furthermore, only a small fraction of the actual surface material is probably represented in the soil sample acquired. For example, a simple test of the surface sampler at JPL on surface tinted soil shows that only 11% of sieve size material 1.5 mm $> d >$ 0.589 mm, 9% of 0.589 mm $> d >$ 0.495 mm and 3% of 0.495 mm $> d$ are of surface origin in the test sample. A distribution of the oxidative capability throughout the entire depth of the trench, not just located on the exposed surface, indicates a history of stability consistent with the concept of the protective iron oxide coat under which is buried the oxygen-producing source. The stability is contrary to having the oxygen coupled surficially on the iron oxide coat because it would be at the mercy of transient changes in the vaporous environment over extremely long periods of time.

Another aspect of the iron oxide coating which bears on the stability question is related to the weak acidic nature of this material. Qualities which are ascribed to acid-like surfaces are the following: the ready emission of CO_2 from the soil in GEX, the small first peak from the pyrolytic release experiment (PR), the initial instant release of $^{14}CO_2$ from the LR and the higher background counts when LR was cooled following heating of the test cell to sterilisation temperatures[1]. The significance of the LR sterilisation regime is that the temperature is sufficient to volatilise such materials as $H_2SO_4{\cdot}2H_2O$ (boiling point 169°C) which can coat surfaces and prevent readsorption of $^{14}CO_2$ at the LR test cell temperature. The latter observation provided us with a basis for speculating that surficial acidity is provided by a thin coat of sulphuric acid from sulphate salts of weak bases (amorphous sulphate). A probable source of this sulphate is SO_3 produced by the visible activity of the past[23]. The SO_3 reacts with any water vapour in the atmosphere to produce sulphuric acid that descends as a mist to cover the iron oxide coat. Such hygroscopic coatings could well preserve the integrity of the highly active inner superoxide coat during episodes of higher water vapour incidences on Mars while explaining durable O_2 production from the soil sample in the test cell. It may explain the observed cohesive nature of the Martian soil. The acid characteristic of the surface ascribed to amorphous sulphate is born out by the high level of sulphur measured by the ICAT (ref. 18) and the oxidised state of the surface as evidenced by the red colouration of the iron oxide outer coat. The acidic surficial characteristics are quickly neutralised by the strongly basic alkali and alkaline earth metal hydroxides formed by the hydrolytic reaction as demonstrated by both GEX and LR.

The above model of the silicate mineral covered with iron oxide requires that ultraviolet light initiate the process of nascent production where adsorbed CO_2 and/or H_2O is concentrated on the surface of the iron oxide particles. To maintain a steady-state concentration of these energy-rich oxygen bonds on the inner mineral surface, the combined dissipative effects of water vapour and heat must be counterbalanced by an equivalent influx of nascent oxygen. When temperatures are low and water vapour is virtually limited, as appears to be the case in the Lander 1 area, the oxygen sources are prevalent. Sites having more water vapour are expected to have less active material and produce less oxygen, but it remains to be determined whether the effective oxidative capacity, that is the capability to oxidise organics, is also reduced. The H_2O_2 product (unstable in the presence of iron oxide) may remain uncontacted by the iron-oxide outer coat.

The degree to which the particle surfaces are re-exposed to the ultraviolet flux to renew the activity, the penetration of nascent oxygen, hydrogen peroxide and/or ozone downward through the soil bed and the concentration of water vapour are factors that must be considered fully to describe and/or predict the potential to produce oxidisers at any site for any depth into the surface of Mars.

At the Viking Lander 1 site, the vista is an ancient one where the soil was formed before the boulders were strewn over it. This suggests that the soil was layered over fluvial scoured rocks well after the episodes of water formed the surface. The oxidisers could have been created by two distinctly different processes. During a period when there was a dense atmosphere, silicate particles could have been lifted into the atmosphere by winds allowing the surfaces to be exposed to solar ultraviolet thus creating oxidisers which have been preserved to this date. Oxidisers could also be formed constantly at a rate through the entire soil bed from atmospheric hydrogen peroxide, ozone or nascent oxygen providing a steady state production of oxidiser which counters the dissipative effects of warmth and humidity. Both ozone and hydrogen peroxide can provide nascent oxygen in our model.

The oxidisers in the Martian soil are good candidates for the reaction observed in the LR experiment[1] as well as some of the observed CO_2 evolution in GEX. They must have the requisite qualities of oxidising the formate in the LR, producing the O_2 in GEX and accounting for some of the CO_2 released in GEX. A reaction that could account for the above observations starts with 534 nmol of the metal superoxide, $M(O_2)_2$:

$$586H_2O + 534M(O_2)_2 \rightarrow 775O_2 + 534M(OH)_2 + 52H_2O_2$$

Thermodynamic calculations proving the validity of these types of reactions in the environment simulating GEX will be tested at a later date.

Thee quivalent amount of oxidiser in the LR (26 nmol per 0.5 ml soil) calculated by one of us (V.I.O.), indicates that the inferred H_2O_2 produced as a result of soil reacting with water is limiting.

Because of the large abundance of water with respect to formate in the LR (ref. 22), 55.56 M per 2.4×10^{-4} M $\sim 2.3 \times 10^5$ l^{-1}, and in the GEX 1.9×10^5 l^{-1}, it is unlikely that the superoxide will oxidise the formate directly, but will provide the requisite H_2O_2 in the reaction with water.

In the GEX, the N_2 changes are small and can be explained from either initial N_2 desorption from soil by water vapour and subsequent resorption in liquid water or biological fixation, while N_2 production could arise from a Van Slyke reaction where α-amino nitrogen in the M4 nutrient reacts on the aforementioned external-acid surfaces with nitrite in Martian soil. Since N_2 exists in the Martian atmosphere and there is postulated NO in the upper atmosphere[24], it would not be inconsistent to assume oxides of nitrogen exist in the Martian soil.

As we have observed from our studies of lunar soils[6,7], metallic iron formed by meteoric bombardment of lunar iron oxides provides a basis for estimating the biologically important, free-water history of a planet. At this time, no H_2 has been found, therefore no metallic iron can be inferred in the Martian sample tested by GEX. This suggests that an amount of water and/or O_2 necessary to oxidise the metallic iron to the depth of the regolith was available at some time for this oxidation. For answers on the question of the state of oxidation of the planetary surface as measured by the presence of metallic iron, we must await the recharge occurring at the start of the second cycle to remove O_2 so as to measure the production of H_2 from the oxidation of Fe° by H_2O, and await the findings of Lander 2. The ultimate sensitivity for H_2 detection will allow us to measure 0.003% of metallic iron in the Martian soil if O_2 is not present in the test cell.

The advantage of starting with metallic iron as a precursor source of iron oxide is that it presents iron in an easily oxidisable form that allows the propagation of the iron oxide product on a broad planetary scale without the need for liquid water. Only water vapour and O_2 are required. The early development of the iron oxide in the history of Mars explains how the dry lubricating qualities of the iron oxide dampens aeolian erosion and removes the need for liquid water to extrude iron oxide deposits from their sources.

We prepared earlier for an expected chemistry of an arid planet[3] and must await the dissipation before life can be determined. At this time, all of the gas changes we have observed in our GEX test cell can be explained most easily by mechanistic processes. Although these results and explanations are preliminary, we find no need to invoke biological processes. While continued careful study is required, it is anticipated that results from Viking Lander 2 will reinforce our conclusions.

This work was done under contract to the National Aeronautics and Space Administration Viking Project Office. We thank M. Lehwalt, E. Merek, M. Rowland, S. Chang, S. Fong, N. Gee, J. Griffin and R. Mendoza of NASA-Ames Research Center for contributions; R. Day, S. Jeanjaquet, F. Brown, R. Gilje, S. Loer, J. Katherler, G. Rosiak and D. McMorris and all the unnamed participants at the TRW Systems Group who helped design, fabricate, test and nurture the GEX; H. Klein, J. Lederberg, A. Rich, J. Hubbard and R. Johnson for their advice; G. Soffen, C. Broome, G. Lee and T. Young for their interest in the GEX; T. Mutch and M. Carr, P. Toulmin, II, H. Moore and R. Hargreaves for their specific contributions in picture interpretations, inorganic analysis, physical properties and magnetics, respectively; C. Reichwein, and D. Buchendahl especially, for development of software specifically for GEX and for precise knowledge of the Viking System, respectively, and finally D. DeVincenxi for encouragement and continued Division support.

Received October 4; accepted November 23, 1976.

1 Klein, H. P., *et al.*, *Science*, **194**, 99–105 (1976).
2 Oyama, V. I., *Icarus*, **16**, 167–184 (1972).
3 Oyama, V. I., Berdahl, B. J., Carle, G. C., Lehwalt, M. E., and Ginoza, H. S., *Origins of Life* (in the press).
4 Merek, E. L., and Oyama, V. I., in *Life Sciences and Space Research*, **VIII**, 108–115 (North-Holland, Amsterdam, 1970).
5 Klein, H. P., Lederberg, J., Rich, A., Horowitz, N. H., Oyama, V. I., and Levin, G. V., *Nature*, **262**, 24–27 (1976).
6 Oyama, V. I., Merek, E. L., Silverman, M. P., and Boylen, C. W., in *Proc. Second Lunar Science Conference*, **2** (edit. by Levinson, A. A.), 1931–1937 (MIT Press, Cambridge, 1971).
7 Oyama, V. I., Berdahl, B. J., Boylen, C. W., and Merek, E. L., in *Third Lunar Science Conference* (edit. by Watkins, C.), 590–592 (Lunar Science Institute, Houston, 1972).
8 Hess, S. L., *et al.*, *Science*, **193**, 788–791 (1976).
9 Clever, H. L., and Battino, R., in *Solutions and Solubilities Part I*, (edit. by Dack, M. R. J.), 380–441, (Wiley, New York, 1975).
10 Morrison, T. J., and Johnstone, N. B., *J. chem. Soc.*, 3441–3446 (1954).
11 Yeh, S. Y., and Peterson, R. E., *J. pharm. Sci.*, **53**, 822–824 (1964).
12 Cox, J. D., and Head, A. J., *Trans. Faraday Soc.*, **58**, 1839–1845 (1962).
13 Austin, W. H., Lacombe, E., Rand, P. W., and Chatterjee, M., *J. appl. Physiol.*, **18**, 301–304 (1963).
14 Morrison, T. J., and Billett, F., *J. chem. Soc.*, 3819–3822 (1952).
15 Owen, T., and Biemann, K., *Science*, **193**, 801–803 (1976).
16 Baver, L. D., *Soil Phys.*, pp 12 and 287 (Wiley, New York, 1961).
17 Fanale, F. P., and Cannon, W. A., *J. geophys. Res.*, **79**, 3397–3402 (1974).
18 Toulmin, P., III, Clark, B. C., Baird, A. K., Keil, K., and Rose, H. J., Jr, *Science*, **194**, 81–83 (1976).
19 Mutch, T. A., *et al.*, *Science*, **194**, 87–91 (1976).
20 Mutch, T. A., *et al.*, *Science*, **193**, 791–801 (1976).
21 Vol'nov, I. I., *Peroxides, Superoxides, and Ozonides of Alkali and Alkaline Earth Metals* (edit. by Petrocelli, A. W.), (Plenum, New York, 1966).
22 Levin, G. V., and Straat, P. A., *Origins of Life* (in the press).
23 Carr, M. H., *et al.*, *Science*, **193**, 766–776 (1976).
24 Nier, A. O., *et al.*, *Science*, **193**, 786–788 (1976).
25 Holmes, H. F., and Gammage, R. B., *Proc. Lunar Science Conf.*, **6**, 3343–3350 (1975).
26 Langmuir, I., *Trans. Faraday Soc.*, **17**, 621 (1921).

Part XIV

THE MOONS OF MARS

Editor's Comments on Paper 36

36 VEVERKA and DUXBURY
Viking Observations of Phobos and Deimos: Preliminary Results

Mariner 9 revealed that Phobos and Deimos, the moons of Mars, were irregular in shape and pocked with craters (Pollack et al. 1972, 1973). Data from the Viking mission have provided more accurate determination of crater densities; improved measurements of the size, shape, and orbital parameters; and the color of the surface (see Paper 36 by Veverka and Duxbury; also Veverka 1977; Tolson et al. 1978; Pang et al. 1978; Pollack et al. 1978). The high resolution images taken from a distance of 950 km (Phobos) and 3000 km (Deimos) show peculiar crater chains and grooves on Phobos, which are described by J. Veverka and T. C. Duxbury (Paper 36).

The crater chains or clusters resemble the shallow secondary craters produced when blocks of debris are ejected from a primary crater by a meteorite impact, although they are not distributed radially around any crater. However, the low gravitational field of Phobos should permit most debris to escape. Therefore the presence of a regolith on Phobos (which was established by photometric, polarimetric, and thermal measurements) came as a surprise (Zellner 1972; Noland and Veverka 1977). Veverka and Duxbury propose some mechanisms of generating the crater clusters.

The nearly parallel sets of grooves or striations are even more puzzling. The February 1977 close encounter of Phobos, at a distance of 120 km shows the grooves to be graben-like rilles rather than crater chains or pits. It is still not certain whether they are caused by tidal stresses or are cracks related to the crater Stickney (the largest impact crater on Phobos, 10 km diameter (Soter and Harris 1977).

36

Reprinted from *J. Geophys. Res.* **82**:4213–4223 (1977)

Viking Observations of Phobos and Deimos: Preliminary Results

J. VEVERKA

Laboratory for Planetary Studies, Cornell University, Ithaca, New York 14853

T. C. DUXBURY

Jet Propulsion Laboratory, California Institute of Technology, Pasadena, California 91103

The improved resolution of the Viking orbiter images has led to the discovery of a number of unusual features on the surface of Phobos: (1) elongated rill-like depressions associated with the crater Stickney (possibly surface fractures), (2) chains of irregular craters which sometimes show a 'herringbone' pattern (possibly secondaries), and (3) sets of almost parallel linear striations of uncertain origin. The crater chains are not randomly oriented but tend to lie parallel to the orbital plane of Phobos. The striations, on the other hand, appear to form arcs of small circles which are normal to the Mars-Phobos direction. With the possible exception of feature (2), similar features have not been recognized on Deimos, possibly because of the coarser resolution of available imagery. The Viking data demonstrate that the surfaces of both satellites are definitely saturated with craters ≥300 m across.

1. INTRODUCTION

During the Viking primary mission (June–November 1976), about three dozen high-resolution images of Phobos and Deimos were obtained by the Viking orbiter cameras [*Duxbury and Veverka*, 1977]. The prime objective was to extend the high-resolution coverage of the satellite surfaces obtained by Mariner 9 in 1971–1972. For example, in the case of Phobos the Mariner 9 coverage of the north polar regions and of the potentially interesting area around 230°W, 10°N, which is antipodal to Stickney (the largest crater on Phobos), was of mediocre quality. In the case of Deimos the quality of the Mariner 9 coverage was generally poor; in fact, so few craters were visible on the images that a statistically meaningful crater density curve could not be constructed for the outer satellite [*Thomas and Veverka*, 1977].

The Viking coverage of the satellites is superior to that obtained by Mariner 9 in several respects but primarily in terms of effective resolution. While Viking obtained images of Phobos and Deimos from distances as small as 880 and 3000 km, respectively, the corresponding minimum ranges in the case of Mariner 9 were 5710 and 5490 km [*Veverka et al.*, 1974]. In addition, the performance of the Viking cameras is superior to that of the Mariner 9 B camera [*Carr et al.*, 1976]. For example, whereas in the Mariner 9 satellite images, limbs are typically 3–4 pixels (picture elements) wide [*Noland and Veverka*, 1976*a*], in the Viking images their width does not exceed 1–2 pixels.

The improved resolution has led to better crater density counts on Phobos and to the first statistically valid measurement of the crater density on Deimos. More important, the improved resolution has led to the discovery of a suite of unusual surface features on Phobos. It is the purpose of this paper to give a preliminary account of these new results. An overview of the satellite coverage during the Viking primary mission is given in a companion paper [*Duxbury and Veverka*, 1977].

2. DISCOVERY OF UNUSUAL SURFACE FEATURES ON PHOBOS

The improved resolution of the Viking images has led to the discovery of a number of unusual surface features on Phobos. On the basis of their morphology these can be divided into three major categories: (1) elongated rill-like depressions associated with the crater Stickney, possibly surface fractures or rows of coalescing secondaries, (2) chains and clusters of irregular elongated craters, possibly secondaries, and (3) parallel linear striations or grooves of enigmatic origin.

With the possible exception of category (2), similar features have not been recognized on the surface of Deimos, possibly because of insufficient resolution. On Phobos, such features become prominent when the surface resolution exceeds 100 m and approaches 50 m.

a. Elongated Rill-Like Depressions Associated With Stickney

A number of elongated rill-like depressions occur in the vicinity of Stickney, the largest crater on Phobos. These features are clearly visible in Figure 1.

The most prominent of the elongated depressions seems to originate at the rim of Stickney and extends some 10 km in almost a straight line. At the resolution of Figure 1 (~100 m) it is difficult to be sure whether this depression is a trough or a string of coalescing irregular pits, although it does look more like a trough which may taper somewhat distally from Stickney. It is conceivable that this feature represents a fracture in the surface of Phobos produced by the severe impact which formed Stickney.

Note that one somewhat similar feature can be followed more or less continuously from the rim of Stickney to the far limb. For most of its trend the feature appears to consist of almost coalescing elongated depressions and terminates in a puzzling arcuate trough near the limb. Clearly, still higher resolution images of these features are needed before their origin can be interpreted more definitively. Nevertheless, the association of at least some of them with Stickney appears certain.

Unfortunately, it was not possible during the nominal mission to examine the neighborhood of Hall, the large 6-km crater near the south pole, at high resolution for similar features. None have been identified in the vicinity of Roche, the 5-km crater near the north pole (Figure 2).

Fig. 1. View of Phobos from Viking Orbiter 1 (87A52). Stickney, the largest crater on Phobos, is seen at left. Note the rill-like depressions trending from left to right which appear to emanate from the rim of this 10-km crater.

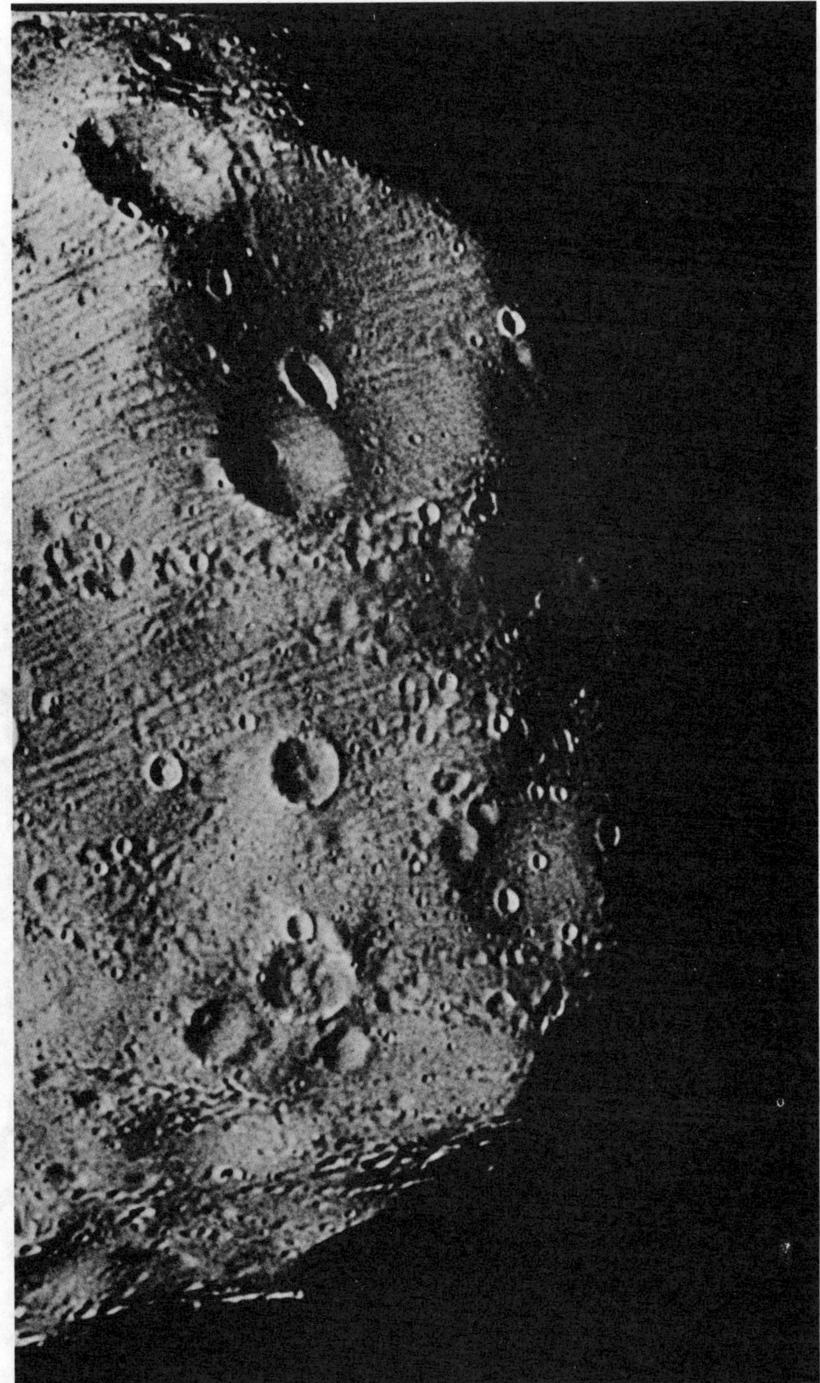

Fig. 2. View of Phobos from Viking Orbiter 2 (039B84). The large crater at top (Roche) lies close to the north pole of Phobos and is about 5 km across. Enlarged segments of this frame are shown in Figures 3, 7, and 8.

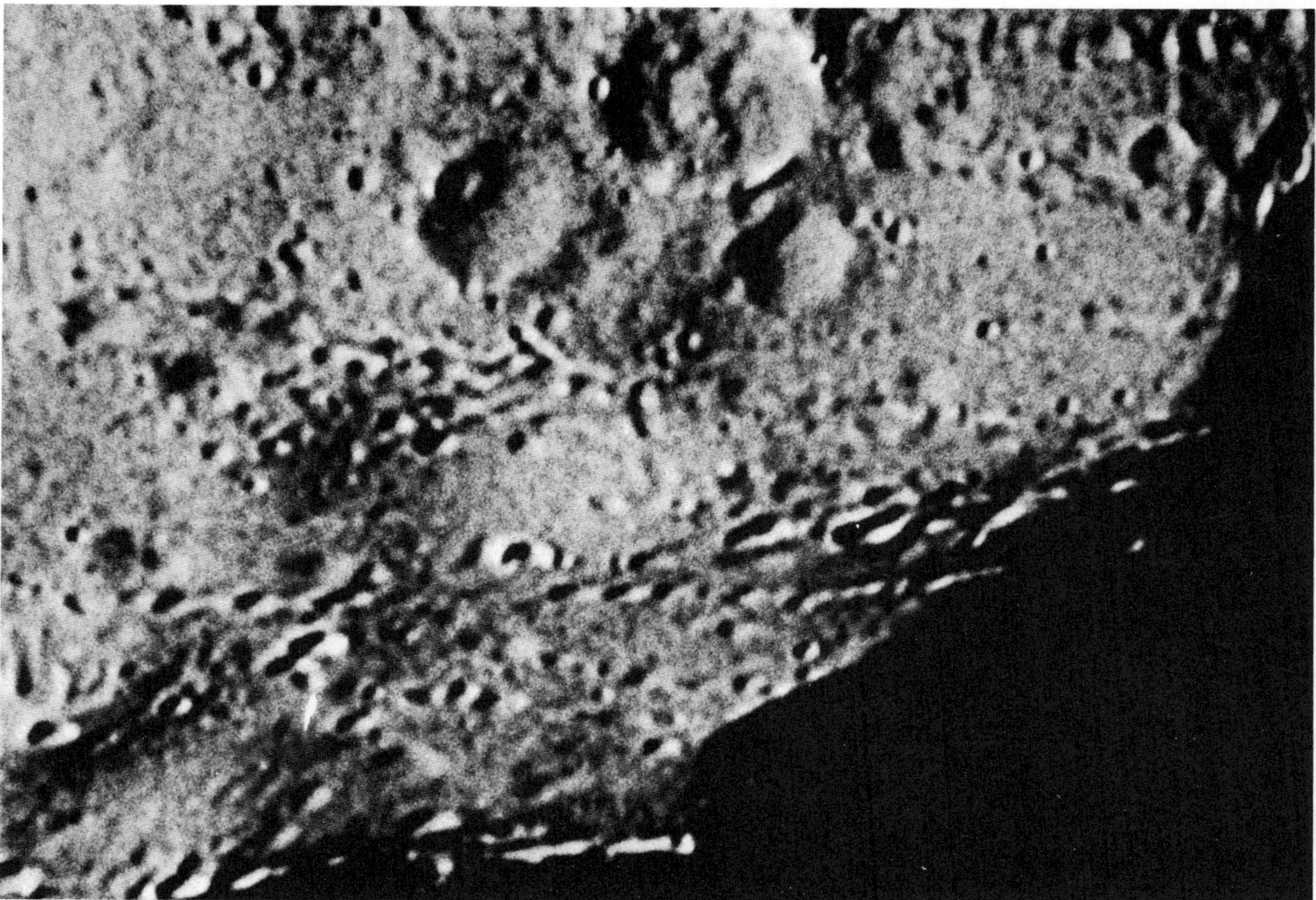

Fig. 3. Enlarged segment of frame 039B84 (Figure 2) showing clusters of irregular craters arranged in herringbone patterns characteristic of secondary impacts. The largest crater in this view (top center) is about 1.5 km across.

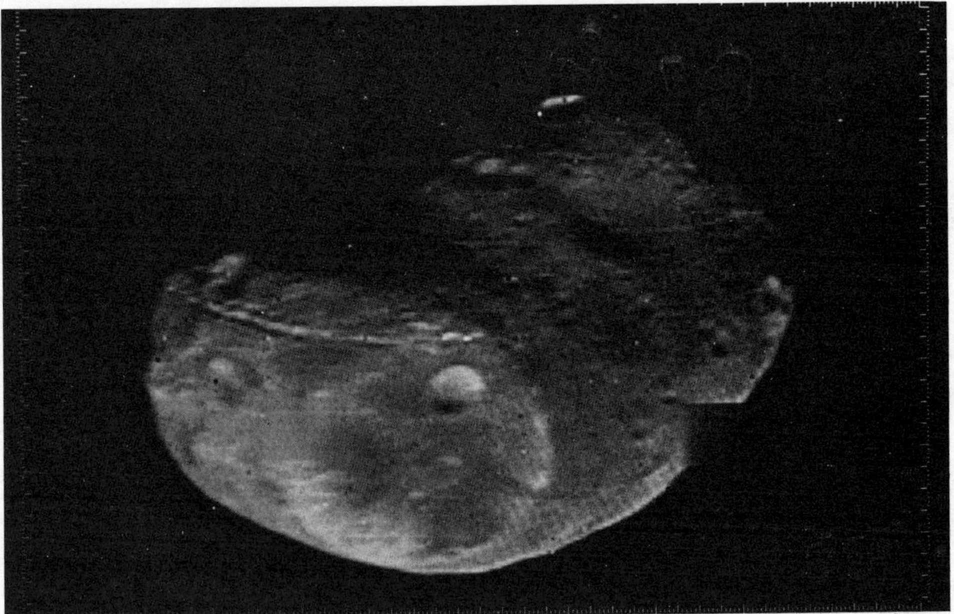

Fig. 4. View of Phobos from Viking Orbiter 2 (081B33) showing a prominent chain of irregular craters superimposed on the crater Stickney.

b. Chains and Clusters of Irregular Elongated Craters

A number of prominent crater chains are seen in Figure 2. Typically, the chains consist of irregular craters ~50–200 m across, which sometimes cluster into the 'herringbone' pattern (Figure 3) characteristic of secondary ejecta [*Oberbeck and Morrison*, 1974]. Such crater chains appear to be common on Phobos (Figure 4) but do not appear to be randomly oriented. Rather, they all seem to be closely parallel to the orbital (equatorial) plane of Phobos (Figure 5).

Traditionally, there has existed the feeling that since the escape velocity from Phobos is so low (~12 m/s), very little ejecta would reimpact the surface and that secondary craters should be absent. Thus when *Pollack et al.* [1973] described a row of several 200-m craters visible in some Mariner 9 images of Phobos, they did not consider the possibility that these were secondaries. Instead, they suggested that such rows of small craters could be produced by localized outgassing along fractures following a severe impact such as that which formed Stickney. However, *Soter* [1971] had noted that some ejecta from Phobos would remain in Mars orbit and would have an opportunity to reimpact Phobos at fairly low velocities.

The evidence that clusters of low-velocity objects do impact the surface of Phobos seems clear. But until detailed calculations are carried out, it is not clear whether any secondaries can be produced on Phobos in single-hop trajectories. It is possible that many of the crater chains are produced by clumps of ejecta which originally were thrown out at slightly more than the escape velocity of Phobos, ended up in Mars orbit, and subsequently reimpacted the surface of the satellite. Although such a mechanism appears to be feasible, the details remain to be worked out. It is also conceivable that some of the crater chains on Phobos were produced by ejecta originating from the outer satellite Deimos.

The nonrandom orientation of the crater chains on Phobos (Figure 5) seems to be more consistent with the 'escape-into-Mars-orbit and reimpact mechanism' than with the idea of a single-hop process. The crater chains on Phobos definitely do not radiate from any particular crater.

Incidentally, at least some of the crater chains on Phobos are younger than the large crater Stickney, as the superposition of the prominent crater chain across the crater seen in Figure 4 demonstrates.

It appears that there may be clusters of irregular craters ('secondaries') on Deimos, although in the case of this satellite we still lack images with adequate resolution to settle the issue. For example, in the upper left of Figure 6 one sees a cluster of irregular craters which may be secondaries. This particular group of craters had been noticed in the Mariner 9 coverage [*Veverka et al.*, 1974, orbit 149] and found to be associated with an area of higher albedo, by about 30% relative to its surroundings [*Noland and Veverka*, 1976*a*].

c. Linear Striations or Grooves

The most enigmatic features discovered on the surface of Phobos are the swarms of almost parallel striations or grooves best seen on frame 039B84 (Figure 2). Typically, the striations are 150–200 m wide, appear to have rather shallow cross-sectional profiles, and are separated from their neighbors by distances comparable to their widths. Individual striations can be followed for distances in excess of 5 km (Figure 2).

The striations seem to occur in at least two sets which are not precisely parallel (Figure 7) but inclined at about 10° in relation to each other. It can be seen from Figures 7 and 8 that elements of the two sets do not cross each other.

The striations do cross the large craters near the north pole

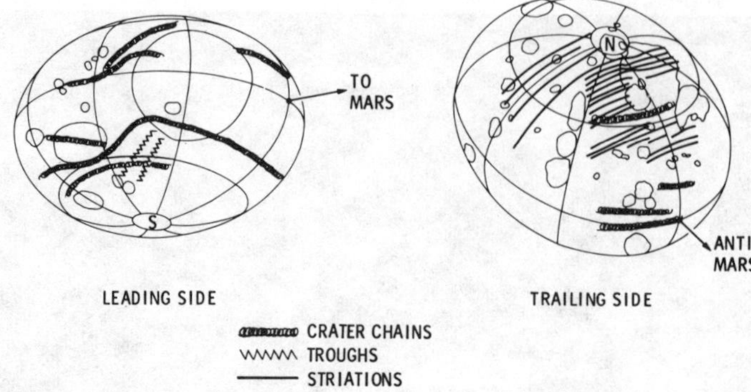

Fig. 5. Sketch map of the distribution of various linear features on the surface of Phobos based on data available from the Viking primary mission. Features such as those shown in Figures 3 and 4 are mapped as crater chains. Rill-like depressions associated with Stickney (Figure 1) are mapped as troughs, while the linear features shown in Figures 7 and 8 are mapped as striations.

visible in Figure 2 but are overlain by a number of smaller craters (Figure 7). Thus they are neither the oldest nor the youngest features visible on the surface of Phobos.

A fundamental question concerns the longitudinal profile of a typical striation. Is a typical striation made up of a chain of coalescing craters, or is it more properly a crack or a gouge? Unfortunately, the evidence obtained during the primary mission is far from conclusive. A few of the striations visible in Figure 7 have segments which have a vaguely pitted appearance and could be interpreted as modified rows of pits. However, many other striations (Figures 7 and 8) look more like grooves at the resolution currently available.

Coverage of Phobos at a resolution adequate to show striations was of very limited extent during the Viking primary mission. The only region of Phobos covered adequately is that shown in Figure 2. The south polar region and the area around Stickney were not imaged with resolution adequate to show striations. The distribution of striations on the surface of Phobos based on the limited standard mission coverage is shown in Figure 5. The striations appear to lie along small circles perpendicular to the Mars-Phobos direction (parallel to the satellite's orbital velocity vector) and show an apparent concentration at high northern latitudes near longitude 180°W.

The origin of the striations remains unclear. Although the striations qualitatively resemble gouges, no reasonable mechanism for gouging the surface of Phobos has been suggested.

The theories proposed to date to account for the striations can be divided into three categories: first, the striations represent layering within Phobos; second, the striations are rows of impact craters; and third, the striations are cracks resulting from tensional stresses.

According to the first hypothesis the striations represent layering within Phobos. The satellite is considered to be a fragment of a much larger parent body which had a radially layered structure. According to this view the layering should pervade Phobos, and expressions of it might be found in other areas of the surface. It is not expected that the striations should be resolvable into rows of craters.

One possible difficulty with this view is that craters should disturb the layering pattern locally, while they evidently do not. Specifically, the formation of the large crater Roche might be expected to have disturbed the layering in some radially symmetric fashion, which is clearly not the case (Figure 8).

According to the second view the striations are actually rows of coalescing impact craters. It is assumed that it is possible (in some unspecified manner) to accumulate strings of debris in Mars orbit with which Phobos subsequently collides. Although a detailed mechanism has not been worked out, the scheme may be possible. It predicts that all striations should be resolvable into rows of small craters (possibly modified by subsequent erosion by micrometeoroids).

It should be noted that this mechanism must be different from that which produces the crater chains discussed in section 2*b*, since the distributions on the surface of Phobos of the two types of features are markedly different (Figure 5).

According to the third hypothesis the striations represent cracks produced by tensional stresses. Specifically, A. W. Harris and S. Soter (private communication, 1977) have shown that tidal forces tend to distort an object at the distance of Phobos from Mars into a triaxial ellipsoid and that tensional stresses would be concentrated in a narrow belt near the poles at right angles to the Mars-Phobos direction. If the striations are simple stress cracks or stretch marks in a regolith, they are not resolvable into rows of craters, and another swarm of striations should be located near the south pole of Phobos.

It should be noted that large stresses in the surface of Phobos can also be produced during the formation of large craters. It is conceivable that the formation of a crater such as Stickney may have initiated failure in the polar regions, which could have been under tension due to the Harris-Soter mechanism. In this event, Stickney would be younger than Roche (compare Figure 8).

It is possible that the striations may have a more complex origin. They could be primarily faults which have, in part, been modified by local outgassing of the sort suggested by *Pollack et al.* [1973]. If we assume that Phobos was originally made of a volatile-rich material similar to that found in type 1 carbonaceous chondrites, severe impacts such as the impact responsible for Stickney would have heated a significant fraction of the satellite's interior to the point where outgassing could have occurred. It is reasonable to expect that such outgassing would have been concentrated along faults which themselves could be the result of impacts. Such a mechanism

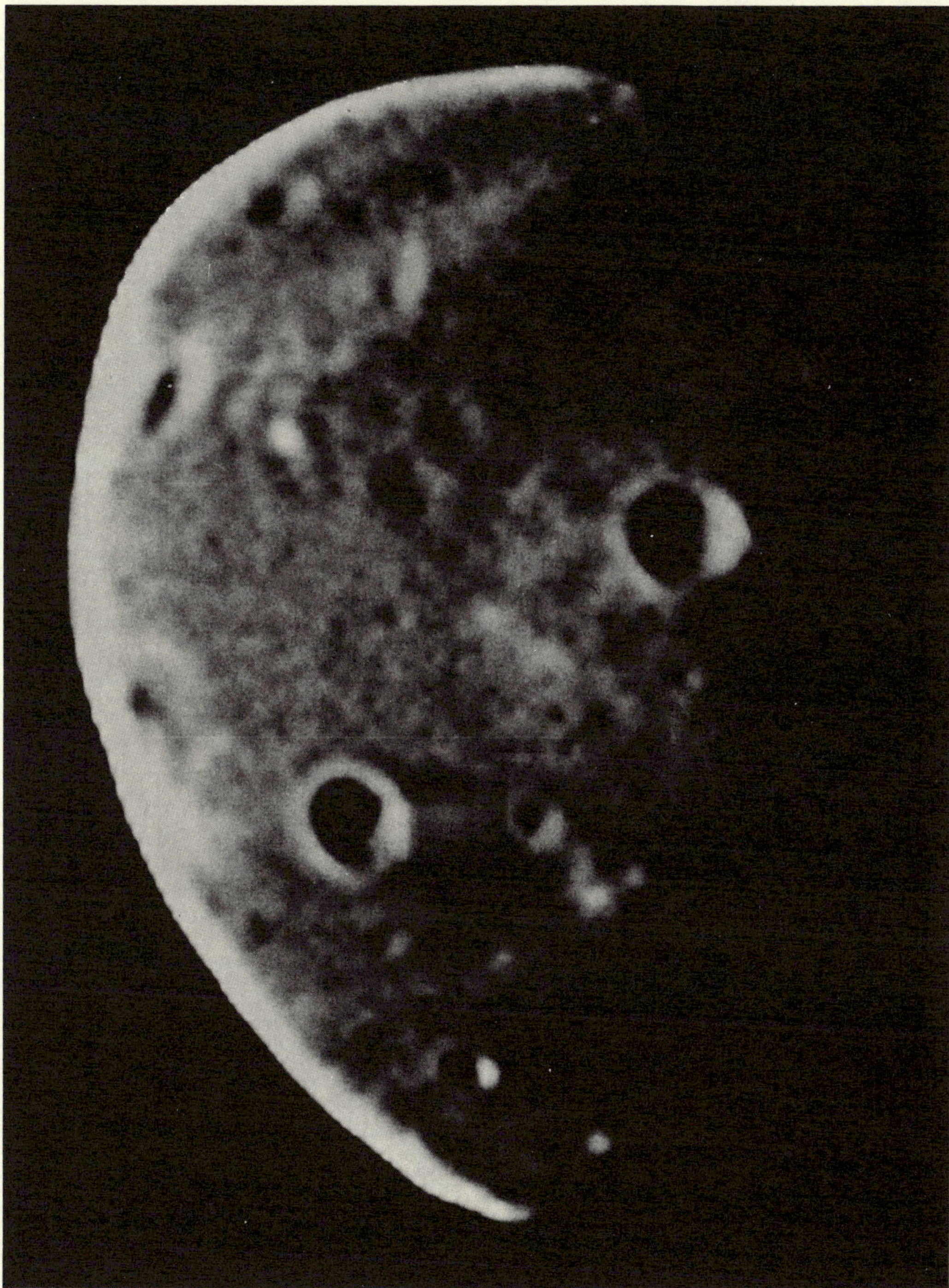

Fig. 6. View of Deimos from Viking Orbiter 1 (056A98). The largest crater is about 1.3 km across. A cluster of irregular depressions occurs near top left.

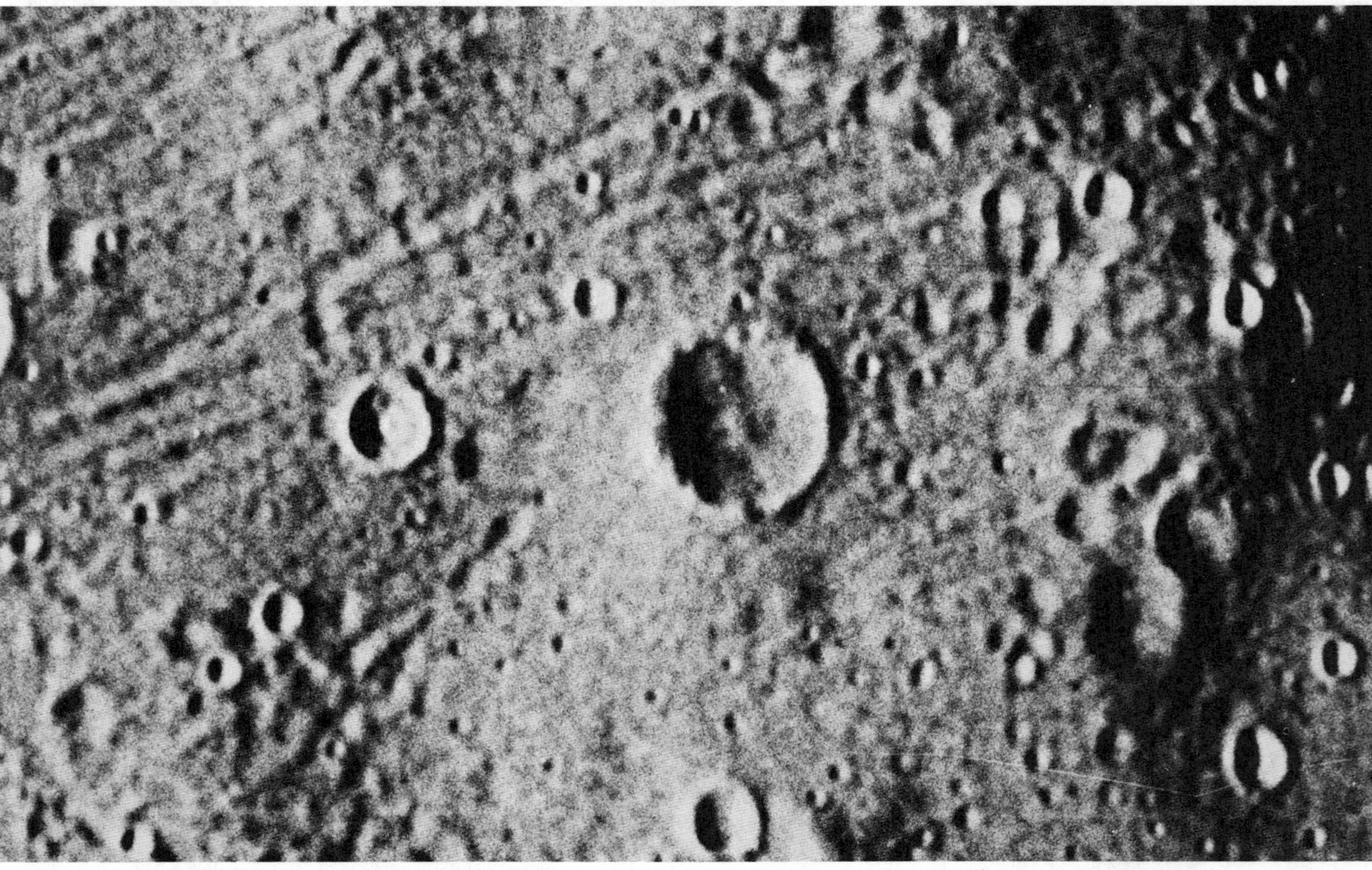

Fig. 7. Enlargement of a segment of frame 039B84 (Figure 2) showing striations. The largest crater in this view is about 900 m across. At least two sets of striations inclined at about 10° to each other can be seen. Some craters are clearly superimposed on the striations.

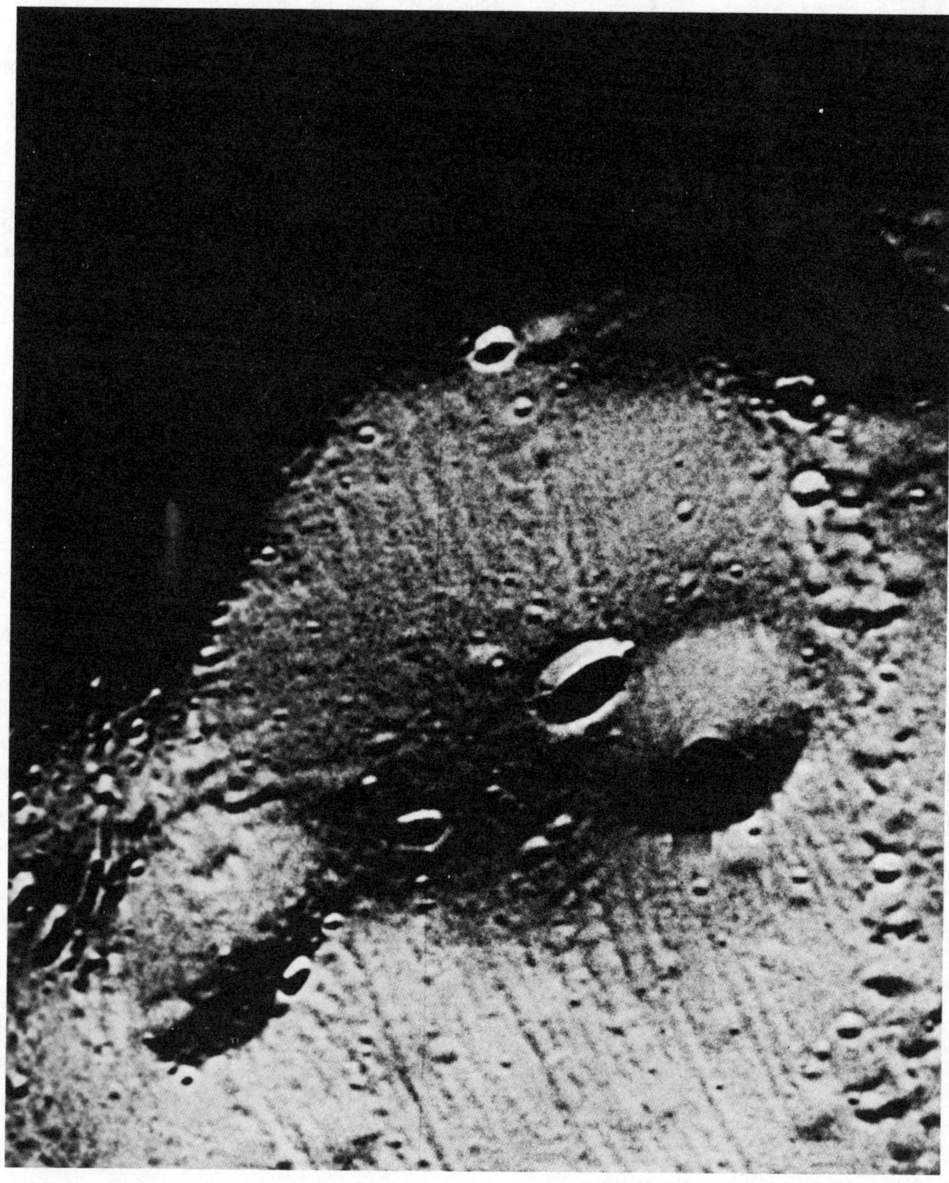

Fig. 8. Enlargement of a segment of frame 039B84 (Figure 2) showing striations crossing the floor of the large crater Roche, which is about 5 km across.

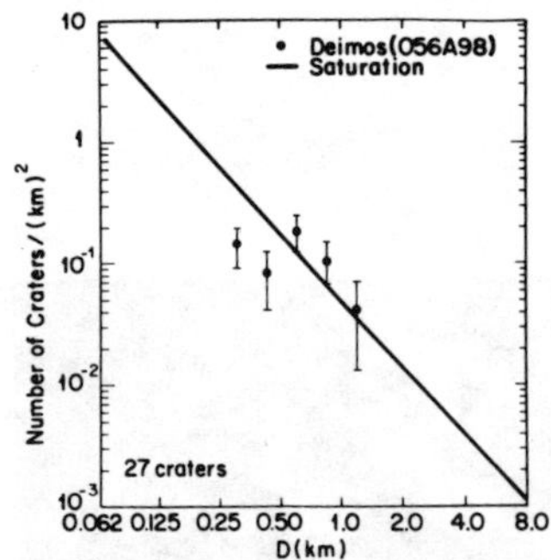

Fig. 9. Crater density on Deimos measured from frame 056A98 (Figure 6).

would account for the pitted appearance of some of the striations.

In order to choose among the above theories one needs additional information. Specifically two important questions must be answered: Do all striations resolve into crater chains given sufficient resolution? What is the distribution of striations on the surface of Phobos? Hopefully, these key questions will be answered during the close encounters with Phobos which will take place during the Viking extended mission. (Preliminary data from the Viking extended mission show that many of the striations are pitted in a manner consistent with the faulting plus outgassing hypothesis outlined above.) The examination of Deimos at very high resolution during the extended mission to see whether similar features exist on the surface of the outer satellite should further limit the scope of speculation.

3. Crater Counts

Crater counts were carried out on the best Deimos frame available (frame 056A98, Figure 6). A total of 27 craters ranging in size from 0.3 to 1.4 km are visible. In Figure 9, the crater counts on Deimos are compared with the 'saturation curve,' an empirical curve based on the most heavily cratered regions of the lunar uplands taken from *Hartmann* [1973]. Hartmann and some other workers hypothesize that the saturation curve represents the limiting crater density that can be reached before crater production and crater obliteration processes balance each other.

From the data in Figure 6 we conclude that the surface of Deimos is saturated with craters in the diameter range of 0.3–1.4 km. Since the flux of impacting bodies over the last 4.6 b.y. at the orbit of Mars is not known accurately, it is not certain how long it takes a fresh surface at the orbit of Mars to reach 'saturation.' *Pollack* [1977] estimated a minimum time of 1.5 b.y. We emphasize that our observation that the satellite surfaces are saturated with small craters provides only a minimum exposure age.

Crater counts were also carried out on frame 039B84 (Figure 2), the best of the Phobos images obtained during the nominal mission. Craters ranging from 50 m to 4.9 km in diameter are visible, and clusters of irregular secondary craters are conspicuous. Only circular craters, not obviously associated with clusters, were included in the crater count. The data for 237 craters are plotted in Figure 10; for crater diameters of ≥0.3 km the data points lie close to the saturation curve. The falloff at smaller sizes probably indicates a progressive loss of resolution as well as obliteration of small primary craters by secondaries.

Over 60 small craters can be counted within the large 4.9-km crater (Roche) near the north pole of Phobos (Figure 2). None of these show any of the characteristics of secondaries. The crater counts for the inside of Roche (Figure 11) do not differ significantly from those for the surrounding terrain or from the saturation curve. The fact that the density of impact craters within Roche is identical to that on the surrounding terrain suggests that erosion mechanisms specific to the interior of large craters (e.g., slumping of walls) are not effective, a fact which is understandable in view of the negligible surface gravity on Phobos ($g \sim 1$ cm/s^2).

4. Other Topics

Three sequences of color coverage, using filter passbands from 0.42 to 0.62 μm, were obtained during the nominal mission. This represents the first time that color measurements have been made of individual surface features on Phobos and Deimos. Preliminary analysis has not revealed any color variations on the surfaces. Between 0.4 and 0.6 μm the satellites appear to be uniformly gray.

The Viking nominal mission has also extended the phase angle coverage of the satellites. Mariner 9 observations were limited to the range 18°–83° [*Veverka et al.*, 1974]. For Phobos, Viking has extended this range to phase angles as large as 125°. Such data are being used to refine the satellite's photometric function and in particular its phase integral [*Noland and Veverka*, 1976*b*].

5. Conclusions

The Viking images revealed a number of unusual features on the surface of Phobos. Prominent chains of noncircular craters suggest that clusters of objects impact the surface at relatively low velocity. It is possible that these clusters are made up of ejecta from either Phobos or Deimos which do not have sufficient velocity to escape Mars [*Soter*, 1971]. The apparent common occurrence of low-velocity impacts is consistent with the observations that both Phobos and Deimos have regoliths [e.g., *Gatley et al.*, 1974; *Noland and Veverka*, 1976*a*, *b*].

The elongated depressions which seem to be associated with the large crater Stickney may be evidence of the local fracturing of a small body by an almost catastrophic impact. This would be the first time that such evidence has been observed in the solar system.

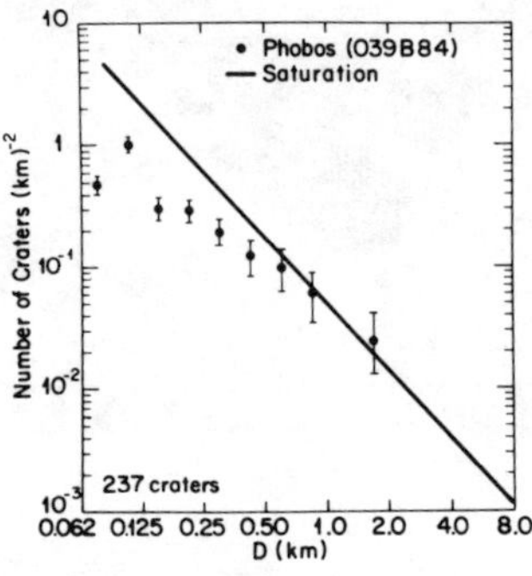

Fig. 10. Crater density on Phobos measured from frame 039B84 (Figure 2).

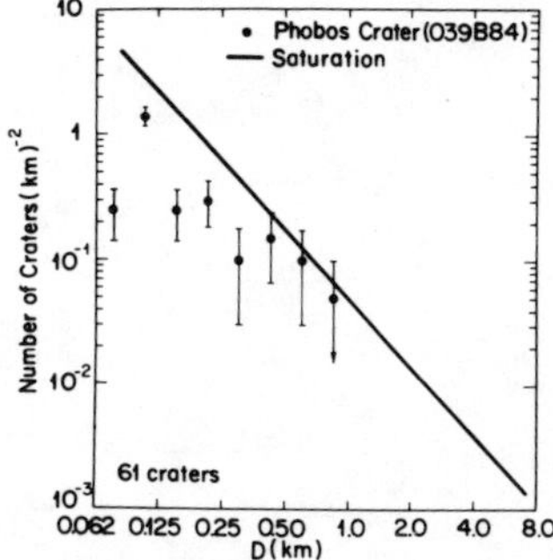

Fig. 11. Crater density within the crater Roche on Phobos measured from frame 039B84 (Figure 8).

Although a definitive explanation of the enigmatic striations is still being worked out, it can be expected that the solution of this puzzle will expand our understanding of small objects in the solar system. This is a case where more coverage at the highest resolution of large extents of Phobos is needed. Such coverage is being obtained during the Viking extended mission.

At the present time we lack adequate imagery of Deimos to say whether similar features occur on the surface of the outer satellite. Clearly, this is another important question which should be answered during the Viking extended mission.

The crater densities on both Phobos and Deimos have now been established reliably. Both surfaces appear to be saturated. Since the flux of impacting objects at the orbit of Mars throughout the history of the solar system is most uncertain, it is difficult to determine how long it takes a surface at the orbit of Mars to become saturated with craters. *Pollack* [1977] estimated at least 1.5 b.y., with time scales comparable to the age of the solar system being quite probable. Thus the surfaces of the satellites are probably very old. If Phobos and Deimos are fragments of larger objects, the fragmentation occurred a long time ago.

Acknowledgments. We thank our colleagues on the Viking Orbiter Imaging Team for helpful discussions, W. Quaide for specific comments, and R. Ruiz and W. Farrel of the Jet Propulsion Laboratory's Image Processing Laboratory for enhancing the satellite images used in this paper. This work was supported by the Viking Project and by NASA's Planetary Geology Program. NASA contract NSG 7156.

REFERENCES

Carr, M. H., et al., Preliminary results from the Viking orbiter imaging experiment, *Science*, *193*, 766–776, 1976.

Duxbury, T. C., and J. Veverka, Viking imaging of Phobos and Deimos: An overview of the primary mission, *J. Geophys. Res.*, *82*, this issue, 1977.

Gatley, I., H. Kieffer, E. Miner, and G. Neugebauer, Infrared observations of Phobos from Mariner 9, *Astrophys. J.*, *190*, 497–503, 1974.

Hartmann, W. K., Martian cratering, 4, Mariner 9 initial analysis of cratering chronology, *J. Geophys. Res.*, *78*, 4096–4116, 1973.

Noland, M., and J. Veverka, The photometric functions of Phobos and Deimos, II, Surface photometry of Deimos, *Icarus*, *29*, 200–211, 1976*a*.

Noland, M., and J. Veverka, The photometric functions of Phobos and Deimos, I, Disc-integrated photometry, *Icarus*, *28*, 405–414, 1976*b*.

Oberbeck, V. R., and R. H. Morrison, Laboratory simulations of the herringbone pattern, associated with lunar secondary crater chains, *Moon*, *9*, 415–455, 1974.

Pollack, J. B., Phobos and Deimos: A review, in *Planetary Satellites*, edited by J. Burns, University of Arizona Press, Tucson, 1977.

Pollack, J. B., et al., Mariner 9 television observations of Phobos and Deimos, 2, *J. Geophys. Res.*, *78*, 4313–4326, 1973.

Soter, S., Studies of the terrestrial planets, Ph.D. thesis, Cornell Univ., Ithaca, N. Y., 1971.

Thomas, P., and J. Veverka, Phobos: Surface density of impact craters, *Icarus*, *30*, 595–597, 1977.

Veverka, J., et al., A Mariner 9 atlas of the moons of Mars, *Icarus*, *23*, 206–289, 1974.

BIBLIOGRAPHY

Adams, J. B., and McCord, T. B. 1969. Mars: Interpretation of spectral reflectivity of light and dark regions. *J. Geophys. Res.* **74**:4851–4856.

Allègre, C. J., and Courtillot, V. E. 1974. Evidence for lateral movements of the Martian crust. *EOS Trans. Am. Geophys. Un.* 55th Annu. Meeting Abstr., p. 341.

Anders, E., and Arnold, J. R. 1965. Age of craters on Mars. *Science* **149**: 1494–1496.

Anders, E., and Owen, T. 1977. Mars and earth: Origin and abundance of volatiles. *Science* **198**:453–465.

Anderson, D. A. 1972. Internal constitution of Mars. *J. Geophys. Res.* **77**: 789–795.

Anderson, D. L., Gaffney, E. S., and Low, P. F. 1967. Frost phenomena on Mars. *Science* **155**:319–322.

Anderson, D. L., Miller, W. F., Latham, G. V., Nakamura, Y., Toksöz, M. N., Dainty, A. M., Duennebier, F. K., Lazarewicz, A. R., Kovach, R. L., and Knight, T. C. D. 1977. Seismology on Mars. *J. Geophys. Res.* **82**: 4524–4546.

Antoniadi, E. M. 1930. *La Planète Mars.* Paris: Libaire Scientifique, Herman et Cie, 239 pp.

Aronson, J. R., and Emslie, A. G. 1975. Composition of the martian dust as derived by infrared spectroscopy from Mariner 9. *J. Geophys. Res.* **80**:4925–4931.

Arvidson, R. E. 1972. Aeolian processes on Mars: Erosive velocities, settling velocities and yellow clouds. *Geol. Soc. Am. Bull.* **83**:1503–1508.

Arvidson, R. E. 1974a. Morphological classification of Martian craters and some implications. *Icarus:* **22**:264–271.

Arvidson, R. E. 1974b. Wind-blown streaks, splotches and associated craters on Mars: Statistical analysis of Mariner 9 photographs. *Icarus* **21**:12–27.

Arvidson, R. E., and Coradini, M. 1975. Crater diameter-frequency distributions for Martian fretted terrain. *Am. Geophys. U. Trans.* **56**: 1014 (abstr.).

Bagnold, R. A. 1941. *The Physics of Blown Sand and Desert Dunes.* London: Methuen and Co., 265 pp.

Baker, V. R., and Milton, D. J. 1974. Erosion by catastrophic floods on Mars and Earth. *Icarus* **23**:27–41.

Baldwin, R. B. 1965. Mars: An estimate of the age of its surface. *Science* **149**:1498–1499.

Barker, E. S., Schorn, R. A., Woszczyk, A., Tull, R. G., and Little, S. J. 1970. Mars: Detection of atmospheric water vapor during the southern hemisphere spring and summer. *Science* **170**:1308–1310.

Barth, C. A., Hord, C. W., Stewart, A. I., Lane. A. L., Dick, M. L. Anderson, G. P. 1973. Mariner 9 UV spectrometer experiment: Seasonal variation of ozone on Mars. *Science* **179**:795–796.

Belcher, O., Veverka, J., and Sagan, C. 1971. Mariner photography of Mars and aerial photography of earth: Some analogies. *Icarus* **15**:241–254.

Belton, M. J. S., and Hunten, D. M. 1966. The abundance and temperature of CO_2 in the Martian atmosphere. *Astrophys. J.* **145**:454–467.

Belton, M. J. S., and Hunten, D. M. 1968. A search for O_2 on Mars, and Venus: A possible detection of oxygen in the atmosphere of Mars. *Astrophys. J.* **153**:963–974.

Biemann, K., Owen, T., Rushneck, D. R., LaFleur, A. L., Howarth, D. W. 1976a. The atmosphere of Mars near the surface: Isotopic ratios and upper limits on the noble gases. *Science* **194**:76–78.

Biemann, K., Oro, J., Toulmin, P., Orgel, L. E., Nier, A. O., Anderson, D. M., Simmonds, P. G., Flory, D., Diaz, A. V., Rushneck, D. R., and Biller, J. E. 1976b. Search for organic and volatile inorganic compounds in two surface samples from the Chryse Planetia region of Mars. *Science* **194**: 72–76.

Biemann, K., Oro, J., Toulmin, P., Orgel, L. E., Nier, A. D., Anderson, D. M., Simmonds, P. G., Flory, D., Diaz, A. V., Rushneck, D. R., Biller, J. E., and LaFleur, A. L. 1977. The search for organic substances and inorganic volatile compounds in the surface of Mars. **82**:4641–4658.

Binder, A. B. 1966. Mariner IV: Analysis of preliminary photographs. *Science* **152**:1053–1055.

Binder, A. B. 1969. Internal structure of Mars. *J. Geophy. Res.* **74**:3110–3118.

Binder, A. B., and Cruikshank, D. P. 1966. Lithologic and mineralogic investigations of the surface of Mars. *Icarus* **5**:521–525.

Binder, A. B., and Davis, D. R. 1973. Internal structure of Mars. *Phys. of Earth and Planet. Interiors* **7**:477–485.

Binder, A. B., and McCarthy, D. W. 1972. Mars: The lineament system. *Science* **176**:279–281.

Birch, F. 1969. Density and composition of the mantle: First approximation as an olivine layer. In *The Earth's Crust and Upper Mantle*, P. J. Hart, ed. AGU Geophys. Monograph 13.

Blasius, K. R. 1976. The record of impact cratering on the great volcanic shields of the Tharsis region of Mars. *Icarus* **29**:343–362.

Blasius, K. R., and Cutts, J. A. 1976. Shield volcanism and lithospheric structure beneath the Tharsis plateau, Mars. *Proc. Lunar Sci. Conf.* 7**3**:3561–3573.

Blasius, K. R., and Cutts, J. A. 1977. Geology of the Valles Marineris: First analysis of imaging from the Viking 1 Orbiter primary mission. *J. Geophys. Res.* **82**:4067–4091.

Booth, M. C., and Kieffer, H. H. 1977. Carbonate formation on Mars. Washington, D. C.: NASA TM-X-3511, p. 200.

Bretz, J. H. 1969. The Lake Missoula floods and the channeled scabland. *J. Geol.* **77**:505–543.

Briggs, G. A. 1974. The nature of the residual martian polar caps. *Icarus* **23**:167–191.

Brinkman, R. T. 1971. Mars: Has nitrogen escaped? *Science* **174**:944–945.

Capen, C. F. 1971. Martian yellow clouds—past and future. *Sky and Telescope* **41**:117–120.

Carleton, N. P., and Traub, W. A. 1972. Detection of molecular oxygen on Mars. *Science* **177**:988–992.

Carr, M. H. 1974a. Tectonism and volcanism of the Tharsis region of Mars. *J. Geophys. Res.* **79**:3943–3949.

Carr, M. H. 1974b. The role of lava erosion in the formation of lunar rilles and martian channels. *Icarus* **22**:1–23.

Carr, M. H. 1975. Geologic map of the Tharsis quadrangle of Mars. *U. S. Geol. Sur. Misc. Inv. Serv. Map I-893.*

Carr, M. H. 1976. The volcanoes of Mars. *Sci. Am.* **234**:32–43.

Carr, M. H., Crumpler, L. S., Cutts, J. A., Greeley, R., Guest, J. E., and Masursky, M. 1977a. Martian impact craters and emplacement of ejecta by surface flow. *J. Geophys. Res.* **82**:4055–4065.

Carr, M. H., Greeley, R., Blasius, K. R., Guest, J. E., and Murray, J. B. 1977b. Some martian volcanic features as viewed from Viking Orbiters. *J. Geophys. Res.* **82**:4039–4054.

Cassini, Jr. 1740. Eléments d'Astronomie. Paris: Imprimerie Royale, 649 pp.

Chamberlain, J. W., and Hunten, D. M. 1965. Pressure and CO_2 content of the martian atmosphere: A critical discussion. *Rev. Geophys.* **3**: 299–317.

Chapman, C. R. 1974. Cratering on Mars 1: Cratering and obliteration history. *Icarus* **22**:272–291.

Chapman, C. R., Pollack, J. B., and Sagan, C. 1969. An analysis of the Mariner 4 cratering statistics. *Astron. J.* **74**:1039–1051.

Christensen, E. J., Born, G. H., Hildebrand, C. E., and Williams, B. G. 1977. The mass of Phobos from Viking flybys. *Geophys. Res. Lett.* **4**:555–557.

Cintala, M. J., Head, J. W., and Mutch, T. A. 1976. Martian crater depth/diameter relationships: Comparisons with the Moon and Mercury. *Proc. Lunar Sci. Conf.* **7**:3575–3587.

Cintala, M. J., Wood, C. A., Head, J. W., and Mutch, T. A. 1977. Interplanetary comparisons of fresh crater morphometry: Preliminary results. Washington, D. C.: NASA TM-X-3511, pp. 94–96.

Clark, B. C., Baird, A. K., Rose, H. J., Toulmin, P., Christian, R. P., Kelliher, W. C., Castro, A. J. Rowe, C. D., Keil, K., and Huss, G. R. 1977. The Viking X-ray fluorescence experiment: Analytical methods and early results. *J. Geophys. Res.* **82**:4577–4594.

Clark, B. C., Toulmin, P., Baird, A. K., Keil, K., and Rose, H. J. 1976a. Argon content of the martian atmosphere at the Viking 1 landing site: Analysis by X-ray fluorescence spectroscopy. *Science* **193**:804–805.

Clark, B. C., Baird, A. K., Rose, H. J., Toulmin, P., Keil, K., Castro, A. J., Kelliher, W. C., Rowe, C. D., and Evans, P. H. 1976b. Inorganic analyses of martian surface samples at the Viking landing sites. *Science* **194**: 1283–1288.

Colthup, N. B. 1961. Identification of aldehyde in Mars vegetation region. *Science* **134**:529.

Conrath, B., Curran, R., Hanel, R., Kunde, V., Maguire, W., Pearl, J., Pirraglia, J., Welker, J., and Burke, T. 1973. Atmospheric and surface properties of Mars obtained by infrared spectroscopy on Mariner 9. *J. Geophys. Res.* **78**:4267–4278.

Corliss, W. R. 1975. *The Viking Mission to Mars,* rev. ed. Washington, D.C.: NASA SP-334, 77 pp.

Courtillot, V. E., Allègre, C. J., and Mattauer, M. 1975. On the existence of lateral relative motion on Mars. *Earth Planet Sci. Lett.* **25**:279–285.

Curran, R. J., Conrath, B. J., Hanel, R. A., Kunde, V. G., and Pearl, J. C. 1973. Mars: Mariner 9 spectroscopic evidence for H_2O ice clouds. *Science* **182**:381–383.

Cutts, J. A. 1973a. Nature and origin of layered deposits of the martian polar regions. *J. Geophys. Res.* **78**:4231–4249.

Cutts, J. A. 1973b. Wind erosion in the martian polar regions. *J. Geophys. Res.* **78**:4211–4221.

Cutts, J. A., and Smith, R. S. U. 1973. Eolian deposits and dunes on Mars. *J. Geophys. Res.* **78**:4139–4154.

Cutts, J. A., Soderblom, L. A., Sharp, R. P., Smith, B. A., and Murray, B. C. 1971. The surface of Mars 3: Light and dark markings. *J. Geophys. Res.* **76**:343–356.

Dalgarno, A., and McElroy, M. B. 1970. Mars: Is nitrogen present? *Science* **170**:167–168.

DeVaucouleurs, G. 1954. *Physics of the Planet Mars—An Introduction to Aerophysics.* London: Faber and Faber, Ltd., 365 pp.

DeVaucouleurs, G., Blunck, J., Davies, M., Dollfus, A., Koval, I. K., Kuiper, G., Masursky, H., Miyamoto, S., Moroz, V. I., Sagan, C., and Smith, B. 1975. The new martian nomenclature of the International Astronomical Union. *Icarus* **26**:85–98.

Dobrovolskis, A., and Ingersoll, A. P. 1975. Carbon dioxide-water clathrate as a reservoir of CO_2 on Mars. *Icarus* **26**:353–357.

Dolginov, S. S., Yeroshenko, Y. G., and Zhugarov, L. N. 1973. Magnetic field in the very close neighborhood of Mars according to data from Mars 2 and Mars 3 spacecraft. *J. Geophys. Res.* **78**:4779–4786.

Dolginov, S. S., Yeroshenko, Y. G., and Zhuzgov, L. N. 1975. Magnetic field of Mars from data of Mars 3 and Mars 5. *Cosmic Res.* **13**:94–106 (trans. from Russian).

Dollfus, A. 1957a. Photometric properties of the desert areas on Mars. *Comptes Rendus.* **244**:162–164.

Dollfus, A. 1957b. Photometric study of the dark areas on the surface of the planet Mars. *Comptes Rendus* **244**(11):1458–1460.

Dollfus, A. 1961. Polarization studies of the planets. In *Planets and Satellites,* Kuiper, G. P., and Middlehurst, B. M., eds, The Solar System III. Chicago: U. Chicago Press, p. 343–397.

Dollfus, A. 1970. The anomalies of the martian surface sòil in the region "Hellas." *Comptes Rendus* Ser. B., **270**:641–644.

Dollfus, A. Focas, J., and Bowell, E. 1969. The planet Mars: The nature of its surface, the characteristics of its atmosphere from the polarimetry of its light, Part II—The nature of the soil. *Astron. and Astrophys.* **2**: 105–121.

Draper, A. L., Adamcik, J. A., and Gibson, E. K. 1964. Comparison of the spectra of Mars and a goethite-hematite mixture in the 1-2 micron region. *Icarus* **3**:63–65.

Evans, D. C. 1965. Ultraviolet reflectivity of Mars. *Science* **149**:969–972.

Fanale, F. P. 1971a. History of martian volatiles: Implications for organic synthesis. *Icarus* **15**:279–303.

Fanale, F. P. 1971b. A case for catastrophic early degassing of the earth. *Chem. Geol.* **8**:79–105.

Fanale, F. P., and Cannon, W. A. 1971. Adsorption on the martian regolith: Implications for the martian volatile budget and diurnal brightening. *Nature* **230**:502–504.

Fanale, F. P., and Cannon, W. A. 1974. Exchange of adsorbed H_2O and CO_2 between the regolith and atmosphere of Mars caused by changes in surface insolation. *J. Geophys. Res.* **79**:3397–3402.

Farmer, C. B. 1976. Liquid water on Mars. *Icarus* **28**:279–289.

Farmer, C. B., Davies, D. W., and LaPorte, D. D. 1976. Mars: Northern summer ice cap-water vapor observations from Viking 2. *Science* **194**: p.1339–1341.

Farmer, C. B., and LaPorte, D. D. 1972. The detection and mapping of water vapor in the martian atmosphere. *Icarus* **16**:34–46.

Focas, J. H. 1962. Seasonal evolution of the fine structure of the dark areas of Mars. *Planet. Space Sci.* **9**:371–381.

Gaiduk, A. R. 1971. Short-wave spectrum of Mars in the region of the NO_2 bands. *Sov. Astron. AJ* **15**:454–456.

Gierasch, P., and Sagan, C. 1971. A preliminary assessment of martian wind regimes. *Icarus* **14**:312–318.

Gifford, F. A. 1964. The martian canals according to a purely aeolian hypothesis. *Icarus* **3**:130–135.

Glasstone, S. 1968. *The Book of Mars.* Washington, D.C.: NASA SP-179, 315 pp.

Golitsyn, G. S. 1973. On the martian dust storms. *Icarus* **18**:113–119.

Greeley, R. 1973. Mariner 9 photographs of small volcanic structures on Mars. *Geology* **1**:175, 180.

Greeley, R., Iversen, J. D., Pollack, J. B., Udovich, N., and White, B. 1974. Wind tunnel simulations of light and dark streaks on Mars. *Science* **183**:847–849.

Greeley, R., Theilig, E., Guest, J. E., Carr, M. H., Masursky, H., and Cutts, J. A. 1977. Geology of Chryse Planitia. *J. Geophys. Res.* **82**:4093–4109.

Guest, J. E., Butterworth, P. S., and Greeley, R. 1977. Geological observations in the Cydonia region of Mars from Viking. *J. Geophys. Res.* **82**:4111–4120.

Hanel, R., Conrath, B. J., Hovis, W. A., Kunde, V. G., Lowman, P. D., Pearl, J. C., Prabhakara, C., Schlachman, B., and Levin, G. V. 1972. IR spectroscopy experiment on the Mariner 9 mission: Preliminary results. *Science* **175**:305–308.

Hanks, T. C., and Anderson, D. L. 1969. The early thermal history of the Earth, *Phys. of Earth and Planet. Interiors* **2**:19–29.

Hargraves, R. B., Collinson, D. W., Arvidson, R. E., and Spitzer, C. R. Viking magnetic properties investigation: Further results. *Science* **194**: 1303–1309.

Hartmann, W. K. 1966. Martian cratering. *Icarus* **5**:565–576.

Hartmann, W. K. 1973a. Martian surface and crust: Review and synthesis. *Icarus* **19**:550–575.

Hartmann, W. K. 1973b. Martian cratering IV: Mariner 9 initial analysis of cratering chronology. *J. Geophys. Res.* **78**:4096–4116.

Hartmann, W. K. 1974a. Martian and terrestrial paleoclimatology: Relevance of solar variability. *Icarus* **22**:301–311.

Hartmann, W. K. 1974b. Geological observations of Martian arroyos. *J. Geophys. Res.* **79**:3951–3957.

Hartmann, W. K. 1977. Relative crater production rates on planets. *Icarus* **31**:260–276.
Hays, J. D., Imbrie, J., and Shackleton, N. J. 1976. Variations in the earth's orbit: Pacemaker of the Ice Ages. *Science* **194**:1121–1132.
Head, J. W. 1976. The significance of substrate characteristics in determining morphology and morphometry of lunar craters. *Lunar Sci.* **7**:354–356.
Herr, K. C., Forney, P. B., and Pimentel, G. C. 1972. Mariner Mars 1969 IR spectrometer. *App. Opt.* **11**:493–501.
Herr, K. C., and Pimentel, G. C. 1969. Infrared absorptions near 3μ recorded over the Polar Cap of Mars. *Science* **166**:496–499
Hess, S. L. 1973. Martian winds and dust clouds. *Planet. Space Sci.* **21**:1549–1557.
Hess, S. L., et al. 1976. Preliminary meteorological results on Mars from the Viking 1 Lander. *Science* **193**:788–791.
Hord, C. W., Barth, C. A., Stewart, A. I., and Lane, A. L. 1972. Mariner 9 UV spectrometer experiment: Photometry and topography of Mars. *Icarus* **17**:443–456.
Horowitz, N. H., and Hobby, G. L. 1977. Viking on Mars: The carbon assimilation experiments. *J. Geophys. Res.* **82**:4659–4662.
Horowitz, N. H., Hobby, G. L., and Hubbard, J. S. 1976. The Viking carbon assimilation experiments: Interim report. *Science* **194**:1321–1322.
Horowitz, N. H., Hubbard, J. S., and Hobby, G. L. 1972. The carbon assimilation experiments: The Viking Mars Lander. *Icarus* **16**:147–152.
Houck, J. R., Pollack, J. B., Sagan, C., Schaack, O., and Dekker, J. A. 1973. High altitude infra-red spectroscopic evidence for bound water on Mars. *Icarus* **18**:470–480.
Hubbard, J. S., Hardy, J. P., and Horowitz, N. H. 1971. Photocatalytic production of organic compounds from CO and H_2O in a simulated martian atmosphere. *Proc. Nat. Acad. Sci. U.S.* **68**:574–578.
Huck, F. O., Jobson, D. J., Park, S. K., Wall, S. D., Arvidson, R. E., Patterson, W. R., and Benton, W. D. 1977. Spectrophotometric and color estimates of the Viking Lander sites. *J. Geophys. Res.* **82**:4401–4411.
Huguenin, R. L. 1976. Surface oxidation: A major sink for water on Mars. *Science* **192**:138–139.
Hunt, G. R., Logan, L. M., and Salisbury, J. W. 1973. Mars: Components of infrared spectra and the composition of the dust cloud. *Icarus* **18**: 459–467.
Hunten, D. M. 1974. Aeronomy of the lower atmosphere of Mars. *Rev. Geophys. and Space Phys.* **12**:529–535.
Ingersoll, A. P. 1974. Mars: The case against permanent CO_2 frost caps. *J. Geophys. Res.* **79**:3403–3410.
Istomin, V. G., and Grechnev, K. V. 1976. Argon in the martian atmosphere: Evidence from the Mars 6 descent module. *Icarus* **28**:155–158.
Istomin, V. G., Grechnev, K. V., Ozerou, L. N., Slutskiy, M. Y., Paulenko, V. A., and Tsvetkov, V. N. 1975. Experiment measuring composition of martian atmosphere on board the descent module of the Mars 6 space station. *Cosmic Research* **13**:13–17 (trans. from Russian).
Iversen, J. D., Pollack, J. B., Greeley, R., and White, B. R. 1976. Saltation threshold on Mars: The effect of interparticle force, surface roughness and low atmosphere density. *Icarus* **29**:381–394.

Jamison, D. 1965. Some speculation on the martian canals. *Astron. Soc. Pac. Publ.* **77**:394–395.

Johnston, D. H., McGetchin, T. R., and Toksöz, M. N. 1974. The thermal state and internal structure of Mars. *J. Geophys. Res.* **79**:3959–3971.

Jones, K. L. 1974. Evidence for an episode of crater obliteration intermediate in martian history. *J. Geophys. Res.* **79**:3917–3931.

Kaplan, L. D., Connes, J., and Connes, P. 1969. Carbon monoxide in the martian atmosphere. *Astrophys. J. Lett.* **157**:L187–L192.

Kaplan, L. D., Connes, P., Connes, J., and Smith, M. A. H. 1975. The spectroscopic evidence for argon on Mars. *EOS Trans. Am. Geophys. Un.* **56**:405.

Katterfeld, G. N., and Hedervari, P. 1969. Ring-shaped and linear structures on Mars. *Sov. Astron. AJ* **12**:863–871.

Kellog, W. W., and Sagan, C. 1961. The atmosphere of Mars and Venus, a report by the ad hoc panel on planetary atmospheres of the Space Sciences Board, Publ. 944. Washington, D. C.: Natl. Acad. Sci.–Nat. Res. Council, 151 pp.

Kepler, J. 1609. *Astronomia Nova, Sue Physica Coelistis Tradita Commentarius de Motibus Stellae Martisex Observationibus.* Prague: G. V. Tychonis Brahe.

Kiess, C. C., Karrer, S., and Kiess, H. K. 1960. A new interpretation of martian phenomena. *Astron. Soc. Pac. Publ.* **72**:256–267.

Klein, H. P. 1976. The Viking biological investigation: Preliminary results. *Science* **194**:99–105.

Klein, H. P. 1978. The Viking biological experiments on Mars. *Icarus* **34**: 666–674.

Klein, M. P., Lederberg, J., and Rich, A. 1972. Biological experiments: The Viking Mars Lander. *Icarus* **16**:139–146.

Kliore, A., Cain, D. L., Levy, G. S., Eshleman, V. R., Fjeldbo, G., and Drake, F. D. 1965. Occulation experiment: Results of the first direct measurement of Mars' atmosphere and ionosphere. *Science* **149**:1243–1248.

Kliore, A., Fjeldbo, G., Seidel, B. L., and Rasool, S. I. 1969. Mariners 6 and 7: Radio occultation measurements of the atmosphere of Mars. *Science* **166**:1393–1397.

Kuiper, G. P. 1952. *The Atmospheres of the Earth and Planets.* Chicago: University of Chicago Press, 434 pp.

Kuiper, G. P. 1955. On the martian surface features. *Astron. Soc. Pac. Publ.* **67**:271–282.

Kuiper, G. P. 1957. Visible observations of Mars, 1956. *Astrophys. J.* **125**: 307–317.

Lachenbruch, A. 1962. Mechanics of thermal contraction cracks and ice-wedge polygons in permafrost. *Geol. Soc. Am. Spec. Paper No. 70*, 69 pp.

Leovy, C. 1966. Mars ice caps. *Science* **154**:1178–1179.

Leovy, C., Smith, B. A., Young, A. T., and Leighton, R. B. 1971. Mariner Mars, 1969: Atmospheric results. *J. Geophys. Res.* **76**:297–312.

Leovy, C., Briggs, G. A., and Smith, B. A. 1973. Mars atmosphere during the extended mission, television results. *J. Geophys. Res.* **78**:4252–4266.

Leovy, C., Briggs, G. A., Young, A. T., Smith, B. A., Pollack, J. B. Shipley, E. N., and Wildey, R. L. 1972. The martian atmosphere: Mariner 9 TV experiment progress report. *Icarus* **17**:373–393.

Levin, G. V. 1972. Detection of metabolically produced labeled gas: The Viking Mars Lander. *Icarus* **16**:153–166.
Levin, G. V., and Straat, P. A. 1976. Viking labeled release biology experiment: Interim report. *Science* **194**:1322–1329.
Levin, G. V., and Straat, P. A. 1977. Recent results from the Viking labeled release experiment on Mars. *J. Geophys. Res.* **82**:4663–4667.
Lewis, J. S. 1972. Low temperature condensation from the solar nebula. *Icarus* **16**:241–252.
Loomis, A. A. 1965. Some geologic problems of Mars. *Geol. Soc. Am. Bull.* **16**:1083–1104.
Lowell, P. 1906. *Mars and Its Canals.* New York: Macmillan Co.
Lowell, P. 1908. *Mars as the Abode of Life.* New York: Macmillan Co.
Lyttleton, R. A. 1965. On the internal structure of the planet Mars. *Mon. Nat. Roy. Astron. Soc.* **129**:21–39.
McCauley, J. F. 1973. Mariner 9 evidence for wind erosion in the equatorial and mid-latitude region of Mars. *J. Geophys. Res.* **78**:4123–4137.
McCord, T. B. 1969. Comparsion of the reflectivity and color of bright and dark regions on the surface of Mars. *Astrophys. J.* **156**:79–86.
McCord, T. B., and Adams, J. B. 1969. Spectral reflectivity of Mars. *Science* **163**:1058–1060.
McCord, T. B., Huguenin, R. L., Mink, D., and Pieters, C. 1977a. Spectral reflectance of Martian areas during the 1973 opposition: Photo-electric filter photometry 0.33-1.10 μ *Icarus* **31**:25–39.
McCord, T. B., Huguenin, R. L., and Johnson, G. L. 1977b. Photometric imaging of Mars during the 1973 opposition. *Icarus* **31**:293–314.
McElroy, M. B. 1972. Mars: An evolving atmosphere. *Science* **175**:443–445.
McGetchin, T. R., and Head, J. W. 1973. Lunar cinder cones. *Science* **180**: 68–70.
McLaughlin, D. B. 1954. Volcanism and aeolian deposition on Mars. *Geol. Soc. Amer. Bull.* **65**:715–717.
Malin, M. C. 1976. Age of martian channels. *J. Geophys. Res.* **81**:4825–4845.
Malin, M. C. 1977. Comparsion of volcanic features of Elysium (Mars) and Tibesti (Earth). *Geol. Soc. Amer. Bull.* **88**:908–919.
Malin, M. C., and Dzurisin, D. 1977. Landform degradation on Mercury, Moon and Mars: Evidence from crater depth/diameter relationships. *J. Geophys. Res.* **82**:376–388.
Marov, M. Y., and Petrov, G. I. 1973. Investigations of Mars from the Soviet automatic stations Mars 2 and 3. *Icarus* **19**:163–179.
Masson, P. 1977. Structure pattern analysis of the Noctis Labryinthus Valles Marineris region of Mars. *Icarus.* **30**:49–62.
Masursky, H., Batson, R. M., McCauley, J. F., Soderblom, L. A., Wildey, R. L., Carr, M. H., Milton, D. J., Wilhelms, D. E., Smith, B. A. Kirby, T. B., Robinson, J. C., Leovy, C. B., Briggs, G. A., Duxbury, T. C., Acton, C. H., Murray, B. C., Cutts, J. A., Sharp, R. P., Smith, S., Leighton, R. B., Sagan, C., Veverka, J., Noland, M., Lederberg, J., Levinthal, E., Pollack, J. B., Moore, J. T., Hartmann, W. K., Shipley, E. N., DeVaucouleurs, G., and Davies, M. 1972. Mariner 9 TV reconnaissance of Mars and its satellites: Preliminary results. *Science* **175**:294–305.
Masursky, H., and Crabill, N. L. 1976a. The Viking landing sites: Selection and certification. *Science* **193**:809–812.

Masursky, H., and Crabill, N. L. 1976b. Search for the Viking 2 landing site. *Science* **194**:62–68.

Maxwell, T. A., Otto, E. P., Picard, M. D., and Wilson, R. C. 1973. Meteorite impact: A suggestion for the origin of some stream channels on Mars. *Geology* **1**:9–10.

Metzger, A. E., Trombka, J. I., Peterson, L. E., Reedy, R. C., and Arnold, J. R. 1973. Lunar surface radioactivity: Preliminary results of the Apollo 15 and Apollo 16 gamma-ray spectrometer experiments. *Science* **179**: 800–803.

Michaux, C. M. 1967. *Handbook of the Physical Properties of the Planets.* Washington, D.C.: NASA SP-3030.

Milankovitch, M. 1941. *K. Serb. Akad. Beogr. Spec. Publ. 132.*

Miller, S. L., and Smythe, W. D. 1970. Clathrate in the martian ice cap. *Science* **170**:531–533.

Milton, D. J. 1973. Water and processes of degradation in the martian landscape. *J. Geophys. Res.* **78**:4037–4047.

Milton, D. J. 1974. Carbon dioxide hydrate and floods on Mars. *Science* **183**:654–656.

Moroz, V. I. 1964. The infrared spectrum of Mars (1.1-4.1μ) *Sov. Astron AJ* **8**:273–281.

Moroz, V. I. 1976. Argon in the martian atmosphere: Do the results of Mars 6 agree with the optical and radio occulation measurements. *Icarus* **28**:159–163.

Moroz, V. I., and Nadzhip, A. E., 1975. Preliminary results of water vapor measurements in the martian atmosphere by the Mars 5 interplanetary probe. *Cosmic Res.* **13**:28–34 (trans. from Russian).

Murray, B. C., and Malin, M. C. 1973a. Polar wandering on Mars? *Science* **179**:997–1000.

Murray, B. C., Soderblom, L. A., Cutts, J. A., Sharp, R. P., Milton, D. J., and Leighton, R. B. 1972. Geological framework of South Polar region of Mars. *Icarus* **17**:328–345.

Murray, B. C., Strom, R. G., Trask, N. J., Gault, D. E. 1975. Surface history of Mercury: Implications for terrestrial planets. *J. Geophys. Res.* **80**: 2508–2514.

Murray, B. C., Ward, W. R., and Yeung, S. C. 1973. Periodic insolation variations on Mars. *Science* **180**:638–640.

Mutch, T. A. 1972. *Geology of the Moon,* rev. ed. Princeton, N. J.: Princeton University Press, 391 pp.

Mutch, T. A., Arvidson, R. E., Head, J. W., III, Jones, K. L., and Saunders, R. S. 1976. *The Geology of Mars.* Princeton, N. J.: Princeton, University Press, 400 pp.

Mutch, T. A., Binder, A. B., Huck, F. D., Levinthal, E. C., Liebes, S., Morris, E. C., Patterson, W. R., Pollack, J. B., Sagan, C., and Taylor, G. R. 1976. The surface of Mars: The view from the Viking Lander. *Science* **193**: 791–801.

Nakamura, K. 1976. Volcanoes as possible indicators of tectonic stress orientation. Unpubl. manuscript, Columbia University, New York.

Neal, J., Langer, A., and Kerr, P. F. 1968. Giant desiccation polygons of Great Basin playas. *Geol. Soc. Amer. Bull.* **79**:69–90.

Neugebauer, G., Münch, G., Chase, S. C., Jr., Hatzenbeler, H., Miner, E., and Schofield, D. 1969. Mariner 1969: Preliminary results of the infrared radiometer experiment. *Science* **166**:98–99.

Nier, A. O., Hanson, W. B., Seiff, A., McElroy, M. B., Spencer, N. W., Duckett, R. J., Knight, T. C. D., and Cook, W. S. 1976a. Composition and structure of the martian atmosphere: Preliminary results from Viking 1. *Science* **193**:786–788.

Nier, A. O., McElroy, M. B., and Yung, Y. L. 1976b. Isotopic composition of the martian atmosphere. *Science* **194**:68–70.

Noland, M., and Veverka, J. 1977a. The photometric function of Phobos and Deimos, II. Surface photometry of Deimos. *Icarus.* **30**:200–211.

Noland, M., and Veverka, J. 1977b. The photometric function of Phobos and Deimos, III. Surface photometry of Phobos. *Icarus* **30**:212–213.

Oberbeck, V. R., Quaide, W. L., Arvidson, R. E., and Aggarwal, H. R. 1977. Comparative studies of lunar, martian and mercurian craters and plains. *J. Geophys. Res.* **82**:1681–1698.

Öpik, E. J. 1966. The martian surface. *Science* **153**:255–265.

Otterman, J., and Bronner, F. E. 1966. Martian wave of darkening: A frost phenomenon? *Science* **153**:56–60.

Owen, T., and Biemann, K. 1976. Composition of the atmosphere at the surface of Mars: Detection of Ar-36 and preliminary analysis. *Science* **193**:801–803.

Owen, T., Biemann, K., Rushneck, D. R., Biller, J. E., Howarth, D. W., La Fleur, A. L. 1976. The atmosphere of Mars: Detection of krypton and zenon. *Science* **194**:1293–1295.

Oyama, V. I. 1972. The gas exchange experiment for life detection: The Viking Mars Lander. *Icarus* **16**:167–184.

Oyama, V. I., and Berdahl, B. J. 1977a. The Viking gas exchange experiment results from Chryse and Utopia surface samples. *J. Geophys. Res.* **82**:4669–4676.

Ozima, M. 1975. Ar isotopes and earth atmosphere evolution models. *Geochim. Cosmochim. Acta* **39**:1127–1134.

Pang, K. D., Pollack, J. B., Veverka, J., Lane, A. L., Ajello, J. M. 1978. The composition of Phobos: Evidence for carbonaceous chrondrite surface from spectral analysis. *Science* **199**:64–66.

Papanastassiou, D. A., and Wasserburg, G. J. 1971. Rb-Sr ages of igneous rocks from the Apollo 14 mission and the age of the Frau Mauro formation. *Earth and Planet. Sci. Lett.* **12**:36–48.

Park, K. D., Pollack, J. B., Veverka, J., Lane, A. L., and Ajello, J. M. 1978. The composition of Phobos: Evidence for carbonaceous chrondrite surface from spectral analysis. *Science* **199**:64–66.

Peale, S. J., Schubert, G., and Lingenfelter, R. E. 1975. Origin of martian channels: Clathrates and water. *Science* **187**:273–274.

Perls, T. A. 1973. Carbon suboxide on Mars: Evidence against formation. *Icarus* **20**:511–512.

Phillips, R. J., Saunders, R. S., and Conel, J. E. 1973. Mars: Crustal structure inferred from Bouguer gravity anomalies. *J. Geophys. Res.* **78**:4815–4820.

Pieri, D. 1976. Distribution of small channels on the martian surface. *Icarus* **27**:25–50.

Pike, R. J. 1976. Simple to complex impact craters: The transition on the Moon. *Lunar Sci.* **7**:700–701.
Pimentel, G. C., Atwood, K. C., Gaffon, H., Hartline, M. K., Jukes, T. H., Pollard, E. C., and Sagan, C. 1966. In *Biology and the Exploration of Mars.* C. Pittendright, W. Vishniac, and J. Pearman, eds. Washington, D.C.: Nat. Acad. Sci.-Nat. Res. Council.
Plummer, W. T., and Carson, R. K. 1969. Mars: Is the surface colored by carbon suboxide? *Science* **166**:1141–1142.
Pollack, J. B., Haberle, R., Greeley, R., and Iverson, J. 1976. Estimates of the wind speeds required for particle motion on Mars. *Icarus* **29**: 395–418.
Pollack, J. B., and Sagan, C. 1967. Secular changes and dark area regeneration on Mars. *Icarus* **6**:434–439.
Pollack, J. B., Toon, O. B., and Sagan, C. 1974. Physical properties of the particles composing the great martian dust storm of 1971. *Bull. Amer. Astron. Soc.* **6**:370.
Pollack, J. B., Veverka, J., Noland, M., Sagan, C., Duxbury, T., Acton, C., Born, G., Hartmann, W., and Smith, B. 1973. Mariner 9 television observations of Phobos and Deimos. *J. Geophys. Res.* **78**:4813–4326.
Pollack, J. B., Veverka, J., Pang, K., Colburn, D., Lane, A. L., and Ajello, J. M. 1978. Multicolor observations of Phobos with the Viking Lander camera: Evidence for a carbonaceous chondritic composition. *Science* **199**:66–69.
Ponnamperuma, C., Shimoyama, A., Yamada, M., Hobo, T., and Pal, R. 1977. Possible surface reactions on Mars: Implications for Viking biology results. *Science* **197**:455–457.
Potter, D. B. 1976. Geologic map of the Hellas quadrangle of Mars. *U. S. Geol. Surv. Misc. Inv. Ser. Map I-941.*
Prokof'eva, V. V., Chuprakova, T. A., Dzyamko, S. S., and Bryzalova, T. V. 1975. The blue clearings on Mars in August and September 1971 and their relation to dust clouds. *Sov. Astron. AJ* **19**:378–384.
Rea, D. G. 1964. The darkening wave on Mars. *Nature* **201**:1014–1015.
Rea, D. G., O'Leary, B. T., and Sinton, W. M. 1965. Mars: The origin of the 3.58 and 3.69μ minima in the infrared specta. *Science* **147**:1286–1288.
Reasenberg, R. D. 1977. The moment of inertia and isostasy of Mars. *J. Geophys. Res.* **82**:369–375.
Richardson, R. S., and Bonestell, C. 1964. *Mars.* New York: Harcourt, Brace and World, 151 pp.
Roland, N. W. 1976. Tektonische Standardnetze und Beanspruchungs pläne für Erde und Mars. *Geol. Rundschau* **65**:17–33.
Sagan, C. 1971. The long winter model of martian biology: A speculation. *Icarus* **15**:511–514.
Sagan, C. 1973a. Sandstorms and eolian erosion on Mars. *J. Geophys. Res.* **78**:4155–4161.
Sagan, C. 1973b. Liquid carbon dioxide and the martian polar laminae. *J. Geophys. Res.* **78**:4250–4251.
Sagan, C. 1977. Reducing greenhouses and the temperature history of Earth and Mars. *Nature* **269**:224–226.
Sagan, C., and Bagnold, R. A. 1975. Fluid transport on earth and aeolian transport on Mars. *Icarus* **26**:209–218.

Sagan, C., and Mullen, G. 1972. Earth and Mars: Evolution of atmospheres and surface temperatures. *Science* **177**:52–55.

Sagan, C., Phaneuf, J. P., and Ihnat, M. 1965. Total reflection spectrophotometry and thermogravimetric analysis of simulated martian surface materials. *Icarus* **4**:43–61.

Sagan, C., Toon, D. B., and Gierasch, P. J. 1973. Climatic change on Mars. *Science* **181**:1045–1049.

Sagan, C., Veverka, J., Fox, P., Dubisch, T., Lederberg, J., Levinthal, E., Quam, L., Tucker, R., Pollack, J. B., and Smith, B. A. 1972. Variable features on Mars: Preliminary Mariner 9 TV results. *Icarus* **17**:346–372.

Sagan, C., Veverka, J., Fox, P., Dubisch, R., French, R., Gierasch, P., Quam, L. Lederberg, J., Levinthal, E., Tucker, R., Eross, B., and Pollack, J. B. 1973. Variable features on Mars, 2: Mariner 9 global results. *J. Geophys. Res.* **78**:4163–4196.

Sagan, C., Veverka, J., and Gierasch, P. 1971. Observational consequences of martian wind regimes. *Icarus* **15**:253–278.

Sagan, C., and Young, A. T. 1973. Solar neutrinos, martian rivers and Praesepe. *Nature* **243**:459–460.

Salisbury, J. W., and Hunt, G. R. 1968. Martian surface materials: Effect of particle size on spectral behavior. *Science* **161**:365–366.

Salisbury, J. W., and Hunt, G. R. 1969. Compositional implications of the spectral behavior of the martian surface. *Nature* **222**:132–136.

Schiaparelli, G. 1893. *Natura ed Arte.* Milano: Casa Editrice Dottor F. Vallardi.

Schumm, S. A. 1974. Structural origin of large martian channels. *Icarus* **22**: 371–384.

Schwartzman, D. W. 1973. Argon degassing models of the Earth. *Nature* **245**:20–21.

Scott, D. M., and Allingham, J. W. 1976. Geologic map of the Elysium quadrangle of Mars. *U. S. Geol. Surv. Mis. Inv. Ser. Map I-935.*

Sengör, A. M. C., and Jones, I. C. 1975. A new interpretation of martian tectonics with special reference to the Tharsis region. *Geol. Soc. Amer.* **7**:1264 (abstract).

Sharonov, V. V. 1961. A lithological interpretation of the photometric and colorimetric studies of Mars. *Sov. Astron. AJ* **5**:199–202.

Sharp, R. P., Soderblom, L. A., Murray, B. C., and Cutts, J. A. 1971. The Surface of Mars, 2: Uncratered terrains. *J. Geophys. Res.* **76**:331–342.

Sharp, R. P. 1973a. Mars, fretted and chaotic terrains. *J. Geophys. Res.* **78**: 4073–4083.

Sharp, R. P. 1973b. Mars, troughed terrain. *J. Geophys. Res.* **78**:4063–4072.

Sharp, R. P. 1973c. Mars, South polar pits and etched terrain. *J. Geophys. Res.* **78**:4221.

Sharp, R. P., and Malin, M. C. 1975. Channels on Mars. *Geol. Soc. Bull.* **86**: 593–609.

Shklovskii, I. S., and Sagan, C. 1966. *Intelligent Life in the Universe,* P. Fern, trans. New York: Dell Publishing Company.

Shoemaker, E. M., and Helin, E. F., 1977. Present impact cratering rates on the terrestrial planets and the moon. Washington, D. C.: NASA TM-X-3511, pp. 74–77.

Sinton, W. M. 1957. Spectroscopic evidence for vegetation on Mars. *Astrophys. J.* **126**:231–239.

Sinton, W. M. 1959. Further evidence of vegetation on Mars. *Science* **130**: 1234–1237.

Sinton, W. M. 1967. On the composition of the martian surface material. *Icarus* **6**:222.

Smith, E. I. 1976. Comparison of the crater morphology size relationships for Mars, Moon and Mercury. *Icarus* **28**:543–550.

Smith, E. I., and Hartnell, J. 1977. The effect of non-gravitational factors on the shape of martian, lunar and mercurian craters: Target effects. Washington, D. C.: NASA TM-X-3511, pp. 91–92.

Smith, E. J., David, L., Coleman, P. J., and Jones, D. E. 1965. Mariner IV measurements near Mars—initial results: Magnetic field measurements near Mars. *Science* **149**:1241–1242.

Smith, J. C., and Born, G. H. 1976. Secular acceleration of Phobos and Q of Mars. *Nature* **27**:51–53.

Smith, S. A., and Smith, B. A. 1972. Diurnal and seasonal behavior of discrete white clouds on Mars. *Icarus* **16**:509–521.

Soderblom, L. A., Kreidler, T. J., and Masursky, H., 1973a. Latitudinal distribution of a debris mantle on the Martian surface. *J. Geophys. Res* **78**: 4117–4122.

Soderblom, L. A., Malin, M. C., Cutts, J. A., and Murray, B. 1973b. Mariner 9 observations of the surface of Mars in the North Polar regions. *J. Geophys. Res.* **78**:4197–4210.

Soffen, G. A., and Snyder, C. W. 1976. The first Viking mission to Mars. *Science* **193**:759–765.

Solomon, S. C., and Chaiken, J. 1976. Thermal expansion and thermal stress in the Moon and terrestrial planets: Clues to early thermal history. *Proc. Lunar Sci. Conf. 7*, pp. 3229–3243.

Soter, S., and Harris, A. 1977. Are striations on Phobos evidence for tidal stress? *Nature* **268**:421–422.

Spinrad, H., Münch, G., and Kaplan, L. D. 1963. The detection of water vapor on Mars. *Astrophys. J.* **137**:1319–1321.

Steinbacher, R. H., Kliore, A., Lorell, J., Hipscher, H., Barth, C. A., Masursky, H., Münch, G., Pearl, J., and Smith, B. 1972. Mariner 9 science experiments: Preliminary results. *Science* **175**:293–294.

Surkov, Y. A. 1977. *Gamma spectrometry in cosmic investigations.* Moscow: Atomizdat.

Surkov, Y. A., and Fedoseyev, G. A. 1977. Radioactivity of the moon, planets and meteorites. In *Soviet-American Conference on Cosmo-Chemistry of the Moon and Planets,* J. H. Pomeroy and N. J. Hubbard, eds. Washington, D. C.: NASA SP-370 **1**:201–218.

Surkov, Y. A., Kirnozov, F. F., Glazov, V. N., Dunchenko, A. G., and Tatsii, L. P. 1976. The content of natural radioactive elements in venusian rocks as determined by Venera 9 and Venera 10. *Cosmic Res.* **14**:618–622.

Surkov, Y. A., Moskalyova, L. P., Kirnozov, F. F., Kharyukova, V. P., Manvelyan, D. S., and Shcheglov, D. P. 1976. Preliminary results of investigations of gamma radiation from Mars from Mars 5 observations. *Space Research,* COSPAR **16**:993–1000.

Tolson, R. H., Duxbury, T. C., Born, G. H., Christensen, E. J., Diehl, R. E., Farless, D., Hildebrand, C. H., Mitchell, R. T., Molko, P. M., Morabito, L. A., Palluconi, F. D., Reichert, R. J., Taraji, H., Veverka, J., Neugebauer,

G., and Findlay, J. T. 1978. Viking first encounter of Phobos: Preliminary results. *Science* **199**:61–64.

Toon, O. B., Pollack, J. B., and Sagan, C. 1977. Physical properties of the particles composing the martian dust storm of 1971–72. *Icarus* **30**: 663–696.

Toon, O. B., Ward, W. R., and Burns, J. A. 1977. Climatic change on Mars: Hot poles at high obliquity. Paper presented at Planet. Sic. Div. Am. Astron. Soc., Honolulu, Jan. 1977.

Toulmin, P. 1976. Preliminary results from the Viking X-ray fluorescence experiment: The first sample from Chryse Planitia, Mars. *Science* **194**: 81–84.

Toulmin, P., Baird, A. K., Clark, B. C., Keil, K., Rose, H. J., Christian, R. P., Evans, P. H., and Kelliher, W. C. 1977. Geochemical and mineralogical interpretation of the Viking inorganic chemical results. *J. Geophys. Res.* **82**:4625–4634.

Tull, R. G. 1966. The reflectivity spectrum of Mars in the near infrared. *Icarus* **5**:505–514.

VanTassel, R. A., and Salisbury, J. W. 1964. The composition of the martian surface. *Icarus* **3**:264–269.

Vasil'chenko, N. V., and Moroz, V. I. 1967. The spectrum of Mars, 1.1–2.5μ, at the 1965 opposition. *Sov. Astron. AJ* **11**:456–461.

Veverka, J. 1975. Mars: Evidence for crater streaks produced by wind erosion. *Icarus* **25**:595–601.

Veverka, J. 1977. Phobos and Deimos. *Sci. Amer.* **236**:30–37.

Veverka, J., and Duxbury, T. C. 1977. Viking observations of Phobos and Deimos: Preliminary results. *J. Geophys. Res.* **82**:4213–4223.

Veverka, J., Thomas, P., and Greeley, R. 1977. A study of the variable features on Mars during the Viking primary mission. *J. Geophys. Res.* **82**: 4167–4187.

Vogt, P. R. 1974. Volcano height and plate thickness. *Earth Planet Sci. Lett.* **23**:337–348.

Ward, W. R. 1974. Climatic variations on Mars, 1: Astronomical theory of insolation. *J. Geophys. Res.* **79**:3375–3386.

Ward, W. R., Murray, B. C., and Malin, M. C. 1974. Climatic variations on Mars, 2: Evolution of carbon dioxide atmosphere and polar caps. *J. Geophys. Res.* **79**:3387–3395.

Wasserburg, G. J., Turner, G., Tera, F., Podosek, F. A., Papanastassiou, D. A., and Huneke, J. C. 1972. Comparison of Rb-Sr, K-Ar, and U-Th-Pb ages, lunar chronology and evolution. *Lunar Sci.* **3**:788–790.

West, M. 1974. Martian volcanism: Additional observations and evidence for pyroclastic activity. *Icarus* **21**:1–11.

Wilson, R. C., Harp, E. L., Picard, M. D., and Ward, S. H. 1973. Chaotic terrain of Mars: A Tectonic interpretation from Mariner 6 imagery. *Geol. Soc. Amer. Bull.* **84**:741–748.

Wise, D. U. 1974. Martian fault pattern and time sequence in relation to volcanism, northern Tharsis ridge area. *Trans. Amer. Geophys. Un.* **55**:341 (abstract).

Wise, D. U. 1977. *Timing of deformational events in the northern Tharsis bulge of Mars.* Washington, D. C.: NASA TM-X-3511, pp. 59–60.

Witting, J., Narin, F., and Stone, C. A. 1965. Mars: Age of its craters. *Science* **149**:1496–1498.
Wood, C. A., Cintala, M. J., and Head, J. W. 1978. Interior morphometry of fresh Martian craters: Preliminary Viking results. *Lunar Sci.* (in press).
Woronow, A. 1977. Crater saturation and equilibrium: A Monte Carlo simulation. *J. Geophys. Res.* **82**:2447–2456.
Younkin, R. L. 1966. A search for limonite near-infrared spectral features on Mars. *Astrophys. J.* **144**:809–818.
Zellner, B. 1972. Minor planets and related objects, VIII Deimos. *Astron. J.* **77**:183.

APPENDIX A:
PHYSICAL PROPERTIES OF MARS AND ITS MOONS

Mars

Mass	6.418×10^{23} kg	
Radius, equatorial average	3396.6 3390 km	
Density	3.933 g/cm	
Surface acceleration due to gravity	370.6 cm/sec^2	
Perihelion distance from sun	206.66×10^6 km	
Aphelion distance from sun	249.22×10^6 km	
Mean distance from sun	227.94×10^6 km	
Inclination of orbit from ecliptic plane	1°51′	
Inclination of equator from orbital plane	23°59′	
Eccentricity of orbit	0.0933	
Period of revolution about sun	686.98 days	
Period of rotation about axis	24 hrs. 37 min. 22 sec.	
Oblateness	0.009 (.0117)	optical .0079 dynamic .0052

Satellites	Phobos	Deimos
Radius, average	10.9 ± 1.5 km {13.5, 10.8, 9.4}	$5.7 \pm .5$ km {7.5, 6.1, 5.5}
Density	2g/cm^3	
Distance from center Mars	9450 km	23,500
Period of revolution about Mars	7 hrs/39 min.	30 hrs/18 min.

Spacecraft	Date launched	Date of encounter	Nearest approach
Mars 1 (USSR)	Nov. 1, 1962	June, 1963	193,000 km
Mariner 4 (USA)	Nov. 28, 1964	July 14, 1965	13,200
Mariner 6 (USA)	Feb. 25, 1969	July 31, 1969	3,390
Mariner 7 (USA)	Mar. 27, 1969	Aug. 5, 1969	3,500
Mars 2 (USSR)	May 19, 1971	Nov. 27, 1971	2,008 (orbiter) Landing site: 44.20°S, 313.2°W
Mars 3 (USSR)	May 28, 1971	Dec. 2, 1971	1,860 (orbiter) Landing site: 44.90°S, 160.08°W
Mariner 9 (USA)	May 30, 1971	Nov. 14, 1971	1,395
Mars 4 (USSR)	July 21, 1973	Feb. 10, 1974	2,200 Flyby
Mars 5 (USSR)	July 25, 1973	Feb. 12, 1974	1,760 Orbiter
Mars 6 (USSR)	May 5, 1973	Mar. 12, 1974	1,600 Orbiter Lander 23°S, 19°SW
Mars 7 (USSR)	Aug. 9, 1973	Mar. 9, 1974	Lander missed Mars by 1300 km
Viking 1 (USA)	Aug. 20, 1975	Orbiter June 19, 1976 Lander July 20, 1976	Orbiter: variable Landing site: Chryse 22.4°N, 47.5°W
Viking 2 (USA)	Sept. 9, 1975	Orbiter Aug. 7, 1976 Lander Sept. 3, 1976	Orbiter: variable Landing site: Utopia 47.89°N, 225.86°W

•ACE MISSIONS TO MARS

Scientific instruments	Major achievements
	Radio contact lost 106 million km from earth
TV camera, magnetometer, cosmic ray telescope, cosmic dust detec-tor, trapped radiation detector, ion chamber, solar plasma detector	22 TV pictures covering 1% of surface; found cratered, moon-like surface; very weak magnetic field; first direct measurement of surface atmospheric pressure, 4–7 mb
'wo TV cameras, infrared and ultraviolet spectrometers, IR radio-neter	76 pictures of surface: Cratered, uncratered, chaotic terrains. Composition and pressure of atmosphere surface temperatures
dentical instrumentation as Mariner 6	126 pictures of surface, 33 of these covering South Polar Cap, showing "quasilinear" features, "etch pits"; data on atmosphere, topography.
ɔrbiter: IR bolometer, 8-40 μ (temperatures) Microwave radiometer 3.37 cm 'R spectrometer 2.06 μ (CO_2) ınterference-polarized photometer (H_2O vapor)	First soft landing on Martian surface Weak magnetic field 1/500 that of earth Water vapor 1/5000 that of earth
/isible and UV photometers, magnetometer, ion detectors. *Lander:* atmospheric spectrometer, TV cameras, instruments for chemical and physical soil analyses, measurement of wind velocity.	TV cameras transmitted only 20 sec (failure probably caused by dust covering during intense dust storm)
'wo TV cameras, IR spectrometer, IR radiometer, UV spectrometer, adio for S-band occultation and celestial mechanics	First spacecraft to orbit Mars Nearly complete TV coverage of surface-over 7300 pictures: Atmospheric composition and pressure; topography; gravity First close-up view of Phobos and Deimos
Magnetometer 'lasma traps Multichannel electrostatic instruments 'V cameras	12 pictures of surface Interplanetary medium
Magnetometer Water vapor photometer (1.38 μ band) Ozone photometer (2600 Å band) Resonance luminescence of Lα line polarimetry (French experiment) TV cameras IR radiometer (8–26μ) Interference filter spectrometer (2–5μ) (2μ) Photometers-narrow band; Gamma radiation	108 TV pictures of surface; measurements of atmospheric pressure and composition Detection of weak magnetic field Gamma-ray spectrometry–inferred radioactivity of surface rocks Localized regions where water vapor content reaches up to 100 ppt.μ
Micrometeorite sensors, cosmic ray detectors Mass spectrometer	Data transmission lasted 2.5 min; mass spectrometer ion pump malfunction led to estimate of 35% Ar in atmosphere
Magnetometer Plama traps Micrometeorite sensors, cosmic ray detectors Solar wave radio experiment (French)	
Orbiter: Two TV cameras, water vapor mapper (MAWD), IR Thermal mapper, entry mass spectrometer Lander: Two imaging facsimile cameras, mass spectrometer, X-ray fluorescence, seismometer, gas chromatograph—mass spectrometer labeled release, gas exchange and pyrolytic release experiment	Detailed imaging of surface; including polar regions, giant channels Detection of N^2, Ar, Xe, Kr in surface atmosphere; N^{15}/N^{14} ratios suggest a denser atmosphere in the past Experimental evidence for frozen water in the permanent polar ice caps
dentical instrumentation as Viking 1	Surface composition compatible with highly weathered basalt Peculiar chemical reactivity of soil—abiological or biological?

APPENDIX C: POTENTIAL RESERVOIRS OF WATER ICE ON MARS

1. Chaotic terrain (includes "fretted" terrain of Sharp 1973).

Area 5.3×10^{16} cm^2
Volume 2.65×10^{21} cm^3 (mean depth 0.5 km)
Volume ice 1.325×10^{21} cm^3 (assuming mean porosity 50%)
Mass ice = $\underline{1.206 \times 10^{21}\ g}$

2. Permanent polar caps

Area	North Polar Cap	6.6×10^{15} cm^2
	South Polar Cap	1.05×10^{15} cm^2
		7.65×10^{15} cm^2
Volume	North Polar Cap	13.2×10^{19} cm^3 (assume mean depth 200m)
	South Polar Cap	2.1×10^{19} cm^3
		15.3×10^{19} cm^3

Mass ice = 0.14×10^{21} g

3. Layered polar deposits (Cutts 1973)

Area 2.6×10^{16} cm^2
Volume 5.2×10^{21} cm^3 (assume mean depth 2 km)
Mass ice = 1.57×10^{21} g (1/3 total volume)

4. Regolith

1×10^3 g/cm^2 (Fanale and Cannon 1974)
$\simeq 1.44 \times 10^{21}$ g

Total
1.21×10^{21} g
0.14
1.57
1.44
4.36×10^{21} g

AUTHOR CITATION INDEX

SUBJECT INDEX

About the Editor

VIVIEN GORNITZ is a research associate and lecturer in the Department of Geological Sciences at Columbia University. She received the B.A. in chemistry from Barnard College in 1963, and the Ph.D. in mineralogy at Columbia University in 1969. From 1971 to 1973, she was granted a National Research Council Associateship in lunar and martian geology. More recently, she has been conducting research in the remote sensing of earth resources from LANDSAT satellite and aircraft.